The Matrix Eigenvalue Problem

The Matrix Eigenvalue Problem

GR and Krylov Subspace Methods

David S. Watkins

Washington State University
Pullman, Washington

siam® Society for Industrial and Applied Mathematics • Philadelphia

Library of Congress Cataloging-in-Publication Data

Watkins, David S.
    The matrix eigenvalue problem : GR and Krylov subspace methods / David S. Watkins.
        p. cm. -- (Other titles in applied mathematics ; 101)
    Includes bibliographical references and index.
    ISBN 978-0-898716-41-2 (alk. paper)
    1. Eigenvalues. 2. Invariant subspaces. 3. Matrices. I. Title.

    QA193.W38 2007
    512.9'436--dc22   2007061800

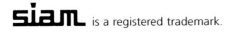 is a registered trademark.

Contents

Preface

Eigenvalue problems are ubiquitous in engineering and science. This book presents a unified theoretical development of the two most important classes of algorithms for solving matrix eigenvalue problems: QR-like algorithms for dense problems, and Krylov subspace methods for sparse problems. I make no claim to completeness. My choice of topics reflects my own interests, a desire to keep the length of the book within reason, and a wish to complete the book within my lifetime.

Prerequisites

Readers of this book are expected to be familiar with the basic ideas of linear algebra and to have had some experience with matrix computations. The student who has absorbed a good chunk of my book *Fundamentals of Matrix Computations* [221] and developed a bit of mathematical maturity will be in a good position to appreciate this book. It is expected that the reader already knows the importance of eigenvalue computations.

How to Read the Book

Chapters 1 and 2 contain background material and are not meant to be read linearly. I suggest that you begin with Chapter 3 and just refer back to the earlier chapters as needed. Chapters 3, 4, 8, and 9 form the heart of the book. Perhaps I should include Chapter 6 on the generalized eigenvalue problem in this list as well. Read Chapter 5 only if you are interested in the details of the convergence theory. Read Chapter 7 if you have an urge to find out what is going on inside the bulge.

Even though I have instructed you to start with Chapter 3, I hope you will end up spending a good chunk of time in Chapter 2, which contains basic theoretical material on eigensystems. I invested a lot of effort in that chapter, and I believe that many readers, including those with very strong backgrounds, will learn new and interesting things there.

A substantial fraction of the book is embedded in the exercises, so please work as many of them as possible. I expect you to spend more time in the exercises than in the text proper. Many proofs of theorems are presented as exercises, which outline my ideas about how you could prove the results. I do not claim that these are the best possible proofs; undoubtedly some of them could be improved. I invite reader feedback.

More on the Contents of the Book

Chapter 3 introduces the tools we need for creating zeros in matrices. Then Chapter 4 introduces and discusses GR algorithms, including the QR algorithm, the differential qd algorithm, and the LR, SR, and HR algorithms, each of which is useful for solving certain special types of eigenvalue problems. Chapter 6 discusses GZ algorithms, including the QZ algorithm, for the generalized eigenvalue problem. Chapter 8 on product eigenvalue problems shows that the QZ algorithm for the generalized eigenvalue problem, the implicit QR algorithm for the singular value decomposition, and periodic QR algorithms for more general product eigenvalue problems are all special cases (not generalizations) of the QR algorithm for the standard eigenvalue problem. Chapter 9 introduces Krylov subspace methods for large, sparse eigenvalue problems. The focus is on short Krylov runs with frequent implicit restarts. A generic Krylov process is presented first, and two methods for making implicit restarts are worked out. Then special cases, including the Arnoldi process and symmetric and unsymmetric Lanczos processes, are discussed. Special Krylov processes that preserve unitary, Hamiltonian, and symplectic structure are also developed. Finally, product Krylov processes (meshing with Chapter 8) and block Krylov processes are considered.

Website

I have compiled a modest collection of MATLAB® programs to go with the book. Some of the exercises refer to them, and you can download them from

www.siam.org/books/ot101

Acknowledgments

I assembled the first notes for this book when I was teaching a course entitled *Eigensystem Computations* at the Technical University of Chemnitz, Germany, in 1999. I thank the German Academic Exchange Service for providing funding. I also thank the attendees for their feedback. (Among them was one undergraduate, a kid named Daniel Kressner.) I thank Volker Mehrmann, who was my host in Chemnitz and has since moved on to Berlin. I thank my many friends who have helped me learn this subject. (I will not name names, as I am afraid of leaving some out.) I thank my home institution, Washington State University, for supporting my research over a period of many years. Heike Fassbender, Nick Higham, Daniel Kressner, and Françoise Tisseur all read a preliminary version of the book and provided me with valuable feedback for which I am most grateful. I claim responsibility for whatever shortcomings the book may have, especially since I did not follow all of their advice.

Chapter 1

Preliminary Material

This book is about matrices with complex entries. The field of complex numbers is denoted $\mathbb{C}$. The set of $n \times 1$ matrices (column vectors) is denoted $\mathbb{C}^n$. The set of $m \times n$ matrices is denoted $\mathbb{C}^{m \times n}$. Most of the time we will be concerned with square matrices, matrices in $\mathbb{C}^{n \times n}$. Sometimes we will restrict our attention to the field of real numbers, denoted $\mathbb{R}$, and the related structures $\mathbb{R}^n$, $\mathbb{R}^{m \times n}$, $\mathbb{R}^{n \times n}$.

Matrices over other fields, e.g., finite fields, are also quite interesting and important in various application areas, but they will not be discussed in this book.

Sometimes we will view our matrices as linear operators on the vector space $\mathbb{C}^n$ or $\mathbb{R}^n$.

1.1   Matrix Algebra

It is assumed that the reader is familiar with the basics of matrix algebra. For example, two $m \times n$ matrices A and B can be added together in the obvious way to form the sum $A + B$. A matrix A can be multiplied by a complex number $\alpha \in \mathbb{C}$ (a *scalar*) to form the product αA. Furthermore, an $m \times n$ matrix A can be multiplied by an $n \times p$ matrix B to form a matrix product AB, which is $m \times p$. The definition of matrix multiplication is less obvious than that of matrix addition. If $C = AB$, we have

$$c_{ij} = \sum_{k=1}^{n} a_{ik} b_{kj}.$$

Here and throughout the book we use the standard convention that c_{ij} denotes the entry in the ith row and jth column of C, a_{ik} denotes the entry in the ith row and kth column of A, and so on. We assume familiarity with the basic properties of these operations, such as the distributive law $(D + E)F = DF + EF$ and the associative law $(GH)K = G(HK)$. Matrix multiplication is not commutative: typically $AB \neq BA$.

The *identity matrix* $I_n \in \mathbb{C}^{n \times n}$ is the $n \times n$ matrix with ones on the main diagonal and zeros elsewhere. For example,

$$I_3 = \begin{bmatrix} 1 & 0 & 0 \\ 0 & 1 & 0 \\ 0 & 0 & 1 \end{bmatrix}.$$

For every $A \in \mathbb{C}^{m \times n}$ we have $A I_n = A$, and for every $A \in \mathbb{C}^{n \times p}$ we have $I_n A = A$. Thus the identity matrix is the (only) matrix that leaves all other matrices unchanged under matrix multiplication. In most cases, when the size of the matrix is not in question, we will drop the subscript and write simply I instead of I_n.

If $A \in \mathbb{C}^{n \times n}$ (A is square!), then the matrix $B \in \mathbb{C}^{n \times n}$ is called the *inverse* of A if

$$AB = I = BA.$$

We write A^{-1} (and say "A inverse") to denote the inverse of A. Thus we have $B = A^{-1}$. If B is the inverse of A, then A is the inverse of B; i.e., $A = B^{-1}$.

Not every square matrix has an inverse. For example, one easily checks that if

$$A = \begin{bmatrix} 1 & 0 \\ 0 & 0 \end{bmatrix},$$

then there is no $B \in \mathbb{C}^{2 \times 2}$ such that $AB = I$. An even simpler example is $A = 0$, the matrix of all zeros. A matrix that has no inverse is called *singular* or *noninvertible*. Conversely, a matrix that does have an inverse is called *nonsingular* or *invertible*. The following theorem lists several characterizations of nonsingularity.

Theorem 1.1.1. *Let $A \in \mathbb{C}^{n \times n}$. The following five conditions are equivalent; if any one of them holds, they all hold.*

(a) *A has an inverse.*

(b) *There is no nonzero $y \in \mathbb{C}^n$ such that $Ay = 0$.*

(c) *The columns of A are linearly independent.*

(d) *The rows of A are linearly independent.*

(e) *$\det(A) \neq 0$.*

(In condition (b), the symbol 0 stands for the vector in $\mathbb{C}^n$ whose entries are all zero. In condition (e), the symbol zero stands for the complex number 0. $\det(A)$ denotes the determinant of A.)

These results are familiar from elementary linear algebra.

We will use determinants only sparingly. They can be avoided altogether, but we will find it convenient to make use of them now and then. In addition to the connection between determinants and invertibility noted in Theorem 1.1.1, we note the following simple connection between determinants and the matrix product.

Theorem 1.1.2. *Let A, $B \in \mathbb{C}^{n \times n}$. Then*

$$\det(AB) = \det(A)\det(B).$$

Powers of matrices will play a big role in this book. We apply exponents to square matrices just as we do to numbers. If $A \in \mathbb{C}^{n \times n}$, we write $A^2 = AA$, $A^3 = AAA = A^2A$, and in general $A^k = A^{k-1}A$. Of course, we also define $A^0 = I$ and $A^1 = A$. Using powers of matrices, we can also define polynomials in matrices. If p is a polynomial, say $p(z) = c_0 + c_1 z + c_2 z^2 + \cdots + c_k z^k$, then, given $A \in \mathbb{C}^{n \times n}$, we define $p(A) \in \mathbb{C}^{n \times n}$ by

$$p(A) = c_0 I + c_1 A + c_2 A^2 + \cdots + c_k A^k.$$

Given $A \in \mathbb{C}^{m \times n}$, the *transpose* of A, denoted A^T, is the matrix $B \in \mathbb{C}^{n \times m}$ such that $b_{ij} = a_{ji}$. The rows of A become the columns of A^T. The *conjugate transpose* of A, denoted A^*, is obtained from A by switching rows and columns *and* taking complex conjugates. Thus if $C = A^* \in \mathbb{C}^{n \times m}$, we have $c_{ij} = \overline{a_{ji}}$.

1.2 Norms and Inner Products

Vector Norms

A *norm* or *vector norm* is a function $x \mapsto \|x\|$ that assigns to each vector $x \in \mathbb{C}^n$ a nonnegative real number $\|x\|$, called the norm of x, such that the following four properties are satisfied:

(a) $\|0\| = 0$;

(b) $\|x\| > 0$ if $x \neq 0$ (*positive definiteness*);

(c) $\|\alpha x\| = |\alpha| \, \|x\|$ for all $\alpha \in \mathbb{C}$ and all $x \in \mathbb{C}^n$ (*absolute homogeneity*);

(d) $\|x + y\| \leq \|x\| + \|y\|$ for all $x, y \in \mathbb{C}^n$ (*triangle inequality*).

Some important examples of norms are the *Euclidean norm* or *vector 2-norm*

$$\|x\|_2 = \left(\sum_{j=1}^{n} |x_j|^2 \right)^{1/2}$$

and the *taxicab norm* or *vector 1-norm*

$$\|x\|_1 = \sum_{j=1}^{n} |x_j|.$$

The norm of a vector should be viewed as a measure of the *length* or *magnitude* of a vector or, equivalently, as the *distance* from tail to tip of the vector in a generalized sense. The Euclidean norm corresponds to our everyday geometric notion of lengths and distances, while the taxicab norm is more relevant to surface travel in Manhattan. Generalizing on these two norms, we have, for $1 \leq p < \infty$, the *vector p-norm*

$$\|x\|_p = \left(\sum_{j=1}^{n} |x_j|^p \right)^{1/p}.$$

Also useful is the *vector ∞-norm*

$$\|x\|_\infty = \max_{1 \leq j \leq n} |x_j|.$$

For our purposes the most useful norm is the Euclidean norm. From this point on, if we write the norm symbol $\| x \|$ with no subscript, we mean the Euclidean or 2-norm, unless otherwise stated.

The Standard Inner Product

Given two vectors $x, y \in \mathbb{C}^n$, the *inner product* (or *dot product* or *scalar product*) of x and y is defined by

$$\langle x, y \rangle = y^* x = \sum_{j=1}^{n} x_j \overline{y_j}. \tag{1.2.1}$$

This is the *standard inner product* on $\mathbb{C}^n$. There are other useful inner products, but the standard one is the only one we will use in this book. The (standard) inner product is closely related to the Euclidean norm. For one thing, it is obvious that

$$\langle x, x \rangle = \| x \|^2.$$

Not so obvious but very important is the CSB (Cauchy–Schwarz–Bunyakovski) inequality.

Theorem 1.2.1 (CSB inequality). *For all $x, y \in \mathbb{C}^n$,*

$$| \langle x, y \rangle | \leq \| x \| \, \| y \|.$$

Equality holds if and only if x and y are linearly dependent.

Here we are exercising our convention that $\| x \|$ means specifically the Euclidean norm. The CSB inequality is proved in Exercise 1.2.4.

Another important fact is the relationship between the standard inner product and the conjugate transpose.

Proposition 1.2.2. *Let $A \in \mathbb{C}^{m \times n}$, $x \in \mathbb{C}^n$, and $y \in \mathbb{C}^m$. Then*

$$\langle Ax, y \rangle = \langle x, A^* y \rangle.$$

Finally we recall the notion of orthogonality. Two vectors x and $y \in \mathbb{C}^n$ are called *orthogonal* if $\langle x, y \rangle = 0$. A set of vectors $u_1, \ldots, u_k \in \mathbb{C}^n$ is called *orthonormal* if $\langle u_i, u_j \rangle = \delta_{ij}$, that is, $\langle u_i, u_j \rangle = 0$ if $i \neq j$ and 1 if $i = j$. Associated with orthogonality, we have the simple but important Pythagorean theorem.

Theorem 1.2.3 (Pythagorean theorem). *Let $x \in \mathbb{C}^n$ and $y \in \mathbb{C}^n$ be orthogonal. Then*

$$\| x + y \|^2 = \| x \|^2 + \| y \|^2.$$

Matrix Norms

A *matrix norm* is a function $A \mapsto \| A \|$ that assigns to each matrix $A \in \mathbb{C}^{n \times n}$ a nonnegative real number $\| A \|$, called the norm of A, such that the following five properties are satisfied:

(a) $\| 0 \| = 0$;

(b) $\|A\| > 0$ if $A \neq 0$ (*positive definiteness*);

(c) $\|\alpha A\| = |\alpha| \|A\|$ for all $\alpha \in \mathbb{C}$ and all $A \in \mathbb{C}^{n \times n}$ (*absolute homogeneity*);

(d) $\|A + B\| \leq \|A\| + \|B\|$ for all $A, B \in \mathbb{C}^{n \times n}$ (*triangle inequality*);

(e) $\|AB\| \leq \|A\| \|B\|$ for all $A, B \in \mathbb{C}^{n \times n}$ (*submultiplicativity*).

The properties are exactly the same as those of a vector norm, except that a property relating the norm to the matrix product has been added.

Each vector norm $\|\cdot\|_v$ induces a matrix norm $\|\cdot\|_M$ by

$$\|A\|_M = \sup_{x \neq 0} \frac{\|Ax\|_v}{\|x\|_v}. \tag{1.2.2}$$

It is easy to verify that the *induced matrix norm* so defined satisfies matrix norm properties (a)–(e). The following simple but useful fact is an immediate consequence of (1.2.2).

Proposition 1.2.4. *Let $\|\cdot\|_v$ be a vector norm on $\mathbb{C}^n$, and let $\|\cdot\|_M$ be its induced matrix norm. Then for all $x \in \mathbb{C}^n$ and all $A \in \mathbb{C}^{n \times n}$,*

$$\|Ax\|_v \leq \|A\|_M \|x\|_v.$$

Each vector p-norm induces a matrix norm, called the *matrix p-norm*. We use the same notation for the matrix p-norm as for the vector p-norm:

$$\|A\|_p = \sup_{x \neq 0} \frac{\|Ax\|_p}{\|x\|_p}.$$

When we take $p = 2$, we get the matrix 2-norm, which is also called the *spectral norm*. This is the matrix norm that will be of greatest importance to us. From this point on, if we write a matrix norm $\|A\|$ with no subscript, we are referring to the spectral norm, the matrix 2-norm, unless otherwise stated.

For our purposes the second most important matrix norm is one that cannot be induced by the construction (1.2.2), namely the *Frobenius norm*

$$\|A\|_F = \left(\sum_{i=1}^{n} \sum_{j=1}^{n} |a_{ij}|^2 \right)^{1/2}. \tag{1.2.3}$$

Each nonsingular matrix A has a *condition number*

$$\kappa(A) = \|A\| \|A^{-1}\|.$$

Different matrix norms give rise to different condition numbers. For example, we will use $\kappa_p(A)$ to denote the condition number given by the matrix p-norm. That is, $\kappa_p(A) = \|A\|_p \|A^{-1}\|_p$. If we write $\kappa(A)$ with no subscript, we mean the *spectral condition number*, that is, the condition number given by the spectral norm, unless otherwise stated.

Perturbations of Nonsingular Matrices

As a first application of matrix norms we will show that if a nonsingular matrix is perturbed slightly, the perturbed matrix is still nonsingular. As a first (huge) step in this direction, we will show that if the identity matrix is perturbed slightly, the perturbed matrix is nonsingular. In other words, if $\| S \|$ is small enough, $I - S$ is nonsingular. It turns out that $\| S \| < 1$ is small enough.

Theorem 1.2.5. *If $\| S \| < 1$, then $I - S$ is nonsingular. Furthermore,*

$$(I - S)^{-1} = \sum_{j=0}^{\infty} S^j$$

and

$$\| (I - S)^{-1} \| \le \frac{1}{1 - \| S \|}.$$

This is just the theory of geometric series extended to matrices. The proof is worked out in Exercise 1.2.11.

Theorem 1.2.6. *Let $A \in \mathbb{C}^{n \times n}$ be nonsingular, and let $E \in \mathbb{C}^{n \times n}$. If $\frac{\|E\|}{\|A\|} < \frac{1}{\kappa(A)}$, then $A + E$ is nonsingular. Furthermore*

$$\| (A + E)^{-1} \| \le \frac{\| A^{-1} \|}{1 - r},$$

where $r = \kappa(A) \frac{\|E\|}{\|A\|} < 1$.

This theorem is an easy consequence of Theorem 1.2.5. See Exercise 1.2.12.

Exercises

1.2.1. Show that the taxicab norm satisfies norm properties (a)–(d).

1.2.2. Show that the vector ∞-norm satisfies norm properties (a)–(d).

1.2.3. Show that for every $x \in \mathbb{C}^n$, $\| x \|_{\infty} = \lim_{p \to \infty} \| x \|_p$.

1.2.4. This exercise leads to a proof of Theorem 1.2.1, the CSB inequality.

(a) Check that the CSB inequality holds if $x = 0$ or $y = 0$. This is part of the "linearly dependent" case.

(b) From here on, assume $x \neq 0$ and $y \neq 0$. The complex number $\langle x, y \rangle$ can be expressed in polar form: $\langle x, y \rangle = r e^{i\theta}$, where $r = |\langle x, y \rangle| \ge 0$. Let $\hat{x} = e^{-i\theta} x$. Show that $\langle \hat{x}, y \rangle = r$ and $\| \hat{x} \| = \| x \|$. Define a function $f(t) = \| \hat{x} + t y \|^2$, where t is a real variable. Show that f is a quadratic polynomial with real coefficients:

$$f(t) = \| y \|^2 t^2 + 2 |\langle x, y \rangle| t + \| x \|^2.$$

(c) Complete the square to obtain

$$f(t) = \left(\| y \| t + \frac{|\langle x, y \rangle|}{\| y \|} \right)^2 + \left(\| x \|^2 - \frac{|\langle x, y \rangle|^2}{\| y \|^2} \right).$$

(d) Use the definition of f to show that $f(t) \geq 0$ for all real t. Then use this fact and the expression from part (c) to show that $|\langle x, y \rangle| \leq \|x\| \|y\|$. This is the CSB inequality.

(e) Suppose $|\langle x, y \rangle| = \|x\| \|y\|$. Use the expression for f from part (c) to show that there is a real $\hat{t}$ at which $f(\hat{t}) = 0$. Refer back to the definition of f to deduce that $\hat{x} + \hat{t}y = 0$ and that therefore x and y are linearly dependent.

(f) Finally, suppose that x and y are linearly dependent. Then either $x = \alpha y$ or $y = \beta x$, for some scalar α or β. Show directly that $|\langle x, y \rangle| = \|x\| \|y\|$.

1.2.5. Prove that the Euclidean vector norm satisfies norm properties (a)–(d). Use the CSB inequality to establish (d).

1.2.6. Let $A \in \mathbb{C}^{m \times n}$, $x \in \mathbb{C}^n$, and $y \in \mathbb{C}^m$. Using the definitions of matrix multiplication and inner product, express $\langle Ax, y \rangle$ as a double sum involving the entries a_{ij}, x_j, and y_i. Do the same for $\langle x, A^* y \rangle$, and show that they are the same. This proves Proposition 1.2.2.

1.2.7. Prove the Pythagorean theorem by expressing the norms as inner products.

1.2.8. Prove that for every vector norm $\| \cdot \|_v$ the induced matrix norm defined by (1.2.2) is indeed a matrix norm; that is, it satisfies matrix norm properties (a)–(e).

1.2.9. Let $\| \cdot \|_v$ be a vector norm on $\mathbb{C}^n$, and let $\| \cdot \|_M$ be its induced matrix norm. Prove the following two obvious but fundamental facts:

(a) $\|I\|_M = 1$.

(b) For all $x \in \mathbb{C}^n$ and all $A \in \mathbb{C}^{n \times n}$, $\|Ax\|_v \leq \|A\|_M \|x\|_v$.

1.2.10.

(a) Using the CSB inequality, show that for all $A \in \mathbb{C}^{n \times n}$ and $x \in \mathbb{C}^n$, $\|Ax\|_2 \leq \|A\|_F \|x\|_2$.

(b) Deduce that $\|A\|_2 \leq \|A\|_F$ for all $A \in \mathbb{C}^{n \times n}$.

(c) Give an example of a matrix $A \in \mathbb{C}^{n \times n}$ ($n \geq 2$) such that $\|A\|_2 < \|A\|_F$. (Look for a very simple example.)

(d) Show that $\|A\|_F^2 = \sum_{k=1}^{n} \|Ae_k\|_2^2$, where e_k is the kth *standard basis vector*, having a 1 in the kth position and zeros elsewhere. Deduce that $\|A\|_F \leq \sqrt{n}\, \|A\|_2$.

1.2.11. The following steps lead to a proof of Theorem 1.2.5:

(a) Let $T_k = \sum_{j=0}^{k} S^j$. Show that

$$T_k(I - S) = (I - S)T_k = I - S^{k+1}.$$

(b) Use matrix norm property (e) repeatedly (technically you should use induction on k) to show that $\|S^{k+1}\| \leq \|S\|^{k+1}$ for all k. Deduce that $\lim_{k \to \infty} S^{k+1} = 0$.

(c) Show that

$$\|T_{k+m} - T_k\| \leq \sum_{j=k+1}^{k+m} \|S\|^j < \frac{\|S\|^{k+1}}{1 - \|S\|}$$

by using matrix norm properties (d) and (e) repeatedly and summing a geometric progression.

(d) Deduce that $\lim_{k \to \infty} \sup_{m \geq 0} \| T_{m+k} - T_k \| = 0$. This means that (T_k) is a Cauchy sequence. Since the space $\mathbb{C}^{n \times n}$ is complete, there is a $T \in \mathbb{C}^{n \times n}$ such that $\lim_{k \to \infty} T_k = T$. Since the matrices T_k are the partial sums of $\sum_{j=0}^{\infty} S^j$, we conclude that $T = \sum_{j=0}^{\infty} S^j$.

(e) Taking a limit of the equation from part (a), show that $T(I - S) = (I - S)T = I$. Thus $T = (I - S)^{-1}$.

(f) Show that $\| T_k \| \leq \frac{1}{1 - \|S\|}$ for all k. Deduce that $\| T \| \leq \frac{1}{1 - \|S\|}$.

All of these steps are valid for any matrix norm, so the theorem is valid for any matrix norm, except for the final claim, which relies on the property $\| I \| = 1$. This property holds for any induced matrix norm.

1.2.12. In this exercise you will prove Theorem 1.2.6.

(a) Show that $A + E = A(I + A^{-1}E)$. Therefore $A + E$ is nonsingular if and only if $I + A^{-1}E$ is. Use Theorem 1.2.5 to show that $I + A^{-1}E$ is nonsingular if $\| A^{-1} \| \| E \| < 1$.

(b) Show that $\| A^{-1} \| \| E \| = r$, and $r < 1$ if and only if $\frac{\|E\|}{\|A\|} < \frac{1}{\kappa(A)}$.

(c) Show that $\| (A + E)^{-1} \| \leq \| (I + A^{-1}E)^{-1} \| \| A^{-1} \| \leq \frac{\|A^{-1}\|}{1 - r}$ if $r < 1$.

(d) Put the pieces together to prove Theorem 1.2.6.

1.3   Matrix Groups and Decompositions

Triangular Matrices

A matrix $T \in \mathbb{C}^{n \times n}$ is called *upper triangular* if $t_{ij} = 0$ whenever $i > j$. T is called *lower triangular* if $t_{ij} = 0$ whenever $i < j$. A *triangular* matrix is one that is either upper or lower triangular. A *diagonal* matrix is a matrix that is both upper and lower triangular.

T is *strictly upper triangular* if $t_{ij} = 0$ whenever $i \geq j$. This means that T is upper triangular, and its main-diagonal entries are all zero as well. We define *strictly lower triangular* matrices analogously.

As an easy consequence of the definition of the determinant, we have the following theorem.

Theorem 1.3.1. *If T is triangular, then* $\det(T) = t_{11}t_{22} \cdots t_{nn}$.

Corollary 1.3.2. *A triangular matrix T is nonsingular if and only if its main-diagonal entries $t_{11}, \ldots, t_{nn}$ are all nonzero.*

A proof of Corollary 1.3.2 that does not rely on determinants is given in Exercise 1.3.1.

Proposition 1.3.3.

(a) *Let $T_1, T_2 \in \mathbb{C}^{n \times n}$. If T_1 and T_2 are both upper (resp., lower) triangular, then $T_1 T_2$ is also upper (resp., lower) triangular. If T_1 and T_2 are nonsingular, then $T_1 T_2$ is also nonsingular.*

(b) *If $T \in \mathbb{C}^{n \times n}$ is upper (resp., lower) triangular and nonsingular, then T^{-1} is also upper (resp., lower) triangular.*

All of these assertions are easily proved by induction on n (Exercise 1.3.2). In the language of abstract algebra, the set of nonsingular matrices in $\mathbb{C}^{n \times n}$, together with the operation of matrix multiplication, forms a group. It is called the general linear group and denoted $GL_n(\mathbb{C})$. Proposition 1.3.3 shows that the set of nonsingular upper triangular matrices is a subgroup of $GL_n(\mathbb{C})$, since it is closed under matrix multiplication and inversion, and so is the set of nonsingular lower triangular matrices. Within these subgroups are other subgroups that will be of interest to us. A triangular matrix is called *unit triangular* if its main-diagonal entries are all ones: $t_{ii} = 1, i = 1, \ldots, n$. One easily checks that the product of two unit lower (upper) triangular matrices is unit lower (upper) triangular, and the inverse of a unit lower (upper) triangular matrix is unit lower (upper) triangular. Thus the unit lower (upper) triangular matrices form a subgroup of the nonsingular lower (upper) triangular matrices. Another useful subgroup is the set of nonsingular upper triangular matrices whose main diagonal entries are all real and positive. Each of these subgroups can be intersected with the set of real, nonsingular $n \times n$ matrices $GL_n(\mathbb{R})$ to form a subgroup of real matrices. For example, the set of real, unit upper triangular matrices is a subgroup of $GL_n(\mathbb{R})$.

Given $A \in \mathbb{C}^{n \times n}$, the kth *leading principal submatrix* of A is the $k \times k$ submatrix A_k formed from the first k rows and columns.

Theorem 1.3.4 (*LR* decomposition). *Let $A \in GL_n(\mathbb{C})$, and suppose that the leading principal submatrices A_k are nonsingular for $k = 1, \ldots, n$. Then there exist unique matrices $L \in GL_n(\mathbb{C})$ and $R \in GL_n(\mathbb{C})$ such that L is unit lower triangular, R is upper triangular, and*

$$A = LR.$$

If A is real, then L and R are real.

Almost every matrix satisfies the conditions on the leading principal submatrices, so almost every matrix has an LR decomposition, which is a decomposition of A into a product of matrices from two complementary subgroups. The LR decomposition can be computed by Gaussian elimination without row or column interchanges. This is discussed in [221] and many other books.

Permutation Matrices

A *permutation matrix* is a matrix that has one 1 in each row and each column, all other entries being zero. The set of $n \times n$ permutation matrices is a finite subgroup of $GL_n(R)$ with $n!$ elements. Permutation matrices effect row and column interchanges (Exercise 1.3.3).

If row interchanges are allowed, then every matrix, invertible or not, can be reduced to upper triangular form by Gaussian elimination.

Theorem 1.3.5 (*LR* decomposition with partial pivoting). *Let $A \in \mathbb{C}^{n \times n}$. Then there exist matrices $P \in \mathbb{C}^{n \times n}$, $L \in \mathbb{C}^{n \times n}$, and $R \in \mathbb{C}^{n \times n}$ such that P is a permutation matrix, L is unit lower triangular and satisfies $|l_{ij}| \le 1$ for all i and j, R is upper triangular, and*

$$A = PLR.$$

If A is real, then P, L, and R can be taken to be real.

The matrices P, L, and R, which are not necessarily unique, can be computed by Gaussian elimination with partial pivoting. See, e.g., [221]. R is singular if and only if A is singular.

Unitary Matrices

A matrix $U \in \mathbb{C}^{n \times n}$ is called *unitary* if $U^*U = UU^* = I$, that is, $U^* = U^{-1}$.

Proposition 1.3.6.

   (a) *If U_1 and $U_2 \in \mathbb{C}^{n \times n}$ are both unitary, then $U_1 U_2$ is also unitary.*

   (b) *If $U \in \mathbb{C}^{n \times n}$ is unitary, then U^{-1} is also unitary.*

Thus the unitary matrices are a subgroup of $GL_n(\mathbb{C})$. A unitary matrix that is real is called an *orthogonal* matrix. The set of orthogonal matrices is a subgroup of $GL_n(\mathbb{R})$. Permutation matrices are orthogonal.

Unitary matrices have numerous desirable properties, some of which are listed in the following proposition.

Proposition 1.3.7. *Let $U \in \mathbb{C}^{n \times n}$ and $V \in \mathbb{C}^{n \times n}$ be unitary. Then the following hold:*

   (a) $\langle Ux, Uy \rangle = \langle x, y \rangle$ *for all $x, y \in \mathbb{C}^n$.*

   (b) $\| Ux \| = \| x \|$ *for all $x \in \mathbb{C}^n$.*

   (c) *The columns of U form an orthonormal set.*

   (d) *The rows of U form an orthonormal set.*

   (e) $\| U \| = 1$.

   (f) $\kappa(U) = 1$.

   (g) $\| UAV \| = \| A \|$ *for all $A \in \mathbb{C}^{n \times n}$.*

   (h) $\| UAV \|_F = \| A \|_F$ *for all $A \in \mathbb{C}^{n \times n}$.*

Parts (a) and (b) of this proposition say that unitary matrices preserve the standard inner product and the Euclidean norm. This means that if we think of a unitary matrix as a linear transformation $x \mapsto Ux$, then this linear transformation preserves (Euclidean) lengths and angles of vectors. Examples of unitary (in fact orthogonal) transformations are rotations and reflections. Part (f) says that unitary matrices are perfectly well conditioned; it is not possible for a matrix to have a condition number less than 1. Algorithms that employ unitary transformations tend, therefore, to be very stable.

Theorem 1.3.8 (QR decomposition). *Let $A \in GL_n(\mathbb{C})$. Then there exist unique matrices $Q \in GL_n(\mathbb{C})$ and $R \in GL_n(\mathbb{C})$ such that Q is unitary, R is upper triangular with positive real entries on the main diagonal, and*

$$A = QR.$$

If A is real, then Q and R are real.

The assumption $A \in GL_n(\mathbb{C})$ means that A is nonsingular. Just as for the LR decomposition, we have a unique decomposition of a matrix into a product of matrices from complementary subgroups.

The next theorem drops the assumption of nonsingularity. Every matrix has a QR decomposition, which is, however, not necessarily unique.

Theorem 1.3.9 (QR decomposition). *Let $A \in \mathbb{C}^{n \times n}$. Then there exist matrices $Q \in \mathbb{C}^{n \times n}$ and $R \in \mathbb{C}^{n \times n}$ such that Q is unitary, R is upper triangular, and*

$$A = QR.$$

If A is real, then Q and R can be taken to be real. Whether A is real or not, the main diagonal entries of R can be taken to be real and nonnegative.

Symplectic Matrices

We now consider matrices of dimension $2n$, beginning with the special matrix

$$J = \begin{bmatrix} 0 & I_n \\ -I_n & 0 \end{bmatrix}. \tag{1.3.1}$$

Clearly $J^2 = -I$. Thus J is nonsingular and $J^{-1} = -J = J^T$.

We will discuss only *real* symplectic matrices. A matrix $S \in \mathbb{R}^{2n \times 2n}$ is called *symplectic* if $S^T J S = J$. The following elementary results are easily verified.

Proposition 1.3.10.

(a) *If S_1 and $S_2 \in \mathbb{R}^{2n \times 2n}$ are both symplectic, then $S_1 S_2$ is also symplectic.*

(b) *If S is symplectic, then $\det(S) = \pm 1$, and thus S is nonsingular. Furthermore, $S^{-1} = -J S^T J$.*

(c) *If $S \in \mathbb{R}^{2n \times 2n}$ is symplectic, then S^{-1} is also symplectic.*

(d) *J is symplectic.*

(e) *If $S \in \mathbb{R}^{2n \times 2n}$ is symplectic, then $-S$ is also symplectic.*

(f) *If $S \in \mathbb{R}^{2n \times 2n}$ is symplectic, then $S^T = -J S^{-1} J$. Moreover, S^T is symplectic.*

From parts (a) and (c) of Proposition 1.3.10 we see that the set of symplectic matrices is a subgroup of $GL_{2n}(\mathbb{R})$. An important subgroup of the symplectic group is the group of orthogonal symplectic matrices. The result $\det(S) = \pm 1$ in part (b) is easy to deduce but not the strongest possible. In fact $\det(S) = 1$ for all symplectic matrices, but this is not so obvious [149].

In the study of symplectic matrices it is often convenient to make use of the *perfect shuffle permutation*

$$\hat{P} = \begin{bmatrix} e_1 & e_3 & \cdots & e_{2n-1} & | & e_2 & e_4 & \cdots & e_{2n} \end{bmatrix}, \tag{1.3.2}$$

where e_j is the jth standard basis vector in $\mathbb{R}^{2n}$; that is, e_j has a 1 in the jth position and zeros elsewhere. Obviously $\hat{P}$ is a permutation matrix. For any $x \in \mathbb{R}^{2n}$, the operation

$x \mapsto \hat{P}x$ performs a perfect shuffle of the entries of x (Exercise 1.3.7). Similarly, given any $A \in \mathbb{R}^{2n \times 2n}$, the operation $A \mapsto \hat{P}A\hat{P}^T$ does a perfect shuffle of the rows and columns of A.

A matrix $R \in \mathbb{R}^{2n \times 2n}$ will be called $\hat{P}$-*triangular* if $\hat{P}R\hat{P}^T$ is upper triangular. It is easy to check that R is $\hat{P}$-triangular if and only if

$$R = \begin{bmatrix} R_{11} & R_{12} \\ R_{21} & R_{22} \end{bmatrix},$$

where each block R_{ij} is $n \times n$ and upper triangular, and R_{21} is strictly upper triangular. The set of nonsingular $\hat{P}$-triangular matrices is a subgroup of $GL_{2n}(\mathbb{R})$.

In analogy with the LR and QR decompositions, we have the following SR decomposition theorem.

Theorem 1.3.11 (SR decomposition). *Let $A \in \mathbb{R}^{2n \times 2n}$. Then A has an SR decomposition,*

$$A = SR,$$

where S is symplectic and R is $\hat{P}$-triangular, if and only if the leading principal submatrices of $\hat{P}A^T J A \hat{P}^T$ of orders $2, 4, 6, \ldots, 2n$ are all nonsingular.

This theorem is proved in [81, Theorem 11] and [49, Satz 4.5.11], for example. The condition on the leading principal submatrices says, "Take $A^T J A$, shuffle its rows and columns, and look at the even leading principal submatrices. If all are nonsingular, then A has an SR decomposition." This is a very mild condition, satisfied by almost every $A \in GL_{2n}(\mathbb{R})$.

Hyperbolic Matrices

A real 2×2 matrix of the form

$$H = \begin{bmatrix} c & s \\ s & c \end{bmatrix}, \qquad c^2 - s^2 = 1,$$

is called a *hyperbolic matrix of type* 1. (The graph of the equation $c^2 - s^2 = 1$ is a hyperbola in the (c, s)-plane.) For our purposes we do not care whether c is positive or negative. However, if c is positive, then there exists a real t such that $c = \cosh t$ and $s = \sinh t$. If we define a matrix D by

$$D = \begin{bmatrix} 1 & 0 \\ 0 & -1 \end{bmatrix}, \tag{1.3.3}$$

then one can easily check that every hyperbolic matrix of type 1 satisfies $H^T D H = D$.

A matrix $D \in \mathbb{R}^{n \times n}$ is called a *signature matrix* if it is a diagonal matrix and each of its main-diagonal entries is either $+1$ or -1. Each signature matrix induces an indefinite "inner product" by $\langle x, y \rangle_D = y^T D x$. A matrix $H \in \mathbb{R}^{n \times n}$ is called *D-orthogonal* if it preserves the D-inner product, i.e., $\langle Hx, Hy \rangle_D = \langle x, y \rangle_D$ for all $x, y \in \mathbb{R}^n$. One easily shows that H is D-orthogonal if and only if $H^T D H = D$.

The set of D-orthogonal matrices is a subgroup of $GL_n(\mathbb{R})$. In the special case $D = I$ (or $D = -I$), the D-orthogonal matrices are exactly the orthogonal matrices. For D as in (1.3.3), the hyperbolic matrices of type 1 are D-orthogonal.

We might hope that almost every matrix $A \in \mathbb{R}^{n \times n}$ has a decomposition $A = HR$, where H is D-orthogonal and R is upper triangular. Unfortunately, this turns out not to be true, except in the special cases $D = \pm I$. Instead we have the following result: The kth *leading principal minor* of a matrix is the determinant of the kth leading principal submatrix. A matrix $A \in \mathbb{R}^{n \times n}$ has a decomposition of the specified type if and only if the kth leading principal minor of $A^T D A$ has the same sign as the kth leading principal minor of D for $k = 1, \ldots, n$. (The easier half of this assertion is proved in Exercise 1.3.10.) The set of matrices satisfying this condition is not dense in $\mathbb{R}^{n \times n}$ except when $D = \pm I$.

To get a useful HR decomposition, we must leave the realm of group theory. Let D and $\hat{D}$ be signature matrices. Then H is $(D, \hat{D})$-*orthogonal* if $H^T D H = \hat{D}$.

Interchanging the rows of a hyperbolic matrix of type 1, we obtain a new matrix

$$ H = \begin{bmatrix} s & c \\ c & s \end{bmatrix}, \qquad c^2 - s^2 = 1, $$

which is called a *hyperbolic matrix of type* 2. It is a simple matter to show that for D as in (1.3.3), every hyperbolic matrix of type 2 satisfies $H^T D H = -D$. Thus hyperbolic matrices of type 2 are $(D, \hat{D})$-orthogonal, where $\hat{D} = -D$.

Theorem 1.3.12 (*HR decomposition*). *Let $A \in \mathbb{R}^{n \times n}$, let D be a signature matrix, and suppose that the leading principal submatrices of $A^T D A$ are all nonsingular. Then there exists a signature matrix $\hat{D}$, a $(D, \hat{D})$-orthogonal matrix H, and an upper triangular matrix R such that*

$$ A = HR. $$

Almost every $A \in \mathbb{R}^{n \times n}$ satisfies the condition that the leading principal submatrices of $A^T D A$ are nonsingular. The HR decomposition theorem is proved in [81, Theorem 9] and [49, Satz 4.5.13], for example.

Exercises

1.3.1. This exercise leads to a proof of Corollary 1.3.2 that does not rely on determinants. We'll do the lower triangular case. By Theorem 1.1.1, we know that T is nonsingular if and only if the only solution of $T y = 0$ is $y = 0$.

(a) Suppose that T is lower triangular and $t_{ii} = 0$ for some i. Let j be the largest index for which $t_{jj} = 0$. Show that the equation $T y = 0$ has a nonzero solution y with $y_j = 1$ and $y_i = 0$ for $i < j$. You may find it helpful to partition T as

$$ T = \begin{bmatrix} T_{11} & 0 & 0 \\ r & t_{jj} & 0 \\ T_{31} & c & T_{33} \end{bmatrix} $$

and show that the equation $Ty = 0$ has a unique solution of the form

$$y = \begin{bmatrix} 0 \\ 1 \\ z \end{bmatrix},$$

where the 1 is in the jth position. (Note: The equation $Ty = 0$ has many other solutions, but only one of this form.)

(b) Conversely, suppose that T is lower triangular with $t_{ii} \neq 0$ for $i = 1, \ldots, n$. Prove by induction on n that the equation $Ty = 0$ has no solutions other than $y = 0$. For the induction step, partition T as

$$T = \begin{bmatrix} t_{11} & 0 \\ c & \hat{T} \end{bmatrix}.$$

1.3.2. Prove Proposition 1.3.3. For both parts you can use partitioned matrices and induction on n. For part (b) let $S = T^{-1}$ and use the equation $ST = I$ or $TS = I$.

1.3.3.

(a) Show that the product of two permutation matrices is a permutation matrix.

(b) Show that if P is a permutation matrix, then $P^T = P^{-1}$. In particular, the inverse of a permutation matrix is a permutation matrix.

(c) Show that if P is a permutation matrix, then PA differs from A only in that its rows have been reordered.

(d) Show that if P is a permutation matrix, then AP differs from A only in that its columns have been reordered.

1.3.4. Prove Proposition 1.3.6.

1.3.5. Prove Proposition 1.3.7.

1.3.6. Prove Proposition 1.3.10.

1.3.7. Let $\hat{P}$ be the permutation matrix given by (1.3.2).

(a) Show that for any $x \in \mathbb{R}^{2n}$, $\hat{P}x = [x_1 \ x_{n+1} \ x_2 \ x_{n+2} \ \cdots \ x_n \ x_{2n}]^T$. Thus $\hat{P}$ performs a perfect shuffle of the entries of x.

(b) Show that the operations $A \mapsto \hat{P}A$ and $B \mapsto B\hat{P}^T$ perfectly shuffle the rows of A and the columns of B, respectively. Deduce that the operation $A \mapsto \hat{P}A\hat{P}^T$ performs a perfect shuffle of the rows and columns of A.

(c) Show that $R \in \mathbb{R}^{2n \times 2n}$ is $\hat{P}$-triangular if and only if

$$R = \begin{bmatrix} R_{11} & R_{12} \\ R_{21} & R_{22} \end{bmatrix},$$

where R_{11}, R_{12}, and R_{22} are upper triangular, and R_{21} is strictly upper triangular.

1.3.8. Show that the set of all hyperbolic matrices of type 1

$$H = \begin{bmatrix} c & s \\ s & c \end{bmatrix}, \qquad c^2 - s^2 = 1,$$

is a subgroup of $GL_2(\mathbb{R})$.

1.3.9.

    (a) Show that $H \in \mathbb{R}^{n \times n}$ is D-orthogonal if and only if $H^T D H = D$.

    (b) Show that if H is D-orthogonal, then $\det(H) = \pm 1$.

    (c) Show that the set of D-orthogonal matrices is a subgroup of $GL_n(\mathbb{R})$.

1.3.10. Suppose $A \in \mathbb{R}^{n \times n}$ has a decomposition $A = HR$, where H is D-orthogonal and R is upper triangular.

    (a) Show that $A^T D A = R^T D R$.

    (b) Show that the kth leading principal minor of $A^T D A$ has the same sign as the kth leading principal minor of D, $k = 1, \ldots, n$.

1.4 Some Other Classes of Matrices

In this section we discuss some important classes of matrices that are not matrix groups.

Normal Matrices

A matrix $A \in \mathbb{C}^{n \times n}$ is called *normal* if $A^* A = A A^*$. A is *Hermitian* if $A^* = A$, and *skew Hermitian* if $A^* = -A$. Recall from Section 1.3 that A is *unitary* if $A^* A = A A^* = I$. Obviously the set of normal matrices contains the Hermitian matrices, the skew-Hermitian matrices, and the unitary matrices.

Now suppose that A is real. Then if A is also Hermitian, it satisfies $A^T = A$ and is called *symmetric*. If A is skew Hermitian, then it satisfies $A^T = -A$ and is called *skew symmetric*.

Let $A \in \mathbb{C}^{n \times n}$ be Hermitian, and let $x \in \mathbb{C}^n$. Then $x^* A x$ is a 1×1 matrix; that is, it is a scalar. Since A is Hermitian, $x^* A x$ is easily seen to be real. A Hermitian matrix is called *positive definite* if $x^* A x > 0$ for all nonzero $x \in \mathbb{C}^n$. Every positive definite matrix is nonsingular. A key result for positive definite matrices is the *Cholesky decomposition theorem*.

Theorem 1.4.1. *Let $A \in \mathbb{C}^{n \times n}$ be a Hermitian positive definite matrix. Then there exists a unique upper triangular $R \in \mathbb{C}^{n \times n}$ such that the main-diagonal entries of R are real and positive, and*

$$A = R^* R.$$

If A is real, then R is real.

The matrix R is called the *Cholesky factor* of A, and the decomposition $A = R^* R$ is called the *Cholesky decomposition*. Theorem 1.4.1 is proved in [221] and many other places.

The Cholesky decomposition is closely related to the LR decomposition (Exercise 1.4.2), and it can be computed by a symmetric variant of Gaussian elimination.

Hamiltonian and Skew-Hamiltonian Matrices

In Section 1.3 we introduced the group of symplectic matrices. Closely related to the symplectic group are the Hamiltonian and skew-Hamiltonian matrices. We will not develop that relationship here; we will just give a few basic facts.

Again we make use of the matrix

$$J = \begin{bmatrix} 0 & I_n \\ -I_n & 0 \end{bmatrix},$$

and we restrict our attention to real matrices. A matrix $H \in \mathbb{R}^{2n \times 2n}$ is called *Hamiltonian* if JH is symmetric, that is, $(JH)^T = JH$. One easily checks that a matrix is Hamiltonian if and only if its block structure is

$$H = \begin{bmatrix} A & K \\ N & -A^T \end{bmatrix}, \qquad \text{where} \qquad N = N^T, \quad K = K^T.$$

Hamiltonian matrices arise in the study of dynamical systems; Exercise 1.4.5 illustrates this in a very simple case.

A matrix $H \in \mathbb{R}^{2n \times 2n}$ is called *skew-Hamiltonian* if JH is skew symmetric, that is, $(JH)^T = -JH$. H is skew-Hamiltonian if and only if its block structure is

$$H = \begin{bmatrix} A & K \\ N & A^T \end{bmatrix}, \qquad \text{where} \qquad N = -N^T, \quad K = -K^T.$$

Proposition 1.4.2. *Suppose that H, H_1, and H_2 are Hamiltonian (resp., skew-Hamiltonian). Then the following hold:*

(a) *$H_1 + H_2$ is Hamiltonian (resp., skew-Hamiltonian).*

(b) *αH is Hamiltonian (resp., skew-Hamiltonian) for all real α.*

(c) *H^T is Hamiltonian (resp., skew-Hamiltonian).*

(d) *If H is nonsingular, then H^{-1} is Hamiltonian (resp., skew-Hamiltonian).*

Parts (a) and (b) show that the set of Hamiltonian (resp., skew-Hamiltonian) matrices is a (vector) subspace of $\mathbb{R}^{2n \times 2n}$.

Proposition 1.4.3. *If H is Hamiltonian, then H^2 is skew-Hamiltonian.*

Proposition 1.4.4. *If S is symplectic, then $S + S^{-1}$ is skew-Hamiltonian, and $S - S^{-1}$ is Hamiltonian.*

Exercises

1.4.1. Let $A \in \mathbb{C}^{n \times n}$, $X \in \mathbb{C}^{n \times m}$, and $x \in \mathbb{C}^n$.

    (a) Show that if A is Hermitian, then X^*AX is Hermitian.

    (b) Show that if A is skew-Hermitian, then X^*AX is skew-Hermitian.

    (c) Show that if A is Hermitian, then x^*Ax is real.

    (d) Show that if A is skew-Hermitian, then x^*Ax is purely imaginary.

1.4.2. Let A be positive definite.

    (a) Show that there is no nonzero vector z such that $Az = 0$. Deduce that A is nonsingular.

    (b) Show that each of the leading principal submatrices of A is positive definite and hence nonsingular. Deduce that A has a unique decomposition $A = LU$, where L is unit lower triangular and U is upper triangular (Theorem 1.3.4).

    (c) Let R be the Cholesky factor of A, and let D be the diagonal matrix built from the main diagonal entries of R: $D = \operatorname{diag}\{r_{11}, \dots, r_{nn}\}$. State and prove formulas for L and U in terms of D and R.

1.4.3. Prove that the matrix $H \in \mathbb{R}^{2n \times 2n}$ is Hamiltonian if and only if it has the block structure

$$H = \begin{bmatrix} A & N \\ K & -A^T \end{bmatrix},$$

where $N = N^T$ and $K = K^T$.

1.4.4. Prove that the matrix $H \in \mathbb{R}^{2n \times 2n}$ is skew-Hamiltonian if and only if it has the block structure

$$H = \begin{bmatrix} A & N \\ K & A^T \end{bmatrix},$$

where $N = -N^T$ and $K = -K^T$.

1.4.5. In the Hamiltonian formulation of classical mechanics [10], the state of a dynamical system is described by $2n$ functions of time:

$$q = \begin{bmatrix} q_1(t) \\ q_2(t) \\ \vdots \\ q_n(t) \end{bmatrix} \quad \text{and} \quad p = \begin{bmatrix} p_1(t) \\ p_2(t) \\ \vdots \\ p_n(t) \end{bmatrix}.$$

The $q_i(t)$ are called *generalized position* coordinates, and the $p_i(t)$ are *generalized momenta*. The dynamics of the system are determined by *Hamilton's equations*,

$$\dot{q} = \frac{\partial \mathcal{H}}{\partial p}, \qquad \dot{p} = -\frac{\partial \mathcal{H}}{\partial q},$$

where $\mathcal{H}$ is the *Hamiltonian* (total energy) of the system. In the simplest dynamical systems the Hamiltonian has the form

$$\frac{1}{2}q^T K q + \frac{1}{2}p^T M p,$$

where K and M are real, symmetric, positive-definite matrices. The first term represents potential energy, and the second represents kinetic energy (something like $\frac{1}{2}mv^2$). Let us consider a slightly more complicated Hamiltonian, adding a term in which both the positions and the momenta appear:

$$\mathcal{H} = \frac{1}{2}q^T K q + \frac{1}{2}p^T M p + p^T A q.$$

Show that for this Hamiltonian, Hamilton's equations take the form

$$\begin{bmatrix} \dot{q} \\ \dot{p} \end{bmatrix} = \begin{bmatrix} A & M \\ -K & -A^T \end{bmatrix} \begin{bmatrix} q \\ p \end{bmatrix}.$$

Notice that the coefficient matrix of this linear system is a Hamiltonian matrix.

1.4.6. Prove Proposition 1.4.2. For the Hamiltonian case of parts (c) and (d), start from the equation $(JH)^T = JH$ or the equivalent $H^T J = -JH$, then multiply on left or right by J, J^T, H^{-1}, and/or H^{-T}, as appropriate. Notice, however, that you cannot use H^{-1} or H^{-T} in part (c), since H might be singular.

1.4.7. Prove Proposition 1.4.3: The square of a Hamiltonian matrix is skew-Hamiltonian.

1.4.8. Prove Proposition 1.4.4.

1.5   Subspaces

Subspaces of the vector spaces $\mathbb{R}^n$ and $\mathbb{C}^n$ play a substantial role in the study of matrices. In discussing complex matrices we normally make use of subspaces of $\mathbb{C}^n$, while in the study of real matrices we sometimes prefer to focus on subspaces of $\mathbb{R}^n$. In order to keep the discussion flexible, let us introduce the symbol $\mathbb{F}$ to mean either the field of real numbers or the field of complex numbers. Then we can speak of subspaces of $\mathbb{F}^n$, the vector space of n-tuples with entries in $\mathbb{F}$. Similarly, $\mathbb{F}^{n \times m}$ will denote the set of $n \times m$ matrices with entries in $\mathbb{F}$.

A *subspace* of $\mathbb{F}^n$ is a nonempty subset that is closed under vector addition and scalar multiplication. In other words, the nonempty subset $\mathcal{S}$ of $\mathbb{F}^n$ is a subspace if the following two conditions hold:

$$v \in \mathcal{S}, \; w \in \mathcal{S} \Rightarrow v + w \in \mathcal{S},$$

$$v \in \mathcal{S}, \; \alpha \in \mathbb{F} \Rightarrow \alpha v \in \mathcal{S}.$$

Given vectors $s_1, \ldots, s_k \in \mathcal{S}$, a vector of the form

$$\sum_{j=1}^k \alpha_j s_j,$$

where $\alpha_1, \alpha_2, \ldots, \alpha_k$ are scalars in $\mathbb{F}$, is called a *linear combination* of the vectors $s_1, \ldots, s_k$. An easy induction argument shows that if S is a subspace of $\mathbb{F}^n$ and $s_1, \ldots, s_k$ are vectors in S, then every linear combination of $s_1, \ldots, s_k$ also lies in S.

Every subspace of $\mathbb{F}^n$ contains the vector 0. The one-element set $\{0\}$ is a subspace of $\mathbb{F}^n$, and $\mathbb{F}^n$ is itself a subspace of $\mathbb{F}^n$. We call these *trivial* subspaces of $\mathbb{F}^n$. We will focus mainly on nontrivial subspaces of $\mathbb{F}^n$.

Given any vectors $s_1, \ldots, s_k \in \mathbb{F}^n$, the set consisting of all linear combinations of $s_1, \ldots, s_k$ is a subspace of $\mathbb{F}^n$. It is denoted span$\{s_1, \ldots, s_k\}$.

Basis and Dimension

Henceforth let S denote a subspace of $\mathbb{F}^n$, and let $s_1, \ldots, s_k$ denote a set of k vectors in S. Every subspace contains the zero vector, and whatever vectors $s_1, \ldots, s_k$ happen to be, we can always express 0 as a linear combination of $s_1, \ldots, s_k$. Indeed, if we take $\alpha_1 = \alpha_2 = \cdots = \alpha_k = 0$, then

$$\alpha_1 s_1 + \alpha_2 s_2 + \cdots + \alpha_k s_k = 0. \tag{1.5.1}$$

In general there might be other choices of $\alpha_1, \ldots, \alpha_k$ for which (1.5.1) holds. If not, $s_1, \ldots, s_k$ are called *linearly independent*. In other words $s_1, \ldots, s_k$ are *linearly independent* if and only if the only solution of (1.5.1) is $\alpha_1 = \cdots = \alpha_k = 0$. The vectors $s_1, \ldots, s_k$ are called a *spanning set* for S if every $v \in S$ is a linear combination of $s_1, \ldots, s_k$. In this case we also say that $s_1, \ldots, s_k$ *span* S. $s_1, \ldots, s_k$ are called a *basis* for S if they are linearly independent *and* they span S. If $s_1, \ldots, s_k$ form a basis for S, then each vector $v \in S$ can be expressed as a linear combination of $s_1, \ldots, s_k$ in exactly one way. That is, for each $v \in S$, there are unique $\alpha_1, \ldots, \alpha_k \in \mathbb{F}$ such that $v = \alpha_1 s_1 + \cdots + \alpha_k s_k$.

Recall that the set $s_1, \ldots, s_k$ is called *orthonormal* if $\langle s_i, s_j \rangle = \delta_{ij}$. Every orthonormal set is linearly independent. (The converse is false; there are numerous linearly independent sets that are not orthonormal.) An orthonormal set that is also a basis of S is called an *orthonormal basis* of S.

We will not work out the theory of bases and dimension, which can be found in any elementary linear algebra text. Every nonzero subspace of $\mathbb{F}^n$ has a basis, indeed many bases. Among these are orthonormal bases. It is not hard to show that any two bases of S have the same number of elements. This number is called the *dimension* of the subspace.[1]

The *standard basis* of $\mathbb{F}^n$ is the set of vectors $e_1, e_2, \ldots, e_n$, where e_i is the vector with a one in the ith position and zeros elsewhere. In other words, $e_1, \ldots, e_n$ are the columns of the identity matrix I_n. Obviously $e_1, e_2, \ldots, e_n$ is an orthonormal basis of $\mathbb{F}^n$. Since $\mathbb{F}^n$ has a basis consisting of n vectors, its dimension is n.

The Projection Theorem

Let S be a subspace of $\mathbb{F}^n$, and let v be a vector that does not lie in S. Is there an element $\hat{s} \in S$ that best approximates v in the sense that $\| v - \hat{s} \|$ (the Euclidean norm) is minimal? If so, how can we characterize s? Our experience with low-dimensional real spaces suggests

---

[1] The subspace $\{0\}$ has the empty set as a basis, and its dimension is zero.

that there is a best s and that it is obtained by projecting v orthogonally onto $\mathcal{S}$. (The shortest distance from a point to a plane is along the perpendicular.) The following theorem shows that this expectation is correct.

Theorem 1.5.1 (projection theorem). *Let $v \in \mathbb{F}^n$, and let S be a subspace of $\mathbb{F}^n$. Then there is a unique $\hat{s} \in S$ that best approximates v in the sense that*

$$\| v - \hat{s} \| = \min\{ \| v - s \| \mid s \in \mathcal{S} \}.$$

$\hat{s}$ is characterized by

$$\langle v - \hat{s}, s \rangle = 0 \quad \text{for all } s \in \mathcal{S}.$$

In other words, $\hat{s}$ is the orthogonal projection of v into S.

The projection theorem is proved in Exercise 1.5.3.

Sums, Intersections, and Orthogonal Complements

There are several simple ways to build new subspaces from old. If $\mathcal{S}$ and $\mathcal{U}$ are subspaces of $\mathbb{F}^n$, then the *intersection* $\mathcal{S} \cap \mathcal{U}$ and the *sum*

$$\mathcal{S} + \mathcal{U} = \{s + u \mid s \in \mathcal{S}, \ u \in \mathcal{U}\}$$

are both subspaces of $\mathcal{S}$. The dimensions of these subspaces are related by a simple rule. See Exercise 1.5.6.

Theorem 1.5.2. *Let S and V be subspaces of $\mathbb{F}^n$. Then*

$$\dim(\mathcal{S} + \mathcal{U}) + \dim(\mathcal{S} \cap \mathcal{U}) = \dim(\mathcal{S}) + \dim(\mathcal{U}).$$

If $\mathcal{S} \cap \mathcal{U} = \{0\}$, then $\mathcal{S} + \mathcal{U}$ is called a *direct sum* and denoted $\mathcal{S} \oplus \mathcal{U}$. By definition, each $v \in \mathcal{S} + \mathcal{U}$ can be written as $v = s + u$ for some $s \in \mathcal{S}$ and $u \in \mathcal{U}$. If the sum is a direct sum $\mathcal{S} \oplus \mathcal{U}$, then this decomposition is unique: for each $v \in \mathcal{S} \oplus \mathcal{U}$, there is a unique $s \in \mathcal{S}$ and $u \in \mathcal{U}$ such that $v = s + u$.

Every subspace $\mathcal{S}$ of $\mathbb{F}^n$ has an *orthogonal complement* $\mathcal{S}^\perp$ defined by

$$\mathcal{S}^\perp = \{w \in \mathbb{F}^n \mid \langle v, w \rangle = 0 \text{ for all } v \in \mathcal{S}\}.$$

Thus $\mathcal{S}^\perp$ consists of all those vectors in $\mathbb{F}^n$ that are orthogonal to every vector in $\mathcal{S}$. Some of the basic facts about the orthogonal complement are catalogued in the following theorem. (See Exercise 1.5.8.)

Theorem 1.5.3. *Let S and $\mathcal{U}$ be subspaces of $\mathbb{F}^n$. Then the following hold:*

(a) $\mathcal{S}^\perp$ *is a subspace of $\mathbb{F}^n$.*

(b) $\mathcal{S} \cap \mathcal{S}^\perp = \{0\}$.

(c) $\mathcal{S} \oplus \mathcal{S}^\perp = \mathbb{F}^n$.

(d) $\dim(\mathcal{S}) + \dim(\mathcal{S}^\perp) = n$.

(e) $\mathcal{S}^{\perp\perp} = \mathcal{S}$.

(f) $(\mathcal{S} + \mathcal{U})^\perp = \mathcal{S}^\perp \cap \mathcal{U}^\perp$.

Range and Null Space

Every matrix $A \in \mathbb{F}^{n \times m}$ can be associated with a linear transformation that maps $\mathbb{F}^m$ into $\mathbb{F}^n$ by the action $x \mapsto Ax$. Viewing A in this way, we can associate two important subspaces with A. The *range* of A, denoted $\mathcal{R}(A)$, is given by

$$\mathcal{R}(A) = \{Ax \mid x \in \mathbb{F}^m\}.$$

This is consistent with the usual use of the word range as applied to a function. One easily checks that $\mathcal{R}(A)$ is a subspace of $\mathbb{F}^n$. The *null space* of A, denoted $\mathcal{N}(A)$, is given by

$$\mathcal{N}(A) = \{x \in \mathbb{F}^m \mid Ax = 0\}.$$

It is easy to check that $\mathcal{N}(A)$ is a subspace of $\mathbb{F}^m$.

The dimensions of $\mathcal{R}(A)$ and $\mathcal{N}(A)$ are called the *rank* of A and the *nullity* of A and denoted $\mathrm{rank}(A)$ and $\mathrm{null}(A)$, respectively. The simple relationship between rank and nullity is given by the following proposition.

Proposition 1.5.4. *If $A \in \mathbb{F}^{n \times m}$, then $\mathrm{rank}(A) + \mathrm{null}(A) = m$.*

Proposition 1.5.4 is proved in Exercise 1.5.10.

If $A \in \mathbb{F}^{n \times m}$, then $A^* \in \mathbb{F}^{m \times n}$. (In the real case, A^* coincides with A^T.) Associated with A^* we have $\mathcal{R}(A^*) \subseteq \mathbb{F}^m$ and $\mathcal{N}(A^*) \subseteq \mathbb{F}^n$.

Proposition 1.5.5. *For every $A \in \mathbb{F}^{n \times m}$, $\mathcal{R}(A)^\perp = \mathcal{N}(A^*)$ and $\mathcal{R}(A^*)^\perp = \mathcal{N}(A)$.*

Proposition 1.5.5 is proved in Exercise 1.5.11. In words, Proposition 1.5.5 says that $\mathcal{R}(A)$ and $\mathcal{N}(A^*)$ are orthogonal complements in the n-dimensional space $\mathbb{F}^n$. It follows that the sum of their dimensions is n. Thus $\mathrm{rank}(A) + \mathrm{null}(A^*) = n$. Similar remarks apply to $\mathcal{R}(A^*)$ and $\mathcal{N}(A)$. We summarize these findings as a corollary.

Corollary 1.5.6. *Let $A \in \mathbb{F}^{n \times m}$. Then $\mathrm{rank}(A) + \mathrm{null}(A^*) = n$ and $\mathrm{rank}(A^*) + \mathrm{null}(A) = m$.*

If we now compare the statement of Proposition 1.5.4 with the second statement of Corollary 1.5.6, we deduce immediately that A and A^* have the same rank.

Theorem 1.5.7. *For every matrix $A \in \mathbb{F}^{n \times m}$, $\mathrm{rank}(A) = \mathrm{rank}(A^*)$.*

If A is real, then A^* is the same as A^T, but if A is complex, there is a difference between A^* and A^T. Since we are sometimes interested in the transpose of a complex matrix, let us pause for a moment to consider the relationship between A^* and A^T. These matrices are just complex conjugates of one another. It follows easily that $\mathcal{R}(A^T)$ is just the complex conjugate of $\mathcal{R}(A^*)$ and therefore has the same dimension. Thus $\mathrm{rank}(A^T) = \mathrm{rank}(A^*)$. Similarly $\mathrm{null}(A^T) = \mathrm{null}(A^*)$. The first of these observations gives us the following corollary.

Corollary 1.5.8. *For every $A \in \mathbb{C}^{n \times n}$, $\mathrm{rank}(A) = \mathrm{rank}(A^T)$.*

In general A and A^* do not have the same nullity. However, in the square case $n = m$, they do. This gives the following result.

Theorem 1.5.9. *For every square matrix A, $\mathrm{null}(A) = \mathrm{null}(A^*) = \mathrm{null}(A^T)$.*

Working with Subspaces

How does one get one's hands on a subspace? The usual way is to pick a spanning set (usually a basis) and work with that. If $s_1, \ldots, s_k$ is a spanning set of the subspace $\mathcal{S}$, then $\mathcal{S}$ is exactly the set of linear combinations of $s_1, \ldots, s_k$; that is, $\mathcal{S} = \mathrm{span}\{s_1, \ldots, s_k\}$. Thus each $s \in \mathcal{S}$ can be written as a linear combination $s = \sum_{j=1}^{k} s_j \alpha_j$. It is convenient to pack the spanning set into a matrix, so let $S = \begin{bmatrix} s_1 & s_2 & \cdots & s_k \end{bmatrix} \in \mathbb{F}^{n \times k}$. Then the linear combination $\sum_{j=1}^{k} s_j \alpha_j$ can be expressed more succinctly as Sx, where $x \in \mathbb{F}^{n \times k}$ is the column vector whose components are $\alpha_1, \ldots, \alpha_k$. Thus each $s \in \mathcal{S}$ can be written in the form $s = Sx$ (and conversely, each Sx is in $\mathcal{S}$), so $\mathcal{S} = \{Sx \mid x \in \mathbb{F}^k\} = \mathcal{R}(S)$.[2] Consequently $\dim(\mathcal{S}) = \mathrm{rank}(S)$. We summarize these findings as in the following proposition.

Proposition 1.5.10. *Let $\mathcal{S}$ be a subspace of $\mathbb{F}^n$, let $s_1, \ldots, s_k$ be vectors in $\mathbb{F}^n$, and let $S = \begin{bmatrix} s_1 & \cdots & s_k \end{bmatrix} \in \mathbb{F}^{n \times k}$. Then $\mathcal{S} = \{Sx \mid x \in \mathbb{F}^k\} = \mathcal{R}(S)$ if and only if $s_1, \ldots, s_k$ lie in $\mathcal{S}$ and form a spanning set for $\mathcal{S}$. In this case, $\dim(\mathcal{S}) = \mathrm{rank}(S)$.*

We will say that the matrix $S \in \mathbb{F}^{n \times k}$ *represents* the subspace $\mathcal{S} \subseteq \mathbb{F}^n$ if $\mathcal{S} = \mathcal{R}(S)$. Normally when we want to do some computations involving a subspace $\mathcal{S}$, we work with a representative matrix S instead. Usually the spanning set $s_1, \ldots, s_k$ is taken to be a basis. In this case the representation of each $s \in \mathcal{S}$ in terms of the basis is unique, so there is a unique $x \in \mathbb{F}^k$ such that $s = Sx$. Furthermore $\mathrm{rank}(S) = \dim(\mathcal{S}) = k$. Since an $n \times k$ matrix cannot possibly have rank greater than k, we say that S has *full rank* in this case.

Each subspace has many different representative matrices. Suppose that $s_1, \ldots, s_k$ and $\hat{s}_1, \ldots, \hat{s}_k$ are two different bases of $\mathcal{S}$, and S and $\hat{S}$ are the matrices whose columns are $s_1, \ldots, s_k$ and $\hat{s}_1, \ldots, \hat{s}_k$, respectively. What is the relationship between S and $\hat{S}$? Conversely, given two matrices $S \in \mathbb{F}^{n \times k}$ and $\hat{S} \in \mathbb{F}^{n \times k}$, both of rank k, under what conditions do they represent the same space? In other words, when is $\mathcal{R}(S) = \mathcal{R}(\hat{S})$? The following proposition, which is proved in Exercise 1.5.12, answers these questions.

Proposition 1.5.11. *Let $S \in \mathbb{F}^{n \times k}$ and $\hat{S} \in \mathbb{F}^{n \times k}$ be matrices of rank k.*

(a) *Then $\mathcal{R}(S) = \mathcal{R}(\hat{S})$ if and only if there is a nonsingular matrix $X \in \mathbb{F}^{k \times k}$ such that $\hat{S} = SX$.*

(b) *Suppose the matrices S and $\hat{S}$ each have orthonormal columns, and suppose $\mathcal{R}(S) = \mathcal{R}(\hat{S})$ and $\hat{S} = SX$. Then X is unitary.*

Exercises

1.5.1. By definition, $\mathrm{span}\{s_1, \ldots, s_k\}$ is the set of all linear combinations of $s_1, \ldots, s_k$. Show that $\mathrm{span}\{s_1, \ldots, s_k\}$ is a subspace of $\mathbb{F}^n$.

1.5.2. Prove that if $s_1, \ldots, s_k$ form a basis for the subspace $\mathcal{S}$, then for every $v \in \mathcal{S}$ there exist unique $\alpha_1, \alpha_2, \ldots, \alpha_k \in \mathbb{F}$ such that $v = \alpha_1 s_1 + \alpha_2 s_2 + \cdots + \alpha_k s_k$. There are two parts to this (existence and uniqueness), but one part is trivial.

---

[2]The *column space* of a matrix is defined to be the space spanned by its columns. The connection $Sx = \sum_{j=1}^{k} s_j \alpha_j$ shows clearly that the terms "column space" and "range" are synonymous.

1.5.3. In this exercise you will prove the projection theorem (Theorem 1.5.1). Let $v \in \mathbb{F}^n$, and let $\mathcal{S}$ be a subspace of $\mathbb{F}^n$.

    (a) We will begin by showing that v has a unique orthogonal projection into $\mathcal{S}$. That is, there is a unique $\hat{s} \in \mathcal{S}$ that satisfies $\langle v - \hat{s}, s \rangle = 0$ for all $s \in \mathcal{S}$. To this end, let $s_1, \ldots, s_k$ be an orthonormal basis of $\mathcal{S}$. Then $\hat{s}$, if it exists, has the form $\hat{s} = \sum_{i=1}^{k} s_i \alpha_i$ for some $\alpha_1, \ldots, \alpha_k \in \mathbb{F}$. Show that $\langle v - \hat{s}, s_j \rangle = 0$ if and only if $\alpha_j = \langle v, s_j \rangle$. Deduce that there is a unique $\hat{s} \in \mathcal{S}$ such that $\langle v - \hat{s}, s \rangle = 0$ for all $s \in \mathcal{S}$ and that this vector is given by

$$\hat{s} = \sum_{i=1}^{n} \langle v, s_i \rangle s_i.$$

        The coefficients $\alpha_i = \langle v, s_i \rangle$ are variously known as *Gram–Schmidt* or *Fourier* coefficients, depending on the context.

    (b) Now we will show that the vector $\hat{s}$ computed in part (a) is the unique best approximation to v from $\mathcal{S}$. Given $s \in \mathcal{S}$, clearly $v - s = (v - \hat{s}) + (\hat{s} - s)$. Show that $v - \hat{s}$ is orthogonal to $\hat{s} - s$, and then use the Pythagorean theorem to show that $\| v - s \| \geq \| v - \hat{s} \|$, with equality if and only if $s = \hat{s}$. This completes the proof of the projection theorem.

1.5.4. Let $\mathcal{V}$ and $\mathcal{W}$ be subspaces of $\mathbb{F}^n$ with $\mathcal{V} \subseteq \mathcal{W}$. Let $v_1, \ldots, v_j$ be a basis for $\mathcal{V}$.

    (a) Show that if $\mathcal{V} \neq \mathcal{W}$, then there is a $w \in \mathcal{W}$ such that $v_1, \ldots, v_j, w$ is a linearly independent set.

    (b) Show that if $\mathcal{V} \neq \mathcal{W}$, then there exist vectors $w_{j+1}, \ldots, w_k \in \mathcal{W}$ such that $v_1, \ldots v_j, w_{j+1}, \ldots w_k$ together form a basis for $\mathcal{W}$.

    (c) Deduce that $\mathcal{V} = \mathcal{W}$ if and only if $\dim(\mathcal{V}) = \dim(\mathcal{W})$.

1.5.5. Let $\mathcal{V}$ and $\mathcal{W}$ be subspaces of $\mathbb{F}^n$ with $\mathcal{V} \subseteq \mathcal{W}$. Let $v_1, \ldots, v_j$ be an orthonormal basis for $\mathcal{V}$.

    (a) Show that if $\mathcal{V} \neq \mathcal{W}$, then there is a $w \in \mathcal{W}$ such that $v_1, \ldots, v_j, w$ is an orthonormal set.

    (b) Show that if $\mathcal{V} \neq \mathcal{W}$, then there exist vectors $w_{j+1}, \ldots, w_k \in \mathcal{W}$ such that $v_1, \ldots v_j, w_{j+1}, \ldots w_k$ together form an orthonormal basis for $\mathcal{W}$.

1.5.6. Let $\mathcal{S}$ and $\mathcal{U}$ be subspaces of $\mathbb{F}^n$.

    (a) Show that $\mathcal{S} \cap \mathcal{U}$ is a subspace of $\mathbb{F}^n$.

    (b) Show that $\mathcal{S} + \mathcal{U}$ is a subspace of $\mathbb{F}^n$.

    (c) Prove Theorem 1.5.2:

$$\dim(\mathcal{S} + \mathcal{U}) + \dim(\mathcal{S} \cap \mathcal{U}) = \dim(\mathcal{S}) + \dim(\mathcal{U}). \tag{1.5.2}$$

Start with a basis for $S \cap U$. From Exercise 1.5.4 you know that you can extend it to form a basis for S. Then you can do a different extension to get a basis for U. From these vectors pick out a basis for $S + U$, and prove that it is a basis. Then compare the number of vectors in the various bases to obtain (1.5.2).

1.5.7. Let S and U be subspaces of $\mathbb{F}^n$ satisfying $S \cap U = \{0\}$. Prove that for each $v \in S \oplus U$ there is a unique $s \in S$ and $u \in U$ such that $v = s + u$. (Show that if $v = s_1 + u_1 = s_2 + u_2$, where $s_1, s_2 \in S$ and $u_1, u_2 \in U$, then $s_1 = s_2$ and $u_1 = u_2$.)

1.5.8. Let S and U be subspaces of $\mathbb{F}^n$.

  (a) Show that $S^\perp$ is a subspace of $\mathbb{F}^n$.

  (b) Show that $S \cap S^\perp = \{0\}$.

  (c) Using Theorem 1.5.1, show that every $v \in \mathbb{F}^n$ can be expressed in the form $v = s + t$, where $s \in S$ and $t \in S^\perp$. Deduce that $S + S^\perp = \mathbb{F}^n$.

  (d) Show that $\dim(S) + \dim(S^\perp) = n$.

  (e) Show that $S \subseteq S^{\perp\perp}$. Then use the result from part (d) together with Exercise 1.5.4, part (d), to deduce that $S^{\perp\perp} = S$.

  (f) Show that $(S + U)^\perp = S^\perp \cap U^\perp$. (Just show the inclusion both ways. This does not rely on the previous parts of the problem.)

1.5.9. Given $A \in \mathbb{F}^{n \times m}$, show that $\mathcal{R}(A)$ is a subspace of $\mathbb{F}^n$ and that $\mathcal{N}(A)$ is a subspace of $\mathbb{F}^m$.

1.5.10. Let $A \in \mathbb{R}^{n \times m}$, and suppose the nullity of A is j. In this exercise you will show that the rank of A is $m - j$, thereby proving Proposition 1.5.4.

  (a) Let $v_1, \ldots, v_j$ be a basis for $\mathcal{N}(A)$, and let $v_{j+1}, \ldots, v_m$ be vectors in $\mathbb{F}^m$ such that $v_1, \ldots v_m$ together form a basis for $\mathbb{F}^m$ (Exercise 1.5.4). Show that $Av_{j+1}, \ldots, Av_m$ form a spanning set for $\mathcal{R}(A)$.

  (b) Show that $Av_{j+1}, \ldots, Av_m$ are linearly independent.

  (c) Deduce that the rank of A is $m - j$.

1.5.11. In this exercise you will prove Proposition 1.5.5.

  (a) Suppose $y \in \mathcal{R}(A)^\perp$. This clearly implies $\langle Ax, y \rangle = 0$ for all $x \in \mathbb{F}^m$. Now make use of Proposition 1.2.2 to show that $\langle A^*y, A^*y \rangle = 0$, from which you can deduce that $y \in \mathcal{N}(A^*)$. This demonstrates that $\mathcal{R}(A)^\perp \subseteq \mathcal{N}(A^*)$.

  (b) Now make an argument that is approximately the reverse of the argument of part (a) to deduce the reverse inclusion $\mathcal{N}(A^*) \subseteq \mathcal{R}(A)^\perp$.

  (c) Show that $\mathcal{R}(A^*)^\perp = \mathcal{N}(A)$. (Use the obvious fact that $A^{**} = A$.)

1.5.12. This exercise proves Proposition 1.5.11. Let $S, \hat{S} \in \mathbb{F}^{n \times k}$.

  (a) Show that if $\hat{S} = SX$, where $X \in \mathbb{F}^{k \times k}$, then $\mathcal{R}(\hat{S}) \subseteq \mathcal{R}(S)$.

(b) Show that if $\hat{S} = SX$ and X is nonsingular, then $S = \hat{S}Y$, where $Y = X^{-1}$. Deduce that $\mathcal{R}(S) \subseteq \mathcal{R}(\hat{S})$. Consequently $\mathcal{R}(\hat{S}) = \mathcal{R}(S)$.

(c) Conversely, suppose that S and $\hat{S}$ have rank k and $\mathcal{R}(S) = \mathcal{R}(\hat{S})$. Show that the inclusion $\mathcal{R}(\hat{S}) \subseteq \mathcal{R}(S)$ implies the existence of a unique matrix $X \in \mathbb{F}^{k \times k}$ such that $\hat{S} = SX$. (Express each column of $\hat{S}$ as a linear combination of columns of S.)

(d) Continuing from part (c), show that the inclusion $\mathcal{R}(S) \subseteq \mathcal{R}(\hat{S})$ implies the existence of a unique $Y \in \mathbb{F}^{k \times k}$ such that $S = \hat{S}Y$.

(e) Show that $XY = I = YX$. Thus $Y = X^{-1}$. In particular, X is nonsingular.

(f) Suppose that S and $\hat{S}$ have orthonormal columns, and $S = \hat{S}X$. Show that $X = \hat{S}^*S$. Show that $X^*X = I$. Thus X is unitary.

1.6 Projectors

Let $\mathcal{U}_1$ and $\mathcal{U}_2$ be two subspaces of $\mathbb{F}^n$ such that $\mathbb{F}^n = \mathcal{U}_1 \oplus \mathcal{U}_2$. Then every $x \in \mathbb{F}^n$ can be expressed as a sum $x = u_1 + u_2$, where $u_1 \in \mathcal{U}_1$ and $u_2 \in \mathcal{U}_2$, in exactly one way. Thus we can unambiguously define two linear transformations on $\mathbb{F}^n$ by

$$P_1 x = u_1 \qquad \text{and} \qquad P_2 x = u_2.$$

P_1 is called the *projector* of $\mathbb{F}^n$ onto $\mathcal{U}_1$ *in the direction of* $\mathcal{U}_2$. Likewise P_2 is the projector of $\mathbb{F}^n$ onto $\mathcal{U}_2$ in the direction of $\mathcal{U}_1$. We have not proved that P_1 and P_2 are linear, but this is easily done, and it follows that P_1 and P_2 can be viewed as matrices. Clearly $P_1 + P_2 = I$. Let us derive expressions for P_1 and P_2. Let $u_1, u_2, \ldots, u_k$ be a basis for $\mathcal{U}_1$ and $u_{k+1}, \ldots, u_n$ a basis for $\mathcal{U}_2$. Then $u_1, \ldots, u_n$ is a basis of $\mathbb{F}^n$. Let $U_1 = \begin{bmatrix} u_1 & \cdots & u_k \end{bmatrix}$, $U_2 = \begin{bmatrix} u_{k+1} & \cdots & u_n \end{bmatrix}$, and $U = \begin{bmatrix} U_1 & U_2 \end{bmatrix} \in \mathbb{F}^{n \times n}$. U is nonsingular. Given any $x \in \mathbb{F}^{n \times n}$, there is a unique $y \in \mathbb{F}^{n \times n}$ such that $x = Uy$. If we partition y as $\begin{bmatrix} y_1 \\ y_2 \end{bmatrix}$, conformably with $U = \begin{bmatrix} U_1 & U_2 \end{bmatrix}$, we see that $x = Uy = U_1 y_1 + U_2 y_2$. Since $\mathcal{U}_i = \mathcal{R}(U_i)$, $i = 1, 2$, we see that $U_1 y_1 \in \mathcal{U}_1$ and $U_2 y_2 \in \mathcal{U}_2$. Thus the decomposition $x = U_1 y_1 + U_2 y_2$ is the unique decomposition of x into a component in $\mathcal{U}_1$ plus a component in $\mathcal{U}_2$. Consequently

$$P_1 x = U_1 y_1 \qquad \text{and} \qquad P_2 x = U_2 y_2.$$

Since $x = Uy$, we have $y = U^{-1}x$. Define a matrix $V \in \mathbb{F}^{n \times n}$ by $V = (U^{-1})^*$, so that $y = V^*x$. Partition V as $\begin{bmatrix} V_1 & V_2 \end{bmatrix}$, where $V_1 \in \mathbb{F}^{n \times k}$. Then $y_1 = V_1^*x$ and $y_2 = V_2^*x$. Thus

$$P_1 x = U_1 V_1^* x \qquad \text{and} \qquad P_2 x = U_2 V_2^* x.$$

Since these hold for all $x \in \mathbb{F}^n$, we have $P_1 = U_1 V_1^*$ and $P_2 = U_2 V_2^*$. We summarize these findings as the next theorem.

Theorem 1.6.1. *Let* $\mathcal{U}_1$ *and* $\mathcal{U}_2$ *be subspaces of* $\mathbb{F}^n$ *such that* $\mathbb{F}^n = \mathcal{U}_1 \oplus \mathcal{U}_2$. *Let* P_1 *(resp.,* P_2*) be the projector of* $\mathbb{F}^n$ *onto* $\mathcal{U}_1$ *(resp.,* $\mathcal{U}_2$*) in the direction of* $\mathcal{U}_2$ *(resp.,* $\mathcal{U}_1$*). Let* $U_1 \in \mathbb{F}^{n \times k}$ *and* $U_2 \in \mathbb{F}^{n \times (n-k)}$ *be matrices such that* $\mathcal{U}_1 = \mathcal{R}(U_1)$ *and* $\mathcal{U}_2 = \mathcal{R}(U_2)$. *Let* $U = \begin{bmatrix} U_1 & U_2 \end{bmatrix} \in \mathbb{F}^{n \times n}$ *and* $V = \begin{bmatrix} V_1 & V_2 \end{bmatrix} = U^{-*}$. *Then*

$$P_1 = U_1 V_1^* \qquad \text{and} \qquad P_2 = U_2 V_2^*.$$

Some fundamental properties of projectors are given in the following theorems.

Theorem 1.6.2. *Let $\tilde{U}$, $\tilde{V} \in \mathbb{F}^{n \times k}$ be matrices such that $\tilde{V}^* \tilde{U} = I_k$. Then the matrix $P = \tilde{U} \tilde{V}^*$ is the projector of $\mathbb{F}^n$ onto $\mathcal{R}(\tilde{U})$ in the direction of $\mathcal{R}(\tilde{V})^\perp$.*

Theorem 1.6.3. *If P is the projector of $\mathbb{F}^n$ onto $\mathcal{U}_1$ in the direction of $\mathcal{U}_2$, then the following hold:*

    (a) $\mathcal{R}(P) = \mathcal{U}_1$ *and* $\mathcal{N}(P) = \mathcal{U}_2$.

    (b) $\mathcal{R}(P^*) = \mathcal{U}_2^\perp$ *and* $\mathcal{N}(P^*) = \mathcal{U}_1^\perp$.

    (c) P^* *is the projector of $\mathbb{F}^n$ onto $\mathcal{U}_2^\perp$ in the direction of $\mathcal{U}_1^\perp$.*

Theorem 1.6.4. *$P \in \mathbb{F}^{n \times n}$ is a projector if and only if $P^2 = P$.*

Orthoprojectors

An important special type of projector is the orthoprojector. Let $\mathcal{U}$ be a subspace of $\mathbb{F}^n$. Then $\mathbb{F}^n = \mathcal{U} \oplus \mathcal{U}^\perp$. The projector of $\mathbb{F}^n$ onto $\mathcal{U}$ in the direction of $\mathcal{U}^\perp$ is also called the *orthoprojector* (or *orthogonal projector*) of $\mathbb{F}^n$ onto $\mathcal{U}$. The projector P of $\mathbb{F}^n$ onto $\mathcal{U}_1$ in the direction of $\mathcal{U}_2$ is an orthoprojector if and only if $\mathcal{U}_2 = \mathcal{U}_1^\perp$. A projector that is not an orthoprojector is called an *oblique projector*.

The developments that we carried out above for arbitrary projectors, in which we introduced the matrices U and V, are all valid for orthoprojectors, of course. However, if we now insist that the vectors $u_1, \ldots, u_k$ and $u_{k+1}, \ldots, u_n$ form orthonormal bases, we then get that U is unitary and $V = U$. The following theorem results.

Theorem 1.6.5. *Let $\mathcal{U}$ be a subspace of $\mathbb{F}^n$, and let P be the orthoprojector of $\mathbb{F}^n$ onto $\mathcal{U}$. Let U_1 be a matrix with orthonormal columns such that $\mathcal{U} = \mathcal{R}(U_1)$. Then*

$$P = U_1 U_1^*.$$

Theorem 1.6.6. *Let P be a projector. Then we have the following:*

    (a) *P is an orthoprojector if and only if $P = P^*$.*

    (b) *If P is an orthoprojector, then $\| P \| = 1$.*

    (c) *If P is an oblique projector, then $\| P \| > 1$.*

Exercises

1.6.1. Let $\mathcal{U}_1$ and $\mathcal{U}_2$ be subspaces of $\mathbb{F}^n$ such that $\mathbb{F}^n = \mathcal{U}_1 \oplus \mathcal{U}_2$.

    (a) Show that if $\dim(\mathcal{U}_1) = k$, then $\dim(\mathcal{U}_2) = n - k$. Show that if $u_1, \ldots, u_k$ is a basis of $\mathcal{U}_1$ and $u_{k+1}, \ldots, u_n$ is a basis of $\mathcal{U}_2$, then $u_1, \ldots, u_n$ is a basis of $\mathbb{F}^n$.

(b) Check the details of the proof of Theorem 1.6.1.

(c) Prove Theorem 1.6.2.

(d) Prove Theorem 1.6.3.

(e) Prove Theorem 1.6.4.

1.6.2. Prove Theorem 1.6.5.

1.6.3.

(a) Use some of the results from Theorem 1.6.3 to prove part (a) of Theorem 1.6.6.

(b) Use the Pythagorean theorem to prove part (b) of Theorem 1.6.6.

(c) Let P be an oblique projector. Thus P is the projector of $\mathbb{F}^n$ onto $\mathcal{U}_1$ along $\mathcal{U}_2$, where $\mathcal{U}_2 \neq \mathcal{U}_1^{\perp}$. Show that there are some $u_1 \in \mathcal{U}_1$ and $u_2 \in \mathcal{U}_2$ such that $\langle u_1, u_2 \rangle < 0$. (The angle between u_1 and u_2 is obtuse.) Letting α denote a real parameter, show that $\|u_1 + \alpha u_2\|^2$ is a quadratic polynomial in α whose minimum occurs when $\alpha > 0$. Let $x = u_1 + \alpha_0 u_2$, where α_0 is taken to be the minimizer of the norm. Show that

$$\frac{\|Px\|}{\|x\|} > 1.$$

Deduce that $\|P\| > 1$.

(d) Draw a picture that illustrates part (c).

1.7 The Singular Value Decomposition (SVD)

Perhaps the most important matrix factorization of all is the singular value decomposition (SVD), which appears in countless applications. Here we describe the SVD and prove its existence and some of its basic properties. In Chapter 2 we will use the SVD to establish some important theoretical results. Specifically, the SVD is crucial to our discussion of angles and distances between subspaces in Section 2.6, and it plays a key role in our development of the Jordan canonical form in Section 2.4 (Exercise 2.4.14). Methods for computing the SVD will be discussed in Chapter 8.

Theorem 1.7.1 (SVD). *Let $A \in \mathbb{F}^{n \times m}$, and suppose* rank$(A) = r$. *Then there exist matrices $U \in \mathbb{F}^{n \times n}$, $\Sigma \in \mathbb{R}^{n \times m}$, and $V \in \mathbb{F}^{m \times m}$ such that*

$$A = U \Sigma V^*, \tag{1.7.1}$$

U and V are unitary, and

$$\Sigma = \begin{bmatrix} \hat{\Sigma} & 0 \\ 0 & 0 \end{bmatrix},$$

where $\hat{\Sigma}$ is a diagonal matrix: $\hat{\Sigma} = \mathrm{diag}\{\sigma_1, \ldots, \sigma_r\}$ *with* $\sigma_1 \geq \sigma_2 \geq \cdots \geq \sigma_r > 0$.

A proof of Theorem 1.7.1 is sketched in Exercise 1.7.1. The positive numbers σ_1, $\sigma_2, \ldots, \sigma_r$ are uniquely determined by A and are called the *singular values* of A. Equation (1.7.1) is called the *SVD* of A. It can be rewritten as $AV = U\Sigma$ and in various other ways, as demonstrated in Theorems 1.7.2 and 1.7.3. The columns of V and U are called *right* and *left singular vectors*, respectively. Taking conjugate transposes, we have $A^* = V\Sigma^T U^*$, which is the SVD of A^*. The relationship between singular values/vectors and eigenvalues/vectors is worked out in Exercise 2.3.12.

Sometimes it is convenient to speak of zero singular values. For example, if $n > m$ and $\text{rank}(A) = m$, we have $A = U\Sigma V^*$ with

$$\Sigma = \begin{bmatrix} \hat{\Sigma} \\ 0 \end{bmatrix} \qquad \text{and} \qquad \hat{\Sigma} = \text{diag}\{\sigma_1, \ldots, \sigma_m\}. \tag{1.7.2}$$

Here $\sigma_1, \ldots, \sigma_m$ are all positive. Now suppose $\text{rank}(A) = r < m$. Then we can still write $A = U\Sigma V^*$ with Σ as in (1.7.2) if we take $\sigma_{r+1} = \cdots = \sigma_m = 0$. In this case we say we have $m - r$ singular values equal to zero. We will make use of zero singular values without further comment whenever it is convenient.

The next few theorems are variants of the SVD theorem.

Theorem 1.7.2 (compact SVD). *Let $A \in \mathbb{F}^{n \times m}$, and suppose $\text{rank}(A) = r > 0$. Then there exist $\hat{U} \in \mathbb{F}^{n \times r}$, $\hat{V} \in \mathbb{F}^{m \times r}$, and $\hat{\Sigma} \in \mathbb{R}^{r \times r}$ such that*

$$A = \hat{U}\hat{\Sigma}\hat{V}^*,$$

$\hat{U}$ and $\hat{V}$ have orthonormal columns, and $\hat{\Sigma} = \text{diag}\{\sigma_1, \ldots, \sigma_r\}$ with $\sigma_1 \geq \sigma_2 \geq \cdots \geq \sigma_r > 0$.

Theorem 1.7.3. *Let $A \in \mathbb{F}^{n \times m}$ with $\text{rank}(A) = r$. Then there exist orthonormal vectors $v_1, \ldots, v_m \in \mathbb{F}^m$ (right singular vectors), orthonormal vectors $u_1, \ldots, u_n \in \mathbb{F}^n$ (left singular vectors), and positive numbers $\sigma_1 \geq \cdots \geq \sigma_r$ such that*

$$Av_i = u_i\sigma_i, \quad i = 1, \ldots, r, \qquad A^*u_i = v_i\sigma_i, \quad i = 1, \ldots, r,$$

$$Av_i = 0, \quad i = r+1, \ldots, m, \qquad A^*u_i = 0, \quad i = r+1, \ldots, n.$$

Moreover,

$$\mathcal{R}(A) = \text{span}\{u_1, \ldots, u_r\}, \qquad \mathcal{N}(A^*) = \text{span}\{u_{r+1}, \ldots, u_n\},$$

$$\mathcal{R}(A^*) = \text{span}\{v_1, \ldots, v_r\}, \qquad \mathcal{N}(A) = \text{span}\{v_{r+1}, \ldots, v_m\}.$$

This gives independent confirmation that $\mathcal{R}(A)^\perp = \mathcal{N}(A^*)$, $\mathcal{R}(A^*)^\perp = \mathcal{N}(A)$, and $\text{rank}(A) = \text{rank}(A^*)$.

Theorem 1.7.4. *Let $A \in \mathbb{F}^{n \times m}$, and suppose $\text{rank}(A) = r > 0$. Then there exist orthonormal vectors $v_1, \ldots, v_m \in \mathbb{F}^m$ (right singular vectors), orthonormal vectors $u_1, \ldots, u_n \in \mathbb{F}^n$ (left singular vectors), and positive numbers $\sigma_1 \geq \cdots \geq \sigma_r$ such that*

$$A = \sum_{j=1}^{r} \sigma_j u_j v_j^*.$$

Since each of $u_j v_j^*$ is an $n \times m$ matrix with rank one, this last theorem shows that A can be expressed as a sum of r rank-one matrices.

There are simple relationships between the singular values and the spectral norm and condition number of a matrix.

Proposition 1.7.5. *Let $A \in \mathbb{F}^{n \times m}$, and let σ_1 denote the largest singular value of A. Then $\|A\| = \sigma_1$.*

If A is square and nonsingular, then its rank is n, so it has n positive singular values.

Proposition 1.7.6. *Let $A \in \mathbb{F}^{n \times n}$ be nonsingular, and let $\sigma_1 \geq \cdots \geq \sigma_n > 0$ be the singular values of A. Then $\|A^{-1}\| = \sigma_n^{-1}$ and $\kappa(A) = \sigma_1/\sigma_n$.*

Exercises

1.7.1. This exercise sketches a proof of the SVD theorem by induction on r, the rank of A. If $r = 0$, take $U \in \mathbb{C}^{n \times n}$ and $V \in \mathbb{C}^{m \times m}$ to be arbitrary unitary matrices, and take $\Sigma = 0 \in \mathbb{R}^{n \times m}$. Then $A = U\Sigma V^*$. Now suppose $r > 0$, and let's consider the induction step. Let $\sigma_1 = \|A\| = \max_{\|v\|=1} \|Av\| > 0$, and let $v_1 \in \mathbb{F}^m$ be a unit vector for which the maximum is attained: $\|Av_1\| = \sigma_1$. Define $u_1 \in \mathbb{R}^n$ by $u_1 = \sigma_1^{-1} Av_1$. Then clearly $\|u_1\| = 1$. Let $U_1 \in \mathbb{F}^{n \times n}$ and $V_1 \in \mathbb{F}^{m \times m}$ be unitary matrices whose first columns are u_1 and v_1, respectively. (It is shown in Section 3.1 that this can always be done.)

(a) Let $A_1 = U_1^* A V_1$. Show that the first column of A_1 consists entirely of zeros, except for a σ_1 in the first position. (Remember that U_1 is unitary, so its columns are orthonormal.) Thus A_1 has the block form

$$A_1 = \begin{bmatrix} \sigma_1 & c^* \\ 0 & \hat{A} \end{bmatrix},$$

where $c \in \mathbb{F}^{m-1}$.

(b) $\|A_1\| = \|A\|$ by part (g) of Proposition 1.3.7. Thus $\|A_1\| = \sigma_1$. Using this fact, show that $c = 0$. To this end, let $x = \begin{bmatrix} \sigma_1 \\ c \end{bmatrix} \in \mathbb{C}^m$, and show that $\frac{\|A_1 x\|}{\|x\|} \geq \sqrt{\sigma_1^2 + \|c\|^2}$. This forces $c = 0$, since otherwise we would have $\|A_1\| \geq \frac{\|A_1 x\|}{\|x\|} > \sigma_1$.

(c) We now have $A = U_1 A_1 V_1^*$, where

$$A_1 = \begin{bmatrix} \sigma_1 & 0^T \\ 0 & \hat{A} \end{bmatrix}.$$

Show that the rank of $\hat{A}$ is $r - 1$, and $\|\hat{A}\| \leq \sigma_1$. Apply the induction hypothesis to $\hat{A}$ to complete the proof.

1.7.2. Deduce Theorem 1.7.2 as a corollary of Theorem 1.7.1.

1.7.3. Deduce Theorem 1.7.3 from Theorem 1.7.1.

1.7.4. Prove Theorem 1.7.4.

1.7.5. Prove Proposition 1.7.5.

1.7.6. Let $A \in \mathbb{F}^{n \times n}$ be nonsingular, and let $\sigma_1 \geq \cdots \geq \sigma_n > 0$ be its singular values.

    (a) Find the SVD of A^{-1} in terms of the SVD of A. What are the singular values and singular vectors of A^{-1}?

    (b) Deduce that $\| A^{-1} \| = \sigma_n^{-1}$ and $\kappa(A) = \sigma_1 / \sigma_n$.

1.8   Linear Transformations and Matrices

Every linear transformation can be represented by a matrix in numerous ways. Here we summarize a few of the basic facts without working out the complete theory. Let V be an n-dimensional vector space over the field $\mathbb{F}$, and let $T : V \to V$ be a linear transformation mapping V to V. Let $v_1, \ldots, v_n$ be a basis of V. Since T is linear, its entire action is determined by its action on a basis. In other words, T is determined by $T v_1, \ldots, T v_n$. For each j there exist unique coefficients $a_{1j}, \ldots, a_{nj}$ such that

$$T v_j = \sum_{i=1}^{n} v_i a_{ij}. \tag{1.8.1}$$

The coefficients a_{ij} can be collected into an $n \times n$ matrix A, which we call the *matrix of T with respect to* the basis $v_1, \ldots, v_n$.

    Now consider a second basis $\hat{v}_1, \ldots, \hat{v}_n$. Then T has a matrix $\hat{A} = (\hat{a}_{ij})$ with respect to $\hat{v}_1, \ldots, \hat{v}_n$. It is easy to establish the relationship between A and $\hat{A}$. Each basis can be written in terms of the other:

$$\hat{v}_j = \sum_{i=1}^{n} v_i s_{ij}, \qquad v_j = \sum_{i=1}^{n} \hat{v}_i \hat{s}_{ij}, \tag{1.8.2}$$

where the coefficients s_{ij} and $\hat{s}_{ij}$ are uniquely determined. Let S and $\hat{S}$ denote the matrices whose (i, j) entry is s_{ij} and $\hat{s}_{ij}$, respectively. Then $\hat{S} = S^{-1}$ (Exercise 1.8.1). Moreover, $\hat{A} = S^{-1} A S$, which means that A and $\hat{A}$ are *similar*; see Section 2.2.

Proposition 1.8.1. *Let $T : V \to V$ be a linear transformation on the n-dimensional vector space V over $\mathbb{F}$, and let A and $\hat{A} \in \mathbb{F}^{n \times n}$ be the matrices of T with respect to the bases $v_1, \ldots, v_n$ and $\hat{v}_1, \ldots, \hat{v}_n$, respectively. Then the following hold:*

    (a) *A and $\hat{A}$ are similar:*

$$\hat{A} = S^{-1} A S,$$

    where $S \in \mathbb{F}^{n \times n}$ is defined by (1.8.2).

    (b) *If V is an inner product space and the bases are orthonormal, then S is unitary; hence A and $\hat{A}$ are unitarily similar.*

    (c) *If $\tilde{A} \in \mathbb{F}^{n \times n}$ is similar to A, then there is a basis $\tilde{v}_1, \ldots, \tilde{v}_n$ of V such that $\tilde{A}$ is the matrix of T with respect to $\tilde{v}_1, \ldots, \tilde{v}_n$.*

Exercises

1.8.1. This exercise refers to the matrices S and $\hat{S}$ defined by (1.8.2).

   (a) Show that $\hat{S} = S^{-1}$.

   (b) Show that if $\mathcal{V}$ is an inner product space and the bases $v_1, \ldots, v_n$ and $\hat{v}_1, \ldots, \hat{v}_n$ are both orthonormal, then S and $\hat{S}$ are unitary.

1.8.2. This exercise proves Proposition 1.8.1.

   (a) Use equations (1.8.2) and (1.8.1) and its $\hat{v}$ counterpart to prove that $\hat{A} = \hat{S}AS$. Combining this result with part (a) of Exercise 1.8.1, we obtain part (a) of Proposition 1.8.1. Part (b) of Proposition 1.8.1 was proved in part (b) of Exercise 1.8.1.

   (b) Prove part (c) of Proposition 1.8.1. Define vectors $\tilde{v}_1, \ldots, \tilde{v}_n$ using a formula like the ones in (1.8.2), then show that $\tilde{v}_1, \ldots, \tilde{v}_n$ are linearly independent (hence a basis) and that $\tilde{A}$ is the matrix of T with respect to this basis.

Chapter 2

Basic Theory of Eigensystems

2.1 Eigenvalues, Eigenvectors, and Invariant Subspaces

Let $A \in \mathbb{C}^{n \times n}$; that is, A is an $n \times n$ matrix with complex entries. A vector $v \in \mathbb{C}^n$ is called an *eigenvector* of A if v is nonzero and Av is a multiple of v. That is, there is a $\lambda \in \mathbb{C}$ such that

$$Av = v\lambda. \tag{2.1.1}$$

The complex scalar λ is called the *eigenvalue* of A associated with the eigenvector v. The pair (λ, v) is called an *eigenpair* of A. The eigenvalue associated with a given eigenvector is unique. However, each eigenvalue has many eigenvectors associated with it: If λ is an eigenvalue associated with eigenvector v, then any nonzero multiple of v is also an eigenvector with associated eigenvalue λ. Indeed the set of all vectors (including the zero vector) satisfying (2.1.1) for a fixed eigenvalue λ is a subspace of $\mathbb{C}^n$. We call it the *eigenspace* of A associated with λ. It consists of all of the eigenvectors of A associated with λ, together with the zero vector.

The set of all eigenvalues of A is called the *spectrum* of A and will be denoted $\lambda(A)$.

Equation (2.1.1) is clearly equivalent to

$$(\lambda I - A)v = 0, \tag{2.1.2}$$

where I is, as usual, the $n \times n$ identity matrix. Thus λ is an eigenvalue of A if and only if (2.1.2) has a nonzero solution.

Proposition 2.1.1. *The following statements are equivalent:*

(a) λ *is an eigenvalue of* A.

(b) $\lambda I - A$ *is a singular matrix.*

(c) $\mathcal{N}(\lambda I - A) \neq \{0\}$.

Recall that $\mathcal{N}(B)$ denotes the *null space* of B, the set of all vectors $w \in \mathbb{C}^n$ such that $Bw = 0$. Notice that if λ is an eigenvalue of A, then $\mathcal{N}(\lambda I - A)$ is exactly the eigenspace associated with λ.

Corollary 2.1.2. *A is nonsingular if and only if 0 is not an eigenvalue of A.*

Proposition 2.1.3. *A and A^T have the same eigenvalues.*

Eigenvectors, as defined in (2.1.1) and (2.1.2), are sometimes called *right eigenvectors* to distinguish them from *left eigenvectors*, which we now define. A nonzero row vector w^T is called a *left eigenvector* of A if there is a scalar λ such that

$$w^T A = \lambda w^T.$$

This equation is clearly equivalent to

$$(\lambda I - A^T)w = 0,$$

which shows that w^T is a left eigenvector of A if and only if w is a right eigenvector of A^T associated with the eigenvalue λ. Since A and A^T have the same eigenvalues, we can conclude that if λ is an eigenvalue of A, then λ has both right and left eigenvectors associated with it. We can define the *left eigenspace* of A associated with an eigenvalue λ in just the same way as we defined the right eigenspace. Clearly, since $\mathcal{N}(\lambda I - A)$ and $\mathcal{N}(\lambda I - A^T)$ have the same dimension, the left and right eigenspaces associated with a given eigenvalue have the same dimension.

We will focus mainly on right eigenvectors, but we will find that left eigenvectors are sometimes useful.

Existence of Eigenvalues

We have now established some of the most basic properties, but we have not yet shown that every matrix has eigenvectors and eigenvalues. For all we know so far, there might be matrices that do not have eigenvalues. Fortunately it turns out that there are no such matrices; the spectrum of every matrix is nonempty. The usual path to this result uses properties of determinants.

Theorem 2.1.4. *λ is an eigenvalue of A if and only if λ satisfies the equation*

$$\det(\lambda I - A) = 0. \tag{2.1.3}$$

This theorem is an immediate consequence of Proposition 2.1.1 and Theorem 1.1.1: $\lambda I - A$ is singular if and only if its determinant is zero. One easily checks that $\det(\lambda I - A)$ is a monic polynomial in λ of degree n, the dimension of A. We call it the *characteristic polynomial* of A, and we call (2.1.3) the *characteristic equation* of A. By the fundamental theorem of algebra, every complex polynomial of degree one or greater has at least one zero. Therefore every complex matrix has at least one eigenvalue. More precisely, $\det(\lambda I - A)$, like every nth degree monic polynomial, has a factorization

$$\det(\lambda I - A) = (\lambda - \lambda_1)(\lambda - \lambda_2) \cdots (\lambda - \lambda_n), \tag{2.1.4}$$

where $\lambda_1, \lambda_2, \ldots, \lambda_n \in \mathbb{C}$, and each of these λ_i is a solution of the characteristic equation and hence an eigenvalue of A. We thus have the following theorem.

Theorem 2.1.5. *Every $A \in \mathbb{C}^{n \times n}$ has n eigenvalues $\lambda_1, \lambda_2, \ldots, \lambda_n \in \mathbb{C}$.*

The numbers $\lambda_1, \lambda_2, \ldots, \lambda_n \in \mathbb{C}$ need not be distinct; for example, the matrices $0 \in \mathbb{C}^{n \times n}$ and $I \in \mathbb{C}^{n \times n}$ have eigenvalues $0, 0, \ldots, 0$ and $1, 1, \ldots, 1$, respectively. The number of times a given eigenvalue appears in the list $\lambda_1, \lambda_2, \ldots, \lambda_n \in \mathbb{C}$, that is, in the factorization (2.1.4), is called the *algebraic multiplicity* of the eigenvalue.

Determinants come in handy now and then, as they have here in the demonstration that every $n \times n$ matrix has n eigenvalues. Theorem 2.1.4 suggests that they might even play a role in the practical solution of eigenvalue problems. The first method for computing eigenvalues learned by students in elementary linear algebra courses is to form the characteristic polynomial and factor it to find the solutions of the characteristic equation. As it turns out, this is a poor method in general, and it is almost never used. Not only is it inefficient (how do you compute that determinant anyway, and how do you factor the polynomial?), it can also be notoriously inaccurate in the face of roundoff errors.

The characteristic polynomial and determinants in general will play only a minor role in this book. My original plan was to avoid determinants completely, and thus to write an entire book on matrix computations that makes no mention of them. Although this is certainly possible, I have decided that it is too extreme a measure. Determinants are just too convenient to be ignored. However, to give an indication of how we could have avoided determinants and the characteristic polynomial, I have outlined below and in the exercises some alternative ways of proving the existence of eigenvalues.

Let $w \in \mathbb{C}^n$ be any nonzero vector, and consider the *Krylov sequence*

$$w, \ Aw, \ A^2 w, \ A^3 w, \ \ldots.$$

If we generate enough vectors in this sequence, we will get a linearly dependent set. Indeed, since $\mathbb{C}^n$ has dimension n, the set of $n + 1$ vectors $w, Aw, \ldots, A^n w$ must be linearly dependent. Let k be the smallest positive integer such that the vectors

$$w, \ Aw, \ A^2 w, \ A^3 w, \ \ldots, \ A^k w$$

are linearly dependent. This number could be as large as n or as small as 1. Notice that if $k = 1$, then Aw must be a multiple of w, that is $Aw = w\lambda$ for some λ. Then w is an eigenvector of A, λ is an eigenvalue, and we are done. Let us assume, therefore, that $k > 1$.

Since $w, Aw, A^2 w, \ldots, A^k w$ are linearly dependent, there exist constants $a_0, \ldots, a_k$, not all zero, such that

$$a_0 w + a_1 A w + a_2 A^2 w + \cdots + a_k A^k w = 0. \tag{2.1.5}$$

Clearly $a_k \neq 0$, since otherwise the smaller set $w, Aw, A^2 w, \ldots, A^{k-1} w$ would be linearly dependent. Let $p(\lambda)$ denote the kth degree polynomial built using the coefficients from (2.1.5):

$$p(\lambda) = a_0 + a_1 \lambda + a_2 \lambda^2 + \cdots + a_k \lambda^k.$$

Then, if we define $p(A)$ by

$$p(A) = a_0 I + a_1 A + a_2 A^2 + \cdots + a_k A^k,$$

(2.1.5) takes the simple form $p(A)w = 0$. By the fundamental theorem of algebra, the equation $p(\lambda) = 0$ has at least one complex root μ. Then $\lambda - \mu$ is a factor of $p(\lambda)$, and we can write $p(\lambda) = (\lambda - \mu)q(\lambda)$, where $q(\lambda)$ is a polynomial of degree exactly $k - 1$. Using this factorization, we can rewrite $p(A)w = 0$ as

$$(A - \mu I)q(A)w = 0. \tag{2.1.6}$$

Let $v = q(A)w$. The polynomial q has the form $q(\lambda) = b_0 + b_1\lambda + \cdots + b_{k-1}\lambda^{k-1}$, with $b_{k-1} = a_k \neq 0$, so

$$v = b_0 w + b_1 A w + \cdots + b_{k-1} A^{k-1} w.$$

Because $w, Aw, \ldots, A^{k-1}w$ are linearly independent, v must be nonzero. Now (2.1.6) shows that $Av = v\mu$, so v is an eigenvector of A with eigenvalue μ. Thus we have proved, without using determinants, that every matrix $A \in \mathbb{C}^{n \times n}$ has an eigenpair. Notice that this argument shows that *every* zero of p is an eigenvalue of A.

   Another proof that uses a Krylov sequence is outlined in Exercise 2.1.3. Yet another approach, using Liouville's theorem of complex analysis, is used in functional analysis [202, Theorem V.3.2].

   Krylov sequences are used frequently in theoretical arguments. In practical computations they are not used so much. However, the *Krylov subspaces* span$\{w, Aw, \ldots, A^{j-1}w\}$ that they span play a huge role, as we will see in Chapters 4 and 9.

Triangular Matrices

The eigenvalues of triangular matrices are obvious.

Proposition 2.1.6. *Let $T \in \mathbb{C}^{n \times n}$ be triangular. Then the main-diagonal entries $t_{11}, \ldots, t_{nn}$ are the eigenvalues of T.*

Proof: Once again determinants come in handy. Since $\lambda I - T$ is triangular, we know from Theorem 1.3.1 that $\det(\lambda I - T) = (\lambda - t_{11})(\lambda - t_{22}) \cdots (\lambda - t_{nn})$. Thus the roots of the characteristic equation are $t_{11}, \ldots, t_{nn}$. $\square$

   A proof that builds eigenvectors and does not use determinants is outlined in Exercise 2.1.4.

   The simplest kind of triangular matrix is the diagonal matrix, which has a string of eigenvalues running down the main diagonal and zeros everywhere else.

   A useful generalization of Proposition 2.1.6 has to do with block triangular matrices.

Proposition 2.1.7. *Suppose that T is block triangular, say*

$$T = \begin{bmatrix} T_{11} & T_{12} & \cdots & T_{1,k-1} & T_{1k} \\ 0 & T_{22} & \cdots & T_{2,k-1} & T_{2k} \\ \vdots & \vdots & \ddots & \vdots & \vdots \\ 0 & 0 & \cdots & T_{k-1,k-1} & T_{k-1,k} \\ 0 & 0 & \cdots & 0 & T_{kk} \end{bmatrix},$$

where the main-diagonal blocks T_{ii} are square. Then the spectrum of T is the union of the spectra of $T_{11}, T_{22}, \ldots, T_{kk}$.

Proof: Since T is block triangular, so is $\lambda I - T$. The determinant of a block triangular matrix is the product of the determinants of the main-diagonal blocks: $\det(\lambda I - T) = \det(\lambda I - T_{11}) \cdots \det(\lambda I - T_{kk})$. The result follows. $\square$

Linear Independence of Eigenvectors

Most matrices have an ample supply of eigenvectors, as the following theorem shows.

Theorem 2.1.8. *Let $A \in \mathbb{C}^{n \times n}$, let $\lambda_1, \ldots, \lambda_k$ be distinct eigenvalues of A, and let $v_1, \ldots, v_k$ be eigenvectors associated with $\lambda_1, \ldots, \lambda_k$, respectively. Then $v_1, \ldots, v_k$ are linearly independent.*

Theorem 2.1.8 is proved in Exercise 2.1.5.

Corollary 2.1.9. *If $A \in \mathbb{C}^{n \times n}$ has n distinct eigenvalues, then A has a set of n linearly independent eigenvectors $v_1, \ldots, v_n$. In other words, there is a basis of $\mathbb{C}^n$ consisting of eigenvectors of A.*

A matrix $A \in \mathbb{C}^{n \times n}$ that has n linearly independent eigenvectors is called *semisimple*. (A synonym for semisimple is *diagonalizable*; see Theorem 2.4.1 below.) Thus Corollary 2.1.9 states that every matrix that has distinct eigenvalues is semisimple. The converse is false; a matrix can have repeated eigenvalues and still be semisimple. A good example is the matrix I, which has the eigenvalue 1, repeated n times. Since every nonzero vector is an eigenvector of I, any basis of $\mathbb{C}^n$ is a set of n linearly independent eigenvectors of I. Thus I is semisimple. Another good example is the matrix 0.

A matrix that is not semisimple is called *defective*. An example is

$$A = \begin{bmatrix} 0 & 1 \\ 0 & 0 \end{bmatrix},$$

which is upper triangular and has the eigenvalues 0 and 0. The eigenspace associated with 0 is one-dimensional, so A does not have two linearly independent eigenvectors. (See Exercise 2.1.6.)

Invariant Subspaces

As in our earlier discussion of subspaces (see Section 1.5), we will allow the symbol $\mathbb{F}$ to denote either $\mathbb{R}$ or $\mathbb{C}$. Then the symbols $\mathbb{F}^n$ and $\mathbb{F}^{n \times n}$ can mean either $\mathbb{R}^n$ and $\mathbb{R}^{n \times n}$ or $\mathbb{C}^n$ and $\mathbb{C}^{n \times n}$, respectively.

Let $A \in \mathbb{F}^{n \times n}$, and let $\mathcal{S}$ be a subspace of $\mathbb{F}^n$. Then we can define another subspace $A(\mathcal{S})$ by

$$A(\mathcal{S}) = \{Ax \mid x \in \mathcal{S}\}.$$

One easily shows that $A(\mathcal{S})$ is indeed a subspace of $\mathbb{F}^n$. The subspace $\mathcal{S}$ is said to be *invariant* under A if $A(\mathcal{S}) \subseteq \mathcal{S}$. This just means that $Ax \in \mathcal{S}$ whenever $x \in \mathcal{S}$. Frequently we will abuse the language slightly by referring to $\mathcal{S}$ as an *invariant subspace of* A, even though it is actually a subspace of $\mathbb{F}^n$, not of A.

The subspaces $\{0\}$ and $\mathbb{F}^n$ are invariant under every $A \in \mathbb{F}^{n \times n}$. We refer to these as *trivial* invariant subspaces. Of greater interest are *nontrivial* invariant subspaces, that is, subspaces that satisfy $\{0\} \neq \mathcal{S} \neq \mathbb{F}^n$ and are invariant under A.

Invariant subspaces are closely related to eigenvectors.

Proposition 2.1.10. *Let* $v_1, v_2, \ldots, v_k$ *be eigenvectors of* A. *Then*

$$\text{span}\{v_1, \ldots, v_k\}$$

is invariant under A.

In particular, a one-dimensional space spanned by a single eigenvector of A is invariant under A, and so is every eigenspace of A. The type of invariant subspace shown in Proposition 2.1.10 is typical for semisimple matrices. However, every defective matrix has one or more invariant subspaces that are not spanned by eigenvectors. For example, consider the space $\mathcal{S} = \mathbb{F}^n$, and see Exercise 2.1.8.

It is convenient to use matrices to represent subspaces. Given a subspace $\mathcal{S}$ of dimension k, let $s_1, \ldots, s_k \in \mathcal{S}$ be any basis of $\mathcal{S}$. Then build an $n \times k$ matrix S using the basis vectors as columns: $S = [s_1 \ s_2 \ \cdots \ s_k]$. Then S represents $\mathcal{S}$, in the sense that $\mathcal{S}$ is the column space of $\mathcal{S}$. In other words, $\mathcal{S} = \mathcal{R}(S)$. Since the columns of S are linearly independent, S has full rank. Two full-rank matrices S and $\hat{S}$ represent the same space if and only if $\mathcal{R}(S) = \mathcal{R}(\hat{S})$. This holds if and only if there is a nonsingular matrix $C \in \mathbb{F}^{k \times k}$ such that $\hat{S} = SC$ (Proposition 1.5.11 and Exercise 1.5.12).

Proposition 2.1.11. *Let* $A \in \mathbb{F}^{n \times n}$, *let* $\mathcal{S}$ *be a* k-*dimensional subspace of* $\mathbb{F}^n$, *and let* $S \in \mathbb{F}^{n \times k}$ *be a matrix such that* $\mathcal{S} = \mathcal{R}(S)$.

(a) *Then* $\mathcal{S}$ *is invariant under* A *if and only if there is a matrix* $B \in \mathbb{F}^{k \times k}$ *such that*

$$AS = SB. \tag{2.1.7}$$

(b) *If* (2.1.7) *holds, then every eigenvalue of* B *is an eigenvalue of* A. *If* v *is an eigenvector of* B *with eigenvalue* λ, *then* Sv *is an eigenvector of* A *with eigenvalue* λ.

(c) *If* (2.1.7) *holds and the columns of* S *are orthonormal, then* $B = S^* A S$.

The proof is left as Exercise 2.1.9. If (2.1.7) holds, then the eigenvalues of B are called the eigenvalues of A *associated with* the invariant subspace $\mathcal{S}$.

When $k = 1$, the equation $AS = SB$ reduces to the eigenvalue equation $Av = v\lambda$. S becomes a single column, and B becomes a 1×1 matrix consisting of a single eigenvalue. When $k > 1$, we can think of S as a "supereigenvector" or "eigenvector packet" and B as a "supereigenvalue" or "eigenvalue packet."

Proposition 2.1.12. *Let* $A \in \mathbb{F}^{n \times n}$, *and let* $\mathcal{S}$ *be invariant under* A. *Then the following hold:*

(a) $\mathcal{S}^\perp$ *is invariant under* A^*.

(b) *Let* $\lambda_1, \ldots, \lambda_n$ *denote the eigenvalues of* A, *and suppose that* $\lambda_1, \ldots, \lambda_k$ *are the eigenvalues associated with the invariant subspace* $\mathcal{S}$. *Then* $\overline{\lambda}_{k+1}, \ldots, \overline{\lambda}_n$ *are the eigenvalues of* A^* *associated with* $\mathcal{S}^\perp$.

Part (a) can be proved using the basic identity $\langle Ax, y \rangle = \langle x, A^*y \rangle$ (Proposition 1.2.2). We defer the proof of part (b) to Exercise 2.2.8

Exercises

2.1.1.

(a) Prove that (2.1.1) and (2.1.2) are equivalent.

(b) Prove Proposition 2.1.1.

2.1.2. Prove Proposition 2.1.3. (Apply Theorem 1.5.9 to $\lambda I - A$, or make a simple argument using determinants.)

2.1.3. This exercise leads to another proof that every $A \in \mathbb{C}^{n \times n}$ has an eigenvalue. Let $w \in \mathbb{C}^n$ be any nonzero vector, and let k be the smallest integer such that $w, Aw, \ldots, A^k w$ are linearly dependent. If $k = 1$, we are done, so assume $k > 1$.

(a) Show that there exist constants $c_0, \ldots, c_{k-1}$ such that

$$A^k w = c_{k-1} A^{k-1} w + \cdots + c_1 Aw + c_0 w. \qquad (2.1.8)$$

(b) Let $W \in \mathbb{C}^{n \times k}$ be the matrix whose columns are $w, Aw, \ldots, A^{k-1}$, in that order. Show that

$$AW = WC, \qquad (2.1.9)$$

where $C \in \mathbb{C}^{k \times k}$ is a *companion matrix*

$$C = \begin{bmatrix} 0 & 0 & 0 & & c_0 \\ 1 & 0 & 0 & & c_1 \\ 0 & 1 & 0 & & c_2 \\ \vdots & \ddots & \ddots & & \vdots \\ 0 & \cdots & 0 & 1 & c_{k-1} \end{bmatrix}. \qquad (2.1.10)$$

The subdiagonal entries of C are all ones, the last column contains the coefficients from (2.1.8), and the rest of the entries are zeros.

(c) Show that if x is a right eigenvector of C with eigenvalue λ, then Wx is a right eigenvector of A with eigenvalue λ. Explain why it is important to this argument that the columns of W are linearly independent. This shows that every eigenvalue of C is an eigenvalue of A. Thus we can complete the proof by showing that C has an eigenvalue.

(d) As it turns out, it is convenient to look for left eigenvectors. Show that $u^T = [u_1 \cdots u_k]$ is a left eigenvector of C with associated eigenvalue λ if and only if $u_2 = \lambda u_1$, $u_3 = \lambda u_2, \ldots, u_k = \lambda u_{k-1}$, and

$$u_1 c_0 + u_2 c_1 + \cdots + u_k c_{k-1} = u_k \lambda. \qquad (2.1.11)$$

(e) Show that u^T is an eigenvector with $u_1 = 1$ (and eigenvalue λ) if and only if $u_j = \lambda^j$ for all j, and

$$\lambda^k - c_{k-1}\lambda^{k-1} - c_{k-2}\lambda^{k-2}\cdots - c_0 = 0. \tag{2.1.12}$$

In particular, λ is an eigenvalue of C if and only if it satisfies (2.1.12).

(f) Invoke the fundamental theorem of algebra to conclude that C, and hence A, has an eigenvalue.

2.1.4. This exercise gives a second proof that the main-diagonal entries of a triangular matrix are its eigenvalues. For the sake of argument, let T be a lower triangular matrix.

(a) Show that if $\lambda \neq t_{ii}$ for $i = 1, \ldots, n$ and v is a vector such that $(\lambda I - T)v = 0$, then $v = 0$. (Use the first equation to show that $v_1 = 0$, the second to show that $v_2 = 0$, and so on.) Conclude that λ is not an eigenvalue.

(b) Suppose $\lambda = t_{kk}$ for some k. If λ occurs more than once in the list $t_{11}, \ldots, t_{nn}$, let t_{kk} denote the last occurrence. Thus $\lambda \neq t_{ii}$ for $i = k + 1, \ldots, n$. Show that there is a vector v of the special form $[0, \ldots, 0, 1, v_{k+1}, \ldots, v_n]^T$ such that $(\lambda I - T)v = 0$. Conclude that λ is an eigenvalue of T with eigenvector v.

(c) How must these arguments be modified for the case when T is upper triangular?

2.1.5. Prove Theorem 2.1.8 by induction on k. For the inductive step, you must show that if

$$c_1 v_1 + c_2 v_2 + \cdots + c_k v_k = 0,$$

then $c_1 = \cdots = c_k = 0$. Multiply this equation by the matrix $A - \lambda_k I$, and use the equations $Av_j = \lambda_j v_j$ to simplify the result. Note that one term drops out, allowing you to use the induction hypothesis to show that $c_1, \ldots, c_{k-1}$ are all zero. Finally, show that $c_k = 0$. Be sure to indicate clearly where you are using the fact that the eigenvalues are distinct.

2.1.6.

(a) Show that every eigenvector of

$$A = \begin{bmatrix} 0 & 1 \\ 0 & 0 \end{bmatrix}$$

is a multiple of $\begin{bmatrix} 1 \\ 0 \end{bmatrix}$. Thus A is defective.

(b) More generally, show that the $n \times n$ matrix

$$B = \begin{bmatrix} \mu & 1 & & & \\ & \mu & 1 & & \\ & & \ddots & \ddots & \\ & & & \mu & 1 \\ & & & & \mu \end{bmatrix}$$

with repeated μ's on the main diagonal, ones on the superdiagonal, and zeros elsewhere, has only one eigenvalue, repeated n times, and the associated eigenspace is one-dimensional. Thus B is defective. Matrices of this form are called *Jordan blocks*.

2.1.7. Prove Proposition 2.1.10.

2.1.8. Show that the matrix

$$A = \left[\begin{array}{cc|c} 2 & 1 & 0 \\ 0 & 2 & 0 \\ \hline 0 & 0 & 1 \end{array}\right]$$

is defective. Show that the space $S = \text{span}\{e_1, e_2\}$ is invariant under A and is not spanned by eigenvectors of A. Here $e_1 = [1\ 0\ 0]^T$ and $e_2 = [0\ 1\ 0]^T$.

2.1.9. Prove Proposition 2.1.11. (Let s_i denote the ith column of S. Show that if S is invariant, then $As_i = Sb_i$ for some $b_i \in \mathbb{F}^k$. Then build the matrix equation $AS = SB$. In part (b), be sure to explain why it matters that the columns of S are linearly independent.)

2.1.10. Use Proposition 1.2.2 to prove part (a) of Proposition 2.1.12.

2.1.11. Let $A \in \mathbb{R}^{2n \times 2n}$, and let v be an eigenvector of A associated with eigenvalue λ. Thus $Av = v\lambda$, where $v \neq 0$.

(a) Show that if A is symplectic (see Section 1.3), then $A^T J v = (Jv)\lambda^{-1}$. Thus Jv is an eigenvector of A^T associated with the eigenvalue $1/\lambda$. Deduce that the eigenvalues of a symplectic matrix occur in $\{\lambda, \lambda^{-1}\}$ pairs. Nonreal eigenvalues appear in quadruples: $\{\lambda, \lambda^{-1}, \bar{\lambda}, \bar{\lambda}^{-1}\}$.

(b) Show that if A is Hamiltonian (see Section 1.4), then $A^T J v = (Jv)(-\lambda)$. Thus Jv is an eigenvector of A^T associated with the eigenvalue $-\lambda$. Deduce that the eigenvalues of a Hamiltonian matrix occur in $\{\lambda, -\lambda\}$ pairs. Eigenvalues that are neither real nor purely imaginary appear in quadruples: $\{\lambda, -\lambda, \bar{\lambda}, -\bar{\lambda}\}$.

(c) Show that if A is skew-Hamiltonian, then $A^T J v = (Jv)\lambda$. Thus Jv is an eigenvector of A^T associated with the eigenvalue λ.

2.2 Similarity, Unitary Similarity, and Schur's Theorem

In the next few sections we will show how eigenvalues and eigenvectors are related to important canonical forms that display the structure of a matrix. Along the way we will begin to speculate on how we might compute the eigenvalues and eigenvectors of a matrix.

Let $A, B \in \mathbb{C}^{n \times n}$. Then B is said to be *similar* to A if there is a nonsingular matrix S such that $B = S^{-1}AS$. The relationship of similarity is clearly an equivalence relation (Exercise 2.2.1). In particular it is a symmetric relationship: A is similar to B if and only if B is similar to A. Thus we can say that A and B are similar matrices without specifying which one comes first. In many contexts it is useful to rewrite the equation $B = S^{-1}AS$ in the equivalent form $AS = SB$. The matrix S is called the *transforming matrix* for the similarity transformation.

Often one wishes to restrict the matrices that are allowed as transforming matrices. A and B are said to be *unitarily similar* if there is a unitary matrix U such that $AU = UB$. If A and B are real, then they are said to be *orthogonally similar* if there is a real, orthogonal matrix U such that $AU = UB$. Clearly both of these relationships are equivalence relations (Exercise 2.2.2).

Proposition 2.2.1. *Let A and B be similar, say* $B = S^{-1}AS$. *Then A and B have the same eigenvalues. If v is an eigenvector of B associated with eigenvalue* λ, *then Sv is an eigenvector of A with eigenvalue* λ.

The proof is straightforward; everything follows from the second assertion.

Propositions 2.1.6 and 2.2.1 suggest a possible path to computing the eigenvalues of a matrix. If we can find a triangular matrix T that is similar to A, then we can read the eigenvalues of A from the main diagonal of T. As we shall later see, some of the most successful methods for computing eigensystems do essentially this.

The key to this program is to be able to find invariant subspaces. Suppose that $\mathcal{S}$ is a nontrivial invariant subspace of dimension k, and let $X_1 \in \mathbb{F}^{n \times k}$ be a matrix that represents $\mathcal{S}$ in the sense that $\mathcal{S} = \mathcal{R}(X_1)$. Then we can use X_1 to build a similarity transformation that block triangularizes A. Specifically, let $X_2 \in \mathbb{F}^{n \times (n-k)}$ be a matrix such that $X = [X_1 \ X_2] \in \mathbb{F}^{n \times n}$ is nonsingular. Then use X to perform a similarity transformation. One easily checks that the matrix $B = X^{-1}AX$ is block triangular:

$$B = \left[\begin{array}{cc} B_{11} & B_{12} \\ 0 & B_{22} \end{array} \right],$$

where B_{11} is $k \times k$. By Proposition 2.1.7, the spectrum of A is just the union of the spectra of B_{11} and B_{22}. Thus the similarity transformation $B = X^{-1}AX$ breaks the eigenvalue problem for A into two smaller subproblems that can be attacked independently. If we can find invariant subspaces for B_{11} and B_{22}, then we can break those problems into even smaller subproblems. Continuing in this manner, we eventually solve the eigenvalue problem for A.

The following proposition summarizes the main idea of the previous paragraph, along with its converse.

Proposition 2.2.2. *Let $A \in \mathbb{F}^{n \times n}$, and let $\mathcal{S}$ be a subspace of $\mathbb{F}^n$ of dimension k, with $1 \leq k \leq n - 1$. Let $X_1 \in \mathbb{F}^{n \times k}$ be any matrix such that $\mathcal{S} = \mathcal{R}(X_1)$, let $X_2 \in \mathbb{F}^{n \times n-k}$ be any matrix such that $X = [X_1 \ X_2] \in \mathbb{F}^{n \times n}$ is nonsingular, and let*

$$B = X^{-1}AX = \left[\begin{array}{cc} B_{11} & B_{12} \\ B_{21} & B_{22} \end{array} \right],$$

where $B_{11} \in \mathbb{F}^{k \times k}$. Then $B_{21} = 0$ if and only if $\mathcal{S}$ is invariant under A.

Proposition 2.2.2 is proved in Exercise 2.2.6. In light of this result, it is fair to make the following claim:

> *The basic task of eigensystem computations is to find a nontrivial invariant subspace.*

Our study of methods for carrying out this task will begin in Chapter 3.

Complex Matrices

Theorem 2.2.3. *Let $A \in \mathbb{C}^{n \times n}$ with $n \geq 2$. Then A has a nontrivial invariant subspace.*

Proof: Let $v \in \mathbb{C}^n$ be any eigenvector of A. Then $\mathcal{S} = \text{span}\{v\}$ is a one-dimensional subspace that is invariant under A. Since $0 < 1 < n$, $\mathcal{S}$ is nontrivial. $\quad\square$

Theorem 2.2.4 (Schur's theorem). *Let $A \in \mathbb{C}^{n \times n}$. Then there exists a unitary matrix U and an upper triangular matrix T such that*

$$T = U^{-1}AU = U^*AU.$$

Proof: We prove Schur's theorem by induction on n. The theorem obviously holds when $n = 1$, so let us move to the inductive step. Assume $A \in \mathbb{C}^{n \times n}$, where $n \geq 2$, and suppose that the theorem holds for matrices that are smaller than $n \times n$. By Theorem 2.2.3, A has a nontrivial invariant subspace $\mathcal{S}$. Let k denote the dimension of $\mathcal{S}$, let $v_1, \ldots, v_k$ be an orthonormal basis of $\mathcal{S}$, and let $v_{k+1}, \ldots, v_n$ be $n - k$ additional orthonormal vectors such that the complete set $v_1, \ldots, v_n$ is an orthonormal basis for $\mathbb{C}^n$ (Exercise 1.5.5). Let $V_1 = [v_1 \cdots v_k] \in \mathbb{C}^{n \times k}$, $V_2 = [v_{k+1} \cdots v_n] \in \mathbb{C}^{n \times (n-k)}$, and $V = [V_1 \ V_2] \in \mathbb{C}^n$. Then V is unitary, since its columns are orthonormal, and $\mathcal{S} = \mathcal{R}(V_1)$. If we let $B = V^{-1}AV$, then by Proposition 2.2.2,

$$B = \begin{bmatrix} B_{11} & B_{12} \\ 0 & B_{22} \end{bmatrix},$$

where $B_{11} \in \mathbb{C}^{k \times k}$. We can now apply the induction hypothesis to both B_{11} and B_{22}, since both are smaller than $n \times n$. Thus there exist unitary W_{11} and W_{22} such that $T_{11} = W_{11}^{-1}B_{11}W_{11}$ and $T_{22} = W_{22}^{-1}B_{22}W_{22}$ are both upper triangular. Let

$$W = \begin{bmatrix} W_{11} & 0 \\ 0 & W_{22} \end{bmatrix},$$

and let $T = W^{-1}BW$. Then clearly

$$T = \begin{bmatrix} T_{11} & T_{12} \\ 0 & T_{22} \end{bmatrix},$$

where $T_{12} = W_{11}^{-1}B_{12}W_{22}$. Of course, the form of T_{12} is unimportant; what matters is that T is upper triangular. Finally let $U = VW$. Then U is unitary, since it is the product of unitary matrices, and $T = U^{-1}AU$. $\quad\square$

Schur's theorem shows that in principle we can triangularize any matrix and hence find all the eigenvalues. We can even arrange to have the transformation done by a unitary matrix, which is highly desirable from the standpoint of stability and accuracy in the face of roundoff errors. Unfortunately our proof of Schur's theorem is not constructive. It does not show how to compute the similarity transformation without knowing some invariant subspaces in advance.

The similarity transformation $T = U^{-1}AU$ can obviously also be written as

$$A = UTU^*.$$

When we write it this way, we call it a *Schur decomposition* of A. The Schur decomposition is not unique; the eigenvalues can be made to appear in any order on the main diagonal of

T (Exercise 2.2.10). The Schur decomposition displays not only the eigenvalues of A, but also a nested sequence of invariant subspaces. Let $u_1, \ldots, u_n$ denote the columns of U, and let $\mathcal{S}_j = \mathrm{span}\{u_1, \ldots, u_j\}$ for $j = 1, \ldots, n-1$. Obviously $\mathcal{S}_1 \subseteq \mathcal{S}_2 \subseteq \cdots \subseteq \mathcal{S}_{n-1}$, and each $\mathcal{S}_j$ is invariant under A.

Real Matrices

In most applications the matrices have real entries. If we happen to be working with a real matrix A, this does not guarantee that the eigenvalues are all real, so we have to be prepared to work with complex numbers at some point.

Notice first, however, that if λ is a real eigenvalue of A, then λ must have a real eigenvector associated with it. This is so because the eigenvectors are the nonzero solutions of $(\lambda I - A)v = 0$. Since $\lambda I - A$ is a real singular matrix, the equation $(\lambda I - A)v = 0$ must have real nonzero solutions.

In the interest of efficiency, we would prefer to keep our computations within the real number system for as long as possible. The presence of a complex eigenvalue would seem to force us out of the real number system. Fortunately, complex eigenvalues of real matrices have a nice property: They occur in pairs. If λ is an eigenvalue of A, then so is its complex conjugate $\bar{\lambda}$. As we shall see, if we treat each complex-conjugate pair as a unit, we can avoid complex arithmetic.

Our objective is to prove a theorem that is as much like Schur's theorem as possible, except that the matrices involved are real. The proof of Schur's theorem relies on the existence of nontrivial invariant subspaces of $\mathbb{C}^n$. If we want to stay within the real number system, we should work with subspaces of $\mathbb{R}^n$ instead. In the real case we have the following theorem instead of Theorem 2.2.3.

Theorem 2.2.5. *Let $A \in \mathbb{R}^{n \times n}$ with $n \geq 3$. Then A has a nontrivial invariant subspace.*

Proof: Let λ be an eigenvalue of A. If λ is real, then let $v \in \mathbb{R}^n$ be a real eigenvector of A associated with λ. Then $\mathcal{S}_1 = \mathrm{span}\{v\}$ is a one-dimensional (hence nontrivial) subspace of $\mathbb{R}^n$ that is invariant under A. Now suppose that λ is not real, and let $v \in \mathbb{C}^n$ be an eigenvector associated with λ. Then v cannot possibly be real. Write $v = v_1 + iv_2$, where $v_1, v_2 \in \mathbb{R}^n$. Then $\mathcal{S}_2 = \mathrm{span}\{v_1, v_2\}$ is a one- or two-dimensional (hence nontrivial) subspace of $\mathbb{R}^n$ that is invariant under A. Hence A has a nontrivial invariant subspace. (See Exercise 2.2.15, in which it is shown that $\mathcal{S}_2$ is invariant and two-dimensional.) $\square$

A matrix $T \in \mathbb{R}^{n \times n}$ is called upper *quasi-triangular* if it is block upper triangular, say

$$T = \begin{bmatrix} T_{11} & T_{12} & \cdots & T_{1k} \\ 0 & T_{22} & \cdots & T_{2k} \\ \vdots & & \ddots & \vdots \\ 0 & 0 & \cdots & T_{kk} \end{bmatrix},$$

where the main-diagonal blocks T_{jj} are all 1×1 or 2×2, and each 2×2 block has complex eigenvalues. Thus the real eigenvalues of T appear in the 1×1 blocks, and the complex conjugate pairs of eigenvalues are carried by the 2×2 blocks. The Wintner–Murnaghan theorem, also known as the real Schur theorem, states that every $A \in \mathbb{R}^{n \times n}$ is orthogonally similar to a quasi-triangular matrix.

Theorem 2.2.6 (Wintner–Murnaghan). *Let $A \in \mathbb{R}^{n \times n}$. Then there exist an orthogonal matrix U and a quasi-triangular matrix T such that*

$$T = U^{-1} A U = U^T A U.$$

The proof is similar to the proof of Schur's theorem and is left as an exercise (Exercise 2.2.16).

If we rewrite the equation in the form

$$A = U T U^T,$$

we have a *Wintner–Murnaghan* or *real Schur decomposition* of A. This decomposition is not unique; the main-diagonal blocks can be made to appear in any order in T.

Exercises

2.2.1. Show that similarity is an equivalence relation on $\mathbb{C}^{n \times n}$, that is that the following are true:

  (a) A is similar to A.

  (b) If B is similar to A, then A is similar to B.

  (c) If A is similar to B and B is similar to C, then A is similar to C.

2.2.2. Show that unitary and orthogonal similarity are equivalence relations on $\mathbb{C}^{n \times n}$ and $\mathbb{R}^{n \times n}$, respectively.

2.2.3. Prove Proposition 2.2.1. (Prove the second assertion, and use the fact that similarity is an equivalence relation.)

2.2.4. Show that if A and B are similar, then $\det(\lambda I - A) = \det(\lambda I - B)$. Thus A and B have the same eigenvalues, and the algebraic multiplicities are preserved.

2.2.5. Suppose that $\mathcal{S}$ is invariant under A, $\mathcal{S} = \mathcal{R}(S) = \mathcal{R}(\hat{S})$, where S and $\hat{S}$ have full rank, $AS = SB$, and $A\hat{S} = \hat{S}\hat{B}$. Show that the matrices B and $\hat{B}$ are similar. Thus they have the same eigenvalues. This shows that the eigenvalues of A *associated with* $\mathcal{S}$ are well defined; they depend on $\mathcal{S}$ alone, not on the representation of $\mathcal{S}$.

2.2.6. In this exercise you will prove Proposition 2.2.2.

  (a) Show that $A X_1 = X_1 B_{11} + X_2 B_{21}$.

  (b) Show that if $\mathcal{S}$ is invariant under A, then $B_{21} = 0$. (Consider the equation from part (a) one column at a time.)

  (c) Show that if $B_{21} = 0$, then $\mathcal{S}$ is invariant under A.

  (d) Show that if $\mathcal{S}$ is invariant under A, then the eigenvalues of B_{11} are the eigenvalues of A associated with $\mathcal{S}$.

2.2.7. Referring to Proposition 2.2.2, show that $B_{12} = 0$ if and only if $\mathcal{R}(X_2)$ is invariant under A.

2.2.8. In this exercise we use the notation from Proposition 2.2.2. Let $\mathcal{S}$ be invariant under A, so that $B_{21} = 0$. Let $Y \in \mathbb{F}^{n \times n}$ be defined by $Y^* = X^{-1}$. Make a partition $Y = \begin{bmatrix} Y_1 & Y_2 \end{bmatrix}$ with $Y_1 \in \mathbb{F}^{n \times k}$.

   (a) Using the equation $Y^* X = I$, show that $\mathcal{R}(Y_2) = \mathcal{S}^{\perp}$.

   (b) Show that $A^* Y = Y B^*$. Write this equation in partitioned form and deduce that $A^* Y_2 = Y_2 B_{22}^*$. This proves (for the second time) part (a) of Proposition 2.1.12: $\mathcal{S}^{\perp}$ in invariant under A^*

   (c) Deduce part (b) of Proposition 2.1.12: The eigenvalues of A^* associated with $\mathcal{S}^{\perp}$ are $\bar{\lambda}_{k+1}, \ldots, \bar{\lambda}_n$.

2.2.9. Check the details of the proof of Schur's theorem (Theorem 2.2.4).

2.2.10. Devise a proof of Schur's theorem that shows that the eigenvalues of A can be made to appear in any order on the main diagonal of T. (Use induction on n. Let λ_1 be the eigenvalue that you would like to have appear as t_{11}. Let v_1 be an eigenvector of A associated with λ_1, normalized so that $\| v_1 \| = 1$. Let V be a unitary matrix whose first column is v_1, and consider the similarity transformation $V^{-1} A V$.)

2.2.11. In Schur's theorem, let $u_1, \ldots, u_n$ denote the columns of U, and let $\mathcal{S}_j = \text{span}\{u_1, \ldots, u_j\}$ for $j = 1, \ldots, n - 1$. Use Proposition 2.2.2 to prove that each of the $\mathcal{S}_j$ is invariant under A.

2.2.12. Given a square matrix $A \in \mathbb{C}^{n \times n}$, the *trace* of A is

$$\text{tr}(A) = \sum_{j=1}^{n} a_{jj}.$$

   Prove the following facts about the trace:

   (a) $\text{tr}(A + B) = \text{tr}(A) + \text{tr}(B)$ for all $A, B \in \mathbb{C}^{n \times n}$.

   (b) $\text{tr}(CE) = \text{tr}(EC)$ for all $C \in \mathbb{C}^{n \times m}$ and $E \in \mathbb{C}^{m \times n}$. Notice that C and E need not be square. CE and EC must both be square but need not be of the same size.

   (c) Show that if $B \in \mathbb{C}^{n \times n}$ is similar to A ($B = S^{-1} A S$), then $\text{tr}(B) = \text{tr}(A)$. (Group the matrices in the product $S^{-1} A S$ and apply part (b).)

   (d) Let $A \in \mathbb{C}^{n \times n}$ with eigenvalues $\lambda_1, \ldots, \lambda_n$. Use Schur's theorem to prove that

$$\text{tr}(A) = \lambda_1 + \cdots + \lambda_n.$$

2.2.13. Use Schur's theorem to show that the determinant of a matrix is the product of its eigenvalues.

2.2.14. Prove that if $A \in \mathbb{R}^{n \times n}$, then $\lambda = \alpha + i\beta$ is an eigenvalue of A if and only if $\bar{\lambda} = \alpha - i\beta$ is. Do it two different ways.

(a) Show that if $v \in \mathbb{C}^n$ is an eigenvector of A associated with eigenvalue λ, then $\bar{v}$ is an eigenvector of A with eigenvalue $\bar{\lambda}$.

(b) Show that if p is the characteristic polynomial of A, then p has real coefficients. Then show that $p(\bar{\lambda}) = \overline{p(\lambda)}$.

2.2.15. Fill in the details of the proof of Theorem 2.2.5.

(a) Show that if λ is not real, then v_1 and v_2 are linearly independent. (Show that if they are dependent, then $Av_1 = \lambda v_1$, $Av_2 = \lambda v_2$, and λ is real.) Hence S_2 is two-dimensional.

(b) Let $V = [v_1 \; v_2]$, and let $\lambda = \lambda_1 + i\lambda_2$, where $\lambda_1, \lambda_2 \in \mathbb{R}$. Show that $AV = VB$, where

$$B = \begin{bmatrix} \lambda_1 & \lambda_2 \\ -\lambda_2 & \lambda_1 \end{bmatrix}.$$

Conclude that $S_2 = \mathcal{R}(V)$ is invariant under A.

(c) Show that the eigenvalues of B are λ and $\bar{\lambda}$.

2.2.16. Prove the Wintner–Murnaghan theorem (Theorem 2.2.6) by induction on n.

(a) Prove the cases $n = 1$ and $n = 2$. This is mostly trivial. The case of a matrix that is 2×2 and has real eigenvalues can be handled by building an orthogonal matrix whose first column is a real eigenvector.

(b) Assume $n \geq 3$, and do the induction step. Use Theorem 2.2.5, and argue as in the proof of Schur's theorem.

2.2.17. (symplectic similarity transformations) Let $A \in \mathbb{R}^{2n \times 2n}$ and $S \in \mathbb{R}^{2n \times 2n}$, and suppose that S is symplectic (see Section 1.3). Let $\hat{A} = S^{-1}AS$.

(a) Show that $\hat{A}$ is Hamiltonian (see Section 1.4) if A is.

(b) Show that $\hat{A}$ is skew-Hamiltonian if A is.

(c) Show that $\hat{A}$ is symplectic if A is.

2.3 The Spectral Theorem for Normal Matrices

For normal matrices, Schur's theorem takes a particularly nice form: The triangular matrix T turns out to be diagonal. This special case of Schur's theorem is called the spectral theorem.

Recall (from Section 1.4) that A is *normal* if $AA^* = A^*A$. The class of normal matrices contains the important subclasses of Hermitian, skew-Hermitian, and unitary matrices.

Proposition 2.3.1. *If A is normal, U is unitary, and $B = U^*AU$, then B is also normal.*

The proof is an easy exercise.

Proposition 2.3.2. *Let $A \in \mathbb{C}^{n \times n}$ be normal, and consider a partition*

$$A = \begin{bmatrix} A_{11} & A_{12} \\ A_{21} & A_{22} \end{bmatrix},$$

where A_{11} and A_{22} are square. Then

$$\| A_{21} \|_F = \| A_{12} \|_F.$$

For the easy proof, see Exercise 2.3.4. As an immediate consequence we have the following.

Proposition 2.3.3. *Let $A \in \mathbb{C}^{n \times n}$ be normal and block triangular:*

$$A = \begin{bmatrix} A_{11} & A_{12} \\ 0 & A_{22} \end{bmatrix},$$

where $A_{11} \in \mathbb{C}^{k \times k}$. Then $A_{12} = 0$, and A_{11} and A_{22} are normal.

Thus block triangular matrices that are normal must be block diagonal. The meaning of this result in terms of invariant subspaces is established in Exercise 2.3.7.

Corollary 2.3.4. *If $A \in \mathbb{C}^{n \times n}$ is normal and triangular, then A is diagonal.*

We can now conclude that every normal matrix is unitarily similar to a diagonal matrix. One easily checks (Exercise 2.3.8) that the converse is true as well. Thus we have the spectral theorem for normal matrices.

Theorem 2.3.5 (spectral theorem). *Let $A \in \mathbb{C}^{n \times n}$. Then A is normal if and only if there is a unitary matrix U and a diagonal matrix Λ such that*

$$A = U \Lambda U^*.$$

Now suppose that A is normal, and let U and Λ be as in Theorem 2.3.5. Let $u_1, \ldots, u_n$ denote the columns of U, and let $\lambda_1, \ldots, \lambda_n$ denote the main-diagonal entries of Λ. Looking at the jth column of the matrix equation $AU = U\Lambda$, we see right away that $Au_j = u_j \lambda_j$. Thus u_j is an eigenvector of A associated with the eigenvalue λ_j. Thus $u_1, \ldots, u_n$ is an orthonormal basis of $\mathbb{C}^n$ consisting of eigenvectors of A. The converse of this result is also true.

Proposition 2.3.6. *$A \in \mathbb{C}^{n \times n}$ is normal if and only if $\mathbb{C}^n$ has an orthonormal basis consisting of eigenvectors of A.*

The details are worked out in Exercise 2.3.11.

As a corollary we see that all normal matrices are semisimple. The converse is false; for semisimplicity a matrix needs to have n linearly independent eigenvectors, but they need not be orthonormal.

The spectral decomposition of Theorem 2.3.5 is closely related to the SVD. Suppose $A = U \Lambda U^*$, as in Theorem 2.3.5. Clearly we can make the eigenvalues $\lambda_1, \ldots, \lambda_n$ appear in any order on the main diagonal of Λ. Suppose we have ordered them so that $|\lambda_1| \geq$

$|\lambda_2| \geq \cdots \geq |\lambda_n|$. Let $|\Lambda|$ denote the diagonal matrix whose main-diagonal entries are $|\lambda_1|, \ldots, |\lambda_n|$, and let $M = \mathrm{diag}\{\mu_1, \ldots, \mu_n\}$ denote the diagonal matrix given by

$$\mu_i = \begin{cases} \lambda_i/|\lambda_i| & \text{if } \lambda_i \neq 0, \\ 1 & \text{if } \lambda_i = 0. \end{cases} \tag{2.3.1}$$

Then $\Lambda = |\Lambda|M$, so $A = U|\Lambda|MU^*$ Let $\Sigma = |\Lambda|$ and $V = UM^*$. Then $A = U\Sigma V^*$. Since M is clearly unitary (Exercise 2.3.10), V is also unitary. Thus the decomposition $A = U\Sigma V^*$ is an SVD of A. We have proved the following theorem.

Theorem 2.3.7. *Let $A \in \mathbb{C}^{n \times n}$ be normal, and let $A = U\Lambda U^*$ be the spectral decomposition of A, as in Theorem 2.3.5, where $|\lambda_1| \geq \cdots \geq |\lambda_n|$. Let $M = \mathrm{diag}\{\mu_1, \ldots, \mu_n\}$ be as defined in (2.3.1), and let $\Sigma = |\Lambda|$ and $V = UM^*$. Then*

$$A = U\Sigma V^*$$

is an SVD of A. In particular, the singular values of A are $|\lambda_1|, \ldots, |\lambda_n|$.

As an immediate corollary we have the following relationships between norms and eigenvalues.

Corollary 2.3.8. *Let $A \in \mathbb{C}^{n \times n}$ be a normal matrix with eigenvalues $\lambda_1, \ldots, \lambda_n$, ordered so that $|\lambda_1| \geq \cdots \geq |\lambda_n|$. Then the following hold:*

(a) $\|A\| = |\lambda_1|$.

(b) $\min_{x \neq 0} \frac{\|Ax\|}{\|x\|} = |\lambda_n|$.

(c) *If A is nonsingular, then $\|A^{-1}\| = |\lambda_n|^{-1}$.*

These results are an immediate consequence of Theorem 2.3.7 and Propositions 1.7.5 and 1.7.6.

Real Normal Matrices

If A is both normal and real, it can be brought nearly to diagonal form by an orthogonal similarity transformation. A matrix is called *quasi-diagonal* if it is block diagonal with 1×1 and 2×2 blocks on the main diagonal, and if the 2×2 blocks have nonreal eigenvalues.

Corollary 2.3.9. *If $A \in \mathbb{R}^{n \times n}$ is normal, then there exist orthogonal $U \in \mathbb{R}^{n \times n}$ and quasi-diagonal $D \in \mathbb{R}^{n \times n}$ such that*

$$A = UDU^T.$$

This is an immediate consequence of Theorem 2.2.6, Proposition 2.3.3, and a simple induction argument. The class of real normal matrices includes the real symmetric, skew symmetric, and orthogonal matrices. For each of these classes of matrices, more refined statements about the blocks in D can be made. For example, if A is symmetric, its eigenvalues are all real. This is a consequence of Exercise 2.3.9, in which it is shown that all eigenvalues of a Hermitian matrix are real. Thus, in the real symmetric case, the blocks in D in Corollary 2.3.9 are all 1×1.

Corollary 2.3.10 (spectral theorem). *Let $A \in \mathbb{R}^{n \times n}$ be symmetric. Then there exist an orthogonal matrix $U \in \mathbb{R}^{n \times n}$ and a diagonal matrix $\Lambda \in \mathbb{R}^{n \times n}$ such that*

$$A = U \Lambda U^T.$$

A few more details are worked out in Exercise 2.3.13.

Exercises

2.3.1. Suppose $B = U^* A U$, where U is unitary.

(a) Prove that B is normal if A is. This is Proposition 2.3.1.

(b) Prove that B is Hermitian if A is.

(c) Prove that B is skew-Hermitian if A is.

(d) Prove that B is unitary if A is.

2.3.2. Show by example that none of the properties discussed in Exercise 2.3.1 is preserved under arbitrary similarity transformations $B = S^{-1} A S$.

2.3.3. In Exercise 2.2.12 we proved some elementary properties of the *trace* of a square matrix. Now we will prove a few more properties that will be used in Exercise 2.3.4 and later. Recall the definition

$$\mathrm{tr}(A) = \sum_{j=1}^{n} a_{jj}.$$

(a) Show that $\mathrm{tr}(BB^*) = \sum_{i=1}^{n} \sum_{j=1}^{m} |b_{ij}|^2 = \|B\|_F^2$ for all $B \in \mathbb{C}^{n \times m}$.

(b) Show that $\mathrm{tr}(BB^*) = 0$ if and only if $B = 0$.

(c) Use the CSB inequality (Theorem 1.2.1) to show that, for all $B, C \in \mathbb{C}^{n \times m}$,

$$|\mathrm{tr}(C^* B)| \le \|B\|_F \|C\|_F.$$

Indeed, this is just a restatement of the CSB inequality, once one realizes that $\mathrm{tr}(C^* B)$ is an inner product.

(d) Show that $|\mathrm{tr}(BC)| \le \|B\|_F \|C\|_F$ for all $B \in \mathbb{C}^{n \times m}$, $C \in \mathbb{C}^{m \times n}$. (CSB inequality)

2.3.4. In this exercise you will prove Proposition 2.3.2.

(a) Using the notation of Proposition 2.3.2, show that

$$A_{11}^* A_{11} + A_{21}^* A_{21} = A_{11} A_{11}^* + A_{12} A_{12}^*.$$

(b) Use properties of the trace functional established in Exercises 2.2.12 and 2.3.3 to show that $\|A_{21}\|_F = \|A_{12}\|_F$.

2.3.5. Prove Proposition 2.3.3.

2.3.6. Prove Corollary 2.3.4 by induction on n.

2.3.7. Let $A \in \mathbb{C}^{n \times n}$ be a normal matrix, and let $\mathcal{S}$ be a subspace of $\mathbb{C}^n$ of dimension k. In this exercise you will prove the following basic fact: $\mathcal{S}$ is invariant under A if and only if the orthogonal complement $\mathcal{S}^\perp$ is invariant under A. Let $u_1, \ldots, u_k$ be an orthonormal basis of $\mathcal{S}$, and let $u_{k+1}, \ldots, u_n$ be additional orthonormal vectors such that $u_1, \ldots, u_n$ is an orthonormal basis of $\mathbb{C}^n$. Let $U_1 = [u_1 \cdots u_k]$, $U_2 = [u_{k+1} \cdots u_n]$, and $U = [u_1 \cdots u_n] \in \mathbb{C}^{n \times n}$.

    (a) Show that $\mathcal{S}^\perp = \text{span}\{u_{k+1}, \ldots, u_n\} = \mathcal{R}(U_2)$.

    (b) Using Proposition 2.2.2, Exercise 2.2.7, and Proposition 2.3.3, prove that $\mathcal{S}$ is invariant under A if and only if $\mathcal{S}^\perp$ is.

2.3.8. Let D be a diagonal matrix, and let U be unitary.

    (a) Prove that D is normal.

    (b) Prove that UDU^* is normal.

2.3.9. Let $A \in \mathbb{C}^{n \times n}$ be normal. Use the spectral theorem (Theorem 2.3.5) to prove the following:

    (a) A is Hermitian ($A^* = A$) if and only if the eigenvalues of A are real.

    (b) A Hermitian matrix A is positive definite if and only if the eigenvalues of A are positive.

    (c) A is skew-Hermitian ($A^* = -A$) if and only if the eigenvalues of A are purely imaginary.

    (d) A is unitary ($A^* = A^{-1}$) if and only if the eigenvalues of A lie on the unit circle ($|\lambda| = 1$) in the complex plane.

2.3.10. Show that the diagonal matrix M defined by (2.3.1) is unitary. Prove it either directly or by use of part (d) of Exercise 2.3.9.

2.3.11. Let $A \in \mathbb{C}^{n \times n}$.

    (a) Prove that if A is normal, then $\mathbb{C}^n$ has an orthonormal basis of eigenvectors of A.

    (b) Conversely, suppose that $u_1, \ldots, u_n$ is an orthonormal basis of $\mathbb{C}^n$ consisting of eigenvectors of A: $Au_j = u_j \lambda_j$, $j = 1, \ldots, n$. Let $U = [u_1 \cdots u_n]$ and $\Lambda = \text{diag}\{\lambda_1, \ldots, \lambda_n\}$. Show that $AU = U\Lambda$. Deduce that A is normal.

2.3.12. Let $A \in \mathbb{C}^{n \times m}$, and let $A = U\Sigma V^*$ be an SVD of A.

    (a) Show that A^*A and AA^* are Hermitian, and hence normal.

    (b) Show that $A^*A = V(\Sigma^*\Sigma)V^*$ and that this equation represents the spectral decomposition of A^*A.

(c) Show that $AA^* = U(\Sigma\Sigma^*)U^*$ and that this equation represents the spectral decomposition of AA^*.

(d) Show that the nonzero eigenvalues of both A^*A and AA^* are $\sigma_1^2, \ldots, \sigma_r^2$, where $\sigma_1, \ldots, \sigma_r$ are the nonzero singular values of A.

(e) Show that the right singular vectors of A are eigenvectors of A^*A, and the left singular vectors are eigenvectors of AA^*.

2.3.13. Let B be a real 2×2 matrix.

(a) Show that if B is normal, then B has one of two forms:

$$\begin{bmatrix} a & b \\ b & c \end{bmatrix} \quad \text{or} \quad \begin{bmatrix} a & b \\ -b & a \end{bmatrix}.$$

In the first case, B has two real eigenvalues and can be transformed to two 1×1 blocks. Show that in the latter case, the eigenvalues of B are $a \pm ib$. State a refined version of Corollary 2.3.9 that is as specific as possible.

(b) Show that if B is skew symmetric, it must have the form

$$\begin{bmatrix} 0 & b \\ -b & 0 \end{bmatrix}.$$

What does a 1×1 skew symmetric matrix look like? State a version of Corollary 2.3.9 for skew symmetric matrices that is as specific as possible.

(c) Show that if $A \in \mathbb{R}^{n \times n}$ is skew symmetric and n is odd, then A is singular.

(d) Show that $B = \begin{bmatrix} a & b \\ -b & a \end{bmatrix}$ is orthogonal if and only if $a^2 + b^2 = 1$. What does a 1×1 orthogonal matrix look like? State a version of Corollary 2.3.9 for orthogonal matrices that is as specific as possible.

2.4   The Jordan Canonical Form

In the world of numerical computing, nontrivial Jordan canonical forms are almost never seen. The theoretical results in this section are nevertheless important. We turn our focus to nonnormal matrices and ask to what extent they can be diagonalized. By now it should be clear that not every matrix can be diagonalized by a similarity transformation; only the semisimple ones can.

Theorem 2.4.1. *Let $A \in \mathbb{C}^{n \times n}$. Then A is semisimple if and only if there exist a nonsingular matrix $S \in \mathbb{C}^{n \times n}$ and a diagonal matrix $\Lambda \in \mathbb{C}^{n \times n}$ such that*

$$A = S\Lambda S^{-1}.$$

The columns of S constitute a basis of $\mathbb{C}^n$ consisting of eigenvectors of A.

The theory of the Jordan canonical form exists because matrices that are not semisimple (defective matrices) exist. How close to diagonal form can a defective matrix A be brought by a similarity transformation? We know from Schur's theorem that every A is similar to an upper triangular matrix. How much additional progress toward diagonal form can we make? Let us begin by considering a simple scenario.

Block Diagonalization

Suppose that A is block triangular, of the form

$$A = \begin{bmatrix} A_{11} & A_{12} \\ 0 & A_{22} \end{bmatrix} \tag{2.4.1}$$

or

$$A = \begin{bmatrix} A_{11} & 0 \\ A_{21} & A_{22} \end{bmatrix}, \tag{2.4.2}$$

where A_{11} and A_{22} are square. Can we get rid of the off-diagonal block A_{12} or A_{21} by a similarity transformation? As we shall see, the answer is yes if A_{11} and A_{22} have no eigenvalues in common. We will study (2.4.2); clearly (2.4.1) can be handled similarly.

We consider similarity transformations by matrices of a very simple form:

$$S = \begin{bmatrix} I & 0 \\ X & I \end{bmatrix}, \quad S^{-1} = \begin{bmatrix} I & 0 \\ -X & I \end{bmatrix}.$$

Similarity transformations of this type will preserve the block lower triangular form of (2.4.2) and leave the blocks A_{11} and A_{22} unchanged. The question, then, is whether there exists an X such that

$$\begin{bmatrix} I & 0 \\ -X & I \end{bmatrix} \begin{bmatrix} A_{11} & 0 \\ A_{21} & A_{22} \end{bmatrix} \begin{bmatrix} I & 0 \\ X & I \end{bmatrix} = \begin{bmatrix} A_{11} & 0 \\ 0 & A_{22} \end{bmatrix}. \tag{2.4.3}$$

An easy calculation shows that (2.4.3) holds if and only if the *Sylvester equation*

$$X A_{11} - A_{22} X = A_{21} \tag{2.4.4}$$

is satisfied. If A_{11} is $k \times k$, then X and A_{21} are both $(n - k) \times k$, and (2.4.4) is a system of $(n - k)k$ linear equations in the $(n - k)k$ unknowns x_{ij}. Letting $x_1, \ldots, x_k$ denote the columns of X, we define

$$\text{vec}(X) = \begin{bmatrix} x_1 \\ x_2 \\ \vdots \\ x_k \end{bmatrix}, \tag{2.4.5}$$

which stacks the unknowns into a long vector in $\mathbb{C}^{(n-k)k}$. The matrix A_{21}, which represents known quantities on the right-hand side of (2.4.4), can be stacked in the same way. If we let $w = \text{vec}(X)$ and $b = \text{vec}(A_{21})$, then (2.4.4) can be rewritten in the form

$$M w = b,$$

where M is some $(n-k)k \times (n-k)k$ coefficient matrix. If M is nonsingular, then (2.4.4) has a unique solution. For the purpose of analyzing M we introduce Kronecker products and develop some of their properties in Exercises 2.4.3 and 2.4.4. We then prove the following theorem in Exercise 2.4.5.

Theorem 2.4.2. *The Sylvester equation (2.4.4) has a unique solution if and only if A_{11} and A_{22} have no eigenvalues in common. If A_{11} and A_{22} do have eigenvalues in common, then (2.4.4) has either no solution (the generic case) or infinitely many solutions.*

Exercise 2.4.6 gives an idea about how to solve Sylvester equations efficiently. In some applications k and $n-k$ are both small. Then the number of equations in (2.4.4) is small, and efficiency is not an issue. However, as k and $n-k$ become larger, efficiency becomes important, and it is worthwhile to look for an efficient method.

The solution of (2.4.3) is related to invariant subspaces. Proposition 2.2.2 shows that (2.4.3) holds if and only if the column space of $\begin{bmatrix} I \\ X \end{bmatrix}$ is invariant under A. This holds, in turn, if and only if

$$\begin{bmatrix} A_{11} & 0 \\ A_{21} & A_{22} \end{bmatrix} \begin{bmatrix} I \\ X \end{bmatrix} = \begin{bmatrix} I \\ X \end{bmatrix} B$$

for some B, by Proposition 2.1.11. A moment's thought shows that this equation also amounts to the Sylvester equation (2.4.4).

We now know how to block diagonalize the block triangular matrices (2.4.1) and (2.4.2) if the main-diagonal blocks do not have common eigenvalues. These results have obvious extensions, for example, the following.

Proposition 2.4.3. *Let A be a block triangular matrix, say*

$$A = \begin{bmatrix} A_{11} & A_{12} & \cdots & A_{1j} \\ 0 & A_{22} & \cdots & A_{2j} \\ \vdots & & \ddots & \vdots \\ 0 & 0 & \cdots & A_{jj} \end{bmatrix},$$

such that no two of the main diagonal blocks A_{ii} have any eigenvalues in common. Then A is similar to the block-diagonal matrix $\mathrm{diag}\{A_{11}, \ldots, A_{jj}\}$.

We can use Proposition 2.4.3 together with Schur's theorem to reduce any matrix to block diagonal form. Given $A \in \mathbb{C}^{n \times n}$, Schur's theorem says that A is unitarily similar to an upper triangular matrix T. We know, furthermore, from Exercise 2.2.10 that the eigenvalues of A can be made to appear in any order on the main diagonal of T. If A is not semisimple, then it must have at least one repeated eigenvalue. Let us order the eigenvalues of T so that repeated eigenvalues appear consecutively on the main diagonal. Then T can be written in the block triangular form

$$T = \begin{bmatrix} T_{11} & T_{12} & \cdots & T_{1j} \\ 0 & T_{22} & \cdots & T_{2j} \\ \vdots & & \ddots & \vdots \\ 0 & 0 & \cdots & T_{jj} \end{bmatrix},$$

where the main-diagonal blocks are upper triangular, each T_{ii} has a single eigenvalue λ_i repeated on its main diagonal, and no two main-diagonal blocks have the same eigenvalue. The blocks may be of any size and number, depending on the form of the spectrum of A. Consider the two extreme cases: If A has n distinct eigenvalues, then we will have $j = n$, and each block T_{ii} will be 1×1. At the other extreme, if A has a single eigenvalue repeated n times, we will have a single block $T = T_{11}$.

Whatever size and number the blocks have, we know from Proposition 2.4.3 that T is similar to the block diagonal matrix $\text{diag}\{T_{11}, \ldots, T_{jj}\}$. We summarize this in the following proposition.

Proposition 2.4.4. *Let $A \in \mathbb{C}^{n \times n}$, and suppose that A has j distinct eigenvalues $\lambda_1, \ldots, \lambda_j$. Then A is similar to a block diagonal matrix $D = \text{diag}\{T_{11}, \ldots, T_{jj}\}$, where each T_{ii} is upper triangular and has the single eigenvalue λ_i.*

Later on we will develop practical procedures for computing the Schur decomposition and for solving Sylvester equations. Thus we will be in a position to calculate the similarity transformation of Proposition 2.4.4 in practice. Notice, however, that the first step of the procedure is on much firmer ground than the second. The unitary similarity transformation guaranteed by Schur's theorem is well conditioned and hence well behaved from a numerical standpoint. In contrast, the transformation to block diagonal form is not unitary and can be ill conditioned (Exercise 2.4.11) and hence numerically unreliable. An ill-conditioned transformation can magnify roundoff (and other) errors disastrously.

Another problem associated with the second step is deciding when two eigenvalues are equal. If A has some repeated eigenvalues to begin with, roundoff errors in the computation of the Schur form will cause each repeated eigenvalue to appear as a tight cluster of nearby eigenvalues. Then it must be decided whether to treat the cluster as a single repeated eigenvalue or as distinct eigenvalues. Attempts to separate close eigenvalues are likely to result in ill-conditioned similarity transformations.

Reduction of a Nilpotent Block

We now focus on a single block. For simplicity we will drop the subscripts. Instead of writing T_{ii}, we will consider a triangular matrix $T \in \mathbb{C}^{k \times k}$ that has a single eigenvalue λ repeated k times. Our mission is to bring T as close to diagonal form as possible.

It is convenient to work with the shifted matrix $N = T - \lambda I$, which is obviously strictly upper triangular; that is, its main-diagonal entries are all zero. Strictly upper triangular matrices have a special property: They are *nilpotent*, which means that $N^d = 0$ for some positive integer d (Exercise 2.4.12). Any similarity transformation we apply to N corresponds to a similarity transformation on T, since $S^{-1}NS = S^{-1}(T - \lambda I)S = S^{-1}TS - \lambda I$. Thus, for any simple form we are able to achieve by a similarity transformation of N, there is a corresponding simple form that can be achieved by applying the same similarity transformation to T.

Nilpotency has a simple characterization in terms of eigenvalues.

Proposition 2.4.5. *Let $N \in \mathbb{C}^{k \times k}$. Then N is nilpotent if and only if all of the eigenvalues of N are 0.*

The easy proof is worked out in Exercise 2.4.13.

We will begin by developing an important compressed form for nilpotent matrices [140] that can be constructed in practice using unitary similarity transformations. The construction does not require that the starting matrix be strictly upper triangular; it just has to be nilpotent.

Proposition 2.4.6. *Let $N \in \mathbb{C}^{k \times k}$ be nilpotent. Then there is a unitary matrix Q and a block strictly upper triangular matrix R such that $N = QRQ^{-1}$. The exact form of R is as follows. There is a positive integer m and positive integers $s_1, \ldots, s_m$ such that $s_1 \geq \cdots \geq s_m$, $s_1 + \cdots + s_m = k$,*

$$
R = \begin{bmatrix}
0 & R_{12} & R_{13} & \cdots & & R_{1m} \\
0 & 0 & R_{23} & & & R_{2m} \\
0 & 0 & & \ddots & \ddots & \vdots \\
\vdots & \vdots & & & 0 & R_{m-1,m} \\
0 & 0 & \cdots & & 0 & 0
\end{bmatrix},
\tag{2.4.6}
$$

the block R_{ij} is $s_i \times s_j$, and the superdiagonal block $R_{j-1,j}$ has full column rank s_j, $j = 2, \ldots, m$.

Proof: We begin by remarking that if $N = 0$, then the result holds with $m = 1$, $Q = I$, and $R = 0$. The proof is by induction on k. If $k = 1$, then N cannot be nilpotent unless $N = 0$, so the result holds. Now suppose $k > 1$, $N \neq 0$, and the proposition holds for nilpotent matrices of dimension less than k.

A *column compression* of N is a transformation

$$
N\tilde{V} = \begin{bmatrix} 0 & \tilde{N} \end{bmatrix},
\tag{2.4.7}
$$

where $\tilde{V}$ is unitary and the columns of $\tilde{N}$ are linearly independent. The number of columns of $\tilde{N}$ is the rank of N, and the number of columns of the zero block is the nullity. Call this number s_1. It is certainly positive, as N is singular. A column compression is always possible; one way to construct one is worked out in Exercise 2.4.14.

Suppose that we have the column compression (2.4.7). If we complete the similarity transformation $\tilde{V}^{-1} N \tilde{V}$, the block of zeros is preserved:

$$
\tilde{V}^{-1} N \tilde{V} = \begin{bmatrix} 0 & M \end{bmatrix} = \begin{bmatrix} 0 & M_1 \\ 0 & M_2 \end{bmatrix}.
$$

The columns of M are linearly independent. In the last expression we have partitioned M into two blocks, where M_2 is square of dimension $k - s_1$. Thus the upper left-hand block of zeros is $s_1 \times s_1$. By Proposition 2.1.7, the eigenvalues of M_2 are all eigenvalues of N, i.e., 0. Thus M_2 is nilpotent.

If $M_2 = 0$, we are done. We have $N = QRQ^{-1}$, where $Q = \tilde{V}$,

$$
R = \begin{bmatrix} 0 & R_{12} \\ 0 & 0 \end{bmatrix},
$$

with $R_{12} = M_1$. R_{12} has linearly independent columns because M does. We have $m = 2$, $s_2 = k - s_1$. The linear independence of the columns of R_{12} implies that $s_1 \geq s_2$. If $M_2 \neq 0$, we can do a column compression of M_2 to push the construction one step further. This is the right course of action if we really want to construct Q and R.

For the purposes of proving the theorem we can ignore the previous paragraph and simply apply the induction hypotheses to M_2: Since M_2 is nilpotent and of order less than k, we have $M_2 = \tilde{Q}\tilde{R}\tilde{Q}^{-1}$, where $\tilde{Q}$ is unitary and $\tilde{R}$ has the form

$$
\tilde{R} = \begin{bmatrix}
0 & R_{23} & & R_{2m} \\
0 & \ddots & \ddots & \vdots \\
\vdots & & 0 & R_{m-1,m} \\
0 & \cdots & 0 & 0
\end{bmatrix}
$$

for some m. The blocks R_{ij} are $s_i \times s_j$, where $s_2 \geq \cdots \geq s_m$ and $s_2 + \cdots + s_m = k - s_1$. Each block $R_{j-1,j}$ has full column rank s_j. Now, if we let $Q = \tilde{V}\begin{bmatrix} I & 0 \\ 0 & \tilde{Q} \end{bmatrix}$, we have

$$
N = QRQ^{-1} = Q\begin{bmatrix} 0 & M_1\tilde{Q} \\ 0 & \tilde{R} \end{bmatrix}Q^{-1}.
$$

If we make a partition $M_1\tilde{Q} = \begin{bmatrix} R_{12} & \cdots & R_{1m} \end{bmatrix}$ conformably with $\tilde{R}$, we find that R has the form (2.4.6), except that we have not yet shown that $s_1 \geq s_2$ and R_{12} has full column rank.

Since M has linearly independent columns, the submatrix of R that occupies the same position must also have linearly independent columns. This is

$$
\begin{bmatrix}
R_{12} & R_{13} & \cdots & R_{1m} \\
0 & R_{23} & & R_{2m} \\
0 & \ddots & \ddots & \vdots \\
\vdots & & 0 & R_{m-1,m} \\
0 & \cdots & 0 & 0
\end{bmatrix}.
$$

The zero pattern of this matrix forces R_{12} to have linearly independent columns. Since R_{12} is $s_1 \times s_2$, we must have $s_1 \geq s_2$. This completes the induction. $\quad\square$

The proof is clearly constructive. One column compression after another is done until nothing but zeros remain in the lower right corner of the matrix. At each step, in order to do the column compression, one needs to know the rank of the remaining nilpotent matrix. This is problematic in practice, i.e., in floating point arithmetic. If the column compression is done using an SVD, as in Exercise 2.4.14, the rank is determined by the number of nonzero singular values. In floating point arithmetic, any singular values that should have been zero will normally be tiny positive numbers because of the effects of roundoff errors. The standard remedy is to treat all tiny singular values as zeros. In practice a threshold on the order of a small multiple of the unit roundoff is set, and any singular values under that threshold are set to zero.

Further Reduction

Any nilpotent matrix can be transformed to the form R of (2.4.6). Now our objective is to eliminate as many of the nonzero entries in R as possible. If we were going to do this in practice, our main tool would be Gaussian elimination without pivoting. Thus the transformations are nonunitary and can be arbitrarily badly conditioned. But we do not intend to carry out the eliminations in practice; this is just a theoretical exercise from here on in. It turns out that we can eliminate all of the R_{ij} except for the superdiagonal entries $R_{j-1,j}$, and these can be transformed to truncated identity matrices.

Proposition 2.4.7. *Let $R \in \mathbb{C}^{k \times k}$ be of the form stated in Proposition 2.4.6. Let m, $s_1, \ldots, s_m$ be as in Proposition 2.4.6. Then there is a nonsingular matrix U of the form*

$$U = \begin{bmatrix} \hat{U} & 0 \\ 0 & I_{s_m} \end{bmatrix} \text{ such that } U^{-1} R U = E, \text{ where } E \text{ has the form}$$

$$E = \begin{bmatrix} 0 & E_{12} & 0 & \cdots & & 0 \\ 0 & 0 & E_{23} & & & 0 \\ 0 & 0 & & \ddots & \ddots & \vdots \\ \vdots & \vdots & & & 0 & E_{m-1,m} \\ 0 & 0 & \cdots & & 0 & 0 \end{bmatrix}, \tag{2.4.8}$$

$$E_{j-1,j} = \begin{bmatrix} I_{s_j} \\ 0 \end{bmatrix} \in \mathbb{C}^{s_{j-1} \times s_j}.$$

Proof: The proof is by induction on m. When $m = 1$, there is nothing to prove; we have $R = 0$ and $E = 0$. It is instructive to work out the cases $m = 2$ and $m = 3$, since they illustrate the two main ideas of the construction. In the case $m = 2$ we have

$$R = \begin{bmatrix} 0 & R_{12} \\ 0 & 0 \end{bmatrix},$$

and we just have to transform R_{12} to the form E_{12}. Recall that R_{12} is $s_1 \times s_2$ with $s_1 \geq s_2$, and R_{12} has full column rank. Since the columns of R_{12} are linearly independent, there is a nonsingular matrix $S_1 \in \mathbb{C}^{s_1 \times s_1}$ such that $S_1^{-1} R_{12} = E_{12} = \begin{bmatrix} I_{s_2} \\ 0 \end{bmatrix}$ (Exercise 2.4.15). One easily checks that

$$\begin{bmatrix} S_1^{-1} & 0 \\ 0 & I \end{bmatrix} \begin{bmatrix} 0 & R_{12} \\ 0 & 0 \end{bmatrix} \begin{bmatrix} S_1 & 0 \\ 0 & I \end{bmatrix} = \begin{bmatrix} 0 & E_{12} \\ 0 & 0 \end{bmatrix}.$$

Notice that the transforming matrix has the stated form $\text{diag}\{S_1, I_{s_2}\}$.

Now consider the case $m = 3$, in which

$$R = \begin{bmatrix} 0 & R_{12} & R_{13} \\ 0 & 0 & R_{23} \\ 0 & 0 & 0 \end{bmatrix}.$$

We need to eliminate R_{13} and normalize R_{12} and R_{23}. These steps can be done in either order; we will do the elimination first. It is a block Gaussian elimination, in which a multiple of the second block row is subtracted from the first to transform R_{13} to zero. The pivotal element for the elimination is R_{23}. Since R_{23} has full column rank, it has a left inverse, a matrix $\hat{W}_{32} \in \mathbb{C}^{s_3 \times s_2}$ such that $\hat{W}_{32} R_{23} = I_{s_3 \times s_3}$ (Exercise 2.4.15). Let $X = R_{13} \hat{W}_{32}$. One easily checks that

$$
\begin{bmatrix} I & -X & 0 \\ 0 & I & 0 \\ 0 & 0 & I \end{bmatrix}
\begin{bmatrix} 0 & R_{12} & R_{13} \\ 0 & 0 & R_{23} \\ 0 & 0 & 0 \end{bmatrix}
\begin{bmatrix} I & X & 0 \\ 0 & I & 0 \\ 0 & 0 & I \end{bmatrix}
=
\begin{bmatrix} 0 & R_{12} & 0 \\ 0 & 0 & R_{23} \\ 0 & 0 & 0 \end{bmatrix},
$$

and the transforming matrices on the left and right are inverses of each other, so this is a similarity transformation. The left-hand multiplication is the elimination step; it subtracts an appropriate multiple of the second row from the first. The right-hand transformation, completing the similarity, adds a multiple of the first column to the second, causing no change. Notice that the transforming matrix has the stated form.

The normalization of the blocks R_{12} and R_{23} is accomplished by a similarity transformation of the form

$$
\begin{bmatrix} S_1^{-1} & & \\ & S_2^{-1} & \\ & & I \end{bmatrix}
\begin{bmatrix} 0 & R_{12} & 0 \\ 0 & 0 & R_{23} \\ 0 & 0 & 0 \end{bmatrix}
\begin{bmatrix} S_1 & & \\ & S_2 & \\ & & I \end{bmatrix}
$$

with suitable chosen S_1 and S_2, to yield

$$
\begin{bmatrix} 0 & S_1^{-1} R_{12} S_2 & 0 \\ 0 & 0 & S_2^{-1} R_{23} \\ 0 & 0 & 0 \end{bmatrix}.
$$

Since R_{23} has full column rank, there is an S_2 such that $S_2^{-1} R_{23} = E_{23}$ (Exercise 2.4.15). Choose such an S_2. Since R_{12} has full column rank, so does $R_{12}' = R_{12} S_2$. Now choose an S_1 such that $S_1 R_{12}' = E_{12}$. Again the transforming matrix has the stated form. This completes the case $m = 3$.

Now consider the general case. We will eliminate the blocks $R_{1m}, R_{2m}, \ldots, R_{m-2,m}$ first. By introducing suitable notation, we can see how to eliminate them all at once. Let

$$
X = \begin{bmatrix} R_{1m} \\ R_{2m} \\ \vdots \\ R_{m-2,m} \end{bmatrix}
\quad \text{and} \quad
Y = \begin{bmatrix} R_{1m-1} \\ R_{2,m-1} \\ \vdots \\ R_{m-2,m-1} \end{bmatrix}.
$$

Then

$$
R = \begin{bmatrix} \tilde{R} & Y & X \\ 0 & 0 & R_{m-1,m} \\ 0 & 0 & 0 \end{bmatrix},
$$

where $\tilde{R}$ is a square submatrix of R of dimension $s_1 + \cdots + s_{m-2}$. We want to transform X to zero. Since $R_{m-1,m}$ has full column rank, it has a left inverse $W_{m,m-1}$. The elimination step is

$$
\begin{bmatrix} I & -X W_{m,m-1} & 0 \\ 0 & I & 0 \\ 0 & 0 & I \end{bmatrix}
\begin{bmatrix} \tilde{R} & Y & X \\ 0 & 0 & R_{m-1,m} \\ 0 & 0 & 0 \end{bmatrix}
=
\begin{bmatrix} \tilde{R} & Y & 0 \\ 0 & 0 & R_{m-1,m} \\ 0 & 0 & 0 \end{bmatrix}.
$$

Completing the similarity transformation, we have

$$
\begin{bmatrix} \tilde{R} & Y & 0 \\ 0 & 0 & R_{m-1,m} \\ 0 & 0 & 0 \end{bmatrix} \begin{bmatrix} I & X W_{m,m-1} & 0 \\ 0 & I & 0 \\ 0 & 0 & I \end{bmatrix} = \begin{bmatrix} \tilde{R} & \tilde{Y} & 0 \\ 0 & 0 & R_{m-1,m} \\ 0 & 0 & 0 \end{bmatrix},
$$

where

$$
\tilde{Y} = Y + \tilde{R} X W_{m,m-1} = \begin{bmatrix} \tilde{R}_{1m-1} \\ \tilde{R}_{2,m-1} \\ \vdots \\ \tilde{R}_{m-2,m-1} \end{bmatrix}.
$$

The form of R implies that the bottom block row of $\tilde{R}$ consists entirely of zeros. Therefore $\tilde{R}_{m-2,m-1} = R_{m-2,m-1}$. In particular it has full column rank.

Now we normalize $R_{m-1,m}$. Let $S_m \in \mathbb{C}^{s_m \times s_m}$ have the property that $S_m^{-1} R_{m-1,m} = E_{m-1,m}$. Then

$$
\begin{bmatrix} I & & \\ & S_m^{-1} & \\ & & I_{s_m} \end{bmatrix} \begin{bmatrix} \tilde{R} & \tilde{Y} & 0 \\ 0 & 0 & R_{m-1,m} \\ 0 & 0 & 0 \end{bmatrix} \begin{bmatrix} I & & \\ & S_m & \\ & & I_{s_m} \end{bmatrix} = \begin{bmatrix} \tilde{R} & \hat{Y} & 0 \\ 0 & 0 & E_{m-1,m} \\ 0 & 0 & 0 \end{bmatrix},
$$

where $\hat{Y} = \tilde{Y} S_m$. The bottom block of $\hat{Y}$ is $\hat{R}_{m-2,m-1} = \tilde{R}_{m-1,m} S_m$, and it has full column rank. Let

$$
\hat{R} = \begin{bmatrix} \tilde{R} & \hat{Y} \\ 0 & 0 \end{bmatrix}.
$$

By the induction hypothesis there is a nonsingular matrix $\hat{U} = \mathrm{diag}\{\check{U}, I_{s_{m-1}}\}$ such that $\hat{U}^{-1} \hat{R} \hat{U} = \hat{E}$ has the form (2.4.8) with m replaced by $m-1$. Then

$$
\begin{bmatrix} \check{U}^{-1} & & \\ & I_{s_{m-1}} & \\ & & I_{s_m} \end{bmatrix} \begin{bmatrix} \tilde{R} & \hat{Y} & 0 \\ 0 & 0 & E_{m-1,m} \\ 0 & 0 & 0 \end{bmatrix} \begin{bmatrix} \check{U} & & \\ & I_{s_{m-1}} & \\ & & I_{s_m} \end{bmatrix}
$$

$$
= \left[\begin{array}{cc|c} & & 0 \\ \multicolumn{2}{c|}{\hat{E}} & E_{m-1} \\ \hline 0 & 0 & 0 \end{array} \right] = E.
$$

The proof is complete. $\square$

Proposition 2.4.8. *Let $N \in \mathbb{C}^{k \times k}$ be a nilpotent matrix. Then N is similar to a matrix of the form of E in Proposition 2.4.7. The integers m and $s_1, \ldots, s_m$ are uniquely determined by N. In fact, m is the smallest integer such that $N^m = 0$, and $s_j = \dim(\mathcal{N}(N^j)) - \dim(\mathcal{N}(N^{j-1}))$, $j = 1, \ldots, m$.*

This result is mostly a corollary of Propositions 2.4.6 and 2.4.7. The characterization of the numbers m and $s_1, \ldots, s_m$ is easily proved by determining which vectors lie in $\mathcal{N}(E)$, $\mathcal{N}(E^2)$, and so on. This is easy because the form of E is so transparent. (See Exercise 2.4.16.) The ordered set of numbers $s_1, \ldots, s_m$ is called the *Weyr characteristic* of the nilpotent matrix N [187].

Jordan Blocks

A *Jordan block* associated with the eigenvalue λ is a matrix with λ's on the main diagonal, ones on the superdiagonal, and zeros elsewhere. For example, a 4×4 Jordan block has the form

$$
\begin{bmatrix}
\lambda & 1 & 0 & 0 \\
0 & \lambda & 1 & 0 \\
0 & 0 & \lambda & 1 \\
0 & 0 & 0 & \lambda
\end{bmatrix}.
\tag{2.4.9}
$$

The matrix $[\ \lambda\ ]$ is a 1×1 Jordan block. A *nilpotent Jordan block* is a Jordan block corresponding to the eigenvalue 0. For example, the 3×3 nilpotent Jordan block is

$$
\begin{bmatrix}
0 & 1 & 0 \\
0 & 0 & 1 \\
0 & 0 & 0
\end{bmatrix}.
$$

Proposition 2.4.9. *Let $N \in \mathbb{C}^{k \times k}$ be a nilpotent matrix, and let m and $s_1, \ldots, s_m$ be the numbers associated with N in Proposition 2.4.8. Then N is similar to a matrix that is a direct sum of nilpotent Jordan blocks $\mathrm{diag}\{J_1, \ldots, J_{s_1}\}$. The number of blocks is s_1. The size of the largest block is $m \times m$. For $k = 1, \ldots, m$, the number of blocks of dimension $k \times k$ is $s_k - s_{k+1}$, where we make the convention $s_{m+1} = 0$.*

Proof: The proof is left as Exercise 2.4.17. Here we just sketch the main idea. The matrix E similar to N in Proposition 2.4.8 is already very close to a direct sum of Jordan blocks. All that is needed is a permutation similarity transformation, as we shall illustrate by example. Consider the matrix

$$
E = \left[
\begin{array}{ccc|cc|cc|c}
0 & 0 & 0 & 1 & 0 & 0 & 0 & 0 \\
0 & 0 & 0 & 0 & 1 & 0 & 0 & 0 \\
0 & 0 & 0 & 0 & 0 & 0 & 0 & 0 \\
\hline
0 & 0 & 0 & 0 & 0 & 1 & 0 & 0 \\
0 & 0 & 0 & 0 & 0 & 0 & 1 & 0 \\
\hline
0 & 0 & 0 & 0 & 0 & 0 & 0 & 1 \\
0 & 0 & 0 & 0 & 0 & 0 & 0 & 0 \\
\hline
0 & 0 & 0 & 0 & 0 & 0 & 0 & 0
\end{array}
\right],
$$

which has $m = 4$, $s_1 = 3$, $s_2 = s_3 = 2$, and $s_4 = 1$. Start at position $(1, 1)$. Move to the right until you hit the 1 at position $(1, 4)$. Now jump down to the main diagonal at $(4, 4)$, then go right until you hit the 1 at position $(4, 6)$. Continuing in this zigzag manner, proceed to $(6, 6)$, $(6, 8)$, and finally $(8, 8)$. You have just outlined a Jordan block. If we perform a permutation similarity transformation that moves rows and columns 1, 4, 6, and 8 to the upper left-hand corner of the matrix, we will have a 4×4 nilpotent Jordan block. A second Jordan block can be built in the same way, starting from the $(2, 2)$ position, traversing rows and columns 2, 5, and 7, and ending at position $(7, 7)$. Since there is no 1 in row 7, the block ends there and is 3×3. A third Jordan block is built starting from the $(3, 3)$ position. Since there is no 1 in row 3, that block is 1×1. After permuting the rows and columns appropriately, we get the result

$$
J = \left[
\begin{array}{cccc|cccc}
0 & 1 & 0 & 0 & 0 & 0 & 0 & 0 \\
0 & 0 & 1 & 0 & 0 & 0 & 0 & 0 \\
0 & 0 & 0 & 1 & 0 & 0 & 0 & 0 \\
0 & 0 & 0 & 0 & 0 & 0 & 0 & 0 \\ \hline
0 & 0 & 0 & 0 & 0 & 1 & 0 & 0 \\
0 & 0 & 0 & 0 & 0 & 0 & 1 & 0 \\
0 & 0 & 0 & 0 & 0 & 0 & 0 & 0 \\
0 & 0 & 0 & 0 & 0 & 0 & 0 & 0
\end{array}
\right],
$$

a direct sum of three nilpotent Jordan blocks.

In general we make a permutation that moves rows and columns $1, s_1 + 1, s_1 + s_2 + 1, \ldots, s_1 + \cdots + s_{m-1} + 1$ to the top of the matrix. This produces an $m \times m$ Jordan block at the upper left-hand corner. The next Jordan block is produced from the rows and columns that were originally in positions $2, s_1 + 2, s_1 + s_2 + 2$, and so on. If $s_m > 1$, another $m \times m$ block is produced. If $s_m = 1$ and $s_{m-1} > 1$, the block is $(m-1) \times (m-1)$. If $s_{m-1} = 1$, the block is smaller. Continuing in this manner, we transform E to a direct sum of s_1 Jordan blocks. $\quad \square$

We now return to the matrix T with which we began this subsection. T is upper triangular and has a single eigenvalue λ. The nilpotent matrix $N = T - \lambda I$ is similar to a direct sum of Jordan blocks: $N = VKV^{-1}$. Therefore $T = N + \lambda I = V(K + \lambda I)V^{-1}$. Since $K + \lambda I$ is clearly a direct sum of Jordan blocks, we have the following corollary.

Corollary 2.4.10. *Let $T \in \mathbb{C}^{k \times k}$ be an upper triangular matrix with a single eigenvalue λ. Then T is similar to a block diagonal matrix $\mathrm{diag}\{J_1, \ldots, J_{s_1}\}$, where each J_i is a Jordan block with eigenvalue λ.*

The Jordan Canonical Form

Putting the pieces together, we obtain the Jordan canonical form.

Theorem 2.4.11 (Jordan canonical form). *Let $A \in \mathbb{C}^{n \times n}$. Then A is similar to a matrix J that is a direct sum of Jordan blocks. J is uniquely determined up to the order of the blocks. That is, the number and size of the blocks associated with each eigenvalue is uniquely determined, but the blocks can appear in any order on the main diagonal. J is called the Jordan canonical form of A. Two matrices are similar if and only if they have the same Jordan canonical form.*

Proof: By Proposition 2.4.4, A is similar to a matrix $D = \mathrm{diag}\{T_1, \ldots, T_j\}$, where each T_i is upper triangular and has a single eigenvalue λ_i. By Corollary 2.4.10, each T_i is similar to a matrix K_i that is a direct sum of Jordan blocks with eigenvalue λ_i. Letting V_i denote the transforming matrix, we have $T_i = V_i K_i V_i^{-1}$. Let $J = \mathrm{diag}\{K_1, \ldots, K_j\}$ and $V = \mathrm{diag}\{V_1, \ldots, V_j\}$. Then J is a direct sum of Jordan blocks that is similar to D: $D = VJV^{-1}$. Hence J is similar to A. J is the Jordan canonical form of A.

The number and size of Jordan blocks associated with each eigenvalue is rigidly determined by the dimensions of certain null spaces, as Propositions 2.4.8 and 2.4.9 show. Therefore the Jordan canonical form is unique up to the order of the blocks. $\quad \square$

Every diagonal matrix is a Jordan canonical form with n Jordan blocks of size 1×1. A matrix is similar to a diagonal matrix if and only if its Jordan canonical form is diagonal. Consequently, a matrix is not semisimple unless its Jordan form is diagonal.

Much more can be said about the Jordan canonical form. For accounts of elegant algebraic and geometric approaches to the Jordan form, see Gantmacher [91] or Lancaster and Tisminetsky [142], for example. Our computational approach seems to have originated with Kublanovskaya [140]. A similar approach is taken by Horn and Johnson [117].

We have refrained from a complete exposition of the Jordan canonical form because nontrivial Jordan forms are more or less invisible in the world of floating point computation. If a matrix has a nondiagonal Jordan form, then it must necessarily have at least one repeated eigenvalue. Any slight perturbation of the matrix (caused by the first roundoff error in the first similarity transformation) is likely to split the repeated eigenvalue into a cluster of nearby eigenvalues. The perturbed matrix is now semisimple; its Jordan form is diagonal.

Nevertheless, it is possible, using algorithms sketched here, to tell whether a given matrix is close to a defective matrix. The triangular Schur form can be computed reliably by the QR algorithm (Chapter 4). If some of the eigenvalues are tightly clustered, they can be be grouped into clusters and each cluster treated as a single eigenvalue of high multiplicity. The Schur form then looks like

$$
T = \begin{bmatrix}
T_{11} & T_{12} & \cdots & T_{1j} \\
0 & T_{22} & \cdots & T_{2j} \\
\vdots & & \ddots & \vdots \\
0 & 0 & \cdots & T_{jj}
\end{bmatrix},
$$

where each T_{ii} has a single eigenvalue (cluster). Each block can be shifted by the average of its eigenvalues to make it approximately nilpotent. Then the compression algorithm sketched in the proof of Proposition 2.4.6 can be applied to ascertain whether the block really is (approximately) nilpotent and to compute the numbers m, $s_1, \ldots, s_m$, which determine the Jordan structure of the (nearby) defective matrix. The QR algorithm and the compression algorithm use only unitary transformations. It is not necessary to use the potentially unstable block diagonalization algorithm because that algorithm does not affect the T_{ii}, which eventually determine the Jordan structure. Nor is it necessary to use the potentially unstable algorithm that produces the Jordan form, since the Jordan structure is already known by the time the matrix R of Proposition 2.4.6 has been produced.

The Real Jordan Form

If A is a real matrix whose eigenvalues are all real, then all of the constructions that we have used in this section to obtain the Jordan form can be carried out entirely in real arithmetic, and Theorem 2.4.11 holds in real arithmetic. That is, $A = SJS^{-1}$, where J is a direct sum of real Jordan blocks, and the transforming matrix S is real. However, real matrices don't necessarily have real eigenvalues, so it is natural to ask what sort of Jordan-like form is attainable, using real similarity transformations, for real matrices that have some nonreal eigenvalues. We know that for each complex eigenvalue $\lambda = \alpha + i\beta$, the complex conjugate $\overline{\lambda} = \alpha - i\beta$ is also an eigenvalue. More generally, for each Jordan block corresponding to λ, there is (Exercise 2.4.21) a Jordan block of the same size corresponding to $\overline{\lambda}$. These two blocks can be combined to form a single *real Jordan block*, whose form is typified by

$$
\begin{bmatrix}
\alpha & \beta & 1 & & & \\
-\beta & \alpha & & 1 & & \\
& & \alpha & \beta & 1 & \\
& & -\beta & \alpha & & 1 \\
& & & & \alpha & \beta \\
& & & & -\beta & \alpha
\end{bmatrix}.
\tag{2.4.10}
$$

This 6×6 block was built from two 3×3 Jordan blocks corresponding to the eigenvalues $\lambda = \alpha + i\beta$ and $\bar{\lambda} = \alpha - i\beta$. The details are worked out in Exercise 2.4.21. The result is the following theorem.

Theorem 2.4.12 (real Jordan canonical form). *Let $A \in \mathbb{R}^{n \times n}$. Then A is similar, via a real similarity transformation, to a matrix J that is a direct sum of Jordan blocks of two types. Corresponding to each real eigenvalue we have a Jordan block (or blocks) of the standard form exemplified by (2.4.9). Corresponding to each complex-conjugate pair of eigenvalues we have a real Jordan block (or blocks) of the form exemplified by (2.4.10). J is uniquely determined up to the order of the blocks. That is, the number and size of the blocks associated with each eigenvalue or conjugate pair of eigenvalues is uniquely determined, but the blocks can appear in any order on the main diagonal. J is called the* real Jordan canonical form *of A. Two real matrices are similar if and only if they have the same real Jordan canonical form.*

Exercises

2.4.1. Prove Theorem 2.4.1.

(a) Let $A \in \mathbb{C}^{n \times n}$ be semisimple, and let $s_1, \ldots, s_n$ be a basis of $\mathbb{C}^n$ consisting of eigenvectors of A. Suppose $As_j = s_j \lambda_j$, $j = 1, \ldots, n$. Let $S = [s_1 \cdots s_n]$, and let $\Lambda = \mathrm{diag}\{\lambda_1, \ldots, \lambda_n\}$. Show that $AS = S\Lambda$. Conclude that A is similar to a diagonal matrix.

(b) Conversely, suppose $A = S\Lambda S^{-1}$, where S is nonsingular, and Λ is a diagonal matrix. Show that $\mathbb{C}^n$ has a basis consisting of eigenvectors of A. Hence A is semisimple.

2.4.2.

(a) Show that

$$
\begin{bmatrix} I & 0 \\ X & I \end{bmatrix}^{-1} = \begin{bmatrix} I & 0 \\ -X & I \end{bmatrix}.
$$

(b) Verify that (2.4.3) holds if and only if the Sylvester equation (2.4.4) holds.

2.4.3. If X is an $\alpha \times \beta$ matrix and Y is a $\gamma \times \delta$ matrix, the *Kronecker product* or *tensor product* $X \otimes Y$ is the $\alpha\gamma \times \beta\delta$ matrix defined by

$$X \otimes Y = \begin{bmatrix} x_{11}Y & x_{12}Y & \cdots & x_{1\beta}Y \\ x_{21}Y & x_{22}Y & \cdots & x_{2\beta}Y \\ \vdots & \vdots & \ddots & \vdots \\ x_{\alpha 1}Y & x_{\alpha 2}Y & \cdots & x_{\alpha\beta}Y \end{bmatrix}.$$

(a) Show that if X, Y, Z, and W are matrices such that the products XZ and YW are defined, then

$$(X \otimes Y)(Z \otimes W) = XZ \otimes YW.$$

(b) Show that if X and Y are square and nonsingular, then so is $X \otimes Y$, and $(X \otimes Y)^{-1} = X^{-1} \otimes Y^{-1}$.

(c) Show that if A_1 and A_2 are square matrices that are similar to B_1 and B_2, respectively, then $A_1 \otimes A_2$ is similar to $B_1 \otimes B_2$.

(d) Suppose that A and B are square matrices. Use Schur's theorem to show that $A \otimes B$ is unitarily similar to an upper triangular matrix that is itself a Kronecker product of upper triangular matrices.

(e) Show that if A has eigenvalues $\lambda_1, \ldots, \lambda_k$ and B has eigenvalues $\mu_1, \ldots, \mu_m$, then $A \otimes B$ has eigenvalues $\lambda_i \mu_j$, $i = 1, \ldots, k$, $j = 1, \ldots, m$.

2.4.4. If Z is an $\alpha \times \beta$ matrix with columns $z_1, \ldots, z_\beta$, we define $\text{vec}(Z) \in \mathbb{C}^{\alpha\beta}$ by

$$\text{vec}(Z) = \begin{bmatrix} z_1 \\ z_2 \\ \vdots \\ z_\beta \end{bmatrix}.$$

Let I_n denote the $n \times n$ identity matrix.

(a) Show that if $A \in \mathbb{C}^{m \times n}$ and $X \in \mathbb{C}^{n \times p}$, then

$$\text{vec}(AX) = (I_p \otimes A)\text{vec}(X).$$

Consequently, the equation $AX = B$ is equivalent to $(I_p \otimes A)\text{vec}(X) = \text{vec}(B)$.

(b) Show that if $X \in \mathbb{C}^{m \times n}$ and $A \in \mathbb{C}^{n \times p}$, then

$$\text{vec}(XA) = (A^T \otimes I_m)\text{vec}(X).$$

Consequently, the equation $XA = B$ is equivalent to $(A^T \otimes I_m)\text{vec}(X) = \text{vec}(B)$.

(c) Suppose $A_1 \in \mathbb{C}^{k \times k}$ and $A_2 \in \mathbb{C}^{m \times m}$. Show that the equation $XA_1 - A_2X = B$ holds if and only if

$$(A_1^T \otimes I_m - I_k \otimes A_2)\text{vec}(X) = \text{vec}(B).$$

(d) Show that $(A_1^T \otimes I_m - I_k \otimes A_2)$ is unitarily similar to an upper triangular matrix of the form $(U_1 \otimes I_m - I_k \otimes U_2)$. Show that if A_1 has eigenvalues $\lambda_1, \ldots, \lambda_k$ and A_2 has eigenvalues $\mu_1, \ldots, \mu_m$, then $(A_1^T \otimes I_m - I_k \otimes A_2)$ has eigenvalues $\lambda_i - \mu_j$, $i = 1, \ldots, k$, $j = 1, \ldots, m$.

2.4.5. Use results from Exercise 2.4.4 to prove Theorem 2.4.2.

(a) Show that the Sylvester equation (2.4.4) holds if and only if $Mw = b$, where $b = \mathrm{vec}(A_{21})$, $w = \mathrm{vec}(X)$, and

$$M = A_{11}^T \otimes I_{n-k} - I_k \otimes A_{22}.$$

(b) Show that the matrix M from part (a) is nonsingular if and only if A_{11} and A_{22} have no eigenvalues in common.

(c) Complete the proof of Theorem 2.4.2.

2.4.6. Exercise 2.4.4 (especially part (d)) suggests a practical method for solving the Sylvester equations by back substitution. Outline such a method. This is a variant of the Bartels–Stewart algorithm (Exercises 4.8.5 and 4.8.6). This variant relies on upper triangular matrices, but of course one could use lower triangular matrices instead. The Bartels–Stewart algorithm uses one upper- and one lower triangular matrix.

2.4.7. Let

$$A = \begin{bmatrix} A_{11} & 0 \\ A_{21} & A_{22} \end{bmatrix} = \begin{bmatrix} 1 & 0 \\ 1 & 1 \end{bmatrix},$$

where $A_{11} = A_{21} = A_{22} = [\, 1 \,] \in \mathbb{C}^{1 \times 1}$. Prove directly that in this case the Sylvester equation (2.4.4) has no solution.

2.4.8.

(a) If $X \in \mathbb{C}^{\alpha \times \beta}$, $Y \in \mathbb{C}^{\gamma \times \delta}$, $u \in \mathbb{C}^{\beta}$, and $v \in \mathbb{C}^{\delta}$, what are the dimensions of $X \otimes Y$, $u \otimes v$, and $Xu \otimes Yv$? Show that $(X \otimes Y)(u \otimes v) = (Xu \otimes Yv)$. (This is actually a special case of part (a) of Exercise 2.4.3. We have made it into a separate exercise simply to draw attention to it.)

(b) Let $X \in \mathbb{C}^{k \times k}$ and $Y \in \mathbb{C}^{m \times m}$. Show that if (λ, u) is an eigenpair of X and (μ, v) is an eigenpair of Y, then $(\lambda\mu, u \otimes v)$ is an eigenpair of $X \otimes Y$.

(c) Continuing the notation from part (b), show that $(\lambda + \mu, u \otimes v)$ is an eigenpair of $I_k \otimes Y + X \otimes I_m$.

2.4.9. In this exercise we consider block diagonalization of matrices of the form (2.4.1).

(a) Show that the equation

$$\begin{bmatrix} I & -X \\ 0 & I \end{bmatrix} \begin{bmatrix} A_{11} & A_{12} \\ 0 & A_{22} \end{bmatrix} \begin{bmatrix} I & X \\ 0 & I \end{bmatrix} = \begin{bmatrix} A_{11} & 0 \\ 0 & A_{22} \end{bmatrix} \tag{2.4.11}$$

holds if and only if X satisfies a Sylvester equation.

(b) Show that the Sylvester equation has a unique solution if and only if A_{11} and A_{22} have no eigenvalues in common.

(c) Show that (2.4.11) holds if and only if the column space of $\begin{bmatrix} X \\ I \end{bmatrix}$ is invariant under A.

2.4.10. Prove Proposition 2.4.3 by induction on j.

2.4.11. Consider a transformation matrix

$$S = \begin{bmatrix} I & 0 \\ X & I \end{bmatrix},$$

of the type that appears in (2.4.3). Show that $\| S \|_2 \geq \| X \|_2$ and $\kappa_2(S) \geq \| X \|_2^2$. Thus S is ill conditioned if $\| X \|_2$ is large.

2.4.12. Let $N \in \mathbb{C}^{k \times k}$ be strictly upper triangular: $n_{ij} = 0$ for $i \geq j$. Let $n_{ij}^{(m)}$ denote the (i, j) entry of N^m.

(a) Prove that $n_{ij}^{(2)} = 0$ if $j - i \leq 1$.

(b) Prove by induction on m that $n_{ij}^{(m)} = 0$ if $j - i \leq m - 1$.

(c) Prove that $N^k = 0$. Hence N is nilpotent.

2.4.13. This exercise shows that a matrix is nilpotent if and only if its eigenvalues are all zero.

(a) Show that if N is nilpotent, then all of its eigenvalues are zero.

(b) Show that if N is a matrix whose eigenvalues are all zero, then N is unitarily similar to a strictly upper triangular matrix. Deduce that N is nilpotent.

2.4.14. This exercise shows one way to build a column compression. N has an SVD $N = U \Sigma V^*$, where U and V are unitary (Theorem 1.7.1). Σ has the form

$$\Sigma = \begin{bmatrix} \hat{\Sigma} & 0 \\ 0 & 0 \end{bmatrix},$$

where $\hat{\Sigma} = \text{diag}\{\sigma_1, \ldots, \sigma_r\}$ is an $r \times r$ nonsingular diagonal matrix. r is the rank of N. Partition V as $V = \begin{bmatrix} V_1 & V_2 \end{bmatrix}$, where V_1 has r columns. Let $\tilde{V} = \begin{bmatrix} V_2 & V_1 \end{bmatrix}$ and $\tilde{\Sigma} = \begin{bmatrix} 0 & \hat{\Sigma} \\ 0 & 0 \end{bmatrix}$.

(a) Show that $N = U \tilde{\Sigma} \tilde{V}^*$.

(b) Show that the equation $N \tilde{V} = U \tilde{\Sigma}$ gives a column compression of N.

We conclude that if we can compute an SVD, we can do a column compression.

2.4.15. This exercise fills in some of the details of the proof of Proposition 2.4.7.

(a) Suppose that $R_{j-1,j} \in \mathbb{C}^{s_{j-1} \times s_j}$ has full column rank. This means that its columns are linearly independent vectors. An additional $s_{j-1} - s_j$ vectors can be adjoined to these vectors to make a basis of $\mathbb{C}^{s_{j-1}}$, a set of s_{j-1} linearly independent vectors. Show that this implies that $R_{j-1,j}$ can be embedded in a square matrix $S_{j-1} = \begin{bmatrix} R_{j-1,j} & M \end{bmatrix} \in \mathbb{C}^{s_{j-1} \times s_{j-1}}$ that is nonsingular. Use the elementary equation $S_{j-1}^{-1} S_{j-1} = I_{s_{j-1}}$ to show that $S_{j-1}^{-1} R_{j-1,j} = \begin{bmatrix} I_{s_j} \\ 0 \end{bmatrix} = E_{j-1,j}$.

(b) Show that $R_{j-1,j}$ has a left inverse, a matrix $W_{j,j-1} \in \mathbb{C}^{s_j \times s_{j-1}}$ such that $W_{j,j-1} R_{j-1,j} = I_{s_j}$. (*Hint:* A well-chosen submatrix of S_{j-1}^{-1} will do the trick.)

2.4.16. This exercise proves Proposition 2.4.8. Let $N \in \mathbb{C}^{k \times k}$ be nilpotent, and let $m, s_1, \ldots, s_m$, and E be related to N as in Propositions 2.4.7 and 2.4.8. Suppose $N = SES^{-1}$.

(a) Show that $x \in \mathcal{N}(N^k)$ if and only if $S^{-1}x \in \mathcal{N}(E^k)$. Deduce that $\dim(\mathcal{N}(N^k)) = \dim(\mathcal{N}(E^k))$ for all k.

(b) Show that $\mathcal{N}(E) = \text{span}\{e_1, \ldots, e_{s_1}\}$, where $e_i \in \mathbb{C}^k$ is the (standard basis) vector with a 1 in the ith position and zeros elsewhere. Thus $\dim(\mathcal{N}(N)) = \dim(\mathcal{N}(E)) = s_1$.

(c) Show that $\mathcal{N}(E^2) = \text{span}\{e_1, \ldots, e_{s_1+s_2}\}$.

(d) Show that for $j = 1, \ldots, m$, $\mathcal{N}(E^j) = \text{span}\{e_1, \ldots, e_k\}$, where $k = s_1 + \cdots + s_j$.

(e) Show that $s_j = \dim(\mathcal{N}(N^j)) - \dim(\mathcal{N}(N^{j-1}))$ for $j = 1, \ldots, m$.

(f) Show that $N^j \neq 0$ if $j < m$, and $N^m = 0$.

2.4.17.

(a) Convince yourself that Proposition 2.4.9 is true. In particular, check that the largest Jordan block is $m \times m$, the number of Jordan blocks is s_1, and the number of $k \times k$ Jordan blocks is $s_k - s_{k+1}$.

(b) Write a careful proof of Proposition 2.4.9 by induction on s_1. If $s_1 = 1$, E is a Jordan block to begin with.

2.4.18. Let $A \in \mathbb{C}^{n \times n}$, and let J be its Jordan canonical form. Recall that the *algebraic multiplicity* of an eigenvalue is defined to be its multiplicity as a root of the characteristic equation.

(a) Show that the algebraic multiplicity of λ is equal to the sum of the dimensions of all of the Jordan blocks associated with λ appearing in J.

(b) The *geometric multiplicity* of λ is defined to be the dimension of $\mathcal{N}(A - \lambda I)$. Show that the geometric multiplicity of λ is equal to the number of Jordan blocks in J associated with the eigenvalue λ.

(c) Show that the geometric multiplicity cannot exceed the algebraic multiplicity of λ and that they are equal if and only if all of the Jordan blocks associated with λ are 1×1.

2.4.19. Jordan blocks of size 2×2 and greater have ones on the superdiagonal. This exercise shows that any other number (except zero) would do as well. The decision to use ones was just a matter of convenience. Let

$$J = \begin{bmatrix} \lambda & 1 & 0 & 0 \\ 0 & \lambda & 1 & 0 \\ 0 & 0 & \lambda & 1 \\ 0 & 0 & 0 & \lambda \end{bmatrix} \quad \text{and} \quad J_\alpha = \begin{bmatrix} \lambda & \alpha & 0 & 0 \\ 0 & \lambda & \alpha & 0 \\ 0 & 0 & \lambda & \alpha \\ 0 & 0 & 0 & \lambda \end{bmatrix},$$

where $\alpha \in \mathbb{C}$ is nonzero. Find a nonsingular diagonal matrix D_α such that $J_\alpha = D_\alpha^{-1} J D_\alpha$.

2.4.20. Theorem 2.4.11 tells us that every A is similar to a matrix in Jordan canonical form. That is, $A = SJS^{-1}$, where J is a direct sum of Jordan blocks. Suppose that A has j distinct eigenvalues $\lambda_1, \ldots, \lambda_j$, and suppose that the Jordan blocks are ordered so that the blocks associated with λ_1 come first, then those from λ_2, and so on. Make the partition $J = \text{diag}\{B_1, B_2, \ldots, B_j\}$, where B_k contains the Jordan blocks associated with λ_k, $k = 1, \ldots, j$. Partition the transforming matrix conformably with the blocks of J: $S = \begin{bmatrix} S_1 & S_2 & \cdots & S_j \end{bmatrix}$. Let $\mathcal{S}_k = \mathcal{R}(S_k)$, $k = 1, \ldots, j$.

(a) Show that $AS_k = S_k J_k$, and deduce that $\mathcal{S}_k$ is invariant under A for $k = 1, \ldots, j$.

(b) Show that $\mathcal{S}_k \cap \mathcal{S}_m = \{0\}$ if $k \neq m$, and show that $\mathbb{C}^n$ is the direct sum of the spaces $\mathcal{S}_1, \ldots, \mathcal{S}_j$:
$$\mathbb{C}^n = \mathcal{S}_1 \oplus \mathcal{S}_2 \oplus \cdots \oplus \mathcal{S}_j.$$
This means that each $x \in \mathbb{C}^n$ can be expressed as a sum
$$x = v_1 + v_2 + \cdots + v_j,$$
$v_k \in \mathcal{S}_k$, $k = 1, \ldots, j$, in exactly one way.

(c) Show that the restricted operator $A|_{\mathcal{S}_k}$ has only one eigenvalue, namely λ_k.

(d) For each k, let d_k denote the dimension of the largest Jordan block associated with λ_k. Show that $\mathcal{S}_k = \mathcal{N}((\lambda I - A)^{d_k})$. This is the *invariant subspace of A associated with* the eigenvalues λ_k.

(e) How do the spaces $\mathcal{S}_1, \ldots, \mathcal{S}_j$ relate to the decomposition given by Proposition 2.4.4?

(f) Suppose that the $n \times n$ matrix A has n distinct eigenvalues $\lambda_1, \ldots, \lambda_n$. Describe the spaces $\mathcal{S}_1, \ldots, \mathcal{S}_n$ in that case.

2.4.21. This exercise sketches a proof of the existence of the real Jordan form. Let $A \in \mathbb{R}^{n \times n}$. Even if you choose not to work out every detail, it is worthwhile to work all the way to the end of this exercise.

(a) Prove the following analogue of Proposition 2.4.4: There exist a real nonsingular matrix W and a real block diagonal matrix $D = \text{diag}\{T_1, \ldots, T_j\}$ such that $A = WDW^{-1}$, and each of the blocks T_i is either upper triangular with a single real eigenvalue λ or block upper triangular with a single complex-conjugate pair of eigenvalues $\lambda, \bar{\lambda}$. Clearly the Wintner–Murnaghan theorem can be used here. Aside from that, it is important that the solution of a Sylvester equation with real data is real.

(b) Show that if T_i is a block with a real eigenvalue λ, then there is a real nonsingular matrix U_i such that $T_i = U_i J_i U_i^{-1}$, where J_i is a direct sum of Jordan blocks. This is a matter of checking that the proof of Propositions 2.4.6 and 2.4.7 can be carried out in $\mathbb{R}^k$ (instead of $\mathbb{C}^k$) if T_i is real.

(c) For the T_i that have complex eigenvalues, we find it convenient to take a detour into the complex plane. Drop the subscript i and consider a matrix $T \in \mathbb{R}^{2k \times 2k}$ that has a single pair of complex-conjugate eigenvalues $\lambda = \alpha + i\beta$ and $\bar{\lambda} = \alpha - i\beta$. Let $\mathcal{S}$ denote the invariant subspace associated with λ. Apply part (d) of Exercise 2.4.20 to T to deduce that the invariant subspace associated with $\bar{\lambda}$ is $\bar{\mathcal{S}}$. Deduce that $\mathcal{S}$ and $\bar{\mathcal{S}}$ both have dimension k.

(d) Let $V_1 \in \mathbb{C}^{n \times k}$ be a matrix such that $\mathcal{R}(V_1) = \mathcal{S}$ and $TV_1 = V_1 J_1$, where J_1 is a direct sum of Jordan blocks associated with λ. Show that $T\bar{V}_1 = \bar{V}_1\bar{J}_1$. Notice that $\bar{J}_1$ is a direct sum of Jordan blocks corresponding to the eigenvalue $\bar{\lambda}$. Moreover, $\mathcal{R}(\bar{V}_1) = \bar{\mathcal{S}}$. Let $V = \begin{bmatrix} V_1 & \bar{V}_1 \end{bmatrix}$ and $J = \mathrm{diag}\{J_1, \bar{J}_1\}$. Show that V is nonsingular and $T = VJV^{-1}$.

(e) Now let $K \in \mathbb{C}^{m \times m}$ denote a single Jordan block from J_1 and $\bar{K}$ its mate from J_2. Let $W \in \mathbb{C}^{n \times m}$ denote the submatrix of V_1 consisting of the m contiguous columns that correspond to the block K. Then $\bar{W}$ is the submatrix of V_2 associated with $\bar{K}$. Then $TW = WK$ and $T\bar{W} = \bar{W}\bar{K}$. Let $w_1, \ldots w_m$ denote the columns of W. Recalling that K is a Jordan block, show that $Tw_1 = w_1\lambda$, and $Tw_k = w_k\lambda + w_{k-1}$ for $k = 2, \ldots, m$.

(f) For $k = 1, \ldots, m$, write $w_k = p_k + iq_k$, where p_k and q_k are real vectors. Write $\lambda = \alpha + i\beta$, where α and β are real numbers. Show that, with the convention $p_0 = q_0 = 0$, we have

$$Tp_k = \alpha p_k - \beta q_k + p_{k-1}$$

and

$$Tq_k = \beta p_k + \alpha q_k + q_{k-1}$$

for $k = 1, \ldots, m$.

(g) Let $U \in \mathbb{R}^{2k \times 2k}$ denote the real matrix whose columns are, in order, p_1, q_1, p_2, $q_2, \ldots, p_k, q_k$. Show that $TU = UM$, where M is a real Jordan block of the form

$$M = \begin{bmatrix} \alpha & \beta & 1 & & & & & \\ -\beta & \alpha & & 1 & & & & \\ & & \alpha & \beta & \ddots & & & \\ & & -\beta & \alpha & & \ddots & & \\ & & & & \ddots & & 1 & \\ & & & & & & & 1 \\ & & & & & & \alpha & \beta \\ & & & & & & -\beta & \alpha \end{bmatrix}.$$

Show that U is nonsingular. Thus $T = UMU^{-1}$.

(h) Put the pieces together to prove Theorem 2.4.12.

2.5 Spectral Projectors

Let $A \in \mathbb{C}^{n \times n}$, and suppose that we break the spectrum of A into two disjoint subsets

$$\Lambda_1 = \{\lambda_1, \ldots, \lambda_k\} \quad \text{and} \quad \Lambda_2 = \{\lambda_{k+1}, \ldots, \lambda_n\}.$$

Then we know from the developments of Section 2.4 that A is similar to a block diagonal matrix that splits the spectrum into two pieces. That is, there is a nonsingular matrix $S = \begin{bmatrix} S_1 & S_2 \end{bmatrix}$ and a block diagonal matrix $B = \text{diag}\{B_1, B_2\}$ such that $\Lambda(B_1) = \Lambda_1$, $\Lambda(B_2) = \Lambda_2$, and $A = SBS^{-1}$. Rewriting the similarity equation as $AS = SB$ and partitioning this equation, we see immediately that $AS_1 = S_1 B_1$ and $AS_2 = S_2 B_2$. These equations signal the by-now familiar fact that the spaces $\mathcal{S}_1 = \mathcal{R}(S_1)$ and $\mathcal{S}_2 = \mathcal{R}(S_2)$ are invariant under A. In fact, $\mathcal{S}_1$ (resp., $\mathcal{S}_2$) is the invariant subspace associated with the eigenvalues in Λ_1 (resp., Λ_2).

Let $U = \begin{bmatrix} U_1 & U_2 \end{bmatrix}$ be defined by $U = S^{-*}$ or $U^* = S^{-1}$, and let $\mathcal{U}_1 = \mathcal{R}(U_1)$ and $\mathcal{U}_2 = \mathcal{R}(U)_2$. Then $A^*U = UB^*$, and one sees straightaway that $\mathcal{U}_1$ and $\mathcal{U}_2$ are the invariant subspaces of A^* associated with $\overline{\Lambda}_1$ and $\overline{\Lambda}_2$, respectively. Since $U^* = S^{-1}$, we have $U^*S = I$, which implies $U_1^*S_1 = I$, $U_2^*S_2 = I$, $U_2^*S_1 = 0$, and $U_1^*S_2 = 0$. It follows easily that

$$\mathcal{U}_1 = \mathcal{S}_2^{\perp} \quad \text{and} \quad \mathcal{U}_2 = \mathcal{S}_1^{\perp}.$$

Turning things around, we have $SU^* = I$, from which we obtain

$$I = S_1 U_1^* + S_2 U_2^*.$$

Let

$$P_1 = S_1 U_1^* \quad \text{and} \quad P_2 = S_2 U_2^*.$$

Then, recalling Theorem 1.6.1 and other basic results from Section 1.6, we see that P_1 and P_2 are complementary projectors. P_1 is the projector of $\mathbb{C}^n$ onto $\mathcal{S}_1$ in the direction of $\mathcal{S}_2$. Since $\mathcal{S}_1$ and $\mathcal{S}_2$ are invariant subspaces of A associated with complementary pieces of the spectrum, we call P_1 a *spectral projector*; it is the spectral projector associated with the invariant subspace $\mathcal{S}_1$ and the spectral subset Λ_1. Similarly P_2 is the spectral projector associated with $\mathcal{S}_2$ and Λ_2. P_1 and P_2 are called complementary because $P_1 + P_2 = I$. This corresponds to the fact that $\mathbb{C}^n = \mathcal{S}_1 \oplus \mathcal{S}_2$ and also that the spectrum of A is the disjoint union of Λ_1 and Λ_2.[1]

The projectors P_1^* and P_2^* are complementary spectral projectors for A^* associated with the invariant subspaces $\mathcal{U}_1$ and $\mathcal{U}_2$, respectively.

The similarity transformation $A = SBS^{-1}$ can be rewritten as $A = SBU^*$, which leads directly to a decomposition of A:

$$A = S_1 B_1 U_1^* + S_2 B_2 U_2^* = A_1 + A_2.$$

---

[1] Using techniques of complex analysis, one can show [202] that the spectral projector P_j associated with Λ_j is given by the contour integral

$$P_j = \frac{1}{2\pi i} \int_{C_j} (\lambda I - A)^{-1} \, d\lambda,$$

where C_j is a closed curve in the complex plane that encloses Λ_j and does not enclose any of the eigenvalues in the complement of Λ_j. This characterization is both elegant and very useful, but we will use other tools.

A_1 is the part of A that acts on $\mathcal{S}_1$ and does nothing on $\mathcal{S}_2$, and A_2 has complementary properties.

Obviously we can obtain finer decompositions by breaking the spectrum into more than two pieces. Suppose that $\Lambda_1, \ldots, \Lambda_k$ are k disjoint subsets whose union is the spectrum of A. Then by Proposition 2.4.3 there is a nonsingular matrix $S = \begin{bmatrix} S_1 & \cdots & S_k \end{bmatrix}$ and a block diagonal matrix $B = \text{diag}\{B_1, \ldots, B_k\}$ such that $A = SBS^{-1}$. The decomposition given by Proposition 2.4.4 is an instance of this, as is the Jordan canonical form.

Defining $U = \begin{bmatrix} U_1 & \cdots & U_k \end{bmatrix}$ by $U^* = S^{-1}$, we have $SU^* = I$, so

$$I = S_1 U_1^* + S_2 U_2^* + \cdots + S_k U_k^*$$

or, letting $P_j = S_j U_j^*$, $j = 1, \ldots, k$,

$$I = P_1 + P_2 + \cdots + P_k. \tag{2.5.1}$$

Each P_j is the spectral projector for A associated with the invariant subspace $\mathcal{S}_j = \mathcal{R}(S_j)$ and eigenvalues Λ_j. Each P_j^* is a spectral projector for A^* associated with the invariant subspace $\mathcal{U}_j = \mathcal{R}(U_j)$ and eigenvalues $\overline{\Lambda}_j$. The equation (2.5.1) and its conjugate transpose correspond to the subspace decompositions

$$\mathbb{C}^n = \mathcal{S}_1 \oplus \mathcal{S}_2 \oplus \cdots \oplus \mathcal{S}_k$$

and

$$\mathbb{C}^n = \mathcal{U}_1 \oplus \mathcal{U}_2 \oplus \cdots \oplus \mathcal{U}_k.$$

The similarity equation $A = SBS^{-1} = SBU^*$ immediately yields a decomposition

$$A = S_1 B_1 U_1^* + S_2 B_2 U_2^* + \cdots + S_k B_k U_k^* \tag{2.5.2}$$
$$= A_1 + A_2 + \cdots + A_k.$$

Each A_j acts as A on $\mathcal{S}_j$ and is zero on $\sum_{m \neq j} \mathcal{S}_m$.

The most extreme case of (2.5.2) occurs when k is taken to be the number of distinct eigenvalues of A and each subset Λ_j is a singleton: $\Lambda_j = \{\lambda_j\}$. If we take B to be a direct sum of Jordan blocks, as Theorem 2.4.11 guarantees we can, then each B_j is the direct sum of Jordan blocks associated with the eigenvalue λ_j.

If A is semisimple, then the Jordan canonical form is diagonal, and each of the blocks B_j is a multiple of the identity matrix: $B_j = \lambda_j I$. Substituting this form for B_j into (2.5.2), we obtain the following theorem.

Theorem 2.5.1. *Let $A \in \mathbb{C}^{n \times n}$ be a semisimple matrix whose distinct eigenvalues are $\lambda_1, \ldots, \lambda_k$. For $j = 1, \ldots, k$, let P_j denote the spectral projector for A associated with the single eigenvalue λ_j. Then $I = P_1 + \cdots + P_k$ and*

$$A = \lambda_1 P_1 + \lambda_2 P_2 + \cdots + \lambda_k P_k.$$

This theorem is the projector version of Theorem 2.4.1. The converse is true as well. The set of spectral projectors $P_1, \ldots, P_k$ is called a *resolution of the identity* associated with A, and the equation $A = \lambda_1 P_1 + \cdots + \lambda_k P_k$ is called the *spectral decomposition* of the semisimple matrix A. It is easy to check that the rank of P_j is equal to the (algebraic and geometric) multiplicity of eigenvalue λ_j. In the generic case, there are n distinct eigenvalues, and each spectral projector has rank one: $P_j = x_j y_j^*$, where x_j and y_j are (right) eigenvectors of A and A^*, respectively, satisfying $y_j^* x_j = 1$.

The semisimple case includes the normal case. If A is normal, we can take the transforming matrix S to be unitary. Then $U = S^{-*} = S$, so we have $P_j = S_j S_j^*$ for $j = 1, \ldots, k$. This means that the spectral projectors are orthoprojectors. The spaces $\mathcal{S}_j$ are pairwise orthogonal, and $\mathcal{S}_j = \mathcal{U}_j$ for $j = 1, \ldots, k$. Theorem 2.5.1 holds for all normal matrices. Restating that result slightly, we have the following theorem.

Theorem 2.5.2. *Let $A \in \mathbb{C}^{n \times n}$ be a normal matrix whose distinct eigenvalues are $\lambda_1, \ldots, \lambda_k$. For $j = 1, \ldots, k$, let P_j denote the spectral projector for A associated with the single eigenvalue λ_j. Then $I = P_1 + \cdots + P_k$, each P_j is an orthoprojector, and*

$$A = \lambda_1 P_1 + \lambda_2 P_2 + \cdots + \lambda_k P_k.$$

This is the projector version of the spectral theorem (Theorem 2.3.5).

Exercises

2.5.1. Using the notation established at the beginning of the section, prove the following:

(a) Show that $A^* U = U B^*$, $A^* U_1 = U_1 B_1^*$, and $A^* U_2 = U_2 B_2^*$. Deduce that $\mathcal{U}_1$ and $\mathcal{U}_2$ are invariant subspaces under A^* associated with $\overline{\Lambda}_1$ and $\overline{\Lambda}_2$, respectively.

(b) Use the equation $U^* S = I$ to show that $U_1^* S_1 = I$, $U_2^* S_2 = I$, $U_2^* S_1 = 0$, and $U_1^* S_2 = 0$. Deduce that $\mathcal{U}_1 = \mathcal{S}_2^\perp$ and $\mathcal{U}_2 = \mathcal{S}_1^\perp$.

(c) Verify that P_1 and P_2 are projectors satisfying $P_1 + P_2 = I$. Show that P_1 (resp., P_2) is the projector of $\mathbb{C}^n$ onto $\mathcal{S}_1$ (resp., $\mathcal{S}_2$) in the direction of $\mathcal{S}_2$ (resp., $\mathcal{S}_1$).

(d) Show that $P_1 P_2 = P_2 P_1 = 0$.

(e) Verify that P_1^* and P_2^* are complementary spectral projectors for A^*, associated with the invariant subspaces $\mathcal{U}_1$ and $\mathcal{U}_2$, respectively.

(f) Show that $A = S_1 B_1 U_1^* + S_2 B_2 U_2^*$. Letting $A_1 = S_1 B_1 U_1^*$ and $A_2 = S_2 B_2 U_2^*$, show that if $x \in \mathcal{S}_1$, then $A_1 x = A x$, and if $x \in \mathcal{S}_2$, then $A_1 x = 0$. Show that $\mathcal{R}(A_1) = \mathcal{S}_1$ and $\mathcal{N}(A_1) = \mathcal{S}_2$. Obviously analogous properties hold for A_2.

2.5.2. Generalize all of the results of Exercise 2.5.1 to the situation where there are k blocks: $B = \text{diag}\{B_1, \ldots, B_k\}$.

2.5.3. Prove Theorem 2.5.1.

2.5.4. Let A be a semisimple matrix with spectral decomposition $A = \lambda_1 P_1 + \cdots + \lambda_k P_k$. Each P_j is a projector, so we know that $P_j^2 = P_j$. Prove each of the following assertions:

(a) $P_i P_j = 0$ if $i \neq j$.

(b) $A^2 = \lambda_1^2 P_1 + \cdots + \lambda_k^2 P_k$.

(c) $A^m = \lambda_1^m P_1 + \cdots + \lambda_k^m P_k$ for $m = 0, 1, 2, 3, \ldots$.

(d) $f(A) = f(\lambda_1) P_1 + \cdots + f(\lambda_k) P_k$ for all polynomials f.

(e) $A^{-1} = \lambda_1^{-1} P_1 + \cdots + \lambda_k^{-1} P_k$ if A is nonsingular.

(f) $(\mu I - A)^{-1} = (\mu - \lambda_1)^{-1} P_1 + \cdots + (\mu - \lambda_k)^{-1} P_k$ if μ is not in the spectrum of A.

2.5.5. Let $A \in \mathbb{C}^{n \times n}$ be block triangular:

$$A = \begin{bmatrix} A_{11} & A_{12} \\ 0 & A_{22} \end{bmatrix},$$

where $\Lambda_1 = \Lambda(A_{11})$ and $\Lambda_2 = \Lambda(A_{22})$ are disjoint. We continue to use notation established in this section. Thus $\mathcal{S}_1$ is the invariant subspace associated with Λ_1, and so on.

(a) (Review) Show that A is similar to $B = \text{diag}\{A_{11}, A_{22}\}$ by a similarity transformation $AS = SB$, where S has the form

$$S = \begin{bmatrix} I & X \\ 0 & I \end{bmatrix}$$

and X satisfies the Sylvester equation

$$X A_{22} - A_{11} X = A_{12}.$$

(b) Show that the spectral projectors are given by

$$P_1 = \begin{bmatrix} I & -X \\ 0 & 0 \end{bmatrix} \quad \text{and} \quad P_2 = \begin{bmatrix} 0 & X \\ 0 & I \end{bmatrix}.$$

(c) Identify bases for the spaces $\mathcal{S}_1, \mathcal{S}_2, \mathcal{U}_1$, and $\mathcal{U}_2$.

(d) What happens when A is normal?

2.5.6. Show that if $A \in \mathbb{C}^{n \times n}$ has a spectral projector that is not an orthoprojector, then A is not normal.

2.6   Angles and Distances between Subspaces

We have seen in Section 2.1 that the ultimate task of eigensystem computations is to compute invariant subspaces. In Chapter 4 we will develop algorithms that produce sequences of subspaces that converge to invariant subspaces. In order to quantify (and prove) convergence of the subspaces, we must develop a notion of distance between two subspaces. Such a notion is also needed for the discussion of perturbations of invariant subspaces in Section 2.7.

Although the distance between two subspaces can be defined very briefly, we will take a more leisurely approach that (we hope) leads to a thorough understanding of the subject. One of the key results in our development is the CS decomposition.

The CS Decomposition (CSD)

Let $Q \in \mathbb{F}^{n \times n}$ be a unitary matrix partitioned as a 2×2 block matrix:

$$Q = \begin{bmatrix} Q_{11} & Q_{12} \\ Q_{21} & Q_{22} \end{bmatrix}. \tag{2.6.1}$$

The blocks Q_{11} and Q_{22} need not be square; they can be of any dimensions whatsoever. Each of the blocks is a matrix in its own right and has its own SVD. The CSD theorem will reveal close relationships among the SVDs of the four blocks.

Theorem 2.6.1 (CS decomposition). *Let $Q \in \mathbb{F}^{n \times n}$ be a unitary matrix, partitioned as in (2.6.1), where $Q_{11} \in \mathbb{F}^{\alpha_1 \times \beta_1}$, $Q_{12} \in \mathbb{F}^{\alpha_1 \times \beta_2}$, $Q_{21} \in \mathbb{F}^{\alpha_2 \times \beta_1}$, and $Q_{22} \in \mathbb{F}^{\alpha_2 \times \beta_2}$. Then there exist unitary matrices $U_1 \in \mathbb{F}^{\alpha_1 \times \alpha_1}$, $U_2 \in \mathbb{F}^{\alpha_2 \times \alpha_2}$, $V_1 \in \mathbb{F}^{\beta_1 \times \beta_1}$, and $V_2 \in \mathbb{F}^{\beta_2 \times \beta_2}$ such that*

$$Q = \begin{bmatrix} Q_{11} & Q_{12} \\ Q_{21} & Q_{22} \end{bmatrix} = \begin{bmatrix} U_1 & \\ & U_2 \end{bmatrix} \begin{bmatrix} \Sigma_{11} & \Sigma_{12} \\ \Sigma_{21} & \Sigma_{22} \end{bmatrix} \begin{bmatrix} V_1^* & \\ & V_2^* \end{bmatrix}, \tag{2.6.2}$$

where each Σ_{ij} is a real "diagonal" matrix of the same dimensions as Q_{ij}. Specifically,

$$\begin{bmatrix} \Sigma_{11} & \Sigma_{12} \\ \Sigma_{21} & \Sigma_{22} \end{bmatrix} = \left[\begin{array}{ccc|ccc} I & & & 0_s^T & & \\ & C & & & -S & \\ & & 0_c & & & -I \\ \hline 0_s & & & I & & \\ & S & & & C & \\ & & I & & & 0_c^T \end{array} \right], \tag{2.6.3}$$

where $C = \mathrm{diag}\{c_1, \ldots, c_m\}$, $S = \mathrm{diag}\{s_1, \ldots, s_m\}$, $1 > c_1 \geq \cdots \geq c_m > 0$, $0 < s_1 \leq \cdots \leq s_m < 1$, and $c_i^2 + s_i^2 = 1$, $i = 1, \ldots, m$.

The c_i and s_i are cosines and sines, respectively. The four identity matrices in (2.6.3) are of different sizes in general, and can be 0×0. The matrices 0_c and 0_s are zero matrices that can have any number of rows and columns, including zero. The matrices C and S are both $m \times m$, where m can be zero.

From (2.6.2) we have $Q_{ij} = U_i \Sigma_{ij} V_j^*$ for $i, j = 1, 2$. Each of these four equations is an SVD or can be turned into an SVD by trivial modifications. Thus there are many relationships between the SVD of the submatrices. For example, Q_{11} and Q_{12} have the same left singular vectors (the columns of U_1), and Q_{11} and Q_{22} have the same singular values $c_1, \ldots, c_m$ between 0 and 1. (In general they have different numbers of singular values equal to 1.)

A proof of Theorem 2.6.1 is worked out in Exercise 2.6.1.

Angles between Subspaces

The relative orientation of two subspaces of $\mathbb{F}^n$ can be described in terms of the *principal angles* between them. These are the angles formed by certain *principal vectors* in the spaces. This concept has numerous applications. We will apply it to the study of distances between subspaces. We will start out with the real case, which is simpler, but ultimately we are interested in the complex case as well.

We begin by considering the concept of *angle* between two vectors in $\mathbb{R}^n$. Recall that for nonzero vectors u and $v \in \mathbb{R}^2$ the formula

$$\langle u, v \rangle = \| u \| \, \| v \| \cos \theta$$

holds, where θ is the angle between u and v. Since $0 \leq \theta \leq \pi$, we can solve this equation for θ to obtain

$$\theta = \arccos \left(\frac{\langle u, v \rangle}{\| u \| \, \| v \|} \right). \tag{2.6.4}$$

This same formula holds in $\mathbb{R}^n$ as well, because it is really just a two-dimensional property: All the action is in the two-dimensional subspace span$\{u, v\}$.

Now let $\mathcal{U} = \mathrm{span}\{u\}$ and $\mathcal{V} = \mathrm{span}\{v\}$ be one-dimensional subspaces of $\mathbb{R}^n$. We will define the *angle* between $\mathcal{U}$ and $\mathcal{V}$ to be just the angle between u and v, with one proviso. The angle between u and v can be right (if $\langle u, v \rangle = 0$), acute (if $\langle u, v \rangle > 0$), or obtuse (if $\langle u, v \rangle < 0$). If the angle is obtuse, then the angle between u and $-v$ is acute, since $\langle u, -v \rangle > 0$. This is easy to picture. Since $-v$ is just as good a representative of $\mathcal{V}$ as v is, we get to decide whether to use the acute or the obtuse angle. Since we are interested in assessing the proximity of two spaces, we will always take the acute one. Thus we define the angle between one-dimensional $\mathcal{U}$ and $\mathcal{V}$ in $\mathbb{R}^n$ as follows: Pick vectors $u \in \mathcal{U}$ and $v \in \mathcal{V}$ such that $\langle u, v \rangle \geq 0$. Then the *angle* between $\mathcal{U}$ and $\mathcal{V}$ is defined to be the angle between u and v. This is given by the formula (2.6.4), and it always satisfies $0 \leq \theta \leq \pi/2$.

Now consider the complex case; suppose $\mathcal{U} = \mathrm{span}\{u\}$ and $\mathcal{V} = \mathrm{span}\{v\}$ are two one-dimensional subspaces of $\mathbb{C}^n$. Now the situation is complicated by the fact that $\langle u, v \rangle$ is a complex number. However, we can easily resolve this issue. Suppose $\langle u, v \rangle = r e^{i\alpha}$, where $r \geq 0$ and $0 \leq \alpha < 2\pi$. This is just the polar representation of a complex number. Now let $\hat{v} = e^{i\alpha} v$. Then $\hat{v}$ represents $\mathcal{V}$ just as well as v does, and we easily see that $\langle u, \hat{v} \rangle = r \geq 0$. Thus it is always possible to choose representatives from $\mathcal{U}$ and $\mathcal{V}$ whose inner product is real and nonnegative. This means that we can define the angle between complex one-dimensional spaces exactly the same as for real spaces: Pick $u \in \mathcal{U}$ and $v \in \mathcal{V}$ such that $\langle u, v \rangle \geq 0$. Then the *angle* between $\mathcal{U}$ and $\mathcal{V}$ is given by (2.6.4). This is always acute or right, never obtuse.

Now consider higher-dimensional subspaces. We could allow them to have different dimensions, but for our purposes it suffices to keep the dimensions equal. Suppose therefore that $\dim(\mathcal{U}) = \dim(\mathcal{V}) = k$. The *smallest angle* between $\mathcal{U}$ and $\mathcal{V}$ is defined to be the smallest angle that can be formed between a vector $u \in \mathcal{U}$ and a vector $v \in \mathcal{V}$. As explained in the previous paragraph, we can restrict our attention to vector pairs that satisfy $\langle u, v \rangle \geq 0$. Moreover, we can simplify the discussion by restricting our attention to vectors that satisfy $\| u \| = \| v \| = 1$, since rescaling the vectors does not change the angle between them. If we make this restriction, the formula (2.6.4) becomes very simple: $\theta = \arccos \langle u, v \rangle$ or $\cos \theta = \langle u, v \rangle$.

We minimize the angle by maximizing the cosine, so let $u_1 \in \mathcal{U}$ and $v_1 \in \mathcal{V}$ be unit vectors such that

$$\langle u_1, v_1 \rangle = \max \{ \langle u, v \rangle \mid u \in \mathcal{U}, \ v \in \mathcal{V}, \ \| u \| = \| v \| = 1, \ \langle u, v \rangle \in \mathbb{R} \}.$$

Then the *smallest angle* between $\mathcal{U}$ and $\mathcal{V}$ is

$$\theta_1 = \arccos \langle u_1, v_1 \rangle.$$

We will call θ_1 the *first principal angle* between $\mathcal{U}$ and $\mathcal{V}$. If $k > 1$, the *second principal angle* is then defined to be the smallest angle that can be formed between a vector in $\mathcal{U}$ that is orthogonal to u_1 and a vector in $\mathcal{V}$ that is orthogonal to v_1. In other words, the second principal angle θ_2 is given by

$$\theta_2 = \arccos \langle u_2, v_2 \rangle,$$

where $u_2 \in \mathcal{U}$ and $v_2 \in \mathcal{V}$ are unit vectors chosen so that

$$\langle u_2, v_2 \rangle = \max\{\langle u, v \rangle \mid u \in \mathcal{U} \cap \mathrm{span}\{u_1\}^\perp, \ v \in \mathcal{V} \cap \mathrm{span}\{v_1\}^\perp,$$
$$\|u\| = \|v\| = 1, \ \langle u, v \rangle \in \mathbb{R}\}.$$

If $k > 2$, the *third principal* angle θ_3 is defined to be the smallest angle that can be formed between a vector in $\mathcal{U}$ that is orthogonal to both u_1 and u_2 and a vector in $\mathcal{V}$ that is orthogonal to both v_1 and v_2, and so on.

In general the principal angles are defined recursively for $i = 1, \ldots, k$ by

$$\theta_i = \arccos \langle u_i, v_i \rangle,$$

where $u_i \in \mathcal{U}$ and $v_i \in \mathcal{V}$ are unit vectors chosen so that

$$\langle u_i, v_i \rangle = \max\{\langle u, v \rangle \mid u \in \mathcal{U} \cap \mathrm{span}\{u_1, \ldots, u_{i-1}\}^\perp,$$
$$v \in \mathcal{V} \cap \mathrm{span}\{v_1, \ldots, v_{i-1}\}^\perp, \ \|u\| = \|v\| = 1, \ \langle u, v \rangle \in \mathbb{R}\}.$$

Obviously $0 \le \theta_1 \le \theta_2 \le \cdots \le \theta_k \le \pi/2$. The orthonormal vectors $u_1, \ldots, u_k \in \mathcal{U}$ and $v_1, \ldots, v_k \in \mathcal{V}$ are called *principal vectors*. Clearly $\mathcal{U} = \mathrm{span}\{u_1, \ldots, u_k\}$ and $\mathcal{V} = \mathrm{span}\{v_1, \ldots, v_k\}$. The principal vectors are not uniquely determined. For example, any pair u_j, v_j can be replaced by $e^{i\alpha}u_j$ and $e^{i\alpha}v_j$ for any phase angle α. Sometimes (when two or more principal angles are equal) the lack of uniqueness is even greater than this. Some extreme examples are considered in Exercise 2.6.2. Although the principal vectors are not uniquely determined, the principal angles *are*, as we shall see. The next few results demonstrate some simple geometric conditions that are satisfied by the principal angles and vectors.

Proposition 2.6.2. *The first principal angle θ_1 and the first principal vectors $u_1 \in \mathcal{U}$ and $v_1 \in \mathcal{V}$ satisfy*

$$\sin \theta_1 = \|u_1 - v_1 \cos \theta_1\| = \min_{v \in \mathcal{V}} \|u_1 - v\| = \min_{\substack{u \in \mathcal{U} \\ \|u\|=1}} \min_{v \in \mathcal{V}} \|u - v\|, \qquad (2.6.5)$$

$$\sin \theta_1 = \|v_1 - u_1 \cos \theta_1\| = \min_{u \in \mathcal{U}} \|v_1 - u\| = \min_{\substack{v \in \mathcal{V} \\ \|v\|=1}} \min_{u \in \mathcal{U}} \|v - u\|. \qquad (2.6.6)$$

Proposition 2.6.2 is easily generalized to yield statements about other principal angles and vectors. For this purpose it is convenient to introduce some new notation. For $j = 1, \ldots, k$, let

$$\mathcal{U}_j = \mathcal{U} \cap \mathrm{span}\{u_1, \ldots, u_{j-1}\}^\perp = \mathrm{span}\{u_j, \ldots, u_k\}$$

and

$$\mathcal{V}_j = \mathcal{V} \cap \mathrm{span}\{v_1, \ldots, v_{j-1}\}^\perp = \mathrm{span}\{v_j, \ldots, v_k\}.$$

Proposition 2.6.3. *For* $i = 1, \ldots, k$,

$$\sin \theta_i = \| u_i - v_i \cos \theta_i \| = \min_{v \in \mathcal{V}_i} \| u_i - v \| = \min_{\substack{u \in \mathcal{U}_i \\ \|u\|=1}} \min_{v \in \mathcal{V}_i} \| u - v \|,$$

$$\sin \theta_i = \| v_i - u_i \cos \theta_i \| = \min_{u \in \mathcal{U}_i} \| v_i - u \| = \min_{\substack{v \in \mathcal{V}_i \\ \|v\|=1}} \min_{u \in \mathcal{U}_i} \| v - u \|.$$

Proposition 2.6.3 is an immediate consequence of Proposition 2.6.2, because θ_i, u_i, and v_i are the first principal angles and vectors associated with the subspaces $\mathcal{U}_i$ and $\mathcal{V}_i$.

The sets $u_1, \ldots, u_k$ and $v_1, \ldots, v_k$ are both orthonormal. In addition there are orthogonality relations between the u_i and the v_j.

Theorem 2.6.4. *Let* $u_1, \ldots, u_k$ *and* $v_1, \ldots, v_k$ *be principal vectors for the subspaces* $\mathcal{U}$ *and* $\mathcal{V}$. *Then*

$$\langle u_i, v_j \rangle = 0 \quad \text{if } i \neq j.$$

This result is an easy consequence of Proposition 2.6.3. (See Exercise 2.6.6.) The next corollary summarizes some of the relationships between principal angles and vectors in matrix form.

Corollary 2.6.5. *Let* $U_1 = \begin{bmatrix} u_1 & \cdots & u_k \end{bmatrix} \in \mathbb{F}^{n \times k}$, $V_1 = \begin{bmatrix} v_1 & \cdots & v_k \end{bmatrix} \in \mathbb{F}^{n \times k}$, *and* $\Gamma_1 = \mathrm{diag}\{\cos \theta_1, \ldots, \cos \theta_k\}$. *Then* $U_1^* V_1 = V_1^* U_1 = \Gamma_1$.

Corollary 2.6.5 has an important converse: Any time vectors satisfy a condition $U_1^* V_1 = \Gamma_1$, as in Corollary 2.6.5, then the vectors must be principal vectors.

Theorem 2.6.6. *Let* $\tilde{u}_1, \ldots, \tilde{u}_k$ *and* $\tilde{v}_1, \ldots, \tilde{v}_k$ *be orthonormal bases for* $\mathcal{U}$ *and* $\mathcal{V}$, *respectively, and let* $\tilde{U} = \begin{bmatrix} \tilde{u}_1 & \cdots & \tilde{u}_k \end{bmatrix}$ *and* $\tilde{V} = \begin{bmatrix} \tilde{v}_1 & \cdots & \tilde{v}_k \end{bmatrix}$. *If* $\tilde{U}^* \tilde{V}$ *is a nonnegative diagonal matrix, say* $\tilde{U}^* \tilde{V} = C = \mathrm{diag}\{c_1, \ldots, c_k\}$, *where* $c_1 \geq c_2 \geq \cdots \geq c_k \geq 0$, *then* $\tilde{u}_1, \ldots, \tilde{u}_k$ *and* $\tilde{v}_1, \ldots, \tilde{v}_k$ *are principal vectors for the spaces* $\mathcal{U}$ *and* $\mathcal{V}$. *The principal angles are* $\theta_i = \arccos c_i$, $i = 1, \ldots, k$.

See Exercise 2.6.7 for the easy proof. This theorem means that

$$\langle \tilde{u}_1, \tilde{v}_1 \rangle = \max\{\langle u, v \rangle \mid u \in \mathcal{U}, \ v \in \mathcal{V}, \ \|u\| = \|v\| = 1, \ \langle u, v \rangle \in \mathbb{R}\},$$

$\langle \tilde{u}_1, \tilde{v}_1 \rangle = c_1 = \cos \theta_1$, and so on.

Principal Angles and the SVD

Theorem 2.6.6 suggests a method for computing principal angles and vectors between two spaces, assuming each of the spaces is given in terms of an orthonormal basis. Let $p_1, \ldots, p_k$ and $q_1, \ldots, q_k$ be orthonormal bases for $\mathcal{U}$ and $\mathcal{V}$, respectively, and let $P_1 = \begin{bmatrix} p_1 & \cdots & p_k \end{bmatrix} \in \mathbb{F}^{n \times k}$ and $Q_1 = \begin{bmatrix} q_1 & \cdots & q_k \end{bmatrix} \in \mathbb{F}^{n \times k}$. Consider an SVD of $P_1^* Q_1$: $P_1^* Q_1 = M_1 \Gamma_1 N_1^*$. Here $M_1, N_1 \in \mathbb{F}^{k \times k}$ are unitary, and $\Gamma_1 \in \mathbb{F}^{k \times k}$ is diagonal, say $\Gamma_1 = \mathrm{diag}\{\gamma_1, \ldots, \gamma_k\}$, with $\gamma_1 \geq \cdots \geq \gamma_k \geq 0$. Let $U_1 = P_1 M_1$ and $V_1 = Q_1 N_1$. Then U_1 and V_1 have orthonormal columns, and furthermore,

$$U_1^* V_1 = M_1^* P_1^* Q_1 N_1 = \Gamma_1.$$

Thus the columns of U_1 and V_1 satisfy the hypotheses of Theorem 2.6.6. Consequently, the angles $\theta_i = \arccos \gamma_i$, $i = 1, \ldots, k$, are the principal angles between $\mathcal{U}$ and $\mathcal{V}$, and the columns of U_1 and V_1 are principal vectors. We summarize all this as a theorem.

Theorem 2.6.7. *Let $\mathcal{U}$ and $\mathcal{V}$ be k-dimensional subspaces of $\mathbb{F}^n$, and let P_1, $Q_1 \in \mathbb{F}^{n \times k}$ be matrices with orthonormal columns such that $\mathcal{U} = \mathcal{R}(P_1)$ and $\mathcal{V} = \mathcal{R}(Q_1)$. Let*

$$P_1^* Q_1 = M_1 \Gamma_1 N_1^* \tag{2.6.7}$$

denote the SVD of $P_1^* Q_1$, where $\Gamma_1 = \mathrm{diag}\{\gamma_1, \ldots, \gamma_k\}$. Let $U_1 = P_1 M_1$ and $V_1 = Q_1 N_1$. Then the columns of U_1 and V_1 are principal vectors associated with $\mathcal{U}$ and $\mathcal{V}$; the principal angles between $\mathcal{U}$ and $\mathcal{V}$ are $\theta_i = \arccos \gamma_i$, $i = 1, \ldots, k$.

We now have a means of computing the principal angles and principal vectors between $\mathcal{U}$ and $\mathcal{V}$, provided that we can compute an SVD: Compute $P_1^* Q_1$. Then compute the SVD: $P_1^* Q_1 = M_1 \Gamma_1 N_1^*$. Compute $U_1 = P_1 M_1$ and $V_1 = Q_1 N_1$. Finally, compute $\theta_i = \arccos \gamma_i$, $i = 1, \ldots, k$, where $\gamma_1, \ldots, \gamma_k$ are the main-diagonal entries of Γ_1.

The computation we have just outlined has a numerical weakness: it cannot compute the small principal angles accurately. This is so because the function $\theta = \arccos \gamma$ is not Lipschitz continuous at $\gamma = 1$ (and $\theta = 0$); its slope approaches ∞ as $\gamma \to 1^-$. Thus a tiny perturbation in γ can make a significant change in θ.

As a consequence of the CSD, there is a different computation, involving the orthogonal complement of one space or the other, that *does* yield accurate values for small principal angles. Let $p_{k+1}, \ldots, p_n$ and $q_{k+1}, \ldots, q_n$ be orthonormal bases for $\mathcal{U}^\perp$ and $\mathcal{V}^\perp$, respectively, and let $P_2 = \begin{bmatrix} p_{k+1} & \cdots & p_n \end{bmatrix}$, $P = \begin{bmatrix} P_1 & P_2 \end{bmatrix}$, $Q_2 = \begin{bmatrix} q_{k+1} & \cdots & q_n \end{bmatrix}$, and $Q = \begin{bmatrix} Q_1 & Q_2 \end{bmatrix}$. Then P and Q are $n \times n$ unitary matrices, and so is $P^* Q$. Associated with the partition

$$P^* Q = \begin{bmatrix} P_1^* Q_1 & P_1^* Q_2 \\ P_2^* Q_1 & P_2^* Q_2 \end{bmatrix}$$

there is a CSD

$$\begin{bmatrix} P_1^* Q_1 & P_1^* Q_2 \\ P_2^* Q_1 & P_2^* Q_2 \end{bmatrix} = \begin{bmatrix} M_1 & \\ & M_2 \end{bmatrix} \begin{bmatrix} \Gamma_1 & -\Sigma_1^T \\ \Sigma_1 & \Gamma_2 \end{bmatrix} \begin{bmatrix} N_1^* & \\ & N_2^* \end{bmatrix}. \tag{2.6.8}$$

We already know that $\Gamma_1 = \mathrm{diag}\{\gamma_1, \ldots, \gamma_k\}$, where the γ_i are the cosines of the principal angles between $\mathcal{U}$ and $\mathcal{V}$. This can also be written as $\Gamma_1 = \mathrm{diag}\{I, C, 0\}$, as in (2.6.3), where the size of the I block is equal to the number of γ_i that are equal to 1 (corresponding to $\theta_i = 0$), and the size of the 0 block, which is square in this case, is equal to the number of γ_i that are equal to zero (corresponding to $\theta_i = \pi/2$). The exact form of Σ_1 depends upon the value of k. The case $k = n/2$ is easiest to describe and understand, so we will begin with it. In this case Σ_1 is square, in fact $\Sigma_1 = \mathrm{diag}\{\sigma_1, \ldots, \sigma_k\}$, where $\sigma_i = \sqrt{1 - \gamma_i^2}$, $i = 1, \ldots, k$. Thus Σ_1 contains the sines of the principal angles. If $k < n/2$, we have a similar situation:

$$\Sigma_1 = \begin{bmatrix} 0 \\ \hat{\Sigma}_1 \end{bmatrix},$$

where $\hat{\Sigma}_1 = \mathrm{diag}\{\sigma_1, \ldots, \sigma_k\}$. This Σ_1 has the form $\mathrm{diag}\{0, S, I\}$, as in (2.6.3), where the size of the I block is equal to the number of σ_i that are equal to 1 (corresponding to

$\theta_i = \pi/2$) and the 0 block has $n - 2k$ rows and 0 columns. The σ_i are the sines of the principal angles between $\mathcal{U}$ and $\mathcal{V}$. They are also the singular values of $P_2^* Q_1$. Thus we have an alternate way of computing principal angles when $k \le n/2$: Let P_2 and Q_1 be matrices with orthonormal columns such that $\mathcal{U}^\perp = \mathcal{R}(P_2)$ and $\mathcal{V} = \mathcal{R}(Q_1)$. Compute the k singular values of the $(n - k) \times k$ matrix $P_2^* Q_1$, some of which may be zero. Call them $\sigma_1 \le \sigma_2 \le \cdots \le \sigma_k$. Then the principal angles are $\theta_i = \arcsin \sigma_i$, $i = 1, \ldots, k$. This computes the small principal angles accurately, since the arcsin function has slope 1 at $\sigma = 0$. However, angles near $\pi/2$ will be computed inaccurately, since $\theta = \arcsin \sigma$ is not Lipschitz continuous at $\sigma = 1$ (and $\theta = \pi/2$).

When $k > n/2$, the situation is a little bit different. In this case $n - k < k$, so the matrix $P_2^* Q_1$ has at most $n - k$ nonzero singular values. It follows that $k - (n - k) = 2k - n$ of the principal angles must be zero. In other words, $\gamma_i = 1$ and $\theta_i = 0$ for $i = 1, \ldots, 2k - n$ (Exercise 2.6.8). The other $n - k$ principal angles are obtained from the $n - k$ singular values of $P_2^* Q_1$: Let $\sigma_{2k-n+1} \le \cdots \le \sigma_k$ denote the $n - k$ singular values of $P_2^* Q_1$, some of which may be zero. Then $\theta_i = \arcsin \sigma_i$, $i = 2k - n + 1, \ldots, k$, are principal angles between $\mathcal{U}$ and $\mathcal{V}$.

In any of these computations we can use $P_1^* Q_2$ in place of $P_2^* Q_1$, since these two matrices have exactly the same singular values, as follows from the CSD (2.6.8). We now summarize these findings as a theorem.

Theorem 2.6.8. *Let $\mathcal{U}$ and $\mathcal{V}$ be k-dimensional subspaces of $\mathbb{F}^n$, and let $\theta_1, \ldots, \theta_k$ denote the principal angles between $\mathcal{U}$ and $\mathcal{V}$. Let P_1, P_2, Q_1, and Q_2 be matrices with orthonormal columns such that $\mathcal{U} = \mathcal{R}(P_1)$, $\mathcal{U}^\perp = \mathcal{R}(P_2)$, $\mathcal{V} = \mathcal{R}(Q_1)$, and $\mathcal{V}^\perp = \mathcal{R}(Q_2)$.*

(a) *If $k \le n/2$, let $\sigma_1 \le \ldots \le \sigma_k$ denote the k singular values of $P_2^* Q_1$ or $P_1^* Q_2$. Then $\theta_i = \arcsin \sigma_i$, $i = 1, \ldots, k$.*

(b) *If $k > n/2$, then $\theta_i = 0$ for $i = 1, \ldots, 2k - n$. Let $\sigma_{2k-n+1} \le \cdots \le \sigma_k$ denote the $n - k$ singular values of $P_2^* Q_1$ or $P_1^* Q_2$. Then $\theta_i = \arcsin \sigma_i$, $i = 2k - n + 1, \ldots, k$.*

Distance between Subspaces

Let $\mathcal{U}$ and $\mathcal{V}$ be two k-dimensional subspaces of $\mathbb{F}^n$. We define $d(\mathcal{U}, \mathcal{V})$, the *distance* between $\mathcal{U}$ and $\mathcal{V}$, by

$$d(\mathcal{U}, \mathcal{V}) = \max_{\substack{u \in \mathcal{U} \\ \|u\| = 1}} d(u, \mathcal{V}) = \max_{\substack{u \in \mathcal{U} \\ \|u\| = 1}} \min_{v \in \mathcal{V}} \| u - v \|. \tag{2.6.9}$$

In words, $d(\mathcal{U}, \mathcal{V})$ is the maximum possible distance to $\mathcal{V}$ from a unit vector in $\mathcal{U}$. As the following theorem shows, there is a simple relationship between the distance between two subspaces and the angles between them.

Theorem 2.6.9. *Let $\mathcal{U}$ and $\mathcal{V}$ be k-dimensional subspaces of $\mathbb{F}^n$, and let P_1, P_2, Q_1, and Q_2 be as in Theorem 2.6.8. Then*

$$d(\mathcal{U}, \mathcal{V}) = \| P_2^* Q_1 \| = \| P_1^* Q_2 \| = \| Q_2^* P_1 \| = \| Q_1^* P_2 \| = \sin \theta_k,$$

where θ_k is the largest principal angle between $\mathcal{U}$ and $\mathcal{V}$.

A proof of Theorem 2.6.9 is worked out in Exercise 2.6.9.

Corollary 2.6.10. *Let $\mathcal{U}$ and $\mathcal{V}$ be k-dimensional subspaces of $\mathbb{F}^n$. Then*

$$d(\mathcal{U}^\perp, \mathcal{V}^\perp) = d(\mathcal{U}, \mathcal{V}).$$

Another characterization of the distance between subspaces involves the orthoprojectors onto the subspaces. Continuing notation from above, let $U \in \mathbb{F}^{n \times n}$ be the unitary matrix given by

$$U = \begin{bmatrix} U_1 & U_2 \end{bmatrix} = \begin{bmatrix} P_1 & P_2 \end{bmatrix} \begin{bmatrix} M_1 & \\ & M_2 \end{bmatrix}, \qquad (2.6.10)$$

where M_1 and M_2 are from the CSD (2.6.8). Similarly, define unitary $V \in \mathbb{F}^{n \times n}$ by

$$V = \begin{bmatrix} V_1 & V_2 \end{bmatrix} = \begin{bmatrix} Q_1 & Q_2 \end{bmatrix} \begin{bmatrix} N_1 & \\ & N_2 \end{bmatrix}. \qquad (2.6.11)$$

Recall that the columns of U_1 and V_1 are principal vectors associated with the spaces $\mathcal{U}$ and $\mathcal{V}$. The CSD (2.6.8) implies that

$$U^* V = \begin{bmatrix} U_1^* V_1 & U_1^* V_2 \\ U_2^* V_1 & U_2^* V_2 \end{bmatrix} = \begin{bmatrix} \Gamma_1 & -\Sigma_1^T \\ \Sigma_1 & \Gamma_2 \end{bmatrix}. \qquad (2.6.12)$$

The following theorem, which is proved in Exercise 2.6.10, makes use of these relationships.

Theorem 2.6.11. *Let $P_\mathcal{U}$ and $P_\mathcal{V}$ denote the orthoprojectors of $\mathbb{F}^n$ onto $\mathcal{U}$ and $\mathcal{V}$, respectively. Then*

(a) $P_\mathcal{V} - P_\mathcal{U} = U \begin{bmatrix} 0 & \Sigma_1^T \\ \Sigma_1 & 0 \end{bmatrix} V^*.$

(b) $\| P_\mathcal{V} - P_\mathcal{U} \| = \sin \theta_k = d(\mathcal{U}, \mathcal{V}).$

Using the characterization of the subspace distance function given in Theorem 2.6.11 (or Theorem 2.6.9), it is easy to show that $d(\cdot, \cdot)$ is a metric on the set of all subspaces of $\mathbb{F}^n$.

Theorem 2.6.12. *The function $d(\cdot, \cdot)$ defined by (2.6.9) is a metric on the set of k-dimensional subspaces of $\mathbb{F}^n$. That is, for any k-dimensional subspaces $\mathcal{U}$, $\mathcal{V}$, and $\mathcal{W}$ of $\mathbb{F}^n$, the following hold:*

(a) $d(\mathcal{U}, \mathcal{V}) > 0$ *if* $\mathcal{U} \neq \mathcal{V}$.

(b) $d(\mathcal{U}, \mathcal{U}) = 0$.

(c) $d(\mathcal{U}, \mathcal{V}) = d(\mathcal{V}, \mathcal{U})$.

(d) $d(\mathcal{U}, \mathcal{W}) \leq d(\mathcal{U}, \mathcal{V}) + d(\mathcal{V}, \mathcal{W})$.

Since subspaces are usually presented in terms of bases, the following result, which is proved in Exercise 2.6.14, comes in handy.

Proposition 2.6.13. *Let $\mathcal{U}$ and $\mathcal{V}$ be two k-dimensional subspaces of $\mathbb{F}^n$, and let $P \in \mathbb{F}^{n \times k}$ be a matrix with orthonormal columns such that $\mathcal{U} = \mathcal{R}(P)$. Then there is a matrix $Q \in \mathbb{F}^{n \times k}$ such that $\mathcal{V} = \mathcal{R}(Q)$ and $\| P - Q \| \leq \sqrt{2} d(\mathcal{U}, \mathcal{V})$.*

As we shall soon see, it is often useful to transform a problem from one coordinate system to another more convenient, one. The following proposition is useful in this connection.

Proposition 2.6.14. *Let $\mathcal{U}$ and $\mathcal{V}$ be k-dimensional subspaces of $\mathbb{F}^n$, let $S \in \mathbb{F}^{n \times n}$ be nonsingular, and let $\tilde{\mathcal{U}} = S^{-1}\mathcal{U}$ and $\tilde{\mathcal{V}} = S^{-1}\mathcal{V}$. Then the following hold:*

(a) $d(\mathcal{U}, \mathcal{V}) \leq \kappa(S) d(\tilde{\mathcal{U}}, \tilde{\mathcal{V}})$.

(b) $d(\tilde{\mathcal{U}}, \tilde{\mathcal{V}}) \leq \kappa(S) d(\mathcal{U}, \mathcal{V})$.

(c) $d(\tilde{\mathcal{U}}, \tilde{\mathcal{V}}) = d(\mathcal{U}, \mathcal{V})$ *if S is unitary.*

This proposition will be proved in Exercise 2.6.15. Part (c) can be deduced from parts (a) and (b), but notice that part (c) is also quite obvious: A unitary transformation preserves lengths and angles. In particular, it preserves the principal angles, so it preserves the distance between the subspaces.

For future reference we make note of one additional small result, which is an easy consequence of definition (2.6.9).

Proposition 2.6.15. *Let $\mathcal{U}$ and $\mathcal{V}$ be subspaces of $\mathbb{F}^n$ of the same dimension. Then $d(\mathcal{U}, \mathcal{V}) < 1$ if and only if $\mathcal{U} \cap \mathcal{V}^\perp = \{0\}$.*

Convenient Coordinates

For the study of invariant subspaces it is often convenient to transform the problem to another coordinate system. Suppose $A \in \mathbb{F}^{n \times n}$ and that $\mathcal{U}$ is a k-dimensional subspace of $\mathbb{F}^n$ that is invariant under A. Let $U = \begin{bmatrix} U_1 & U_2 \end{bmatrix}$ be a unitary matrix such that $\mathcal{U} = \mathcal{R}(U_1)$, and let $B = U^{-1}AU$. This similarity transformation represents a change of coordinate system: A and B are two representations of the same linear transformation in different coordinate systems (Proposition 1.8.1). A vector that is represented by x in the original (A) coordinate system is represented by $U^{-1}x$ in the new (B) coordinate system. Thus the subspace $\mathcal{U}$ in the original coordinate system corresponds to $U^{-1}\mathcal{U}$ in the new coordinate system. The columns of U_1 are an orthonormal basis for $\mathcal{U}$. To get an orthonormal basis for $U^{-1}\mathcal{U}$, we take the columns of $U^{-1}U_1 = \begin{bmatrix} I_k \\ 0 \end{bmatrix} = \begin{bmatrix} e_1 & \cdots & e_k \end{bmatrix}$. Thus $U^{-1}\mathcal{U} = \text{span}\{e_1, \ldots, e_k\}$. Let us call this space $\mathcal{E}_k$. Since $\mathcal{U}$ is invariant under A, it must be true that $\mathcal{E}_k$ is invariant under B. Indeed, one can easily check that this is true (Exercise 2.6.17). This finding is consistent with Proposition 2.2.2, which states that the invariance of $\mathcal{U}$ under A implies that B is block triangular:

$$B = \begin{bmatrix} B_{11} & B_{12} \\ 0 & B_{22} \end{bmatrix},$$

where $B_{11} \in \mathbb{F}^{k \times k}$. The block triangularity is a consequence of the invariance of $\mathcal{E}_k$ under B (Exercise 2.6.18).

It is convenient to work in the coordinate system of B, in which the invariant subspace of interest has the very simple form $\mathcal{E}_k = \text{span}\{e_1, \ldots, e_k\}$. Letting $E_k = \begin{bmatrix} I \\ 0 \end{bmatrix} \in \mathbb{F}^{n \times k}$, we have $\mathcal{E}_k = \mathcal{R}(E_k)$. Our fundamental objective throughout this book is to find invariant subspaces. More realistically, we wish to find subspaces that approximate invariant subspaces as well as possible. If $\mathcal{S}$ is a subspace that approximates the invariant subspace $\mathcal{U}$, then $\hat{\mathcal{S}} = U^{-1}\mathcal{S}$ will be a space that approximates the invariant (under B) subspace $\mathcal{E}_k$. Indeed, $d(\hat{\mathcal{S}}, \mathcal{E}_k) = d(\mathcal{S}, \mathcal{U})$ by Proposition 2.6.14.

Let $S_1 \in \mathbb{F}^{n \times k}$ be a matrix with orthonormal columns such that $\hat{\mathcal{S}} = \mathcal{R}(S_1)$, and consider the partition $S_1 = \begin{bmatrix} S_{11} \\ S_{21} \end{bmatrix}$, where $S_{11} \in \mathbb{F}^{k \times k}$. Then, by Theorem 2.6.7, the singular values of $E_k^* S_1 = S_{11}$ are the cosines of the principal angles between $\mathcal{E}_k$ and $\hat{\mathcal{S}}$. It follows immediately that S_{11} is nonsingular if and only if the largest principal angle θ_k satisfies $\cos \theta_k > 0$, or equivalently $\theta_k < \pi/2$ and $\sin \theta_k < 1$. Recalling that $d(\mathcal{E}_k, \hat{\mathcal{S}}) = \sin \theta_k$, we see that S_{11} is nonsingular if and only if $d(\mathcal{E}_k, \hat{\mathcal{S}}) < 1$. Thus S_{11} will be nonsingular if $\hat{\mathcal{S}}$ approximates $\mathcal{E}_k$ at all well. This allows us to write down a different basis for $\hat{\mathcal{S}}$ that has a certain appeal. Letting $X = S_{21} S_{11}^{-1}$, we see immediately that

$$S_1 = \begin{bmatrix} S_{11} \\ S_{21} \end{bmatrix} = \begin{bmatrix} I \\ X \end{bmatrix} S_{11}.$$

By Proposition 1.5.11, the columns of $\begin{bmatrix} I \\ X \end{bmatrix}$ form a (nonorthonormal) basis for $\hat{\mathcal{S}}$. Comparing this basis with the basis $E_k = \begin{bmatrix} I \\ 0 \end{bmatrix}$ of $\mathcal{E}_k$, we suspect that $\| X \|$ must be closely related to $d(\mathcal{E}_k, \hat{\mathcal{S}})$. Indeed this is so.

Proposition 2.6.16. *Let $\hat{\mathcal{S}}$ be a k-dimensional subspace of $\mathbb{F}^n$, and suppose $\hat{\mathcal{S}} = \mathcal{R}(\hat{S})$, where $\hat{S} = \begin{bmatrix} I \\ X \end{bmatrix}$. Then*

$$\| X \| = \tan \theta_k,$$

where θ_k is the largest principal angle between $\mathcal{E}_k$ and $\hat{\mathcal{S}}$. Consequently

$$d(\mathcal{E}_k, \hat{\mathcal{S}}) = \frac{\| X \|}{\sqrt{1 + \| X \|^2}}.$$

This proposition is easily proved using the SVD of X; see Exercise 2.6.19.

If $\hat{\mathcal{S}}$ is quite close to $\mathcal{E}_k$, then θ_k is small, and we have

$$d(\mathcal{E}_k, \hat{\mathcal{S}}) = \sin \theta_k \approx \theta_k \approx \tan \theta_k = \| X \|, \tag{2.6.13}$$

so all of the quantities listed in (2.6.13) are equally good measures of the proximity of $\hat{\mathcal{S}}$ to $\mathcal{E}_k$.

Notes and References

The study of angles between subspaces goes back at least to an 1875 paper by Jordan [121] and a 1936 paper by the statistician Hotelling [118]. The basic ideas associated with the

CSD are implicit in papers by Davis and Kahan [66, 67] around 1970, but the first explicit statement and proof of (a special case of) the CSD was given in Stewart's 1977 paper [197]. Paige and Saunders [164] proved the CSD in the generality given here. Paige and Wei [166] give a good overview of the history and uses of the CSD. The proof of the CSD outlined in Exercise 2.6.1 is similar to the one outlined in [166]. Björck and Golub [37] showed how to compute principal angles and vectors using the SVD.

Exercises

2.6.1. In this exercise you will prove Theorem 2.6.1. There are many steps, but each step is easy.

(a) Show that $\| Q_{ij} \| \leq 1$ for $i, j = 1, 2$. Deduce that the singular values of Q_{ij} are all less than or equal to 1.

(b) Show that
$$Q_{11}^* Q_{11} + Q_{21}^* Q_{21} = I_{\beta_1} \tag{2.6.14}$$
and
$$Q_{21} Q_{21}^* + Q_{22} Q_{22}^* = I_{\alpha_2}. \tag{2.6.15}$$

(c) Let $j = \mathrm{rank}(Q_{21})$, and let $s_1 \leq s_2 \leq \cdots \leq s_j$ denote the (positive) singular values of Q_{21} in ascending order. None of these exceed 1, but some of them may equal 1. Suppose $s_m < 1$ and $s_{m+1} = \cdots = s_j = 1$. Show that the eigenvalues of $Q_{21}^* Q_{21}$ are $s_1^2, \ldots, s_m^2$, 1 repeated $j - m$ times, and 0 repeated $\beta_1 - j$ times. (See Exercise 2.3.12.)

(d) Let $c_i = \sqrt{1 - s_i^2}$, $i = 1, \ldots, m$. Then $1 > c_1 \geq c_2 \geq \cdots \geq c_m > 0$. Use (2.6.14) to show $Q_{11}^* Q_{11}$ has the same eigenvectors as $Q_{21}^* Q_{21}$, and that the eigenvalues of $Q_{21}^* Q_{21}$ are 1 repeated $\beta_1 - j$ times, $c_1^2 \geq \cdots \geq c_m^2$, and 0 repeated $j - m$ times. Deduce that the nonzero singular values of Q_{11} are 1 repeated $\beta_1 - j$ times and $c_1, \ldots, c_m$.

(e) Proceeding as in part (d) and using (2.6.15), show that the nonzero singular values of Q_{22} are 1 repeated $\alpha_2 - j$ times and $c_1, \ldots, c_m$.

(f) Pick unitary U_1 and V_1 and diagonal Σ_{11} so that the equation $Q_{11} = U_1 \Sigma_{11} V_1^*$ is an SVD of Q_{11}. Similarly let $Q_{22} = \tilde{U}_2 \Sigma_{22} \tilde{V}_2^*$ be an SVD of Q_{22}. Deduce from parts (d) and (e) that $\Sigma_{11} = \mathrm{diag}\{I_{\beta_1 - j}, C, 0_c\}$ and $\Sigma_{22} = \mathrm{diag}\{I_{\alpha_2 - j}, C, 0_c^T\}$, where $C = \mathrm{diag}\{c_1, \ldots, c_m\}$ and 0_c is a zero matrix with $\alpha_1 - \beta_1 + j - m$ rows and $j - m$ columns. Show that

$$
\begin{bmatrix} U_1^* & \\ & \tilde{U}_2^* \end{bmatrix}
\begin{bmatrix} Q_{11} & Q_{12} \\ Q_{21} & Q_{22} \end{bmatrix}
\begin{bmatrix} V_1 & \\ & \tilde{V}_2 \end{bmatrix}
=
\begin{bmatrix} \Sigma_{11} & Y \\ X & \Sigma_{22} \end{bmatrix}
$$

$$
=
\left[
\begin{array}{cc|cc}
I & 0 & Y_{11} & Y_{12} \\
0 & \tilde{C} & Y_{21} & Y_{22} \\
\hline
X_{11} & X_{12} & I & 0 \\
X_{21} & X_{22} & 0 & \tilde{C}^T
\end{array}
\right], \tag{2.6.16}
$$

where $\tilde{C} = \mathrm{diag}\{C, 0_c\}$ and the form of X and Y is yet to be determined.

(g) The matrix in (2.6.16) is unitary. Use this fact to show that the submatrices X_{11}, X_{21}, X_{12}, Y_{11}, Y_{21}, and Y_{12} are all zero.

(h) Now focus on the submatrix

$$\begin{bmatrix} \tilde{C} & Y_{22} \\ X_{22} & \tilde{C}^T \end{bmatrix}. \tag{2.6.17}$$

Show that this matrix is unitary. Show that X_{22} and Y_{22} are square (but not necessarily of the same size). Let $X_{22} = QR$ and $-Y_{22}^* = \tilde{Q}\tilde{R}$ be QR decompositions of X_{22} and $-Y_{22}^*$, respectively, where the main diagonal entries of the triangular matrices R and $\tilde{R}$ are nonnegative (Theorem 1.3.9). Show that

$$\begin{bmatrix} I & \\ & Q^* \end{bmatrix} \begin{bmatrix} \tilde{C} & Y_{22} \\ X_{22} & \tilde{C}^T \end{bmatrix} \begin{bmatrix} I & \\ & \tilde{Q} \end{bmatrix} = \begin{bmatrix} \tilde{C} & -L \\ R & Z \end{bmatrix}, \tag{2.6.18}$$

where $L = \tilde{R}^*$ is lower triangular and the form of Z is to be determined. Show that the matrix in (2.6.18) is unitary.

(i) Show that $R^*R = I - \tilde{C}^*\tilde{C} = \tilde{S}^2$, where $\tilde{S} = \text{diag}\{S, I\}$, $S = \text{diag}\{s_1, \ldots, s_m\}$. Using Theorem 1.4.1, deduce that $R = \tilde{S}$.

(j) Show that $LL^* = I - \tilde{C}\tilde{C}^* = \check{S}^2$, where $\check{S} = \text{diag}\{S, I\}$. (The difference between $\check{S}$ and $\tilde{S}$ is that the I blocks need not be the same size.) Deduce that $L = \check{S}$.

(k) We now know that the unitary matrix in (2.6.18) has the form

$$\begin{bmatrix} \tilde{C} & -L \\ R & Z \end{bmatrix} = \left[\begin{array}{cc|cc} C & & -S & \\ & 0 & & -I \\ \hline S & & Z_{11} & Z_{12} \\ & I & Z_{21} & Z_{22} \end{array} \right].$$

Show that the blocks Z_{21}, Z_{12}, and Z_{22} are all zero. Use the orthogonality of the columns to show that $-CS + SZ_{11} = 0$. Deduce that $Z_{11} = C$.

(l) Now put the pieces together. Let

$$U_2 = \tilde{U}_2 \begin{bmatrix} I & \\ & Q \end{bmatrix} \quad \text{and} \quad V_2 = \tilde{V}_2 \begin{bmatrix} I & \\ & \tilde{Q} \end{bmatrix}.$$

Show that

$$U_2^* Q V_2^* = \left[\begin{array}{cc|cc} I & & 0 & \\ & C & & -S \\ & 0 & & -I \\ \hline 0 & & I & \\ & S & & C \\ & I & & 0 \end{array} \right].$$

Now check that we have completed the proof of Theorem 2.6.1.

2.6.2. Let $\mathcal{U}$ and $\mathcal{V}$ be two subspaces of $\mathbb{C}^n$ of dimension k, and let $\theta_1, \ldots, \theta_k$ be the principal angles between $\mathcal{U}$ and $\mathcal{V}$.

(a) Show that if $\mathcal{U} = \mathcal{V}$, then $\theta_1 = \cdots = \theta_k = 0$. Show that the principal vectors satisfy $v_j = u_j$, $j = 1, \ldots, k$, where $u_1, \ldots, u_k$ can be taken to be any orthonormal basis of $\mathcal{U}$.

(b) Show that if $\dim(\mathcal{U} \cap \mathcal{V}) = j < k$, then $\theta_1 = \cdots = \theta_j = 0$ and $\theta_{j+1} > 0$.

(c) Show that if $\mathcal{U}$ is orthogonal to $\mathcal{V}$, then $\theta_1 = \cdots = \theta_k = \pi/2$. Show that $u_1, \ldots, u_k$ and $v_1, \ldots, v_k$ can be taken to be any orthonormal basis of $\mathcal{U}$ and $\mathcal{V}$, respectively, in this case.

2.6.3. Let $\mathcal{U}$ and $\mathcal{V}$ be two-dimensional subspaces of $\mathbb{R}^3$.

(a) Show that $\theta_1 = 0$.

(b) What does θ_2 represent geometrically?

2.6.4. This exercise leads to a proof of Proposition 2.6.2. It suffices to prove (2.6.5).

(a) Show by direct computation that $\| u_1 - v_1 \cos\theta_1 \|^2 = \sin^2\theta_1$. (Write the norm as an inner product.)

(b) Let $u \in \mathcal{U}$ and $v \in \mathcal{V}$, with $\| u \| = 1$. We will now complete the proof by showing that $\| u - v \| \geq \sin\theta_1$. Let $\hat{v}$ be the best approximation to u from $\mathcal{V}$. Since $\| u - v \| \geq \| u - \hat{v} \|$, it now suffices to show that $\| u - \hat{v} \| \geq \sin\theta_1$. Let θ denote the angle between u and $\hat{v}$. By the projection theorem (Theorem 1.5.1), $\langle u - \hat{v}, \hat{v} \rangle = 0$. Deduce that $\langle u, \hat{v} \rangle = \| \hat{v} \|^2$ and, moreover, $\cos\theta = \| \hat{v} \|$. Apply the Pythagorean theorem to the sum $u = (u - \hat{v}) + \hat{v}$ to obtain $\| u - \hat{v} \| = \sin\theta$. Show that $\sin\theta \geq \sin\theta_1$. Conclude that $\| u - v \| \geq \sin\theta_1$.

2.6.5. This exercise refers to objects discussed in Exercise 2.6.4.

(a) Draw a picture that shows that the relationship $\| u_1 - v_1 \cos\theta_1 \| = \sin\theta_1$ is obvious.

(b) Draw a picture that shows that the relationships $\| \hat{v} \| = \cos\theta$ and $\| u - \hat{v} \| = \sin\theta$ are obvious. (Don't forget that $\langle u - \hat{v}, \hat{v} \rangle = 0$.)

2.6.6. This exercise proves Theorem 2.6.4.

(a) Proposition 2.6.3 shows that $v_i \cos\theta_i$ is the best approximation to u_i from $\mathcal{V}_i$. Use the Projection theorem to show that $\langle u_i - v_i \cos\theta_i, v_j \rangle = 0$ for $j = i + 1, \ldots, k$.

(b) Use the result from part (a) to show that $\langle u_i, v_j \rangle = 0$ if $i < j$.

(c) Show that $\langle u_i, v_j \rangle = 0$ for $i > j$ (or, equivalently, $\langle v_i, u_j \rangle = 0$ for $i < j$).

(d) Prove Corollary 2.6.5.

2.6.7. In this exercise you will prove Theorem 2.6.6. Assume notation as in the statement of the theorem.

(a) For every $u \in \mathcal{U}$ there is a unique $x \in \mathbb{F}^k$ such that $u = \tilde{U}x$. Show that $\| u \| = \| x \|$. Similarly, every $v \in \mathcal{V}$ can be written as $v = \tilde{V}y$ for some unique $y \in \mathbb{F}^k$ satisfying $\| v \| = \| y \|$. Show that if we represent u and v this way, we have $\langle u, v \rangle = \sum_{i=1}^{k} c_i x_i \overline{y_i}$, where the c_i are the main-diagonal entries of $C = \tilde{V}^* \tilde{U} = \tilde{U}^* \tilde{V}$.

(b) Make simple estimates including the CSB inequality (Theorem 1.2.1) to show that

$$\max\{| \langle u, v \rangle | \mid u \in \mathcal{U}, \ v \in \mathcal{V}, \| u \| = \| v \| = 1\} \leq \max_i c_i = c_1.$$

(c) Show that $\langle \tilde{u}_1, \tilde{v}_1 \rangle = c_1$. Conclude that the first principal angle between $\mathcal{U}$ and $\mathcal{V}$ is $\arccos c_1$, and that $\tilde{u}_1$ and $\tilde{v}_1$ are principal vectors.

(d) Show that for $i = 1, \ldots, k$ the ith principal angle is $\arccos c_i$ with associated principal vectors $\tilde{u}_i$ and $\tilde{v}_i$. This proves Theorem 2.6.6.

2.6.8. Suppose $\mathcal{U}$ and $\mathcal{V}$ have dimension $k > n/2$.

(a) Use the CSD (2.6.8) to show that the principal angles $\theta_1, \ldots, \theta_{2k-n}$ are all zero.

(b) Use the fundamental dimension relationship

$$\dim(\mathcal{U} \cap \mathcal{V}) + \dim(\mathcal{U} + \mathcal{V}) = \dim(\mathcal{U}) + \dim(\mathcal{V})$$

(Theorem 1.5.2) to obtain a lower bound on the dimension of $\mathcal{U} \cap \mathcal{V}$. Use this bound to derive a second proof that $\theta_i = 0$, $i = 1, \ldots, 2k - n$.

2.6.9. This exercise proves Theorem 2.6.9.

(a) Let $u \in \mathcal{U}$ with $\| u \| = 1$. Then u can be expressed as a sum $u = v + v^{\perp}$, where $v \in \mathcal{V}$ and $v^{\perp} \in \mathcal{V}^{\perp}$. Use Projection Theorem 1.5.1 to show that $\| v^{\perp} \| = d(u, \mathcal{V})$.

(b) Show that $\| v^{\perp} \|^2 = \| Q^* v^{\perp} \|^2 = \| Q_1^* v^{\perp} \|^2 + \| Q_2^* v^{\perp} \|^2 = \| Q_2^* v^{\perp} \|^2$.

(c) From parts (a) and (b) and the fact that $Q_2^* v = 0$, deduce that $d(u, \mathcal{V}) = \| Q_2^* u \|$.

(d) Each $u \in \mathcal{U}$ with $\| u \| = 1$ can be expressed uniquely as $u = P_1 x$ for some $x \in \mathbb{F}^k$ with $\| x \| = 1$. Deduce that

$$\max_{\substack{u \in \mathcal{U} \\ \| u \| = 1}} d(u, \mathcal{V}) = \max_{\substack{x \in \mathbb{F}^k \\ \| x \| = 1}} \| Q_2^* P_1 x \| = \| Q_2^* P_1 \|.$$

(e) Using Theorem 2.6.8, deduce that $\| P_2^* Q_1 \| = \| P_1^* Q_2 \| = \| Q_2^* P_1 \| = \| Q_1^* P_2 \| = \sin \theta_k$.

2.6.10. This exercise proves Theorem 2.6.11. Recall that $P_{\mathcal{U}} = U_1 U_1^*$ and $P_{\mathcal{V}} = V_1 V_1^*$.

(a) Show that

$$P_{\mathcal{U}} = U \begin{bmatrix} I_k & \\ & 0 \end{bmatrix} U^* = U \begin{bmatrix} I_k & \\ & 0 \end{bmatrix} (U^* V) V^* = U \begin{bmatrix} \Gamma_1 & -\Sigma_1^T \\ 0 & 0 \end{bmatrix} V^*.$$

(b) Similarly, show that

$$P_{\mathcal{V}} = U \begin{bmatrix} \Gamma_1 & 0 \\ \Sigma_1 & 0 \end{bmatrix} V^*.$$

(c) Deduce that $P_{\mathcal{V}} - P_{\mathcal{U}} = U \begin{bmatrix} 0 & \Sigma_1^T \\ \Sigma_1 & 0 \end{bmatrix} V^*.$

(d) Deduce that $\| P_{\mathcal{V}} - P_{\mathcal{U}} \| = \sin \theta_k = d(\mathcal{U}, \mathcal{V}).$

2.6.11. Show that $d(\mathcal{U}^{\perp}, \mathcal{V}^{\perp}) = d(\mathcal{U}, \mathcal{V}).$

2.6.12. Prove Theorem 2.6.12.

2.6.13. Let $U = \begin{bmatrix} U_1 & U_2 \end{bmatrix}$ and $V = \begin{bmatrix} V_1 & V_2 \end{bmatrix}$ be defined by (2.6.10) and (2.6.11), respectively. Recall that the columns of U_1 and V_1 are the principal vectors associated with the subspaces $\mathcal{U} = \mathcal{R}(U_1)$ and $\mathcal{V} = \mathcal{R}(V_1)$, respectively. The *direct rotator* from $\mathcal{V}$ to $\mathcal{U}$ is defined to be the unitary matrix $W = UV^*$.

(a) Show $W(\mathcal{V}) = \mathcal{U}$. More precisely, show that $W v_i = u_i$ for $i = 1, \ldots, k$.

(b) Show that if $X \in \mathbb{F}^{n \times n}$ is a unitary matrix such that $X(\mathcal{V}) = \mathcal{U}$, then X must have the form

$$X = U \begin{bmatrix} Y_1 & \\ & Y_2 \end{bmatrix} V^*,$$

where $Y_1 \in \mathbb{F}^{k \times k}$ and $Y_2 \in \mathbb{F}^{(n-k) \times (n-k)}$ are unitary. (W has this form with $Y_1 = I_k$ and $Y_2 = I_{n-k}$.)

(c) Using (2.6.12), show that $W = U(V^*U)U^* = U \begin{bmatrix} \Gamma_1 & -\Sigma_1^T \\ \Sigma_1 & \Gamma_2 \end{bmatrix} U^*.$ This shows that W is unitarily similar to a product of k rotators through angles $\theta_1, \ldots, \theta_k$.

(d) Show that every unitary X such that $X(\mathcal{V}) = \mathcal{U}$ has the form

$$X = U \begin{bmatrix} Y_1 \Gamma_1 & -Y_1 \Sigma_1^T \\ Y_2 \Sigma_1 & Y_2 \Gamma_2 \end{bmatrix} U^*,$$

where Y_1 and Y_2 are unitary.

(e) Show that

$$\| W - I \| = \min \| X - I \|,$$

where the minimum is taken over all unitary X such that $X(\mathcal{V}) = \mathcal{U}$. Thus of all X in this class, the direct rotator from $\mathcal{V}$ to $\mathcal{U}$ is the one that is closest to the identity matrix.

2.6.14. This exercise proves Proposition 2.6.13.

(a) Let $U, V \in \mathbb{F}^{n \times k}$ be matrices whose columns are principal vectors for the spaces $\mathcal{U}$ and $\mathcal{V}$. Show that the columns of $U - V$ are mutually orthogonal. Then use the Pythagorean theorem to show that, for any $x \in \mathbb{F}^k$,

$$\| (U - V)x \|^2 = \sum_{j=1}^{k} \| u_j - v_j \|^2 |x_j|^2 \leq \| u_k - v_k \|^2 \| x \|^2.$$

(b) Show that $\| U - V \| = \| u_k - v_k \|$.

(c) Show that $\| u_k - v_k \| = 2 \sin(\theta_k/2) \leq \sqrt{2} \sin \theta_k = \sqrt{2} \, d(\mathcal{U}, \mathcal{V})$, where θ_k denotes the largest principal angle between $\mathcal{U}$ and $\mathcal{V}$. (Notice, by the way, that if θ_k is small, $2 \sin(\theta_k/2) \approx \sin \theta_k$, so the factor $\sqrt{2}$ gives a significant overestimate.)

(d) Since $\mathcal{R}(U) = \mathcal{R}(P)$, there is a unitary $M \in \mathbb{F}^{k \times k}$ such that $P = UM$. Let $Q = VM$. Then $\mathcal{R}(Q) = \mathcal{R}(V) = \mathcal{V}$. Show that $\| P - Q \| = \| U - V \| \leq \sqrt{2} d(\mathcal{U}, \mathcal{V})$.

2.6.15. This exercise proves Proposition 2.6.14. Pick $u \in \mathcal{U}$ such that $\| u \| = 1$ and the maximum is attained in (2.6.9), i.e., $d(\mathcal{U}, \mathcal{V}) = d(u, \mathcal{V})$. Then let $\hat{u} = S^{-1}u \in \tilde{\mathcal{U}}$, $\alpha = \| \hat{u} \| > 0$, and $\tilde{u} = \alpha^{-1}\hat{u} \in \tilde{\mathcal{U}}$. Pick $\tilde{v} \in \tilde{\mathcal{V}}$ such that $\| \tilde{u} - \tilde{v} \| = d(\tilde{u}, \tilde{\mathcal{V}})$. Let $v = \alpha S\tilde{v} \in \mathcal{V}$.

(a) Show that $d(\mathcal{U}, \mathcal{V}) \leq \| u - v \|$.

(b) Show that $\alpha \leq \| S^{-1} \|$. Then show that $\| u - v \| \leq \kappa(S) \| \tilde{u} - \tilde{v} \|$.

(c) Show that $\| \tilde{u} - \tilde{v} \| \leq d(\tilde{\mathcal{U}}, \tilde{\mathcal{V}})$.

(d) Put the pieces together to obtain part (a) of Proposition 2.6.14.

(e) Deduce part (b) of Proposition 2.6.14 from part (a) by reversing the roles of S and S^{-1}.

(f) Deduce part (c) of Proposition 2.6.14 from parts (a) and (b).

2.6.16. Use (2.6.9), the Pythagorean theorem, and other elementary tools to prove Proposition 2.6.15.

2.6.17. Suppose A, B, $U \in \mathbb{F}^{n \times n}$ and $B = U^{-1}AU$. Show that the subspace $\mathcal{S}$ is invariant under A if and only if $U^{-1}\mathcal{S}$ is invariant under B.

2.6.18. Let $B = \begin{bmatrix} B_{11} & B_{12} \\ B_{21} & B_{22} \end{bmatrix}$, where $B_{11} \in \mathbb{F}^{k \times k}$. Show that $B_{21} = 0$ if and only if $\mathcal{E}_k = \mathrm{span}\{e_1, \dots, e_k\}$ is invariant under B.

2.6.19. This exercise proves Proposition 2.6.16.

(a) It is convenient to work the case $k \leq n/2$ first. Since $X \in \mathbb{F}^{(n-k) \times k}$, the assumption $k \leq n/2$ implies that X has at least as many rows as columns. Let the SVD of X be given by $X = UTV^*$, where U and V are unitary, and $T = \begin{bmatrix} \hat{T} \\ 0 \end{bmatrix} \in \mathbb{F}^{(n-k) \times k}$, $\hat{T} = \mathrm{diag}\{t_1, \dots, t_k\}$. The singular values of X are $t_1, \dots, t_k$, some of which may be zero. Let us assume for convenience that they are ordered so that $0 \leq t_1 \leq \cdots \leq t_k$.

(b) The matrix $I + X^*X$ is symmetric and positive definite. Show that $I + X^*X = V(I + \hat{T}^2)V^* = VDV^*$, where $D = \mathrm{diag}\{1 + t_1^2, \dots, 1 + t_k^2\}$.

(c) $I + X^*X$ has a unique positive definite square root given by $(I + X^*X)^{1/2} = VD^{1/2}V^*$, and this has an inverse $(I + X^*X)^{-1/2} = VD^{-1/2}V^*$, where $D^{-1/2} = \mathrm{diag}\{(1 + t_1^2)^{-1/2}, \dots, (1 + t_k^2)^{-1/2}\}$. Define $S_1 \in \mathbb{F}^{n \times k}$ by

$$S_1 = \begin{bmatrix} S_{11} \\ S_{21} \end{bmatrix} = \begin{bmatrix} I \\ X \end{bmatrix} (I + X^*X)^{-1/2}.$$

Show that $S_1^* S_1 = I_k$. Deduce that the columns of S_1 form an orthonormal basis of $\hat{\mathcal{S}}$.

(d) Apply Theorem 2.6.8 with $\begin{bmatrix} P_1 & P_2 \end{bmatrix} = I$ and $Q_1 = S_1$ to show that the singular values of $X(I + X^*X)^{-1/2}$ are $\sin\theta_1, \ldots, \sin\theta_k$, where $\theta_1, \ldots, \theta_k$ are the principal angles between $\mathcal{E}_k$ and $\hat{\mathcal{S}}$.

(e) Show that $\sin\theta_i = t_i / \sqrt{1 + t_i^2}$, $i = 1, \ldots, k$. Then deduce that $t_i = \tan\theta_i$, $i = 1, \ldots, k$, and $\|X\| = \tan\theta_k$.

(f) Show that $d(\mathcal{E}_k, \hat{\mathcal{S}}) = \|X\| / \sqrt{1 + \|X\|^2}$.

(g) Modify the preceding arguments to take care of the case $k > n/2$. What is the form of T in the SVD of X in this case? Show that $T^*T = \mathrm{diag}\{t_1^2, \ldots, t_k^2\}$, where $t_i = 0$ for $i = 1, \ldots, 2k - n$. (This is consistent with part (b) of Theorem 2.6.8.) Show that $I + X^*X = V D V^*$, where $D = \mathrm{diag}\{1 + t_1^2, \ldots, 1 + t_k^2\}$. Show that the rest of the arguments carry through as in the case $k \le n/2$.

2.6.20. This exercise builds on Exercise 2.6.19 and uses some of the same notation. Let $\hat{\mathcal{S}}$ be a k-dimensional subspace of $\mathbb{F}^n$, and suppose $\hat{\mathcal{S}} = \mathcal{R}(\hat{S}_1)$, where $\hat{S}_1 = \begin{bmatrix} I \\ X \end{bmatrix}$. Assume $k \le n/2$ for convenience. Let $\hat{S}_2 = \begin{bmatrix} -X^* \\ I \end{bmatrix}$. Let $\theta_1, \ldots, \theta_k$ denote the principal angles between $\hat{\mathcal{S}}$ and $\mathcal{E}_k$.

(a) Show that $\hat{S}_2^* \hat{S}_1 = 0$. Thus the columns of $\hat{S}_2$ are orthogonal to those of $\hat{S}_1$. Deduce that $\hat{\mathcal{S}}^\perp = \mathcal{R}(\hat{S}_2)$.

(b) Let

$$U = \begin{bmatrix} U_{11} & U_{12} \\ U_{21} & U_{22} \end{bmatrix} = \begin{bmatrix} I & -X^* \\ X & I \end{bmatrix} \begin{bmatrix} (I + X^*X)^{-1/2} & \\ & (I + XX^*)^{-1/2} \end{bmatrix}.$$

Show that U is a unitary matrix.

(c) Show that the singular values of $U_{11} = (I + X^*X)^{-1/2}$ are $\cos\theta_1, \ldots, \cos\theta_k$, those of $U_{21} = X(I + X^*X)^{-1/2}$ and $U_{12} = -X^*(I + XX^*)^{-1/2}$ are $\sin\theta_1, \ldots, \sin\theta_k$, and those of $U_{22} = (I + XX^*)^{-1/2}$ are 1, repeated $n - 2k$ times, and $\cos\theta_1, \ldots, \cos\theta_k$. You can prove all of these assertions directly, using the SVD of X, or you can save some work by using the CSD of U.

(d) How are the assertions of part (c) changed when $k > n/2$?

2.7   Perturbation Theorems

When we compute the eigenvalues of a matrix to solve a real-world problem, there is certain to be some error in the matrix to begin with. We may have wanted to compute the eigenvalues of A, but in fact we computed those of a perturbed matrix $A + E$, where E

represents measurement or modeling errors. It is therefore of fundamental importance to ask whether a small perturbation in a matrix changes the eigenvalues/vectors by a great deal or by a small amount. Are the eigenvalues of $A + E$ guaranteed to be close to those of A, or are they not? As we shall see, the eigenvalues of a matrix are not all equally sensitive. Each eigenvalue has a *condition number*, which is a measure of its sensitivity to perturbations. A relatively insensitive eigenvalue has a small condition number and is called well conditioned. A small perturbation in A will cause that eigenvalue to be changed only slightly. An extremely sensitive eigenvalue has a large condition number and is called ill conditioned.[2] A small perturbation in A can cause that eigenvalue to change drastically.

A related question is that of roundoff errors in the computational process. In the face of roundoff errors it is usually impossible to guarantee that a given algorithm computes, for every A, numbers that are close to the true eigenvalues of A. The best algorithms at our disposal are the *backward-stable* ones, which have a weaker guarantee: The computed eigenvalues are those of a nearby matrix $A + E$, where E is tiny compared with A. (E is called the *backward error*.) Backward stability does not guarantee accuracy. As in the previous paragraph, the accuracy depends upon the condition numbers of the eigenvalues.

We begin with a basic theoretical result.

Theorem 2.7.1. *The eigenvalues of a matrix are continuous functions of the matrix entries.*

Proof: We outline the proof without going into the details of the analysis. The eigenvalues of A are the zeros of the characteristic polynomial $\det(\lambda I - A)$. Writing this polynomial in standard form, we have $\det(\lambda I - A) = \lambda^n + c_{n-1}\lambda^{n-1} + \cdots + c_1\lambda + c_0$. As an easy consequence of the definition of the determinant, one checks that the coefficients $c_0, c_1, \ldots, c_{n-1}$ are polynomials in the entries of A. Therefore, they depend continuously on the entries of A. We can now prove the theorem by showing that the zeros of a polynomial depend continuously on its coefficients. To this end we use a special case of the *argument principle* of complex variable theory: Let p be a polynomial, and let C be a simple closed curve in the complex plane that does not pass through any of the zeros of p. Then the number of zeros of p enclosed by C is

$$\frac{1}{2\pi i} \int_C \frac{p'(\lambda)}{p(\lambda)} \, d\lambda, \tag{2.7.1}$$

where p' denotes the derivative of p. For a proof, see any book on complex analysis, for example, [3]. Now let μ be a zero of p of multiplicity k, and let $\epsilon > 0$ be smaller than the distance from μ to the nearest zero distinct from μ. Let C be the circle about μ of radius ϵ. Then the value of the integral in (2.7.1) is k. This is true, regardless of how small ϵ is, as long as ϵ remains positive. We will now show that if the coefficients of p are perturbed by a sufficiently small amount, the number of zeros of the perturbed polynomial within ϵ of μ will be exactly k. This will prove continuity of the zeros of p. The integral in (2.7.1) depends continuously on the integrand $p'(\lambda)/p(\lambda)$, whose values on C depend continuously on the coefficients of p. Therefore, we can guarantee that if p is perturbed by a sufficiently small amount, the value of the integral in (2.7.1) will be changed by less than one. But since that integral must take on an integer value, a change of less than one implies that it remains constant. Thus, for sufficiently small perturbations of the coefficients of p, the value of

---

[2]If a matrix has very ill conditioned eigenvalues, then pseudoeigenvalues [204] are more relevant than eigenvalues.

the integral will be k, which means that k zeros of the perturbed polynomial lie within ϵ of μ. □

The continuity of the eigenvalues is an important, indispensable fact. However, in some matrices the continuity can be quite weak.

Example 2.7.2. *Consider the Jordan block*

$$A = \begin{bmatrix} \mu & 1 & 0 \\ 0 & \mu & 1 \\ 0 & 0 & \mu \end{bmatrix},$$

which has eigenvalues μ, μ, and μ. Let ϵ be a small positive number, and let

$$E_\epsilon = \begin{bmatrix} 0 & 0 & 0 \\ 0 & 0 & 0 \\ \epsilon & 0 & 0 \end{bmatrix}.$$

If we perturb A by E_ϵ, we obtain a matrix $A_\epsilon = A + E_\epsilon$, whose characteristic equation is easily seen to be $(\lambda - \mu)^3 - \epsilon = 0$. The eigenvalues are therefore $\lambda = \mu + \epsilon^{1/3}\omega^j$, $j = 0, 1, 2$, where $\omega = e^{2\pi i/3}$, a primitive cube root of unity. Thus the three eigenvalues of A_ϵ satisfy

$$|\lambda - \mu| = \epsilon^{1/3}.$$

For small values ϵ, the cube root of ϵ is much bigger than ϵ itself. Therefore the perturbation in the eigenvalues is much greater than the perturbation in the matrix. To take an extreme example, suppose $\epsilon = 10^{-15}$. Then $\epsilon^{1/3} = 10^{-5}$, which is 10^{10} times larger. Thus a tiny perturbation in the matrix causes a perturbation in the eigenvalues that is ten billion times as big. If we take a smaller ϵ, the situation is even worse. If we take a larger ϵ, such as 10^{-6}, the situation is much better but still bad.

There are lots of variations on Example 2.7.2. Larger Jordan blocks are worse (Exercise 2.7.1). The situation can also be worsened or improved by replacing the superdiagonal ones in A by larger or smaller numbers, respectively (Exercise 2.7.2).

Jordan blocks are not semisimple. As we shall soon see, semisimple matrices are in some sense better behaved. First we must introduce Gerschgorin disks. A matrix $B \in \mathbb{C}^{n \times n}$ has n Gerschgorin disks associated with it. The ith *Gerschgorin disk* is the closed disk in the complex plane with center b_{ii} and radius

$$r_i = \sum_{j \neq i} |b_{ij}|.$$

That is,

$$D_i = \{z \in \mathbb{C} \mid |z - b_{ii}| \leq r_i\}. \tag{2.7.2}$$

Theorem 2.7.3 (Gerschgorin disk theorem[3]). *Let $B \in \mathbb{C}^{n \times n}$. Then every eigenvalue of B lies within one of the Gerschgorin disks (2.7.2). Each disk D_i that is disjoint from the*

---

[3]For a detailed account of the Gerschgorin disk theorem and its many extensions, see [209].

other disks contains exactly one eigenvalue of B. More generally, if the union of j disks is disjoint from the other n − j disks, then that union contains exactly j eigenvalues of B.

Proof: Let λ be an eigenvalue of B, and let v be an associated eigenvector. Let v_i be a component of v that has maximal magnitude: $|v_i| = \max_j |v_j|$. The ith component of the equation $Bv = v\lambda$ is

$$\sum_{j=1}^{n} b_{ij} v_j = v_i \lambda,$$

which can be rewritten as

$$(\lambda - b_{ii})v_i = \sum_{j \neq i} b_{ij} v_j.$$

Taking absolute values, we have

$$|\lambda - b_{ii}| \, |v_i| \leq \sum_{j \neq i} |b_{ij}| \, |v_j| \leq |v_i| \sum_{j \neq i} |b_{ij}| = |v_i| \, r_i.$$

Dividing both sides of the inequality by $|v_i|$, we find that λ lies in D_i, the ith Gerschgorin disk of B.

The rest of the proof requires a continuity argument. Let

$$D = \mathrm{diag}\{b_{11}, \ldots, b_{nn}\}$$

and $E = B - D$, so that $B = D + E$, D contains the main diagonal of B, and E contains the off-diagonal part. For $t \in [0, 1]$, let $B_t = D + tE$, and let $D_i(t)$ denote the ith Gerschgorin disk associated with $B(t)$. For all t, the center of $D_i(t)$ is b_{ii}, and the radius is $t r_i$. Thus $D_i(t) \subseteq D_i(1) = D_i$ for all $t \in [0, 1]$. As t runs from 0 to 1, the disks grow from points to the full disks D_i. The eigenvalues of $B_0 = D$ are $b_{11}, \ldots, b_{nn}$, the centers of the Gerschgorin disks. By continuity of the eigenvalues, as t runs from 0 to 1, the eigenvalues of B_t trace out continuous paths (connected sets) from $b_{11}, \ldots, b_{nn}$ to the eigenvalues of B. Let D_i be a disk that is disjoint from the others. Let $\lambda_i(t)$ denote the eigenvalue path that starts at b_{ii}. For all $t \in [0, 1]$, $\lambda_i(t)$ must lie within one of the disks $D_j(t)$, $j = 1, \ldots, n$. In fact, it must lie in $D_i(t)$, since its appearance in some other $D_j(t)$ at any time would violate connectedness of the path $\lambda_i(t)$. Hence the entire path must lie in D_i, including the endpoint, the eigenvalue $\lambda_i = \lambda_i(1)$. Clearly D_i cannot contain any other eigenvalues, since each of the other eigenvalue paths must lie completely within the other disks. No path can jump into D_i without violating continuity.

The argument is easily extended to each union of j overlapping disks. If the union does not intersect any of the other disks, then that union must contain exactly j eigenvalue paths and exactly j eigenvalues of B. $\square$

Using Gerschgorin disks, we can prove the following version of the Bauer–Fike theorem.

Theorem 2.7.4. *Let $A \in \mathbb{C}^{n \times n}$ be a semisimple matrix, and let $A = S \Lambda S^{-1}$, where Λ is a diagonal matrix. Let ν be an eigenvalue of the perturbed matrix $A + E$. Then A has an eigenvalue λ_i such that*

$$|\nu - \lambda_i| \leq \kappa_\infty(S) \| E \|_\infty.$$

Proof: $A + E = S(\Lambda + F)S^{-1}$, where $F = S^{-1}ES$. The eigenvalues of A are $\lambda_1, \ldots, \lambda_n$, the main-diagonal entries of Λ, and the eigenvalues of $A + E$ are the eigenvalues of $\Lambda + F$. Since ν is an eigenvalue of $\Lambda + F$, it must lie in one of the Gerschgorin disks associated with $\Lambda + F$, say the ith one. This disk is centered at $\lambda_i + f_{ii}$, and its radius is $r_i = \sum_{j \neq i} |f_{ij}|$. Therefore

$$|\nu - \lambda_i| \leq r_i + |f_{ii}| = \sum_{j=1}^{n} |f_{ij}| \leq \|F\|_{\infty}.$$

Since $F = S^{-1}ES$, we see immediately that $\|F\|_{\infty} \leq \kappa_{\infty}(S)\|E\|_{\infty}$, from which the theorem follows. $\square$

Theorem 2.7.4 implies that each eigenvalue of $A + E$ must be near an eigenvalue of A if $\|E\|_{\infty}$ is sufficiently small. As stated, it does not exclude the possibility that several or even all of the eigenvalues of $A + E$ are clustered near a single eigenvalue of A. However, we can see from the proof that this cannot happen. If the Gerschgorin disks of $\Lambda + F$ are disjoint, then each disk contains exactly one eigenvalue, and each eigenvalue of A has exactly one eigenvalue of $A + E$ nearby. If disks overlap, each overlapping cluster of j disks contains exactly j eigenvalues of $A + E$. Thus each cluster of j eigenvalues of A has j eigenvalues of $A + E$ nearby.

The Bauer–Fike theorem is valid not just for the infinity norm but for any p-norm, and it can be proved without the aid of the Gerschgorin theorem. We chose the Gerschgorin approach because it gives a more precise picture of the location of the eigenvalues.

Theorem 2.7.5 (see Bauer and Fike [24]). *Let $A \in \mathbb{C}^{n \times n}$ be a semisimple matrix, and let $A = S\Lambda S^{-1}$, where Λ is a diagonal matrix. Let ν be an eigenvalue of the perturbed matrix $A + E$. Then A has an eigenvalue λ_j such that*

$$|\nu - \lambda_j| \leq \kappa_p(S)\|E\|_p,$$

$1 \leq p \leq \infty$.

For a proof that does not use Gerschgorin disks, see Exercise 2.7.3. The infinity norm and condition number mesh best with the Gerschgorin disk theorem, but now we will return our focus to the 2-norm and its associated condition number $\kappa(S) = \kappa_2(S)$. Theorem 2.7.5 shows that $\kappa(S)$ puts a bound on how badly the eigenvalues can behave. For example, if $\kappa(S) = 100$, the movement of the eigenvalues cannot be more than 100 times the norm of the perturbing matrix. Thus $\kappa(S)$ serves as an upper bound for the condition numbers for the eigenvalues. If $\kappa(S)$ is small, all of the eigenvalues must be well conditioned, that is, not overly sensitive to perturbations in A. If, on the other hand, $\kappa(S)$ is very large, then some of the eigenvalues of A may be ill conditioned. Recall that the columns of S are eigenvectors of A. The geometric meaning of a large condition number is that the eigenvectors of A are "nearly" linearly dependent.

We must bear in mind also that S is not uniquely determined; the columns can be scaled up and down arbitrarily. This is compensated by a reciprocal scaling of the rows of S^{-1}. We can make $\kappa(S)$ arbitrarily bad by taking some eigenvectors (columns) with huge norms and some with tiny norms. If we wish to use $\kappa(S)$ to gauge the sensitivity of the eigenvalues, we must choose an S whose columns are reasonably well scaled. Scaling to minimize the condition number is discussed in [114, §7.3].

If A is normal, the transforming matrix S can be taken to be unitary, as we know from Theorem 2.3.5. Thus $\kappa_2(S) = 1$, and we have the following very satisfactory corollary.

Corollary 2.7.6. *Let $A \in \mathbb{C}^{n \times n}$ be normal, and let ν be an eigenvalue of the perturbed matrix $A + E$. Then A has an eigenvalue λ_j such that*

$$|\nu - \lambda_j| \le \|E\|_2.$$

This says that the eigenvalues of a normal matrix are very well behaved: A tiny perturbation in A causes an equally tiny perturbation in the eigenvalues.

A continuous function f is said to be *Lipschitz continuous* at x if there is a constant κ (called a *Lipschitz constant*) such that $|f(x + h) - f(x)| \le \kappa |h|$ for all h. The Bauer–Fike theorem tells us that the eigenvalues of a semisimple matrix are Lipschitz continuous functions of the entries of the matrix, where the Lipschitz constant is $\kappa(S)$. We can now contrast this with the situation of the Jordan block of Example 2.7.2. There we saw that a perturbation of norm ϵ can cause the eigenvalues to move by $\epsilon^{1/3}$. In that example there is no constant κ such that the perturbation satisfies $|\lambda - \mu| \le \kappa\epsilon$ for all $\epsilon > 0$, because the function $f(x) = x^{1/3}$ is not Lipschitz continuous at zero (Exercise 2.7.5). For an $n \times n$ Jordan block, the relevant function is $f(x) = x^{1/n}$ (Exercise 2.7.1), which is not Lipschitz continuous at zero if $n > 1$. Graphically, the function has a vertical tangent at zero. It is therefore reasonable to assign a condition number of infinity to the eigenvalues of a Jordan block of size 2×2 or larger. In this sense the eigenvalues of a Jordan block are more sensitive than those of a semisimple matrix.

Condition Number of a Simple Eigenvalue

The Bauer–Fike theorem gives an upper bound on the sensitivities of the all of the eigenvalues, but this often turns out to be overly pessimistic. We now turn our attention to individual condition numbers for eigenvalues of algebraic multiplicity 1, which we call *simple*.

Let λ be a simple eigenvalue of A. By continuity of eigenvalues, if $\|E\|$ is small enough, $A + E$ will have a single eigenvalue $\lambda + \delta$ near λ. Our objective is to determine how large $|\delta|$ is relative to $\|E\|$. Let v be an eigenvector of A associated with λ: $Av = v\lambda$. Then we would expect $A + E$ to have an eigenvector $v + z$ associated with $\lambda + \delta$, with $\|z\|$ small. Indeed this is the case, and we will assume it here, but we will not prove it until later in the section. Let $\epsilon = \|E\|/\|A\|$. Then we also have $|\delta| = O(\epsilon)$ and $\|z\| = O(\epsilon)$. The equation $(A + E)(v + z) = (v + z)(\lambda + \delta)$ implies

$$Av + Ev + Az = v\lambda + z\lambda + v\delta + O(\epsilon^2). \tag{2.7.3}$$

The term $O(\epsilon^2)$ is the sum of all terms obtained by multiplying together two $O(\epsilon)$ quantities and is negligible if ϵ is sufficiently small. Since $Av = v\lambda$, we can cancel two terms from (2.7.3).

Let w^T be a left eigenvector of A associated with λ ($w^T A = \lambda w^T$), scaled so that $w^T v = 1$. This scaling is always possible, as it cannot happen that $w^T v = 0$ (Exercise 2.7.6). Multiplying (2.7.3) on the left by w^T, canceling out two more terms, and using the normalization $w^T v = 1$, we find that

$$\frac{\delta}{\|A\|} = \frac{w^T E v}{\|A\|} + O(\epsilon^2).$$

Thus

$$\frac{|\delta|}{\|A\|} \le \|w\|\,\|v\|\,\epsilon + O(\epsilon^2).$$

This inequality is sharp in the sense that there exists a matrix E such that $w^T E v = \|v\|\,\|w\|\,\|E\|$: Take

$$E = c\overline{w}v^*,$$

where the choice $c = \epsilon/(\|w\|\,\|v\|\,\|A\|)$ gives $\|E\|/\|A\| = \epsilon$.

This result shows that $\|w\|\,\|v\|$ is a condition number for the eigenvalue: If A is perturbed by a small amount, then the amplification factor for the resulting perturbation in λ can be as much as $\|w\|\,\|v\|$ and no more (ignoring the $O(\epsilon^2)$ term). We write

$$\kappa(\lambda) = \|w\|\,\|v\|. \tag{2.7.4}$$

We summarize these findings as a theorem.

Theorem 2.7.7 (condition number of simple eigenvalue). *Let $A \in \mathbb{C}^{n\times n}$, and let λ be a simple eigenvalue of A. Let v and w^T be right and left eigenvectors of A, respectively, normalized so that $w^T v = 1$. Let E be a small perturbation of A satisfying $\|E\|/\|A\| = \epsilon \ll 1$. Then $A + E$ has an eigenvalue $\lambda + \delta$ satisfying*

$$\frac{|\delta|}{\|A\|} \le \kappa(\lambda)\,\epsilon + O(\epsilon^2),$$

where $\kappa(\lambda) = \|w\|\,\|v\|$. This inequality is sharp.

Let v and w be as in the theorem, and let $P = vw^T \in \mathbb{C}^{n\times n}$. Then the equation $w^T v = 1$ implies that P is a projector. In fact, it is the spectral projector for A associated with the simple eigenvalue λ (Exercise 2.7.7). Moreover, $\|P\| = \|v\|\,\|w\|$. This gives us a second characterization of the condition number: $\kappa(\lambda)$ is the norm of the spectral projector associated with λ.

Condition Number of an Invariant Subspace

Now that we have condition numbers for simple eigenvalues, we would also like to get condition numbers for the associated eigenvectors. Each simple eigenvector is just a representative (that is, a basis) of the one-dimensional eigenspace that it spans, so what we are really getting is a condition number for the eigenspace.

Our approach to these results will be to obtain first a more general result, a condition number for invariant subspaces, using the approach of Stewart [194, 196], and Stewart and Sun [199]. We will then get condition numbers for simple eigenspaces as a byproduct.

Let $\mathcal{S}$ be a subspace of $\mathbb{C}^n$ that is invariant under A. Suppose we perturb A slightly to form $\hat{A} = A + E$. Does $\hat{A}$ have an invariant subspace that is close to $\mathcal{S}$? This is only our first question, and the answer is no in general but "usually" yes. In those cases where there is a nearby invariant subspace, how far from $\mathcal{S}$ is it? Can we establish a condition number for $\mathcal{S}$?

Let us set the scene to answer these questions. Suppose that the dimension of $\mathcal{S}$ is k, let $q_1, \ldots, q_k$ be an orthonormal basis for $\mathcal{S}$, and let $q_{k+1}, \ldots, q_n$ be additional orthonormal vectors such that $q_1, \ldots, q_n$ is an orthonormal basis for $\mathbb{C}^n$. Let $Q = [\, q_1 \; q_2 \; \cdots \; q_n \,]$ and $B = Q^{-1}AQ$. Then Q is unitary, and B is unitarily similar to A. Furthermore B is block upper triangular:

$$B = \begin{bmatrix} B_{11} & B_{22} \\ 0 & B_{22} \end{bmatrix},$$

where B_{11} is $k \times k$, by Proposition 2.2.2. Corresponding to the invariant subspace $\mathcal{S}$, B has the invariant subspace $\mathcal{E}_k = Q^{-1}\mathcal{S} = \mathrm{span}\{e_1, \ldots, e_k\}$, where $e_1, \ldots, e_k$ are standard basis vectors in $\mathbb{C}^n$. If we let $E_k = [\, e_1 \; \cdots \; e_k \,]$, then $\mathcal{E}_k = \mathcal{R}(E_k)$. E_k can be written in the block form $E_k = \begin{bmatrix} I \\ 0 \end{bmatrix}$, where I denotes the $k \times k$ identity matrix, and the invariance of $\mathcal{E}_k$ is expressed by the equation (cf. Proposition 2.1.11)

$$\begin{bmatrix} B_{11} & B_{12} \\ 0 & B_{22} \end{bmatrix} \begin{bmatrix} I \\ 0 \end{bmatrix} = \begin{bmatrix} I \\ 0 \end{bmatrix} B_{11}.$$

It is convenient to work with $\mathcal{E}_k$ instead of $\mathcal{S}$. Since the unitary similarity transformation preserves lengths and angles, any information we obtain about perturbations of $\mathcal{E}_k$ translates directly into information about $\mathcal{S}$. We will therefore study $\mathcal{E}_k$. In the interest of nonproliferation of notation, let us assume from the outset that

$$A = \begin{bmatrix} A_{11} & A_{12} \\ 0 & A_{22} \end{bmatrix} \tag{2.7.5}$$

and that the invariant subspace of interest is $\mathcal{E}_k$.

Our first question is this: If $\hat{A} = A + E$, where $\| E \| / \| A \|$ is small, does $\hat{A}$ have an invariant subspace that is close to $\mathcal{E}_k$? The following example shows that the answer is no in general.

Example 2.7.8. Let $A = \begin{bmatrix} 1 & 0 \\ 0 & 1 \end{bmatrix}$, which has the invariant subspace $\mathcal{E}_1 = \mathrm{span}\{e_1\}$. Now let

$$\hat{A} = \begin{bmatrix} 1 & 0 \\ \epsilon & 1 \end{bmatrix},$$

where ϵ is positive but as small as you please. Then $\hat{A}$ is extremely close to A but has no invariant subspace that is close to $\mathcal{E}_1$. One easily checks that the only one-dimensional subspace that is invariant under $\hat{A}$ is $\mathrm{span}\{e_2\}$, which is orthogonal to $\mathcal{E}_1$. Notice that $\hat{A}$ is a defective matrix; it is essentially a Jordan block.

Are there simple conditions on $\hat{A}$ that guarantee that it has an invariant subspace near $\mathcal{E}_k$? Yes! As we shall see, a sufficient condition is that A_{11} and A_{22} have no eigenvalues in common, where A has the form (2.7.5). Since the eigenvalues of A_{11} are exactly the eigenvalues of A associated with the invariant subspace $\mathcal{E}_k$, we can restate the condition as follows: The eigenvalues associated with the invariant subspace are disjoint from the other eigenvalues of A. Notice that this condition is violated in the matrix in Example 2.7.8.

As a first step toward this result, let us suppose that the perturbation E has the special form

$$E = \begin{bmatrix} 0 & 0 \\ A_{21} & 0 \end{bmatrix},$$

so that

$$\hat{A} = \begin{bmatrix} A_{11} & A_{12} \\ A_{21} & A_{22} \end{bmatrix}.$$

If $\hat{A}$ has an invariant subspace $\hat{\mathcal{E}}_k$ near $\mathcal{E}_k$, then that subspace must have a basis $\hat{e}_1, \ldots, \hat{e}_k$ that is close to $e_1, \ldots, e_k$. If we build an $n \times k$ matrix whose columns are $\hat{e}_1, \ldots, \hat{e}_k$, then that matrix, which we write in partitioned form as $\begin{bmatrix} \hat{X}_1 \\ \hat{X}_2 \end{bmatrix}$, will be close to $\begin{bmatrix} I \\ 0 \end{bmatrix}$. Since $\hat{X}_1$ is close to I, it must be nonsingular. If we let $X = \hat{X}_2 \hat{X}_1^{-1}$, then the columns of

$$\begin{bmatrix} I \\ X \end{bmatrix} = \begin{bmatrix} \hat{X}_1 \\ \hat{X}_2 \end{bmatrix} \hat{X}_1^{-1}$$

are also a basis of $\hat{\mathcal{E}}_k$. This shows that it suffices to look for a matrix of the form

$$\begin{bmatrix} I \\ X \end{bmatrix} \tag{2.7.6}$$

whose columns span an invariant subspace of $\hat{A}$. If we can find such a matrix, and if $\| X \|$ is small, then we will have found an invariant subspace for $\hat{A}$ that is close to the invariant subspace of A. Recall from Proposition 2.6.16 that $\| X \|_2$ is the tangent of the largest angle between the two subspaces.

By Proposition 2.1.11, a matrix of the form (2.7.6) represents a subspace that is invariant under $\hat{A}$ if and only if there is a $B \in \mathbb{C}^{k \times k}$ such that

$$\begin{bmatrix} A_{11} & A_{12} \\ A_{21} & A_{22} \end{bmatrix} \begin{bmatrix} I \\ X \end{bmatrix} = \begin{bmatrix} I \\ X \end{bmatrix} B.$$

Writing the two parts of this block equation separately and eliminating B, we easily obtain the following theorem.

Theorem 2.7.9. *The column space of* $\begin{bmatrix} I \\ X \end{bmatrix}$ *is invariant under*

$$\hat{A} = \begin{bmatrix} A_{11} & A_{12} \\ A_{21} & A_{22} \end{bmatrix}$$

if and only if

$$X A_{11} - A_{22} X = A_{21} - X A_{12} X. \tag{2.7.7}$$

Equation (2.7.7) is called an *algebraic Riccati equation*. It is identical to the Sylvester equation (2.4.4), except that it has an additional quadratic term $X A_{12} X$. When $A_{12} = 0$, the Riccati equation reduces to the Sylvester equation.

The left-hand side of (2.7.7) is a linear transformation of X; let us call it φ. Thus $\varphi : \mathbb{C}^{(n-k) \times k} \to \mathbb{C}^{(n-k) \times k}$ is defined by

$$\varphi(X) = X A_{11} - A_{22} X. \tag{2.7.8}$$

The content of Theorem 2.4.2 can be rephrased as follows: φ is nonsingular if and only if A_{11} and A_{22} have no eigenvalues in common.

From this point on, let us assume that A_{11} and A_{22} have no common eigenvalues. Then φ^{-1} exists, and we can rewrite (2.7.7) as

$$X = \varphi^{-1}\left(A_{21} - XA_{12}X\right) \tag{2.7.9}$$

or

$$X = \psi(X),$$

where ψ is the nonlinear map given by

$$\psi(X) = \varphi^{-1}\left(A_{21} - XA_{12}X\right). \tag{2.7.10}$$

Before we prove anything, let us pause to map out our plan. It may happen that $\|\varphi^{-1}\|$ or $\|A_{12}\|$ is large, but whether they be large or small, they are fixed constants. Recall that A_{21} is the small perturbation that changes A to $\hat{A}$. If $\|A_{21}\|$ is sufficiently small, ψ should map small X to small $\psi(X)$. This expectation rests partly on the observation that if $\|X\|$ is small, say $\|X\| = \delta$, then the quadratic term $XA_{12}X$ will satisfy $\|XA_{12}X\| = O(\delta^2)$ and will therefore be negligible. Then, if $\|A_{21}\|$ is small enough, $\|\psi(X)\|$ will be small. Furthermore, we expect ψ to be a contraction mapping, at least for small values of X, since a small change in X will make a negligible change in $XA_{12}X$. We will establish below that ψ is indeed a contraction, mapping small X to small $\psi(X)$. Then the contraction-mapping theorem guarantees that the equation $X = \psi(X)$ has a unique small-norm solution, which is then a solution of (2.7.9) and (2.7.7). The existence of a small-norm solution to (2.7.7) implies that $\hat{A}$ has an invariant subspace $\hat{\mathcal{E}}_k$ that is near $\mathcal{E}_k$.

Let us now recall our original question. How does a perturbation in A perturb the invariant subspaces? Let $\epsilon = \|A_{21}\|_F / \|A\|_F$, the relative magnitude of the perturbation. (We use Frobenius norms for reasons that will be explained shortly.) The subspace $\hat{\mathcal{E}}_k$ is the perturbed version of $\mathcal{E}_k$, so our question now is how does the distance from $\mathcal{E}_k$ to $\hat{\mathcal{E}}_k$ depend on ϵ. If there is a constant κ such that the distance is no greater than $\kappa\,\epsilon + O(\epsilon^2)$, then the smallest such κ is a condition number for the invariant subspace $\mathcal{E}_k$.

It is not hard to guess what the condition number might be. Given that there is a small solution X, the quadratic term $XA_{12}X$ in (2.7.9) will be negligible, as we have already remarked. Thus $X \approx \varphi^{-1}(A_{21})$, and

$$\|X\|_F \le \|\varphi^{-1}\|\,\|A_{21}\|_F + O(\epsilon^2) = \|\varphi^{-1}\|\,\|A\|_F\,\epsilon + O(\epsilon^2).$$

This inequality is sharp, as there exist perturbations A_{21} for which $\|\varphi^{-1}(A_{21})\|_F = \|\varphi^{-1}\|\,\|A_{21}\|_F$. This shows that

$$\|\varphi^{-1}\|\,\|A\|_F \tag{2.7.11}$$

is a condition number for the invariant subspace $\mathcal{E}_k$.

Now that we have our plan, we just need to fill in some details. We begin with a word about norms. The symbol $X \in \mathbb{C}^{(n-k)\times k}$ appearing in (2.7.8) and throughout this section functions as a vector that is acted on by the linear transformation φ. Therefore, when we take the norm of X, it is convenient to use the Euclidean vector norm, which is the same as the Frobenius matrix norm:

$$\|X\|_F = \sqrt{\sum_{i=1}^{n-k}\sum_{j=1}^{k}|x_{ij}|^2}.$$

If we use this norm on X, we have

$$\| \varphi \| = \max_{X \neq 0} \frac{\| \varphi(X) \|_F}{\| X \|_F} = \max_{X \neq 0} \frac{\| X A_{11} - A_{22} X \|_F}{\| X \|_F}$$

and

$$\| \varphi^{-1} \| = \max_{X \neq 0} \frac{\| X \|_F}{\| \varphi(X) \|_F} = \max_{X \neq 0} \frac{\| X \|_F}{\| X A_{11} - A_{22} X \|_F}.$$

The nonnegative number

$$r = 4 \| \varphi^{-1} \|^2 \| A_{21} \|_F \| A_{12} \|_F \tag{2.7.12}$$

is a key quantity in the following lemma and subsequent results. If $r < 1$, we can define

$$s = \frac{2 \| \varphi^{-1} \| \, \| A_{21} \|_F}{1 + \sqrt{1 - r}} \tag{2.7.13}$$

and

$$\mathcal{B} = \{ X \in \mathbb{C}^{(n-k) \times k} \mid \| X \|_F \leq s \}. \tag{2.7.14}$$

Lemma 2.7.10. *Let ψ be the nonlinear map defined by (2.7.10). If $r < 1$, then ψ maps $\mathcal{B}$ into $\mathcal{B}$.*

Proof: We have to show that if $\| X \|_F \leq s$, then $\| \psi(X) \|_F \leq s$. From (2.7.9) we see immediately that

$$\| \psi(X) \|_F \leq \| \varphi^{-1} \| \left(\| A_{21} \|_F + \| A_{12} \|_F \| X \|_F^2 \right).$$

Letting $a = \| \varphi^{-1} \| \, \| A_{12} \|_F$ and $c = \| \varphi^{-1} \| \, \| A_{21} \|_F$, we can rewrite this inequality as

$$\| \psi(X) \|_F \leq c + a \| X \|_F^2. \tag{2.7.15}$$

The related quadratic equation

$$z = c + az^2$$

has two solutions

$$z = \frac{1 \pm \sqrt{1 - 4ac}}{2a} = \frac{2c}{1 \mp \sqrt{1 - 4ac}}. \tag{2.7.16}$$

Since $4ac = r < 1$, both solutions are real and positive. The smaller of the two solutions is easily seen to be s. Thus

$$s = c + as^2.$$

Subtracting this equation from the inequality (2.7.15), we obtain

$$\| \psi(X) \|_F - s \leq a \left(\| X \|_F^2 - s^2 \right).$$

Since $\| X \|_F^2 \leq s^2$, we conclude that $\| \psi(X) \|_F \leq s$. $\square$

Theorem 2.7.11. *Let r and s be given by (2.7.12) and (2.7.13), respectively, and suppose $r < 1$. Then (2.7.7) has exactly one solution that satisfies $\| X \|_F \leq s$.*

Proof: Lemma 2.7.10 shows that ψ maps $\mathcal{B}$ into $\mathcal{B}$. From the definitions of r, s, ψ, and $\mathcal{B}$, one easily shows (Exercise 2.7.11) that for all Y, $Z \in \mathcal{B}$,

$$\| \psi(Y) - \psi(Z) \|_F \leq r \| Y - Z \|_F.$$

Since $r < 1$, this shows that ψ is a contraction mapping on $\mathcal{B}$. Therefore, by the contraction mapping theorem (Exercise 2.7.12), ψ has a unique fixed point in $\mathcal{B}$. That is, there is a unique $X \in \mathcal{B}$ such that $X = \psi(X)$. Since the equation $X = \psi(X)$ is equivalent to (2.7.7), we conclude that (2.7.7) has a unique solution satisfying $\| X \|_F \leq s$. $\square$

We now restate Theorem 2.7.11 as a perturbation theorem for invariant subspaces. Recall that the matrix

$$A = \left[\begin{array}{cc} A_{11} & A_{12} \\ 0 & A_{22} \end{array} \right]$$

has $\mathcal{E}_k = \mathrm{span}\{e_1, \ldots, e_k\}$ as an invariant subspace, and we are interested in investigating how much this subspace is perturbed under a small perturbation in A. Specifically, if

$$\hat{A} = \left[\begin{array}{cc} A_{11} & A_{12} \\ A_{21} & A_{22} \end{array} \right],$$

where $\| A_{21} \|_F / \| A \|_F$ is small, what is the distance from $\mathcal{E}_k$ to the nearest k-dimensional subspace that is invariant under $\hat{A}$? Recall that we get no useful result unless the submatrices A_{11} and A_{22} have no eigenvalues in common. Stewart [194, 196] defines the *separation* between A_{11} and A_{22} by

$$\mathrm{sep}(A_{11}, A_{22}) = \min_{X \neq 0} \frac{\| X A_{11} - A_{22} X \|_F}{\| X \|_F}.$$

Here $\mathrm{sep}(A_{11}, A_{22})$ is nonzero if and only if A_{11} and A_{22} have no eigenvalues in common. In fact, it is the reciprocal of $\| \varphi^{-1} \|$, which figured prominently above. Define

$$\kappa(\mathcal{E}_k) = \| \varphi^{-1} \| \, \| A \|_F = \frac{\| A \|_F}{\mathrm{sep}(A_{11}, A_{22})}. \tag{2.7.17}$$

This is the condition number that we identified earlier (2.7.11).

Corollary 2.7.12. *Let*

$$A = \left[\begin{array}{cc} A_{11} & A_{12} \\ 0 & A_{22} \end{array} \right] \quad and \quad \hat{A} = \left[\begin{array}{cc} A_{11} & A_{12} \\ A_{21} & A_{22} \end{array} \right],$$

and let $\epsilon = \| A_{21} \|_F / \| A \|_F$. Assume that A_{11} and A_{22} have no eigenvalues in common. If $\epsilon < 1 / \left[4 \kappa(\mathcal{E}_k)^2 (\| A_{12} \|_F / \| A \|_F) \right]$, then there is a unique subspace $\hat{\mathcal{E}}_k$ that is both invariant under $\hat{A}$ and near $\mathcal{E}_k$. Moreover,

$$\mathrm{d}(\hat{\mathcal{E}}_k, \mathcal{E}_k) < 2 \kappa(\mathcal{E}_k) \epsilon.$$

Proof: The condition $\epsilon < 1/\left[4\kappa(\mathcal{E}_k)^2(\|A_{12}\|_F/\|A\|)\right]$ just means that $r < 1$. Therefore Theorem 2.7.11 guarantees that there is a unique X that satisfies the algebraic Riccati equation (2.7.7) and has $\|X\|_F \leq s$. Let $\hat{\mathcal{E}}_k$ denote the column space of $\begin{bmatrix} I \\ X \end{bmatrix}$. This space is invariant under $\hat{A}$, and it is the only invariant subspace that is near $\mathcal{E}_k$, in the sense that $\|X\|_F \leq s$. We know from Proposition 2.6.16 that $\|X\|_2 = \tan\theta_k$, where θ_k is the largest principal angle between $\mathcal{E}_k$ and $\hat{\mathcal{E}}_k$, and furthermore $\mathrm{d}(\hat{\mathcal{E}}_k, \mathcal{E}_k) = \sin\theta_k$. Thus $\mathrm{d}(\hat{\mathcal{E}}_k, \mathcal{E}_k) = \sin\theta_k \leq \tan\theta_k = \|X\|_2 \leq \|X\|_F \leq s < 2\|\varphi^{-1}\|\,\|A_{21}\|_F = 2\kappa(\mathcal{E}_k)\epsilon.$ $\square$

Because of the restriction on ϵ in Corollary 2.7.12, we get a bound on the perturbation of the invariant subspace only if the perturbation A_{21} is sufficiently small. How small ϵ needs to be depends upon A_{12} and $\kappa(\mathcal{E}_k)$. In the best case, $A_{12} = 0$, there is no restriction on ϵ. Even if $A_{12} \neq 0$, fairly large values of ϵ are allowed if $\kappa(\mathcal{E}_k)^2$ is not too large. On the other hand, if A_{12} is neither zero nor close to zero, and $\kappa(\mathcal{E}_k)^2$ is large, the bound holds only for very small values of ϵ. Thus we hope that $\kappa(\mathcal{E}_k)$ is not large. Since $\kappa(\mathcal{E}_k)$ is also the pivotal quantity in the bound $\mathrm{d}(\hat{\mathcal{E}}_k, \mathcal{E}_k) < 2\kappa(\mathcal{E}_k)\epsilon$, we now have two reasons for wanting $\kappa(\mathcal{E}_k)$ to be small.

So far we have considered only a special type of perturbation. Now consider a more general situation: $\hat{A} = A + E$, where E has small norm but is perfectly general. Thus

$$\hat{A} = \begin{bmatrix} \hat{A}_{11} & \hat{A}_{12} \\ \hat{A}_{21} & \hat{A}_{22} \end{bmatrix} = \begin{bmatrix} A_{11} + E_{11} & A_{12} + E_{12} \\ E_{21} & A_{22} + E_{22} \end{bmatrix}.$$

We continue to assume, for convenience, that A is block triangular. Thus $\mathcal{E}_k$ is invariant under A. We want to find a nearby subspace $\hat{\mathcal{E}}_k$ that is invariant under $\hat{A}$ and show (hopefully) that $\hat{\mathcal{E}}_k$ is not too far from $\mathcal{E}_k$.

Our approach is simple. The matrix

$$\tilde{A} = \begin{bmatrix} \hat{A}_{11} & \hat{A}_{12} \\ 0 & \hat{A}_{22} \end{bmatrix}$$

also has $\mathcal{E}_k$ as an invariant subspace, so we can get a result for our more general $\hat{A}$ by applying Theorem 2.7.11 and Corollary 2.7.12 to $\tilde{A}$ instead of A. We just have to take into account that we are now working with the perturbed quantities $\hat{A}_{11} = A_{11} + E_{11}$, and so on.

Theorem 2.7.13. *Let*

$$A = \begin{bmatrix} A_{11} & A_{12} \\ 0 & A_{22} \end{bmatrix},$$

where A_{11} and A_{22} have no eigenvalues in common. Let $\hat{A} = A + E$, and let $\epsilon = \|E\|_F/\|A\| > 0$. If ϵ is sufficiently small, then there is a unique subspace $\hat{\mathcal{E}}_k$ that is both invariant under $\hat{A}$ and near $\mathcal{E}_k$. Specifically,

$$\mathrm{d}(\hat{\mathcal{E}}_k, \mathcal{E}_k) < 4\kappa(\mathcal{E}_k)\epsilon.$$

We could have made a precise statement about how small ϵ needs to be, but that would just have been a distraction. As the proof of the theorem will show, the bound on ϵ is of the same order of magnitude as that given in Corollary 2.7.12.

Proof: We wish to apply Theorem 2.7.11 to $\tilde{A}$. We have

$$r = 4 \, \| \hat{\varphi}^{-1} \| \, \| \hat{A}_{21} \|_F \, \| \hat{A}_{12} \|_F,$$

where $\hat{\varphi}$ is the linear operator given by

$$\hat{\varphi}(X) = X \hat{A}_{11} - \hat{A}_{22} X.$$

Let us check that we can guarantee that $r < 1$ by taking ϵ sufficiently small. The main problem is to get a bound on $\| \hat{\varphi}^{-1} \|$. Recall that φ (without the hat) is represented by the matrix $M = A_{11}^T \otimes I - I \otimes A_{22}$. Similarly, $\hat{\varphi}$ is represented by $\hat{M} = \hat{A}_{11}^T \otimes I - I \otimes \hat{A}_{22}$. Clearly $\hat{M} = M + F$, where $F = E_{11}^T \otimes I - I \otimes E_{22}$, and $\| F \| \leq 2 \| E \|$. By Theorem 1.2.6, if $\| F \| \leq 1/(2 \| M \|)$, we get $\| \hat{M}^{-1} \| \leq 2 \| M^{-1} \|$, and hence $\| \hat{\varphi}^{-1} \| \leq 2 \| \varphi^{-1} \|$. We can ensure this by taking $\epsilon \leq 1/(4 \kappa(\mathcal{E}_k))$. This takes care of the bound on $\| \hat{\varphi}^{-1} \|$. Now looking at the other factors in r, we see that $\| \hat{A}_{21} \|_F = \| E_{21} \|_F \leq \epsilon \| A \|_F$ and $\| \hat{A}_{12} \|_F \leq \| A_{12} \|_F + \| E_{12} \|_F \leq \| A_{12} \|_F + \epsilon \| A \|_F$, so it is clear that we can make $r < 1$ by taking ϵ sufficiently small. Under these conditions the conclusions of Theorem 2.7.11 and Corollary 2.7.12 hold. There is a unique $\hat{\mathcal{E}}_k$ that is invariant under $\hat{A}$ and close to $\mathcal{E}_k$. Arguing as in the proof of Corollary 2.7.12, we have $d(\hat{\mathcal{E}}_k, \mathcal{E}_k) \leq s < 2 \| \hat{\varphi}^{-1} \| \, \| A_{21} \|_F \leq 4 \| \varphi^{-1} \| \, \| A_{21} \|_F = 4 \kappa(\mathcal{E}_k) \epsilon$. We had to admit an extra factor of 2, because φ^{-1} was replaced by $\hat{\varphi}^{-1}$ in the definition of s. $\quad \square$

Theorem 2.7.13 shows that $\kappa(\mathcal{E}_k)$ serves as a condition number for $\mathcal{E}_k$ under arbitrary perturbations in A, not just the special perturbations of the type considered in Theorem 2.7.11 and Corollary 2.7.12.

Since $\kappa(\mathcal{E}_k)$ is large when $\text{sep}(A_{11}, A_{22})$ is small, we get the best results when A_{11} and A_{22} are well separated, in the sense that $\text{sep}(A_{11}, A_{22})$ is large. Computing $\text{sep}(A_{11}, A_{22})$ is an expensive proposition: Since φ acts on $\mathbb{C}^{(n-k) \times k}$, the task of computing $\| \varphi^{-1} \|$ is essentially that of determining the singular values of an $m \times m$ matrix, where $m = k(n-k)$. Fortunately, the exact value of $\text{sep}(A_{11}, A_{22})$ is not needed; a good estimate will suffice. LAPACK [8] includes an excellent and relatively economical sep estimator. See also [16].

There is a loose relationship between sep and the gap between the eigenvalues of A_{11} and A_{22}.

Proposition 2.7.14.

$$\text{sep}(A_{11}, A_{22}) \leq \min\{| \lambda - \mu | \mid \lambda \in \lambda(A_{11}), \ \mu \in \lambda(A_{22})\}.$$

The proof is worked out in Exercise 2.7.13. Proposition 2.7.14 shows that if the spectra of A_{11} and A_{22} are not well separated from each other, then $\text{sep}(A_{11}, A_{22})$ is small and the invariant subspace $\mathcal{E}_k$ is ill conditioned. The converse is unfortunately false: It can happen that $\text{sep}(A_{11}, A_{22})$ is small even though the spectra of A_{11} and A_{22} are well separated. This can happen when A is far from normal; for normal matrices, as we shall see below, Proposition 2.7.14 holds with equality, so $\mathcal{E}_k$ is well conditioned if and only if the spectra of A_{11} and A_{22} are well separated.

We have established a condition number for the special invariant subspace $\mathcal{E}_k$ of a block triangular matrix. Now, by way of summary and review, we define the condition number of a nearly arbitrary nontrivial invariant subspace S of an arbitrary $A \in \mathbb{C}^{n \times n}$. We assume only that the eigenvalues of A associated with S are disjoint from the other eigenvalues of A. Let $q_1, \ldots, q_k$ be an orthonormal basis of S, and let $q_{k+1}, \ldots, q_n$ be additional orthonormal vectors such that $q_1, \ldots, q_n$ together form an orthonormal basis of $\mathbb{C}^n$. Let Q be the unitary matrix whose columns are $q_1, \ldots, q_n$, and let $B = Q^{-1}AQ$. Then B is block upper triangular:

$$B = \begin{bmatrix} B_{11} & B_{12} \\ 0 & B_{22} \end{bmatrix}.$$

The eigenvalues of B_{11} are the eigenvalues of A associated with S. Define

$$\kappa(S) = \| B \|_F / \text{sep}(B_{11}, B_{22}). \tag{2.7.18}$$

This is the condition number of S. This is finite, since B_{11} and B_{22} have no common eigenvalues. It is easy to check that $\kappa(S)$ really is a function of S only and does not depend upon the choice of basis $q_1, \ldots, q_n$ (Exercise 2.7.14). The following theorem is an immediate consequence of Theorem 2.7.13.

Theorem 2.7.15. *Let $A \in \mathbb{C}^{n \times n}$, and let S be a nontrivial invariant subspace of A such that the eigenvalues associated with S are disjoint from the other eigenvalues of A. Let $\hat{A} = A + E$, and let $\epsilon = \| E \|_F / \| A \|_F > 0$. If ϵ is sufficiently small, then there is a unique subspace $\hat{S}$ that is both invariant under $\hat{A}$ and near S. Specifically,*

$$d(\hat{S}, S) < 4\kappa(S)\epsilon.$$

Condition Number of a Simple Eigenvector

An eigenvalue is called *simple* if its algebraic multiplicity is 1. Then its geometric multiplicity is also 1, since the latter can never exceed the former (Exercise 2.4.18). This means that the eigenspace S associated with λ is one-dimensional. This is an invariant subspace whose only associated eigenvalue is λ, which is distinct from the other eigenvalues of A. Thus Theorem 2.7.15 pertains to this space. The condition number $\kappa(S)$ can also be thought of as a condition number for the eigenvectors that belong to S.

What does $\kappa(S)$ look like in this case? Assuming, without loss of generality, that A is block triangular, we have

$$A = \begin{bmatrix} \lambda & A_{12} \\ 0 & A_{22} \end{bmatrix},$$

where A_{11} is now the 1×1 matrix $[\lambda]$. The (essentially) unique eigenvector is the standard unit vector e_1. If we perturb A slightly, the perturbed matrix has an eigenvector $\begin{bmatrix} 1 \\ x \end{bmatrix}$, where $\| x \|$ is small. As before, we have $\kappa(S) = \| \varphi^{-1} \| \| A \|_F$, where φ is given by $\varphi(X) = X A_{11} - A_{22} X$. In the present context we have $A_{11} = [\lambda]$, and the matrix X has become the vector x, so we have $\varphi(x) = x\lambda - A_{22}x = (\lambda I - A_{22})x$. Thus, in this simple case, $\varphi = (\lambda I - A_{22})$ and $\| \varphi^{-1} \| = \| (\lambda I - A_{22})^{-1} \|$. Thus the condition number for the eigenvector(s) associated with λ is simply

$$\| (\lambda I - A_{22})^{-1} \| \| A \|_F.$$

The Normal Case

The perturbation theory is simpler and much more satisfactory when A is normal. This is partly due to the fact that any normal matrix that is block triangular must be block diagonal (Proposition 2.3.3). The following result is a version of Corollary 2.7.12 appropriate for normal matrices.

Proposition 2.7.16. *Let*

$$A = \begin{bmatrix} A_{11} & 0 \\ 0 & A_{22} \end{bmatrix}$$

be a normal matrix such that A_{11} and A_{22} have no eigenvalues in common; let $\hat{A} = A + E$, where

$$E = \begin{bmatrix} 0 & E_{12} \\ E_{21} & 0 \end{bmatrix};$$

and let $\epsilon = \|E\|_F / \|A\|_F$. If $\epsilon < 1/[2\,\kappa(\mathcal{E}_k)]$, then there is a unique subspace $\hat{\mathcal{E}}_k$ that is both invariant under $\hat{A}$ and near $\mathcal{E}_k$. Moreover,

$$d(\hat{\mathcal{E}}_k, \mathcal{E}_k) < 2\,\kappa(\mathcal{E}_k)\,\epsilon.$$

The major difference between this proposition and Corollary 2.7.12 is that the restriction on ϵ is much less stringent here, as it depends on $1/\kappa(\mathcal{E}_k)$ instead of $1/\kappa(\mathcal{E}_k)^2$. This is so because $\|\hat{A}_{21}\|_F / \|A\|_F$ and $\|\hat{A}_{12}\|_F / \|A\|_F$ are both small in the normal case. This carries over to Theorems 2.7.13 and 2.7.15 as well: In the normal case, the restriction on ϵ is much less stringent.

The most important benefit of normality is that we have equality in Proposition 2.7.14.

Proposition 2.7.17. *Let A_{11} and A_{22} be normal. Then*

$$\mathrm{sep}(A_{11}, A_{22}) = \min\{|\lambda - \mu| \mid \lambda \in \lambda(A_{11}), \ \mu \in \lambda(A_{22})\}.$$

The proof is worked out in Exercise 2.7.17. The following corollary is immediate.

Corollary 2.7.18. *Let $A \in \mathbb{C}^{n \times n}$ be normal, and let S be a subspace $\mathbb{C}^n$ that is invariant under A. Let Λ_1 denote the set of eigenvalues of A associated with S, let Λ_2 denote the other eigenvalues of A, and assume that Λ_1 is disjoint from Λ_2. Then the condition number of S is given by*

$$\kappa(S) = \frac{\|A\|_F}{\min\{\,|\lambda_1 - \lambda_2| \mid \lambda_1 \in \Lambda_1, \ \lambda_2 \in \Lambda_2\,\}}.$$

In other words, the sensitivity of an invariant subspace of a normal matrix depends entirely upon how well the eigenvalues associated with that subspace are separated from the other eigenvalues of the matrix. Good separation implies low sensitivity, and poor separation implies high sensitivity.

For the special case of a subspace of dimension 1, we have the following result.

Corollary 2.7.19. *Let $A \in \mathbb{C}^{n \times n}$ be normal, and let μ be a simple eigenvalue of A with associated eigenvector v. Then the condition number of v (i.e., the condition number of the one-dimensional invariant subspace S spanned by v) is*

$$\kappa(S) = \frac{\|A\|_F}{\min\{\,|\mu - \lambda| \mid \lambda \neq \mu, \ \lambda \in \lambda(A)\,\}}.$$

Finally, what can we say about the eigenvalues? Well, there is nothing new to say; we just have to recall that we already showed in Corollary 2.7.6 that the eigenvalues of a normal matrix are perfectly conditioned: A small perturbation in A makes an equally small perturbation of the eigenvalues.

A second confirmation of this fact comes from the realization that the left eigenvectors of a normal matrix are essentially the same as the right eigenvectors (Exercise 2.7.18). Specifically, if v is a right eigenvector, then v^* is an associated left eigenvector. Thus we can take $w^T = v^*$ (with $\|v\| = 1$ to ensure $w^T v = 1$) in (2.7.4) to find that $\kappa(\lambda) = 1$ for each simple eigenvalue of a normal matrix. This can also be understood in the language of spectral projectors. As we noted in Section 2.5, all of the spectral projectors of a normal matrix are orthoprojectors. This implies (Theorem 1.6.6) $\kappa(\lambda) = \|P\| = 1$.

We end by pointing out the now obvious fact that the eigenvalue condition number can be much better than the eigenvector condition number. Let λ be a simple eigenvalue of a normal matrix A. Then λ is well conditioned no matter what. However, the associated eigenvector (or, if you prefer, invariant subspace) is ill conditioned if A has an eigenvalue that is extremely close to λ. A specific example is discussed in Exercise 2.7.19.

Residuals, Conditioning, and Backward Stability

Suppose that we have computed an eigenpair (λ, v) of the matrix A with $\|v\| = 1$ by some method. Because of roundoff or other errors, (λ, v) will never be an exact eigenpair in practice, but we can assess its quality by computing the residual,

$$r = Av - v\lambda.$$

If (λ, v) is an exact eigenpair, r will be zero. In practice r will not be zero, but if $\|r\|$ is tiny, we can feel some satisfaction. The following question remains: If $\|r\|$ is tiny, does that guarantee that (λ, v) is close to a true eigenpair of A?

We can answer this question as follows. Since $\|v\| = 1$, we have $v^*v = 1$, so $r = rv^*v$. Thus the equation $r = Av - v\lambda$ can be written as $(A - rv^*)v = v\lambda$ or

$$(A + E)v = v\lambda,$$

where $E = -rv^*$. It is easy to check that $\|E\| = \|r\|$. Thus a tiny residual implies that (λ, v) is an exact eigenpair of a matrix $A + E$ that is very close to A. For future reference we record the following result, which we have just proved.

Proposition 2.7.20. *Let $A \in \mathbb{C}^{n \times n}$, $v \in \mathbb{C}^n$ with $\|v\| = 1$, $\lambda \in \mathbb{C}$, and*

$$Av = v\lambda + r.$$

Then there is a matrix $E \in \mathbb{C}^{n \times n}$ with $\|E\| = \|r\|$ such that

$$(A + E)v = v\lambda.$$

Now the question of whether A has an eigenpair that is close to (λ, v) is simply a question of conditioning. If (λ, v) is a well-conditioned eigenpair of $A + E$, then (λ, v) is close to an eigenpair of $A = (A + E) - E$. In practice the matrix we have in hand might be A, not $A + E$. But if $\|E\|$ is tiny, the difference is slight. We rely on a result of Demmel

[72] that says roughly that the condition number of the condition number is the condition number. If we have a well-conditioned problem and we perturb it slightly, the condition number will change only slightly. Thus the perturbed problem is also well conditioned. In conclusion, a tiny residual and a good condition number together imply an accurate result.

This discussion can be generalized to invariant subspaces. Let $\mathcal{U} \subseteq \mathbb{C}^n$ be a subspace, and let U be a matrix with orthonormal columns such that $\mathcal{U} = \mathcal{R}(U)$. Although $\mathcal{U}$ could be any subspace, let us think of it as an approximation to an invariant subspace of $A \in \mathbb{C}^{n \times n}$. Let $B = U^*AU$ and $R = AU - UB$. If $\mathcal{U}$ is invariant under A, then the residual R will be zero (Proposition 2.1.11). If $\| R \|$ is nonzero but tiny, then $\mathcal{U}$ is invariant under a matrix that is a tiny perturbation of A.

Proposition 2.7.21. *With all terms as defined above, let $E = -RU^*$. Then $\| E \| = \| R \|$, and*

$$(A + E)U = UB.$$

Thus $\mathcal{U}$ is invariant under $A + E$.

The easy proof consists mainly of showing that $\| E \| = \| R \|$, as the rest is obvious (Exercise 2.7.22). We draw the following conclusion. If a (near) invariant subspace has a tiny residual R with $\| R \| = \epsilon$ and the condition number κ as given by (2.7.18) is not too large, then the distance to the nearest true invariant subspace is no more than about $\kappa\epsilon$.

A numerical algorithm is called *backward-stable* if it returns a result that is the true solution of a nearby problem. For example, an algorithm for computing an invariant subspace of A is backward-stable if it returns a $\mathcal{U} = \mathcal{R}(U)$ such that $(A + E)U = UB$ for some E such that $\| E \|$ is tiny. The above discussion shows that an algorithm is backward-stable if and only if it returns an answer that has a tiny residual. The term backward stability is used in a couple of different ways. First, one sometimes can prove that an algorithm is backward-stable (with no modifiers), meaning that it always returns an answer that has a tiny residual. But even if one cannot prove that an algorithm is (unconditionally) backward-stable, and even if it is not (always) backward-stable, one can compute the residual after a given computation and check whether it is tiny. If it is, then we can say that the algorithm performed this particular computation in a backward-stable way.

Backward stability alone does not imply accuracy; backward stability and a good condition number together imply accuracy.

Exercises

2.7.1. Consider the situation of Example 2.7.2, except that the matrices are $n \times n$ instead of 3×3. Thus A is an $n \times n$ Jordan block, and E_ϵ has a lone nonzero entry ϵ in the $(n, 1)$ position. Let $A_\epsilon = A + E_\epsilon$.

    (a) Show that the characteristic equation of A_ϵ is $(\lambda - \mu)^n - \epsilon = 0$; the eigenvalues of A_ϵ are $\lambda = \mu + \epsilon^{1/n}\omega^j$, $j = 0, 1, \dots, n - 1$, where $\omega = e^{2\pi i/n}$; and the eigenvalues satisfy $|\lambda - \mu| = \epsilon^{1/n}$.

    (b) Find the ratio $|\lambda - \mu|/\epsilon$ when $n = 10$ and $\epsilon = 10^{-10}$.

    (c) Find values of n and ϵ for which $|\lambda - \mu|/\epsilon > 10^{20}$.

2.7.2. Let E_ϵ be as in Example 2.7.2, and let

$$A = \begin{bmatrix} \mu & \alpha & 0 \\ 0 & \mu & \alpha \\ 0 & 0 & \mu \end{bmatrix},$$

where $\alpha > 0$. Let $A_\epsilon = A + E_\epsilon$.

(a) Find the eigenvalues of A_ϵ as a function of ϵ and α.

(b) Suppose $\epsilon = 10^{-6}$. Find $|\lambda - \mu|$ in three cases: (i) $\alpha = 1$, (ii) $\alpha = 10^6$, and (iii) $\alpha = 10^{-6}$.

2.7.3. This exercise leads to a proof of the Bauer–Fike theorem.

(a) Let v be an eigenvector of $A + E$ associated with the eigenvalue ν. Let $w = S^{-1}v$ and $F = S^{-1}ES$ (using the notation of Theorem 2.7.5). Show that $(\Lambda + F)w = \nu w$ and $\|F\| \leq \kappa(S)\|E\|$.

(b) Show that $w = (\nu I - \Lambda)^{-1}Fw$ and, taking norms of both sides,

$$\|(\nu I - \Lambda)^{-1}\|_p \|F\|_p \geq 1.$$

(c) Show that $\|(\nu I - \Lambda)^{-1}\|_p = 1/\min_{1 \leq j \leq n}|\nu - \lambda_j|$, where $\lambda_1, \ldots, \lambda_n$ are the eigenvalues of A.

(d) Put the pieces together to prove Theorem 2.7.5.

2.7.4. This exercise proves a variant of the Bauer–Fike theorem that is also due to Bauer and Fike [24]. Let $A \in \mathbb{C}^{n \times n}$ be a semisimple matrix with k distinct eigenvalues $\lambda_1, \ldots, \lambda_k$. By Theorem 2.5.1 A has a spectral decomposition $A = \lambda_1 P_1 + \cdots + \lambda_k P_k$, where $P_1, \ldots, P_k$ are spectral projectors. Let $A + E$ be a perturbation of A with $\|E\|_p = \epsilon$. Consider the k disks $M_j(\epsilon) = \{\lambda \mid |\lambda - \lambda_j| \leq k\|P_j\|\epsilon\}$. The jth disk is centered on λ_j and has radius $k\|P_j\|\epsilon$. We will prove that every eigenvalue of $A + E$ lies in one of the disks $M_j(\epsilon)$.

(a) Show that if ν is an eigenvalue of $A + E$ with eigenvector x, then $x = (\nu I - A)^{-1}Ex$. Deduce that $1 \leq \|(\nu I - A)^{-1}\|_p \|E\|_p = \|(\nu I - A)^{-1}\|_p \epsilon$.

(b) Show that $(\nu I - A)^{-1} = \sum_{j=1}^{k}(\nu - \lambda_j)^{-1}P_j$ (cf. Exercise 2.5.4).

(c) Show that $1 \leq \left(\sum_{j=1}^{k}|\nu - \lambda_j|^{-1}\|P_j\|_p\right)\epsilon$. Then show that there must be some m such that $1 \leq k|\nu - \lambda_m|^{-1}\|P_m\|_p\epsilon$, and deduce that ν must lie in the disk $M_m(\epsilon)$. This proves that every eigenvalue of $A + E$ must lie in one of the disks.

(d) What can we say about the disks $M_j(\epsilon)$ when ϵ is very small? How many eigenvalues must each disk contain? Discuss the relationship between this result and Theorem 2.7.7.

(e) Show that if $A = S\Lambda S^{-1}$, where Λ is diagonal, then $\|P_j\|_p \leq \|S\|_p$ for $j = 1, \ldots, k$. Discuss the relationship of this result to the Bauer–Fike theorem (Theorem 2.7.5).

2.7.5.

(a) Suppose that f satisfies the Lipschitz condition $|f(x+h) - f(x)| \le \kappa |h|$ for all h. Show that if f is differentiable at x, then $|f'(x)| \le \kappa$. Thus the Lipschitz constant gives an upper bound on the slope of f.

(b) Let $f(x) = x^{1/3}$. Sketch the graph of f for $0 \le x \le 1$, paying special attention to what happens near $x = 0$.

(c) Prove that the function $f(x) = x^{1/3}$ is not Lipschitz continuous at 0.

(d) Prove that for every integer $n > 1$ the function $f(x) = x^{1/n}$ is not Lipschitz continuous at 0.

2.7.6. Let λ be a simple eigenvalue of the matrix A. Let $A = VJV^{-1}$ be a Jordan decomposition of A (Theorem 2.4.11). Assume that the Jordan matrix J is arranged so that the 1×1 block $[\lambda]$ is in the upper left-hand corner. Define $W \in \mathbb{C}^{n \times n}$ by $W^T = V^{-1}$. Let v and w denote the first columns of V and W, respectively.

(a) Show that v and w are right and left eigenvectors of A, respectively, associated with the eigenvalue λ. Show that $w^T v = 1$.

(b) Show that every right (resp., left) eigenvector of A associated with λ is a (nonzero) multiple of v (resp., w).

(c) Show that if $\tilde{v}$ and $\tilde{w}$ are right and left eigenvectors of A associated with λ, then $\tilde{w}^T \tilde{v} \ne 0$.

2.7.7. Let $A \in \mathbb{C}^{n \times n}$, suppose that λ is a simple eigenvalue of A, and suppose $Av = v\lambda$ and $w^T A = \lambda w^T$ with $w^T v = 1$. Let $P = vw^T$.

(a) Show that P is the spectral projector for A associated with the eigenvalue λ.

(b) Show that for every $x \in \mathbb{C}^n$, $\|Px\| \le \|v\| \, \|w\| \, \|x\|$. Deduce that $\|P\| \le \|v\| \, \|w\|$.

(c) Show that $Pw = v\|w\|^2$. Deduce that $\|Pw\| = \|v\| \, \|w\| \, \|w\|$ and $\|P\| = \|v\| \, \|w\|$. Thus $\|P\|$ is the condition number of λ.

2.7.8. Define a 6×6 upper triangular matrix A by

$$
A = \begin{bmatrix}
6 & 9 & & & & \\
 & 5 & 9 & & & \\
 & & 4 & 9 & & \\
 & & & 3 & 9 & \\
 & & & & 2 & 9 \\
 & & & & & 1
\end{bmatrix}.
$$

Obviously the eigenvalues of A are $6, 5, 4, 3, 2, 1$.

(a) Use MATLAB's `eig` function to compute a nonsingular S and diagonal Λ such that $A = S\Lambda S^{-1}$. The columns of S are right eigenvectors of A.

(b) Use MATLAB's `inv` function to deduce the left eigenvectors of A. Use the left and right eigenvectors to compute a condition number $\kappa(\lambda) = \|w\| \|v\|$ for each eigenvalue λ.

(c) Use MATLAB's `condeig` function to check your work from part (b).

(d) Use MATLAB's `cond` command to compute $\kappa(S) = \|S\| \|S^{-1}\|$. Recall that this is the upper bound for the condition numbers of the eigenvalues given by (Bauer–Fike) Theorem 2.7.5. Observe that $\kappa(S)$ significantly overestimates all of the individual condition numbers.

(e) Let A_ϵ denote the matrix obtained from A by changing the $(6, 1)$ entry from 0 to ϵ. Use MATLAB's `eig` command to compute the eigenvalues of A_ϵ for $\epsilon = 10^{-6}$. For each eigenvalue λ show that $\kappa(\lambda)\epsilon$ gives a good order-of-magnitude estimate of how far the eigenvalue was moved by the perturbation ϵ.

Note: To get more extreme results, replace the 9's in A by larger numbers or build larger versions of A.

2.7.9. Work out the details of Example 2.7.8.

2.7.10. Prove Theorem 2.7.9.

2.7.11. In this exercise you will show that the map ψ given by (2.7.10) is a contraction mapping on $\mathcal{B}$ if $r < 1$. (The quantities r, s, and $\mathcal{B}$ are defined by (2.7.12), (2.7.13), and (2.7.14), respectively.) Let $Y, Z \in \mathcal{B}$.

(a) Using the definition of ψ (2.7.10), show that

$$\psi(Y) - \psi(Z) = \varphi^{-1}(ZA_{12}(Z - Y) + (Z - Y)A_{12}Y).$$

(b) Show that $\|\psi(Y) - \psi(Z)\|_F \le 2s\|\varphi^{-1}\| \|A_{12}\|_F \|Y - Z\|_F$.

(c) Show that $2s\|\varphi^{-1}\| \|A_{12}\|_F < r$, and hence

$$\|\psi(Y) - \psi(Z)\|_F \le r\|Y - Z\|_F.$$

2.7.12. In this exercise we outline a proof of the contraction mapping theorem. Let $\psi : \mathcal{B} \to \mathcal{B}$, and suppose that there is a constant $r < 1$ such that $\|\psi(Y) - \psi(Z)\| \le r\|Y - Z\|$ for all $Y, Z \in \mathcal{B}$. We will show that there is a unique $X \in \mathcal{B}$ such that $X = \psi(X)$. Define a sequence $(X_m) \subseteq \mathcal{B}$ by $X_0 = 0$ and $X_{m+1} = \psi(X_m)$, $m = 0, 1, 2, \ldots$.

(a) Show that $\|X_{k+1} - X_k\| \le r\|X_k - X_{k-1}\|$ for all k, and consequently

$$\|X_{m+1} - X_m\| \le r^m\|X_1 - X_0\| \le r^m s$$

for all m.

(b) Show that for all j and k,

$$\|X_{k+j} - X_k\| \le \sum_{m=k}^{k+j-1} \|X_{m+1} - X_m\| \le s \sum_{m=k}^{k+j-1} r^m < \frac{sr^k}{1 - r}.$$

(c) Show that $\lim_{k\to\infty} \sup_j \| X_{k+j} - X_k \| = 0$. This implies that (X_m) is a Cauchy sequence in $\mathcal{B}$. Since $\mathcal{B}$ is a closed subset of the complete space $\mathbb{C}^{(n-k)\times k}$, $\mathcal{B}$ is itself complete. Therefore there is an $X \in \mathcal{B}$ such that $\lim_{m\to\infty} X_m = X$.

(d) Show that $X = \psi(X)$. Thus ψ has a fixed point in $\mathcal{B}$.

(e) Show that if $\hat{X} \in \mathcal{B}$ and $\hat{X} = \psi(\hat{X})$, then $\| X - \hat{X} \| \le r \| X - \hat{X} \|$. Conclude that $X = \hat{X}$. Thus the fixed point is unique.

2.7.13.

(a) (Review) Show that the linear transformation φ of (2.7.8) can be represented by the matrix $A_{11}^T \otimes I_{n-k} - I_k \otimes A_{22}$, which has eigenvalues $\nu_{ij} = \lambda_i - \mu_j$, where $\lambda_1, \ldots, \lambda_k$ and $\mu_1, \ldots, \mu_{n-k}$ are the eigenvalues of A_{11} and A_{22}, respectively (cf. Exercises 2.4.5 and 2.4.4).

(b) Show that for any matrix (or linear operator) B, if λ is an eigenvalue of B, then $|\lambda| \le \| B \|$.

(c) Show that $\| \varphi^{-1} \| \ge \max\{ |\lambda_i - \mu_j|^{-1} \mid \lambda_i \in \lambda(A_{11}), \ \mu_j \in \lambda(A_{22}) \}$.

(d) Prove Proposition 2.7.14.

2.7.14. Use the invariance of the spectral and Frobenius norms under unitary transformations to prove that the definition of condition number (2.7.18) is independent of the choice of basis. Specifically, suppose that $q_1, \ldots, q_n$ and $\tilde{q}_1, \ldots, \tilde{q}_n$ are two orthonormal bases of $\mathbb{C}^n$, each of whose first k columns is a basis of $\mathcal{S}$. Let Q and $\tilde{Q}$ be the corresponding unitary matrices, and let $B = Q^* A Q$ and $\tilde{B} = \tilde{Q}^* A \tilde{Q}$.

(a) Show that $\tilde{Q} = QU$, where U is a block diagonal unitary matrix:

$$U = \begin{bmatrix} U_{11} & 0 \\ 0 & U_{22} \end{bmatrix}.$$

Derive equations linking B_{11} to $\tilde{B}_{11}$ and B_{22} to $\tilde{B}_{22}$.

(b) Use unitary invariance of the Frobenius norm to show that that $\operatorname{sep}(B_{11}, B_{22}) = \operatorname{sep}(\tilde{B}_{11}, \tilde{B}_{22})$.

(c) Deduce that the definition (2.7.18) does not depend on the choice of basis.

2.7.15. Prove Theorem 2.7.15.

2.7.16. Use Theorem 2.7.11 to prove Proposition 2.7.16.

2.7.17. Prove Proposition 2.7.17. (Use Proposition 2.3.3, then show that $A_{11}^T \otimes I_{n-k} - I_k \otimes A_{22}$ is normal. Refer to Exercises 2.4.5 and 2.7.13 for the connection between the matrix $A_{11}^T \otimes I_{n-k} - I_k \otimes A_{22}$, the operator $\varphi(X) = X A_{11} - A_{22} X$, and the sep function. Refer to Exercises 2.4.3 and 2.4.4 for properties of the Kronecker product. Finally, apply Corollary 2.3.8 to $A_{11}^T \otimes I_{n-k} - I_k \otimes A_{22}$ to finish the proof.)

2.7.18. Let $A \in \mathbb{C}^{n\times n}$ be a normal matrix, let $\lambda \in \mathbb{C}$, and let $v \in \mathbb{C}^n$.

(a) Show that $\| (\lambda I - A)v \|^2 = \| (\bar{\lambda} I - A^* v) \|^2$. (Use the relationship between the Euclidean norm and the standard inner product, along with the basic identity $\langle Ax, y \rangle = \langle x, A^* y \rangle$.)

(b) Deduce that v is a right eigenvector of A associated with λ if and only if v^* is a left eigenvector associated with λ.

2.7.19. Let $A_t = \operatorname{diag}\{1 + t, 1 - t, 2\}$, where t is a real parameter. We are interested in values of t that are near zero.

(a) Show that A_t is normal for all $t \in \mathbb{R}$.

(b) Find the eigenvalues and associated eigenspaces of A_t when $t \neq 0$.

(c) Compute the condition numbers of the eigenvectors e_1 and e_2, and note that these eigenvectors become increasingly ill conditioned as $t \to 0$. Meanwhile the eigenvalues $1 + t$ and $1 - t$ are perfectly well conditioned.

(d) Let

$$
\hat{A}_t = \left[\begin{array}{ccc} 1 + t & \sqrt{3}t & \\ \sqrt{3}t & 1 - t & \\ & & 2 \end{array} \right].
$$

Compute $\| \hat{A}_t - A_t \|$, and note that $\| \hat{A}_t - A_t \| \to 0$ as $t \to 0$. Compute the three eigenspaces of $\hat{A}_t$, and note that two of them are nowhere near the eigenspaces of A_t.

(e) What eigenvectors does A_0 have that A_t does not have when $t \neq 0$?

(f) Compute the condition number of the two-dimensional invariant subspace $\operatorname{span}\{e_1, e_2\}$, and note that it is well conditioned for all t, even though the eigenvectors within it become ill conditioned as $t \to 0$.

2.7.20. Consider the situation in Theorem 2.7.13. The matrix A is block triangular and has $\mathcal{E}_k$ as an invariant subspace. The theorem gives a bound on the perturbation in $\mathcal{E}_k$ due to a perturbation of A. We might also ask about perturbation bounds for the eigenvalues $\lambda_1, \ldots, \lambda_k$ associated with $\mathcal{E}_k$. We already have the expression (2.7.4) for the condition numbers of individual simple eigenvalues. In this exercise you will derive a perturbation bound for the mean of the eigenvalues $\lambda_1, \ldots, \lambda_k$. It sometimes happens that the mean is much better behaved than the individual eigenvalues are.

We will freely use notation from Theorem 2.7.13 and the results leading up to it. The perturbed matrix $\hat{A} = A + E$ has a unique invariant subspace $\hat{\mathcal{E}}_k$ near A. It is the column space of the matrix of the form $\left[\begin{array}{c} I \\ X \end{array} \right]$ that satisfies

$$
\left[\begin{array}{cc} \hat{A}_{11} & \hat{A}_{12} \\ \hat{A}_{21} & \hat{A}_{22} \end{array} \right] \left[\begin{array}{c} I \\ X \end{array} \right] = \left[\begin{array}{c} I \\ X \end{array} \right] B
$$

for some B. The eigenvalues of B are the eigenvalues of $\hat{A}$ associated with $\hat{\mathcal{E}}_k$. Let μ and $\hat{\mu}$ denote the means of the eigenvalues associated with $\mathcal{E}_k$ and $\hat{\mathcal{E}}_k$, respectively. We want to get an upper bound on $|\hat{\mu} - \mu|$.

(a) Show that $\mu = \frac{1}{k}\operatorname{tr}(A_{11})$ and $\hat{\mu} = \frac{1}{k}\operatorname{tr}(B)$ (cf. Exercise 2.2.12).

(b) Show that $B = \hat{A}_{11} + \hat{A}_{12}X = A_{11} + E_{11} + A_{12}X + E_{12}X$. Deduce that

$$\hat{\mu} - \mu = \frac{1}{k}\left(\operatorname{tr}(E_{11}) + \operatorname{tr}(A_{12}X) + \operatorname{tr}(E_{12}X)\right). \tag{2.7.19}$$

(c) As in Theorem 2.7.13, let $\epsilon = \|E\|_F / \|A\|_F$. Show that $\frac{1}{k}\operatorname{tr}(E_{11}) \le \epsilon\|A\|_F$ and $\frac{1}{k}\operatorname{tr}(E_{12}X) = O(\epsilon^2)$. Thus the interesting term in (2.7.19) is $\frac{1}{k}\operatorname{tr}(A_{12}X)$.

(d) Show that X satisfies the algebraic Riccati equation

$$X\hat{A}_{11} - \hat{A}_{22}X = \hat{A}_{21} - X\hat{A}_{12}X,$$

and deduce that
$$XA_{11} - A_{22}X = E_{21} + O(\epsilon^2). \tag{2.7.20}$$

(Recall that $A_{21} = 0$, so $\hat{A}_{21} = E_{21}$.)

(e) Show that

$$\begin{bmatrix} A_{11} & A_{12} \\ & A_{22} \end{bmatrix}\begin{bmatrix} I & Z \\ & I \end{bmatrix} = \begin{bmatrix} I & Z \\ & I \end{bmatrix}\begin{bmatrix} A_{11} & \\ & A_{22} \end{bmatrix}$$

if and only if Z satisfies the Sylvester equation

$$ZA_{22} - A_{11}Z = A_{12} \tag{2.7.21}$$

(cf. Exercise 2.5.5). The spectral projector for A associated with $\mathcal{E}_k$ is

$$P = \begin{bmatrix} I & -Z \\ 0 & 0 \end{bmatrix}$$

(cf. Exercise 2.5.5).

(f) Recall from part (c) that our object is to estimate $\frac{1}{k}\operatorname{tr}(A_{12}X)$. Substitute (2.7.21) into the expression $A_{12}X$, then use (2.7.20) to obtain the equation

$$A_{12}X = (ZX)A_{11} - A_{11}(ZX) - ZE_{21} + O(\epsilon^2).$$

Deduce that
$$\operatorname{tr}(A_{12}X) = -\operatorname{tr}(ZE_{21}) + O(\epsilon^2).$$

(Do not forget Exercise 2.2.12.)

(g) Recalling properties of the trace from Exercise 2.3.3, show that

$$\frac{|\hat{\mu} - \mu|}{\|A\|} \le \epsilon + \frac{1}{k}\|Z\|_F\,\epsilon + O(\epsilon^2).$$

From part (e), we know that Z is related to the spectral projector P associated with $\mathcal{E}_k$.

(h) We can clean up the result of part (g) by expressing it in terms of $\| P \|_F$. Show that $\| P \|_F = \sqrt{k + \| Z \|_F^2}$. Using the CSB inequality (Theorem 1.2.1), show that $k + \| Z \|_F = \sqrt{k}\sqrt{k+1}\,\| Z \|_F \le \sqrt{k+1}\,\| P \|_F$. Deduce that

$$\frac{|\hat{\mu} - \mu|}{\| A \|} \le \epsilon + \frac{\sqrt{k+1}}{k}\| P \|_F \epsilon + O(\epsilon^2).$$

Thus the eigenvalue mean μ is well conditioned if the norm of the spectral projector P is not large.

2.7.21. Give an example of a matrix whose eigenvalues are all ill conditioned that has a cluster whose mean is well conditioned. (Don't get too elaborate here; there are nice 2×2 examples.)

2.7.22. Prove Proposition 2.7.21.

(a) In particular, show that for any $x \in \mathbb{R}^n$, $\| Ex \| \le \| R \|\| x \|$. Deduce that $\| E \| \le \| R \|$.

(b) There is a $y \in \mathbb{R}^k$ with $\| y \| = 1$ such that $\| Ry \| = \| R \|$. Use this y to build an $x \in \mathbb{R}^k$ with $\| x \| = 1$ and $\| Ex \| = \| r \|$. Deduce that $\| E \| = \| R \|$.

(c) Show that $(A + E)U = UB$.

Chapter 3
Elimination

The topic of this chapter is *how to create zeros in matrices*. We begin with vectors: Given a nonzero $x \in \mathbb{F}^n$, construct a nonsingular $G \in \mathbb{F}^{n \times n}$ so that either Gx or $G^{-1}x$ has as many zeros as possible. Then we move to matrices: Given $A \in \mathbb{F}^{n \times n}$, construct a nonsingular $G \in \mathbb{F}^{n \times n}$ such that $G^{-1}A$ has as many zeros as possible. We will always be able to construct G so that $G^{-1}A = R$ is upper triangular. Then $A = GR$, and we have a GR *decomposition* of A. Finally we consider similarity transformations of matrices: Given $A \in \mathbb{F}^{n \times n}$, construct a nonsingular $G \in \mathbb{F}^{n \times n}$ so that $G^{-1}AG$ has as many zeros as possible.

3.1   Matrices That Create Zeros in Vectors

The fundamental question of this chapter is this: Given a nonzero vector $x \in \mathbb{F}^n$, construct a nonsingular matrix $G \in \mathbb{F}^{n \times n}$ such that $Gx = \alpha e_1$ for some $\alpha \in \mathbb{F}$. Here e_1 denotes, as usual, the standard basis vector with a 1 in the first position and 0's elsewhere. Thus G maps x to a vector with zeros in all but one position. In other words, "G creates many zeros when applied to x." Zero-creating matrices will also be called *elimination matrices*.

This is an easy problem in the sense that it has many solutions. After all, the equation $Gx = \alpha e_1$ is a system of only n equations in $n^2 + 1$ unknowns (α and the entries of G). But we may wish to put additional restrictions on G. The point of finding such a G is to use it in a computational algorithm. Once we have a G, we will wish to use it in further computations, so we would like G to have a form such that we can compute Gz cheaply for any vector z. Furthermore, we would like G to be well conditioned in the interest of numerical stability. This could be ensured by making G unitary, for example [230]. Sometimes a problem will have a certain special structure that we want to preserve, and this will determine the structure of G. For example, we might like G to be symplectic.

It is often convenient to phrase the problem in a slightly different way: Construct a nonsingular G such that $G^{-1}x = \alpha e_1$. This is just a cosmetic change. For all of the classes of zero-creating transformations that we will discuss, the relationship between G and G^{-1} is very simple, and it is trivial to produce G^{-1} from G and vice versa. Notice that $G^{-1}x = \alpha e_1$ if and only if $Ge_1 = \alpha^{-1}x$, which means that the first column of G is proportional to x.

Elementary Matrices

An *elementary matrix* is a matrix of the form $G = I - uv^*$, where $u, v \in \mathbb{F}^n$. Since uv^* is a rank-one matrix (Exercise 3.1.1), $I - uv^*$ is also known as a *rank-one modification of the identity*. Elementary matrices are computationally efficient. If G is elementary, then we can compute Gz for any $z \in \mathbb{F}^n$ with only $O(n)$ arithmetic (Exercise 3.1.2). Recall that it costs $O(n^2)$ arithmetic to compute Gz if G is a generic matrix stored in the conventional way. Elementary matrices are also efficient from the standpoint of storage. If we store just the vectors u and v instead of computing the entries of G and storing G in the conventional way, we get by with storing $2n$ numbers instead of n^2.

Before we consider the question of building an elementary matrix G such that $Gz = \alpha e_1$, we consider the conditions under which an elementary matrix is nonsingular. Notice that while uv^* is an $n \times n$ rank-one matrix, v^*u is a scalar.

Theorem 3.1.1. *The elementary matrix $G = I - uv^*$ is nonsingular if and only if $v^*u \neq 1$. If G is nonsingular, then $G^{-1} = I - \beta uv^*$, where $\beta = 1/(v^*u - 1)$. Thus G^{-1} is also an elementary matrix.*

For the proof, see Exercise 3.1.4. As a practical matter, if we want G to be well conditioned, then v^*u should not just be different from 1; it should be significantly different from 1, since otherwise β will be large, implying that G is ill conditioned (Exercise 3.1.6).

Now we consider the question of using elementary matrices to introduce zeros into vectors.

Proposition 3.1.2. *Let $x \in \mathbb{F}^n$ be nonzero. Then the elementary matrix $G = I - uv^*$ satisfies $Gx = \alpha e_1$ if and only if*

$$u = \gamma^{-1}(x - \alpha e_1), \qquad where \qquad \gamma = v^*x \neq 0. \tag{3.1.1}$$

The proof is straightforward. Proposition 3.1.2 shows that it is easy to build a zero-creating elementary matrix. The nonzero scalar α can be chosen freely, as can the vector v, subject only to the easily satisfied condition $v^*x \neq 0$. Then u has to be the appropriate multiple of $x - \alpha e_1$. There are many ways to choose α and v, so there are many solutions to the problem.

Gauss Transforms

Continuing to use the notation established above, suppose that the vector x satisfies $x_1 \neq 0$, and we want to find $G = I - uv^*$ such that $Gx = \alpha e_1$. Suppose that we choose $\alpha = x_1 \neq 0$ and $v = e_1$. Then $v^*x = x_1 \neq 0$, so by (3.1.1) we should take

$$u = x_1^{-1}(x - x_1 e_1) = \begin{bmatrix} 0 \\ x_2/x_1 \\ x_3/x_1 \\ \vdots \\ x_n/x_1 \end{bmatrix}. \tag{3.1.2}$$

This particular type of elementary matrix is called a *Gauss transform* because it effects a Gaussian elimination operation (Exercise 3.1.8). Gauss transforms are unit lower triangular.

Notice that if x happens to be real, then so are u and G. This is true of all of the classes of zero-creating transformations that will be introduced in this section: If x is real, then G is real.

The condition number of a Gauss transform is closely related to $\|u\|$ (Exercise 3.1.10): G is ill conditioned if and only if $\|u\|$ is large, which in turn occurs if and only if $|x_j/x_1|$ is large for some j. Thus we have ill conditioning exactly when x_1 is small relative to at least one other component of x.

A simple modification of the Gauss transform yields a well-conditioned transformation. Suppose $x \neq 0$ (we allow $x_1 = 0$), and suppose that the largest entry of x lies in the jth position, i.e., $|x_j| = \max_k |x_k|$. Let P be the permutation matrix obtained by interchanging the first and jth rows of the identity matrix. Let $\hat{x} = Px$. Then $\hat{x}$ is the same as x, except that the first and jth entries have been interchanged. Thus $\hat{x}$ has its largest entry in its first position. Let $\hat{G}$ be the Gauss transform such that $\hat{G}\hat{x} = \alpha e_1$. Then $\hat{G}$ is well conditioned. Define G by $G = \hat{G}P$. Then G is well conditioned, and $Gx = \alpha e_1$. This type of transformation is called a *Gauss transform with pivoting*.

Elementary Reflectors

Unitary matrices are especially desirable for computational purposes because they are perfectly conditioned ($\kappa(U) = 1$). Let us now consider the conditions under which an elementary matrix $G = I - uv^*$ is unitary. Recall that G is unitary if $G^*G = I$. Since $G^* = I - vu^*$, it is a simple matter to compute G^*G and determine the conditions under which $G^*G = I$. One finds right away that if G is unitary, then v is a multiple of u (Exercise 3.1.11). This implies that we can write G in the form $G = I - \beta uu^*$, where β is a nonzero scalar. If we insist that β be real, then G is unitary if and only if $\beta = 2/\|u\|^2$ (Exercise 3.1.11).

A unitary elementary matrix $Q = I - \beta uu^*$ ($\beta = 2/\|u\|^2$) is called an *elementary reflector* or *Householder transformation*. The name *reflector* refers to the fact that Q effects a reflection of $\mathbb{F}^n$ through the hyperplane consisting of vectors orthogonal to u (Exercise 3.1.12). For a geometric treatment of reflectors (real case), see [221, § 3.2].

We now consider the question of constructing zero-creating elementary reflectors: Given $x \neq 0$, find an elementary reflector Q such that $Qx = \alpha e_1$ for some α. We can simplify the development by transforming x to a standard form. Define a constant c satisfying $|c| = 1$ by $c = x_1/|x_1|$ if $x_1 \neq 0$ and $c = 1$ if $x_1 = 0$. Then the vector $\hat{x} = \bar{c}x/\|x\|$ satisfies $\|\hat{x}\| = 1$ and $\hat{x}_1 \geq 0$. If we can find a Q such that $Q\hat{x} = \hat{\alpha}e_1$ for some $\hat{\alpha}$, then $Qx = \alpha e_1$, where $\alpha = \hat{\alpha}c\|x\|$. This shows that it suffices to work with vectors with norm 1 that have $\hat{x}_1$ real and nonnegative.

Assume therefore that x itself satisfies $\|x\| = 1$ and $x_1 \geq 0$. We want to find an elementary reflector $Q = I - \beta uu^*$ such that $Qx = \alpha e_1$ for some α. We know from Proposition 3.1.2 that u must be proportional to $x - \alpha e_1$. Once we have specified α, Q will be completely determined. Since such a Q is unitary, it preserves norms, so we must have $|\alpha| = \|x\| = 1$. This does not determine α completely. In the interest of numerical accuracy (Exercise 3.1.14) we choose $\alpha = -1$. This means that u must be proportional to $x + e_1$. Not bothering to scale u, we simply take $u = x + e_1$. Then we must have $\beta = 2/\|x + e_1\|^2$. An easy computation shows that $\|x + e_1\|^2 = 2(1 + x_1)$, so $\beta = 1/(1 + x_1) = 1/u_1$.

Theorem 3.1.3 (zero-creating elementary reflector). *Given a vector $x \in \mathbb{C}^n$ satisfying $\|x\| = 1$ and $x_1 \geq 0$, let $u = x + e_1$, $\beta = 1/u_1$, and $Q = I - \beta uu^*$. Then Q is an elementary reflector satisfying $Qx = -e_1$. If x is real, then u and Q are real.*

A few details are worked out in Exercise 3.1.13. For a far-reaching generalization of reflectors, see [151].

Rotators

Rotators are another important class of unitary transformations. These do not fall into the category of elementary transformations as defined above. We begin with the 2×2 case.

Let c and s be two complex numbers satisfying $|c|^2 + |s|^2 = 1$, and define a 2×2 matrix

$$Q = \begin{bmatrix} \bar{c} & \bar{s} \\ -s & c \end{bmatrix}.$$

One easily checks that $Q^*Q = I$, and thus Q is unitary. If c and s are real, then there is an angle θ such that $c = \cos\theta$ and $s = \sin\theta$. The transformation $x \to Qx$ rotates x through the angle $-\theta$. The inverse of Q rotates vectors through the angle $+\theta$. Therefore we will refer to this class of transformations as *rotators* or *plane rotators*.

Given a nonzero vector $x = \begin{bmatrix} a \\ b \end{bmatrix}$ with $\|x\| = r > 0$, it is easy to build a rotator Q such that

$$\begin{bmatrix} \bar{c} & \bar{s} \\ -s & c \end{bmatrix} \begin{bmatrix} a \\ b \end{bmatrix} = \begin{bmatrix} r \\ 0 \end{bmatrix}.$$

Simply take $c = a/r$ and $s = b/r$. One easily checks that the resulting Q is unitary and does the job. Zero-creating rotators of this type are called *Givens transformations*.

Now let us consider $n \times n$ matrices. An $n \times n$ matrix Q is called a *plane rotator acting in the (i, j) plane* if it has the form

$$Q = \begin{bmatrix} I_{i-1} & & & & \\ & \bar{c} & & \bar{s} & \\ & & I_{j-i-1} & & \\ & -s & & c & \\ & & & & I_{n-j} \end{bmatrix},$$

where $|c|^2 + |s|^2 = 1$. This looks like an identity matrix, except that a rotator has been inserted in rows and columns i and j. Given any $x \in \mathbb{F}^n$, the transformation $x \to Qx$ leaves all entries of x unchanged except x_i and x_j, which undergo the transformation

$$\begin{bmatrix} x_i \\ x_j \end{bmatrix} \to \begin{bmatrix} \bar{c} & \bar{s} \\ -s & c \end{bmatrix} \begin{bmatrix} x_i \\ x_j \end{bmatrix}.$$

Thus one can easily construct a rotator in the (i, j) plane that transforms x_j to zero while leaving all other entries of x, except x_i, fixed.

It follows that we can build a zero-creating matrix G from a product of $n - 1$ plane rotators. Given a nonzero vector $x \in \mathbb{F}^n$, let Q_{n-1} be a plane rotator acting in the $(n-1, n)$ plane such that $Q_{n-1}x$ has a zero in the nth position. Then let Q_{n-2} be a rotator acting in the $(n-2, n-1)$ plane such that $Q_{n-2}(Q_{n-1}x)$ has a zero in position $(n-1)$. The zero in the nth position is preserved by this transformation. Then let Q_{n-3} be a rotator in the $(n-3, n-2)$ plane such that $Q_{n-3}(Q_{n-2}Q_{n-1}x)$ has a zero in position $n-2$. The zeros created on the previous two steps are preserved by this transformation. Continuing in this manner, we can construct plane rotators $Q_{n-4}, \ldots, Q_1$ such that $Q_1 \cdots Q_{n-1}x = \alpha e_1$, where $\alpha = \|x\|$. Let $G = Q_1 \cdots Q_{n-1}$. Then G is unitary, since it is a product of unitary matrices, and it satisfies $Gx = \alpha e_1$.

When we build a zero-creating product of rotators in this way, we never compute G explicitly; rather, we work with the factored form $Q_1 \cdots Q_{n-1}$. This is more efficient and requires very little storage. Only the $2(n-1)$ numbers that define the $n-1$ plane rotators need to be stored.

Complex Orthogonal Matrices

Reflectors and rotators that operate on real data are real and orthogonal: $Q^T Q = I$. There are some applications in which complex symmetric matrices arise. These are matrices $A \in \mathbb{C}^{n \times n}$ that satisfy $A^T = A$ (*not* $A^* = A$). In these applications complex orthogonal elimination matrices are needed. These $Q \in \mathbb{C}^{n \times n}$ satisfy $Q^T Q = I$ (*not* $Q^* Q = I$). An example of such a matrix is

$$Q = \begin{bmatrix} c & -s \\ s & c \end{bmatrix},$$

where c and s are complex numbers satisfying $c^2 + s^2 = 1$. We have opted not to develop this class of matrices and the associated algorithms in this book, but it does constitute one more example that is covered by the general theory. Some works that discuss complex symmetric and orthogonal matrices are [15, 19, 20, 59, 60].

Symplectic Elimination Matrices

Recall that a matrix $S \in \mathbb{R}^{2n \times 2n}$ is called *symplectic* if $S^T J S = J$, where J is defined by

$$J = \begin{bmatrix} 0 & I_n \\ -I_n & 0 \end{bmatrix}.$$

The following proposition identifies several useful classes of symplectic matrices.

Proposition 3.1.4. *The following classes of matrix are symplectic:*

(a) *If $B \in \mathbb{R}^n$ is nonsingular, then*

$$\begin{bmatrix} B & 0 \\ 0 & B^{-T} \end{bmatrix}$$

   is symplectic.

(b) *If $Q \in \mathbb{R}^{n \times n}$ is orthogonal, then*

$$\begin{bmatrix} Q & 0 \\ 0 & Q \end{bmatrix}$$

   is symplectic.

(c) *If $C \in \mathbb{R}^{n \times n}$ and $S \in \mathbb{R}^{n \times n}$ are diagonal matrices with $C^2 + S^2 = I$, then*

$$\begin{bmatrix} C & S \\ -S & C \end{bmatrix}$$

is symplectic.

(d) *If $X \in \mathbb{R}^{n \times n}$ is symmetric, then*

$$\begin{bmatrix} I & 0 \\ X & I \end{bmatrix} \quad and \quad \begin{bmatrix} I & X \\ 0 & I \end{bmatrix}$$

are symplectic. These are sometimes called symplectic shears *[150].*

All of these assertions are easily verified by direct computation.

A matrix of type (c) in Proposition 3.1.4 is a product of n nonoverlapping plane rotators. If $C = \text{diag}\{c_1, \ldots, c_n\}$ and $S = \text{diag}\{s_1, \ldots, s_n\}$, then

$$\begin{bmatrix} c_j & s_j \\ -s_j & c_j \end{bmatrix}$$

effects a rotation in the $(j, n+j)$ plane, $j = 1, \ldots, n$. Some of the rotations can be trivial: $c_i = 1$, $s_i = 0$. Thus we can effect rotations in some of the planes $(j, n+j)$ while leaving others untouched.

Matrices of types (b) and (c) in Proposition 3.1.4 are not only symplectic but also orthogonal. There are various ways of building zero-creating matrices from matrices of this type. For example, consider the following construction. Let $x = \begin{bmatrix} y \\ z \end{bmatrix}$ be a nonzero vector in $\mathbb{R}^{2n}$. For each j, choose c_j and s_j so that $c_j^2 + s_j^2 = 1$ and

$$\begin{bmatrix} c_j & s_j \\ -s_j & c_j \end{bmatrix} \begin{bmatrix} y_j \\ z_j \end{bmatrix} = \begin{bmatrix} w_j \\ 0 \end{bmatrix},$$

where $w_j = (y_j^2 + z_j^2)^{1/2}$. Then, letting $C = \text{diag}\{c_1, \ldots, c_n\}$, $S = \text{diag}\{s_1, \ldots, s_n\}$, and $w = [\, w_1 \ldots w_n \,]^T$, we have

$$\begin{bmatrix} C & S \\ -S & C \end{bmatrix} \begin{bmatrix} y \\ z \end{bmatrix} = \begin{bmatrix} w \\ 0 \end{bmatrix}.$$

Now let $Q \in \mathbb{R}^{n \times n}$ be an elementary reflector such that $Qw = \alpha e_1$. Then

$$\begin{bmatrix} Q & 0 \\ 0 & Q \end{bmatrix} \begin{bmatrix} w \\ 0 \end{bmatrix} = \begin{bmatrix} \alpha e_1 \\ 0 \end{bmatrix}.$$

If we now let

$$V = \begin{bmatrix} Q & 0 \\ 0 & Q \end{bmatrix} \begin{bmatrix} C & S \\ -S & C \end{bmatrix},$$

then V is clearly symplectic and orthogonal, and $Vx = \alpha e_1$.

Unfortunately we will not be able to meet all of our symplectic computational needs with orthogonal, symplectic transformations. For example, if we wish to compute an SR decomposition, as described in Theorem 1.3.11, we need to use nonorthogonal transformations of type (d) (Proposition 3.1.4) at crucial points in the algorithm. (See [53].) Exercise 3.1.18 illustrates how this type of transformation can be used to perform eliminations.

Hyperbolic Elimination Matrices

In our treatment of hyperbolic matrices, it is convenient to seek hyperbolic H such that $H^{-1}x = \alpha e_1$. Recall that a hyperbolic matrix of type 1 in $\mathbb{R}^{2\times 2}$ is a matrix of the form

$$H = \begin{bmatrix} c & s \\ s & c \end{bmatrix}, \tag{3.1.3}$$

where $c^2 - s^2 = 1$. One easily checks that H^{-1} is also hyperbolic of type 1:

$$H^{-1} = \begin{bmatrix} c & -s \\ -s & c \end{bmatrix}. \tag{3.1.4}$$

Given a nonzero vector $x = \begin{bmatrix} a \\ b \end{bmatrix} \in \mathbb{R}^2$, it may or may not be possible to build a hyperbolic H of type 1 such that

$$\begin{bmatrix} c & -s \\ -s & c \end{bmatrix} \begin{bmatrix} a \\ b \end{bmatrix} = \begin{bmatrix} r \\ 0 \end{bmatrix} \tag{3.1.5}$$

for some r. This would clearly require $sa = cb$, which means that we must choose $c = \gamma a$ and $s = \gamma b$ for some real scalar γ. The condition $c^2 - s^2 = 1$ then forces $\gamma^2(a^2 - b^2) = 1$ or $\gamma = \pm 1/\sqrt{a^2 - b^2}$. Clearly this is possible only if $a^2 - b^2 > 0$.

If $a^2 - b^2 < 0$, we can do a row interchange, interchanging the roles of a and b, and then apply a hyperbolic transformation of type 1 to effect the elimination. This is exactly the same as applying a hyperbolic transformation of type 2 to the original vector.

If $a^2 - b^2 = 0$, there is no hyperbolic matrix of type 1 or 2 that creates the desired zero. Perhaps more importantly, if $a^2 - b^2$ is nonzero but near zero, the matrix H^{-1} that effects (3.1.5) will be ill conditioned (Exercise 3.1.19). Thus we can safely apply a transformation of this type only if a^2 and b^2 are not too nearly equal.

We now turn our attention to $n \times n$ matrices. Given a signature matrix $D \in \mathbb{R}^{n\times n}$ and a nonzero $x \in \mathbb{R}^n$, we seek an $H \in \mathbb{R}^{n\times n}$ such that $H^{-1}x = \alpha e_1$ for some α and $H^T DH = \hat{D}$ for some signature matrix $\hat{D}$. We can do this using two reflectors and a plane hyperbolic transformation, provided $x^T Dx \neq 0$.

Theorem 3.1.5. *Let $D \in \mathbb{R}^{n\times n}$ be a signature matrix, and let $x \in \mathbb{R}^{n\times n}$ be a nonzero vector. Then there is a nonsingular matrix $H \in \mathbb{R}^{n\times n}$ such that $H^{-1}x = \alpha e_1$ for some α and $H^T DH = \hat{D}$ for some signature matrix $\hat{D}$ if and only if $x^T Dx \neq 0$.*

Proof: Exercise 3.1.20 shows that no such H exists if $x^T Dx = 0$. We now assume $x^T Dx \neq 0$ and outline a construction of H. We begin by dealing with a trivial case. If all of the main diagonal entries of D are the same, either 1 or -1, the task can be done by a single reflector.

Now assume that the main diagonal of D contains both 1's and -1's. Let P be a permutation matrix that moves the 1's to the top, that is, $P^T DP = \tilde{D}$, where $\tilde{d}_{ii} = +1$ for $i = 1, \ldots, k$ and $\tilde{d}_{ii} = -1$ for $i = k+1, \ldots, n$. This step is not strictly necessary in practice, but it simplifies the description. With this transformation, the vector x is now replaced by $P^{-1}x = P^T x$, which we partition as $P^{-1}x = \begin{bmatrix} y \\ z \end{bmatrix}$, where $y \in \mathbb{R}^k$ and $z \in \mathbb{R}^{n-k}$. Let $U_1 \in \mathbb{R}^{k\times k}$ and $U_2 \in \mathbb{R}^{n-k\times n-k}$ be reflectors such that $U_1 y = \beta e_1 \in \mathbb{R}^k$ and

$U_2 z = \gamma e_1 \in \mathbb{R}^{n-k}$. Let $U = \mathrm{diag}\{U_1, U_2\} \in \mathbb{R}^{n \times n}$. Then $U P^{-1} x = U^{-1} P^{-1} x$ has only two nonzero entries, a β and a γ, in positions 1 and $k+1$, respectively. Furthermore,

$$
U^T \tilde{D} U = \begin{bmatrix} U_1^T U_1 & \\ & -U_2^T U_2 \end{bmatrix} = \begin{bmatrix} I_k & \\ & -I_{n-k} \end{bmatrix} = \tilde{D}.
$$

It is not hard to show that $\beta^2 - \gamma^2 = x^T D x$ and is therefore nonzero. Indeed, letting $z = U^{-1} P^{-1} x$, we have $x^T D x = z^T U^T P^T D P U z = z^T U^T \tilde{D} U z = z^T \tilde{D} z = \beta^2 - \gamma^2$.

The next step depends upon the sign of $\beta^2 - \gamma^2$. If $\beta^2 - \gamma^2 > 0$, let $\begin{bmatrix} c & s \\ s & c \end{bmatrix}$ be a hyperbolic transformation of type 1 such that

$$
\begin{bmatrix} c & -s \\ -s & c \end{bmatrix} \begin{bmatrix} \beta \\ \gamma \end{bmatrix} = \begin{bmatrix} \alpha \\ 0 \end{bmatrix},
$$

where $\alpha = \sqrt{\beta^2 - \gamma^2}$. Now define $K \in \mathbb{R}^{n \times n}$ to be a matrix that effects the appropriate hyperbolic transformation in the $(1, k+1)$ plane:

$$
K = \begin{bmatrix} c & & s & \\ & I_{k-1} & & \\ s & & c & \\ & & & I_{n-k-1} \end{bmatrix}.
$$

Letting $z = U^{-1} P^{-1} x = [\, \beta\ 0\ \cdots\ 0\ \gamma\ 0\ \cdots\ 0\,]^T$, we easily see that $K^{-1} z = \alpha e_1$ and $K^T \tilde{D} K = \tilde{D}$. Now let $H = P U K$ and $\hat{D} = \tilde{D}$. Then $H^{-1} x = \alpha e_1$ and $H^T D H = \hat{D}$.

If $\beta^2 - \gamma^2 < 0$, we use a hyperbolic transformation of type 2. In other words, we interchange rows 1 and $k+1$, then apply a hyperbolic transformation of type 1. We get $H^{-1} x = \alpha e_1$ and $H^T D H = \hat{D}$, where $\hat{D}$ is the signature matrix obtained from $\tilde{D}$ by interchanging the $(1, 1)$ and $(k+1, k+1)$ entries. $\square$

We note that whenever $\beta^2 \approx \gamma^2$, the matrix K will be ill conditioned, and so will H. The construction outlined here has been used in [42] and [203], among others.

Exercises

3.1.1. Prove that $A \in \mathbb{F}^{n \times n}$ has rank one if and only if there exist nonzero vectors $u, v \in \mathbb{F}^n$ such that $A = uv^*$. To what extent is there flexibility in the choice of u and v?

3.1.2.

    (a) If $G \in \mathbb{F}^{n \times n}$ and $z \in \mathbb{F}^n$, show that the computation of Gz by the standard matrix multiplication procedure requires about n^2 multiplications and n^2 additions.

    (b) Show that if $G = I - uv^*$, where $u, v \in \mathbb{F}^n$, then $Gz = z - u(v^* z)$. Notice that $v^* z$ is a scalar. Deduce that Gz can be computed using approximately $2n$ multiplications and $2n$ additions.

3.1.3. Let $A \in \mathbb{F}^{n \times k}$, $B \in \mathbb{F}^{k \times n}$, and $C \in \mathbb{F}^{n \times k}$, where $k \ll n$.

(a) What are the dimensions of the product ABC? How much memory is needed to store the matrices A, B, C, and the product ABC? (How many numbers need to be stored?)

(b) Matrix multiplication is associative: $(AB)C = A(BC)$. This suggests two different ways to compute ABC. Show that one way takes much less arithmetic and intermediate storage space than the other does. Part (b) of Exercise 3.1.2 illustrates a special case of this fact.

3.1.4.

(a) Recall that a matrix G is singular if and only there is a nonzero y such that $Gy = 0$. Prove that if $Gy = 0$, where $G = I - uv^*$, then y is a multiple of u. Thus, if $G = I - uv^*$ is singular, then its null space ($\mathcal{N}(G) = \{y \mid Gy = 0\}$) is the one-dimensional space spanned by u.

(b) Prove that G is singular if and only if $v^*u = 1$.

(c) Suppose $v^*u \neq 1$, and let $H = I - \beta uv^*$ for some β. Show that $GH = HG = I$ if and only if $\beta = 1/(v^*u - 1)$. Thus $H = G^{-1}$ if we take $\beta = 1/(v^*u - 1)$.

3.1.5. Prove that an elementary matrix is singular if and only if it is a projector.

3.1.6. In this exercise we prove that $G = I - uv^*$ is ill conditioned if $v^*u \approx 1$. Let $\epsilon = |1 - v^*u|$, and assume $0 < \epsilon \ll 1$.

(a) Show that there are nonzero $x \in \mathbb{F}^n$ (and lots of them) such that $Gx = x$. Thus 1 is an eigenvalue of G. Deduce that $\|G\| \geq 1$.

(b) Show that u is an eigenvector of G with eigenvalue $-\beta^{-1} = 1 - v^*u$. Conclude that $\|G^{-1}\| \geq |\beta| = 1/\epsilon$.

(c) Deduce that $\kappa(G) \geq 1/\epsilon$. Thus G is ill conditioned.

3.1.7. Prove Proposition 3.1.2.

3.1.8. Let G be a Gauss transform that effects the transformation $Gx = \alpha e_1$. Thus $G = I - ue_1^*$, where u is given by (3.1.2).

(a) Show that for any vector y, the computation Gy can be accomplished with $n - 1$ multiplications and $n - 1$ additions (or subtractions). Show that the computation amounts to subtracting multiples of the first entry from each of the other entries of y.

(b) Show that if $Y \in \mathbb{F}^{n \times k}$, the computation GY can be accomplished with $k(n - 1)$ multiplications and $k(n - 1)$ additions. Describe the transformation $Y \to GY$ in terms of operations on the rows of Y. This justifies the name *Gauss transform*. Notice that if the first column of Y is x, then the first column of GY will contain a lot of zeros.

(c) Show that the inverse of G is given by $G^{-1} = I + ue_1^*$.

(d) Write out G and G^{-1} as $n \times n$ matrices. Observe that Gauss transforms are unit lower triangular.

3.1.9.

(a) Show that $Gx = \alpha e_1$ for some α if and only if the first column of G^{-1} is proportional to x. This gives us a simple (theoretical) method of generating G of the desired form: Just write down any nonsingular matrix whose first column is proportional to x, and take its inverse.

(b) In at least one simple case, the method of part (a) can be put into practice. Suppose $x_1 \neq 0$, and build a matrix H as follows: The main diagonal entries of H are all 1's, and H looks exactly like an identity matrix, except that its first column is proportional to x (*and* its first entry is 1). Show that H is nonsingular.

(c) Let $G = H^{-1}$. Compute G, and note that it is a Gauss transform. We have constructed a Gauss transform whose inverse's first column is proportional to x.

3.1.10. Let $G = I - ue_1^*$ be a Gauss transform.

(a) Show that $\|u\| - 1 \leq \|G\| \leq \|u\| + 1$.

(b) In Exercise 3.1.8 you showed that $G^{-1} = I + ue_1^*$. Show that $\|u\| - 1 \leq \|G^{-1}\| \leq \|u\| + 1$.

(c) Show that $(\|u\| - 1)^2 \leq \kappa(G) \leq (\|u\| + 1)^2$.

3.1.11.

(a) Prove that if $G = I - uv^*$ is unitary, then u and v are proportional; i.e., $v = \rho u$ for some nonzero scalar ρ. Prove that we can write G as $I - \beta \hat{u}\hat{u}^*$, where $\|\hat{u}\| = 1$ ($\hat{u} = u/\|u\|$) and β is a complex scalar.

(b) Let $G = I - \beta \hat{u}\hat{u}^*$, where $\|\hat{u}\| = 1$. Prove that G is unitary if and only if β lies on the circle of radius 1 centered at $1 + 0i$ in the complex plane. Show that if G is unitary and β is real and nonzero, then $\beta = 2$.

(c) Show that, regardless of whether or not u has norm 1, $G = I - \beta uu^*$ is unitary if $\beta = 2/\|u\|^2$.

3.1.12. Consider an elementary reflector $Q = I - \beta uu^*$, where $\beta = 2/\|u\|^2$.

(a) Show that $Q = Q^* = Q^{-1}$.

(b) Show that $Qu = -u$.

(c) Show that $Qw = w$ if $\langle u, w \rangle = 0$.

(d) The set of $w \in \mathbb{F}^n$ such that $\langle u, w \rangle = 0$ is an $(n-1)$-dimensional subspace of $\mathbb{F}^n$ called the *hyperplane* orthogonal to u. Let's call it S for now. Every $x \in \mathbb{F}^n$ has a unique representation $x = cu + w$, where $c \in \mathbb{F}$ and $w \in S$. The component of x in the direction of u is cu, and the component orthogonal to u is w. Show that $Q(cu + w) = -cu + w$. Here $-cu + w$ is the reflection of $cu + w$ in the hyperplane S.

3.1.13. This exercise checks some of the details of Theorem 3.1.3.

    (a) Check that $\| x + e_1 \|^2 = 2(1 + x_1)$, so that the formula $\beta = 1/(1 + x_1) = 1/u_1$ is correct.

    (b) Check by direct calculation that if Q is constructed as described in Theorem 3.1.3, then $Qx = -e_1$, as desired.

3.1.14. This exercise justifies the choice $\alpha = -1$ in the development leading up to Theorem 3.1.3.

    (a) Verify that cancellation cannot occur in any of the additions in the computation of u as specified in Theorem 3.1.3. In other words, show that at no point is a negative number added to a positive number. This is a consequence of the choice $\alpha = -1$.

    (b) Show that if we had made the choice $\alpha = 1$, then one cancellation would occur in the computation of u. Since cancellations sometimes result in loss of accuracy in floating point computations, we prefer the choice that results in no cancellations.

    (c) Show further that if we choose $\alpha = 1$, the algorithm can break down. Exactly when does this happen?

    (d) The condition that Q is unitary forces $|\alpha| = 1$. This seems to leave open the possibility that we could have taken α to be complex. A closer study of the restraints imposed by Proposition 3.1.2 reveals that this is not a possibility; α is perforce real. To reconcile our reflector notation with that of Proposition 3.1.2, write $Q = I - uv^*$, where $v = \beta u$. From Proposition 3.1.2 we have $u = \gamma^{-1}(x - \alpha e_1)$, where $\gamma = v^* x = \beta u^* x$. Substitute the first of these equations into the second and simplify to get $|\gamma|^2 = \beta(1 - \bar{\alpha}x_1)$. Recalling that x_1 is real, we see that this equation forces α to be real when $x_1 \neq 0$.

3.1.15. Let $G = I - uv^*$ be an $n \times n$ elementary matrix with $n > 1$.

    (a) Show that 1 is an eigenvalue of G with geometric multiplicity $n - 1$, and the corresponding eigenspace is the hyperplane $S = \{w \in \mathbb{C}^n \mid \langle v, w \rangle = 0\}$ consisting of vectors orthogonal to v.

    (b) Show that if u is not orthogonal to v, then the other eigenvalue of G is $1 - v^* u$ with eigenspace span$\{u\}$.

    (c) Show that if u is orthogonal to v, then G is unitarily similar to a Gauss transform. (Let Q be a reflector or other unitary matrix such that $Qv = \alpha e_1$. Then show that QGQ^{-1} is a Gauss transform.)

    (d) Show that a Gauss transform has only 1 as an eigenvalue with algebraic multiplicity n. Since the geometric multiplicity is only $n - 1$, Gauss transforms are defective.

    (e) Show that the eigenvalues of an elementary reflector are $1, 1, \ldots, 1, -1$.

3.1.16. Verify all of the assertions about plane rotators that were made in this section.

3.1.17. Prove Proposition 3.1.4. Show that the matrices in parts (b) and (c) are orthogonal, and those of part (d) are nonorthogonal if $X \neq 0$.

3.1.18. This exercise studies matrices of type (d) in Proposition 3.1.4.

(a) Let x, y, and z be real numbers. Show that if $z \neq 0$, there is a unique real a such that

$$
\left[\begin{array}{ccc|cc}
1 &   &   & 0 & a \\
  & 1 &   & a & 0 \\
  &   & 1 &   &   \\
\hline
  &   &   & 1 &   \\
  &   &   &   & 1
\end{array}\right]
\left[\begin{array}{c}
x \\ y \\ z \\ 0 \\ 0
\end{array}\right]
=
\left[\begin{array}{c}
x \\ 0 \\ z \\ 0 \\ 0
\end{array}\right].
$$

This matrix is clearly of type (d).

(b) Show that the transforming matrix from part (a) is ill conditioned if $|z| \ll |y|$ and well conditioned otherwise.

(c) Let $x = [\, y_1 \ \cdots \ y_j \, y_{j+1} \, 0 \cdots 0 \mid z_1 \ \cdots \ z_j \, 0 \, 0 \cdots 0 \,]^T \in \mathbb{R}^{2n}$ with $z_j \neq 0$. Construct a matrix S of type (d) such that $Sx = [\, y_1 \ \cdots \ y_j \, 0 \, 0 \cdots 0 \mid z_1 \ \cdots \ z_j \, 0 \, 0 \cdots 0 \,]^T$.

3.1.19.

(a) Show that if λ is an eigenvalue of the matrix A, then $|\lambda| \le \|A\|$.

(b) Show that if H is a hyperbolic matrix of the form (3.1.3), then the eigenvalues of H satisfy $|\lambda_1| = |c| + |s|$ and $|\lambda_2| = |c| - |s|$. Deduce that $\kappa(H) \ge (|c| + |s|)/(|c| - |s|)$. As an optional exercise, you could show that equality holds.

(c) Show that if H is as in (3.1.5), then $\kappa(H) \ge (|a| + |b|)/||a| - |b||$, which is large if $|a|$ and $|b|$ are nearly equal.

3.1.20. Let D be a signature matrix, and suppose that $x \in \mathbb{R}^n$ is a nonzero vector such that $x^T D x = 0$. Show that there is no $H \in \mathbb{R}^{n \times n}$ such that $H^{-1}x = \alpha e_1$ for some α and $H^T D H = \hat{D}$ for some signature matrix $\hat{D}$.

3.2 Creating Zeros in Matrices, *GR* Decompositions

This is a book about eigenvalue problems, so our primary interest is in square matrices. However, matrices with fewer columns than rows arise in the analysis of the methods, so let us consider a matrix $A \in \mathbb{F}^{n \times m}$, where $n \ge m$. Any of the types of transformations discussed in the previous section can be used to transform A to upper triangular form by the systematic introduction of zeros. The first step in such a process introduces zeros into the first column of A, the second step produces zeros in the second column, and so on.

At the beginning of the previous section we mentioned that it is sometimes convenient to call our zero-creating transformations G^{-1} instead of G. We will do that here. For the first step let x denote the first column of A. Construct a nonsingular matrix $G_1 \in \mathbb{F}^{n \times n}$ such that $G_1^{-1}x = \alpha_1 e_1$ for some α_1. Then $G_1^{-1}A$ has $n - 1$ zeros in the first column.

$$
G_1^{-1}A = \begin{bmatrix} \alpha_1 & s^T \\ 0 & \tilde{A}_1 \end{bmatrix}.
$$

Now let us consider the kth step. After $k-1$ steps, A has been transformed to

$$A_{k-1} = G_{k-1}^{-1} \cdots G_2^{-1} G_1^{-1} A = \begin{bmatrix} R_{k-1} & S_{k-1} \\ 0 & \tilde{A}_{k-1} \end{bmatrix},$$

where R_{k-1} is upper triangular of dimension $(k-1) \times (k-1)$. Let $\tilde{x}$ be the first column of $\tilde{A}_{k-1}$, and let $\tilde{G}_k$ be a nonsingular matrix of dimension $n-k+1$ such that $\tilde{G}_k^{-1} \tilde{x} = \alpha_k e_1$. Let

$$G_k = \begin{bmatrix} I_{k-1} & 0 \\ 0 & \tilde{G}_k \end{bmatrix},$$

and let $A_k = G_k^{-1} A_{k-1}$. Then

$$A_k = \begin{bmatrix} I_{k-1} & 0 \\ 0 & \tilde{G}_k^{-1} \end{bmatrix} \begin{bmatrix} R_{k-1} & S_{k-1} \\ 0 & \tilde{A}_{k-1} \end{bmatrix} = \begin{bmatrix} R_{k-1} & S_{k-1} \\ 0 & \tilde{G}_k^{-1} \tilde{A}_{k-1} \end{bmatrix}.$$

Since $\tilde{G}_k^{-1} \tilde{A}_{k-1}$ has $n-k$ new zeros in its first column, A_k has the form

$$A_k = \begin{bmatrix} R_k & S_k \\ 0 & \tilde{A}_k \end{bmatrix},$$

where R_k is $k \times k$ and upper triangular. Its last column consists of the first column of S_{k-1} with the scalar α_k adjoined.

After m steps we will have reduced the matrix to upper triangular form:

$$A_m = \begin{bmatrix} \tilde{R} \\ 0 \end{bmatrix},$$

where $\tilde{R}$ is $m \times m$ and upper triangular. Let $R = A_m = G_m^{-1} G_{m-1}^{-1} \cdots G_1^{-1} A$ and $G = G_1 G_2 \cdots G_m$. Then we have $R = G^{-1} A$, which can be rewritten as

$$A = GR.$$

Thus the process of creating zeros in A is equivalent to producing a so-called *GR decomposition*, in which $A \in \mathbb{F}^{n \times m}$ is decomposed into a product of a nonsingular matrix $G \in \mathbb{F}^{n \times n}$ and an upper triangular matrix $R \in \mathbb{F}^{n \times m}$.

If $m = n$, that is, A is square, we can skip the nth step, as triangular form is reached after just $n-1$ steps. The equations above remain correct if we take $G_m = I$. In this case R is a square upper triangular matrix.

Examples of *GR* Decompositions

As we have seen, there are many ways to produce the elimination matrices G_k. If we take each G_k to be a reflector or other unitary transformation (e.g., product of rotators), then the product G is unitary, and we usually call it Q instead of G. Thus we have a QR decomposition $A = QR$ with Q unitary and R upper triangular. By now we have assembled all of the elements of a constructive proof of Theorem 1.3.9. We apparently have a bit more, since we now have the result for $n \times m$ matrices.

Theorem 3.2.1. *Let $A \in \mathbb{F}^{n \times m}$ with $n \geq m$. Then there is a unitary matrix $Q \in \mathbb{F}^{n \times n}$ and an upper triangular matrix $R \in \mathbb{F}^{n \times m}$ such that*

$$A = QR.$$

Q can be constructed as a product of m reflectors.

We are also within a few technicalities of a proof of Theorem 1.3.8.

If we take the G_k to be Gauss transforms instead of reflectors, we get a decomposition $A = GR$, in which G is a product of Gauss transforms. Since each of the Gauss transforms is unit lower triangular (Exercise 3.1.8), the product G is itself unit lower triangular. In this case we generally use the symbol L instead of G and speak of an LR decomposition. The reduction of a matrix by Gauss transforms is not always possible, since it depends upon certain numbers being nonzero. With a bit of technical work we could figure out the exact conditions under which the reduction succeeds. We would then have a constructive proof of Theorem 1.3.4.

While the decomposition by Gauss transforms sometimes breaks down, a decomposition by Gauss transforms with pivoting is always possible. Thus we can always obtain a decomposition $A = MR$, where M is a product of Gauss transforms and permutation matrices. With a bit more work we can untangle the permutation matrices from the Gauss transforms to prove Theorem 1.3.5.

If $A \in \mathbb{R}^{2n \times 2n}$, we can try doing a reduction using symplectic elimination matrices. If we manage such a reduction, we obtain an SR decomposition as in Theorem 1.3.11. This takes a bit more effort than the QR and LR decompositions. We omit the details.

If $A \in \mathbb{R}^{n \times m}$, we can do the reduction using hyperbolic matrices of the type described in the proof of Theorem 3.1.5. We then have an HR decomposition. This procedure can break down because it relies on certain quantities being nonzero. If we analyze precisely the conditions under which the procedure succeeds, we obtain a constructive proof of Theorem 1.3.12.

There is nothing to prevent us from doing a GR decomposition using a mixture of different kinds of transformations. For example, one can follow a procedure that does the elimination by a Gauss transform when "safe," and a reflector otherwise.

Computational Cost of a *GR* Decomposition

How much arithmetic is involved in the computation of a GR decomposition? The exact amount depends on the details. We will content ourselves with a rough count. Let us consider the square case $A \in \mathbb{F}^{n \times n}$ first. Suppose that the matrices G_k used in the decomposition are all elementary matrices. Then each G_k^{-1} is an elementary matrix as well. We have $G_1^{-1} = I - uv^*$ for some u and v. The cost of determining u and v depends on the nature of the transformation, but in any case it is $O(n)$. The expensive part is the computation $A_1 = G_1^{-1} A$. From Exercise 3.1.2 we know that the cost of multiplying G_1^{-1} by a vector is about $2n$ multiplications and $2n$ additions. The cost of computing $G_1^{-1} A$ is n times that, since this amounts to multiplying G_1^{-1} by each of the columns of A. Each column is a vector, and there are n of them.[1] Thus the cost of computing $A_1 = G_1^{-1} A$ is about $2n^2$ multiplications and $2n^2$ additions.

---

[1] In actuality we do not need to multiply G_1^{-1} by the first column of A, since we already know the outcome of that computation. This observation makes a negligible change in the operation count.

The second step, which computes $A_2 = G_2^{-1} A_1$, is just like the first step, except that G_2 is effectively a smaller matrix. Recall that G_2 has the form

$$G_2 = \begin{bmatrix} 1 & \\ & \tilde{G}_2 \end{bmatrix},$$

where $\tilde{G}_2 \in \mathbb{F}^{n-1 \times n-1}$ is an elementary matrix. The computation does not touch the first row of A_1, nor does it need to touch the first column, which is (and remains) entirely zero from the second position on down. Thus the second step is the same as the first, except that it acts on an $(n-1) \times (n-1)$ matrix. The cost is thus $2(n-1)^2$ multiplications and as many additions.

Obviously the cost of the third step must be $2(n-2)^2$ multiplications and $2(n-2)^2$ additions, and so on. Thus the total number of multiplications is about

$$2(n^2 + (n-1)^2 + \cdots + 1) = 2 \sum_{j=1}^{n} j^2.$$

Approximating the sum by an integral, we get

$$2 \sum_{j=1}^{n} j^2 \approx 2 \int_0^n j^2 \, dj = \frac{2n^3}{3}.$$

We conclude that the total computational cost of a GR decomposition of an $n \times n$ matrix is about $(2/3)n^3$ multiplications and as many additions.

Depending on the details, there may be other operations as well, notably divisions and square roots. We glossed over the operations involved in setting up the elementary transformations, because those costs are relatively insignificant. However, those operations may well include divisions and square roots. A division takes longer than a multiplication does, and so does a square root, but if there are only $O(n)$ (or even $O(n^2)$) of them, we can ignore them.

The term *flop* is often used as an abbreviation for "floating point operation." We can summarize our operation count for the GR decomposition briefly by saying that it costs about $(4/3)n^3$ flops. These are split almost exactly fifty-fifty between multiplications and additions (including subtractions), with an insignificant number of other operations thrown in. Most matrix computation algorithms have this property. Any time we summarize the computational cost by saying that it costs a certain number of flops, we will mean that the overwhelming majority of the operations are multiplications and additions, and the numbers of the two types of operations are about equal. Any time this is not the case, we will say so explicitly.

The $(4/3)n^3$ flop count should not be taken too seriously; the exact count depends on the details. For example, if the elementary matrices happen to be Gauss transforms, we have $v = e_1$ for each transform, which has the consequence that the total arithmetic is cut roughly in half. If the reduction is done by nonelementary transformations, products of rotators for example, the count comes out differently.[2] However, in any event it is $O(n^3)$. There is no point in getting too precise about the operation counts, since the flop count gives

---

[2]In this case, the number of multiplications may be greater than the number of additions, depending on how the rotators are implemented.

only a rough indication of the execution time of an algorithm on a computer anyway. Other factors such as the amount of movement of data between cache and main memory are at least as important. This will be discussed immediately below.

A similar analysis for nonsquare matrices shows that the GR decomposition of an $n \times m$ matrix takes about $2nm^2 - 2m^3/3$ flops (Exercise 3.2.3). In the important case $n \gg m$, we can abbreviate this to $2nm^2$.

Blocking for Efficient Cache Use

Every computer has a memory hierarchy. The arithmetic is done in a few very fast registers, which communicate with the high-speed cache [110]. The cache is large compared to the number of registers but small when compared with the main memory. Transfer of data between cache and main memory is relatively slow. Still larger and slower is the hard drive or other bulk storage device.[3] In order to operate on some data, the computer must first move it from main memory to cache, then from cache into registers. If the entire matrix can fit into cache, then there is no problem; all of the data is readily available. However, if the matrix is so large that only a small part of it can fit into cache at once, then pieces of it must be swapped in and out periodically in order to complete the computation, whether it be a GR decomposition or whatever. If the computer wants to operate on some data that is not currently in cache, then there is a significant delay while that data is being moved into cache. This is called a *cache miss*.

The key to efficient cache use is to avoid cache misses. Suppose we can move some data from main memory into cache and then do a large amount of arithmetic on that data before it has to be moved out and replaced by other data. If the data that is being operated on occupies only about half of the cache, then, while the arithmetic is being done, a second set of data can be transferred into cache for whatever operations need to be done next. Once the first set of operations is done, the new operations can commence immediately. While these operations are being done, the old data can be transferred out of cache and replaced by new data. If the processor has enough arithmetic to do on each set of data, there will seldom be delays due to having to wait for data to be transferred into cache. On the other hand, if only a small amount of arithmetic is done on each set of data, there will be frequent delays, causing a significant degradation of performance.

The GR decomposition algorithm, as we have described it, is not able to make efficient use of cache (Exercise 3.2.4), so it is natural to ask whether we can improve things by rearranging the algorithm. The most common way to use cache efficiently is to arrange the operations into large matrix-matrix multiplies, where none of the matrix dimensions are small (Exercise 3.2.5). In the GR decomposition algorithm we can achieve this by applying the transformations in blocks instead of one at a time. At each step we generate an elementary transformation. Instead of applying it to the matrix immediately, we save it. We do just enough work on the matrix to figure out what the next elementary transformation should be, and so on for several steps. Once we have generated sufficiently many elementary transformations, we apply them all at once in a block. The details are worked out in

---

[3]Furthermore, the cache may be more complex than we have indicated here. Many machines have two or more levels of larger and successively slower caches. We can get the main ideas across without going into so much detail.

Exercise 3.2.6. The algorithms in LAPACK are arranged this way whenever possible. (See [8, 36, 186].)

Nested Subspaces and *GR* Decompositions

In a decomposition $A = GR$, if $A \in \mathbb{F}^{n \times m}$ has more rows than columns, then R has the form

$$R = \begin{bmatrix} \hat{R} \\ 0 \end{bmatrix},$$

where $\hat{R}$ is $m \times m$ and upper triangular. If we write $G = \begin{bmatrix} \hat{G} & H \end{bmatrix}$, where $\hat{G} \in \mathbb{F}^{n \times m}$, then clearly $A = \hat{G}\hat{R}$. We call this a *condensed GR* decomposition.

Let us now simplify the notation by dropping the hats. Consider a condensed decomposition $A = GR$, where $G \in \mathbb{F}^{n \times m}$ has linearly independent columns, and $R \in \mathbb{F}^{m \times m}$ is upper triangular. Let us assume that A has linearly independent columns as well, which implies that R is nonsingular. Let $a_1, a_2, \ldots, a_m$ denote the columns of A and $g_1, g_2, \ldots, g_m$ the columns of G. Then the equation $A = GR$ implies that

$$\text{span}\{a_1, \ldots, a_k\} = \text{span}\{g_1, \ldots, g_k\}, \qquad k = 1, \ldots, m.$$

Conversely, any time these subspace conditions hold, there must be an R such that $A = GR$. We summarize these facts as a theorem.

Theorem 3.2.2. *Let $A \in \mathbb{F}^{n \times m}$ and $G \in \mathbb{F}^{n \times m}$ have linearly independent columns. Then the subspace equalities*

$$\text{span}\{a_1, \ldots, a_k\} = \text{span}\{g_1, \ldots, g_k\}, \qquad k = 1, \ldots, m, \tag{3.2.1}$$

hold if and only if there is an upper triangular matrix $R \in \mathbb{F}^{n \times n}$ such that $A = GR$. R is uniquely determined and nonsingular.

The details of the proof are worked out in Exercise 3.2.7.

Relationship to the Gram–Schmidt Process

Given any m linearly independent vectors $a_1, \ldots, a_m$, the Gram–Schmidt process produces orthonormal $q_1, \ldots, q_m$ such that

$$\text{span}\{a_1, \ldots, a_k\} = \text{span}\{q_1, \ldots, q_k\}, \qquad k = 1, \ldots, m.$$

Let us briefly recall the Gram–Schmidt process. In the first step we let $\hat{q}_1 = a_1$ and then compute

$$r_{11} = \| \hat{q}_1 \| > 0, \qquad q_1 = \hat{q}_1 r_{11}^{-1}. \tag{3.2.2}$$

Clearly $\| q_1 \| = 1$ and $\text{span}\{q_1\} = \text{span}\{a_1\}$. Now suppose we have computed orthonormal vectors $q_1, \ldots, q_{k-1}$ such that $\text{span}\{q_1, \ldots, q_{k-1}\} = \text{span}\{a_1, \ldots, a_{k-1}\}$, and consider the computation of q_k. First let

$$\hat{q}_k = a_k - \sum_{j=1}^{k-1} q_j r_{jk}, \tag{3.2.3}$$

where each of the coefficients r_{jk} is chosen so that $\hat{q}_k$ is orthogonal to q_j. One easily checks
(Exercise 3.2.8) that $\hat{q}_k$ is orthogonal to $q_1, \ldots, q_{k-1}$ if and only if

$$r_{jk} = q_j^* a_k, \quad j = 1, \ldots, k-1. \tag{3.2.4}$$

Since $a_1, \ldots, a_k$ are linearly independent, $\hat{q}_k \neq 0$. The kth step concludes with the normalization

$$r_{kk} = \|\hat{q}_k\| > 0, \qquad q_k = \hat{q}_k r_{kk}^{-1}. \tag{3.2.5}$$

Then clearly $\|q\|_k = 1$, and $q_1, \ldots, q_k$ are orthonormal. Combining (3.2.4) and (3.2.5), we
easily find that

$$a_k = \sum_{j=1}^{k} q_j r_{jk}, \tag{3.2.6}$$

from which it follows that $\mathrm{span}\{a_1, \ldots, a_k\} = \mathrm{span}\{q_1, \ldots, q_k\}$. Equations (3.2.2), (3.2.3),
(3.2.4), and (3.2.5), assembled in the logical order, constitute the *classical Gram–Schmidt
process*.

Equation (3.2.6) is valid for $k = 1, \ldots, m$. If we combine these m equations into a
single matrix equation, we get an interesting result. Let $A \in \mathbb{F}^{n \times m}$ and $Q \in \mathbb{F}^{n \times m}$ be the
matrices whose columns are $a_1, \ldots, a_m$ and $q_1, \ldots, q_m$, respectively, and let $R \in \mathbb{F}^{m \times m}$ be
the upper triangular matrix

$$R = \begin{bmatrix} r_{11} & \cdots & r_{1m} \\ & \ddots & \vdots \\ & & r_{mm} \end{bmatrix}$$

built from the coefficients r_{jk}. Then R is not only upper triangular but also nonsingular,
since its main-diagonal entries are all positive. With this notation established, one easily
sees that the equations (3.2.6), $k = 1, \ldots, m$, are equivalent to the matrix equation

$$A = QR. \tag{3.2.7}$$

This is an instance of Theorem 3.2.2. Since the columns of Q are orthonormal, this equation
is a condensed QR decomposition. We conclude that *the Gram–Schmidt process computes
a condensed QR decomposition*.

We now have two methods for computing a QR decomposition, the Gram–Schmidt
process and the elimination process, using reflectors, outlined earlier in this section (Theorem 3.2.1). In theory the two methods produce equivalent results, but in floating point
arithmetic the elimination process is far superior. It delivers vectors (the columns of Q) that
are orthonormal to working precision, while the classical Gram–Schmidt process sometimes
produces vectors that are far from orthogonal due to roundoff errors. This does not make
the Gram–Schmidt process useless in practice, but it must be used with care. The *modified* Gram–Schmidt process is a rearrangement of Gram–Schmidt that has better numerical
properties. Another remedy that is widely used is to apply the process twice. Specifically,
once $\hat{q}_k$ has been computed in (3.2.3), if $\hat{q}_k$ is (because of roundoff errors) not (sufficiently
close to) orthogonal to $q_1, \ldots, q_{k-1}$, then apply the process again, this time to $\hat{q}_k$ instead
a_k, to improve the orthogonality. In fact you could repeat the process several times until
satisfied, but it turns out that two steps are generally enough. These issues are discussed in
[221], for example.

Finally, we note that there must surely be a connection between the Gram–Schmidt process and orthoprojectors. This connection is worked out in Exercise 3.2.9

Exercises

3.2.1. Write down a constructive, inductive proof of Theorem 3.2.1.

3.2.2. Consider a reduction of $A \in \mathbb{F}^{n \times n}$ to upper triangular form by a product of Gauss transforms $G_1^{-1}, \ldots, G_{n-1}^{-1}$. Then $A = LR$, where $L = G_1 \ldots G_{n-1}$.

(a) Show that for $k = 1, 2, \ldots, n - 1$, G_k is a unit lower triangular matrix

$$G_k = \begin{bmatrix} 1 & & & & & \\ & \ddots & & & & \\ & & 1 & & & \\ & & l_{jk} & 1 & & \\ & & \vdots & & \ddots & \\ & & l_{nk} & & & 1 \end{bmatrix}$$

($j = k + 1$), where the entries in the kth column are the only nonzero entries below the main diagonal.

(b) Verify the identity

$$\begin{bmatrix} 1 & \\ v & I \end{bmatrix} \begin{bmatrix} 1 & \\ 0 & \tilde{L} \end{bmatrix} = \begin{bmatrix} 1 & \\ v & \tilde{L} \end{bmatrix}.$$

(c) Prove by induction that the product $L = G_1 \ldots G_{n-1}$ is unit lower triangular, and its (i, k) entry ($i > k$) is l_{ik}. (Start with G_{n-1}, then consider $G_{n-2}G_{n-1}$, and so on.)

3.2.3. Suppose we reduce the matrix $A \in \mathbb{F}^{n \times m}$ to upper triangular form by applying m elementary transformations.

(a) Show that the first step takes about $2nm$ multiplications and as many additions, the second step takes $2(n - 1)(m - 1)$, and the total for m steps is about $2 \sum_{k=0}^{m}(n - k)(m - k)$.

(b) Approximate the sum by an integral to deduce that the total number of multiplications (and additions) is about $nm^2 - m^3/3$. Thus the total flop count is $2nm^2 - 2m^3/3$.

3.2.4. In this exercise you will show that the conventional version of the GR decomposition algorithm is unable to make efficient use of cache. The main work consists of updates of the form $B \to G^{-1}B$, where $G^{-1} = I - vw^*$. Thus the transformation is $B \to B - v(w^*B)$. The effective size of the elementary matrix G changes from step to step. Let's say it is k. Then B has k rows. Suppose that B is too big to fit into cache. If we can fit in j columns at a time, then we can process B in $k \times j$ chunks. Let us assume therefore that B is $k \times j$. Show that the operation $B \to B + v(w^*B)$ requires $4kj$ flops, and the amount of data involved is $k(j+2)$ numbers. Deduce that the ratio of arithmetic to data can never exceed 4, regardless of how large k and j are. Thus for each number that is moved into cache, only at most

4 operations are done before that data needs to be moved out and replaced by other data. Under these conditions efficient cache use is impossible.

3.2.5. In Exercise 3.2.4, about half of the arithmetic in the update $B \to G^{-1}B$ is contained in the operation w^*B. Although this is a matrix-matrix multiply, the matrix w^* is extremely thin in one dimension. In this exercise we will see that if the matrices are sufficiently fat in all directions, we can make efficient use of cache during a matrix-matrix multiply. Suppose we want to multiply B by C to get A. Suppose further that t is an integer such that $t \gg 1$, and the dimensions of B, C, and A are $mt \times nt$, $nt \times pt$, and $mt \times pt$, respectively. Then B can be expressed as a block matrix

$$B = \begin{bmatrix} B_{11} & \cdots & B_{1n} \\ \vdots & & \vdots \\ B_{m1} & \cdots & B_{mn} \end{bmatrix},$$

where each block B_{ij} is $t \times t$, and C and A can be blocked in the same way. These very restrictive assumptions are made purely to keep the arguments simple. In practice we do not need to have all of the blocks square and of the same size. The important thing is that they be roughly square and of roughly the same size. Thus the dimensions of the matrices do not need to be exact multiples of t. What matters is that each of the dimensions is at least t, so none of the matrices is excessively narrow in any one dimension. Now how big should t be? It should be as big as possible, subject to the restriction that six $t \times t$ blocks should fit into cache at once.[4]

(a) Show that if A, B, and C are blocked as described, then $A_{ij} = \sum_{k=1}^{n} B_{ik}C_{kj}$. Deduce that the computation of A can be achieved by a sequence of updates $S \leftarrow S + B_{ik}C_{kj}$, where each of the blocks is $t \times t$.

(b) Show that the update $S \leftarrow S + B_{ik}C_{kj}$ requires $2t^3$ flops on $3t^2$ data items. Deduce that the ratio of flops to data is $O(t)$. Thus the amount of arithmetic per item of data is large if t is large.

(c) Explain how the result of part (b) solves the efficient-cache-use problem if t is sufficiently large. Why do we insist that six blocks should fit into cache at once?

3.2.6. This exercise shows how to make efficient use of cache by applying elementary transformations in blocks. Suppose we want to compute a GR decomposition of an $n \times m$ matrix A that is much too large to fit into cache. Let t be an integer such that $1 \ll t \ll m$, and partition A as $A = \begin{bmatrix} A_1 & A_2 \end{bmatrix}$, where A_1 consists of the first t columns. All of the information needed for the first t steps of a GR decomposition is contained in A_1. In fact, let us begin by computing the GR decomposition of A_1. We have $R_1 = G_t^{-1} \cdots G_2^{-1} G_1^{-1} A_1$, where $G_i^{-1} = I + v_i w_i^*$, $i = 1, \ldots, t$. (We find it convenient to use plus signs here.) We do not apply the G_i^{-1} individually to A_2. The objective of this exercise is to combine them into a single transformation, which is then applied to A_2 in a cache-efficient way. For $j = 1, \ldots, t$, let V_j and W_j denote the $n \times j$ matrices whose columns are $v_1, \ldots, v_j$ and $w_1, \ldots, w_j$, respectively.

---

[4] Since there are often multiple levels of cache, it can be useful to do two or more layers of blocking (blocks within blocks). The blocking can even be done recursively [80].

(a) Show that $G_2^{-1} G_1^{-1} = I + V_2 T_2 W_2^*$, where $T_2 = \begin{bmatrix} 1 & 0 \\ w_2^* v_1 & 1 \end{bmatrix}$.

(b) Show that if $G_{j-1}^{-1} \cdots G_1^{-1} = I + V_{j-1} T_{j-1} W_{j-1}^*$, then $G_j^{-1} G_{j-1}^{-1} \cdots G_1^{-1} = I + V_j T_j W_j^*$, where

$$T_j = \begin{bmatrix} T_{j-1} & 0 \\ x_j^* & 1 \end{bmatrix}, \qquad x_j^* = w_j^* V_{j-1} T_{j-1}.$$

Deduce that $G_t^{-1} \cdots G_1^{-1} = I + V_t T_t W_t^*$, where T_t is $t \times t$ and unit lower triangular.

(c) We have seen in Exercise 3.2.3 that the cost of the GR decomposition of the $n \times t$ matrix A_1 is $nt^2 - t^3/3 \approx nt^2$ multiplications. We must pay this price, regardless of whether or not we are going to do the GR decomposition by blocks. If we are going to do it by blocks, we need to pay the additional price of computing T_t. This is a matter of computing $x_j^* = w_j^* V_{j-1} T_{j-1}$ at each step. In which order should this product be computed? Show that this costs about $n(j-1) + (j-1)^2/2 \approx nj$ multiplications and as many additions. Deduce that the total cost of computing T_t is about nt^2 flops.

(d) In order to continue the GR decomposition, we must now transform A_2 to $(I + V_t T_t W_t^*) A_2 = A_2 + V_t T_t W_t^* A_2$. Here we have a product of four matrices, whose dimensions are $n \times t$, $t \times t$, $t \times n$, and $n \times (m-t)$, respectively. Show that if the matrix-matrix multiplications are done in the most efficient order, the total operation count for the update is $(4nt + 2t^2)(m-t) \approx 4nt(m-t)$ flops. (Notice that the cost of computing T_t is small compared to this.) The important thing is that each of the matrices involved in the update is $t \times t$ or larger, so all of the matrix-matrix multiplies can be done blockwise, as explained in Exercise 3.2.5, in a cache-efficient way, provided that t is sufficiently large.

(e) Show that if the update of A_2 had been done by applying the transformations $I + v_i w_i^*$ one at a time, the operation count would have been $(4nt - 2t^2)(m-t)$ flops. Thus the block update costs slightly more flops, but this extra work is far more than offset by the benefits of efficient cache use.

We have described the first block step of a GR decomposition. Once A_2 has been updated, we use its first t columns to do another block step, and so on until we are done.

3.2.7. In this exercise you will prove Theorem 3.2.2.

(a) Suppose $A = GR$, where R is upper triangular. Show that $a_k \in \text{span}\{g_1, \ldots, g_k\}$ for $k = 1, \ldots, m$. Deduce that $\text{span}\{a_1, \ldots, a_k\} \subseteq \text{span}\{g_1, \ldots, g_k\}$ for $k = 1, \ldots, m$.

(b) Show that if A has linearly independent columns, then $r_{kk} \neq 0$ for $k = 1, \ldots, m$. (For example, assume $r_{kk} = 0$ and get $a_k \in \text{span}\{g_1, \ldots, g_{k-1}\}$, which leads to a contradiction.) Deduce that R is nonsingular.

(c) Show that $G = AR^{-1}$. Recalling from Proposition 1.3.3 that R^{-1} is upper triangular, deduce that $\text{span}\{g_1, \ldots, g_k\} \subseteq \text{span}\{a_1, \ldots, a_k\}$ for $k = 1, \ldots, m$. This proves (3.2.1).

(d) Conversely, suppose that (3.2.1) holds. Show that for each k there are uniquely determined scalars $r_{1k}, \ldots, r_{kk} \in \mathbb{F}$ such that $a_k = g_1 r_{1k} + \cdots + g_k r_{kk}$. Show that $r_{kk} \neq 0$.

(e) Show that there is a unique upper triangular $R \in \mathbb{F}^{m \times m}$ such that $A = GR$. Show that R is nonsingular.

3.2.8. In this exercise you will verify the assertions made in the development of the Gram–Schmidt process.

(a) Check that in (3.2.2), $r_{11} > 0$, $\| q_1 \| = 1$, and $\mathrm{span}\{q_1\} = \mathrm{span}\{a_1\}$.

(b) Verify that the vector $\hat{q}_k$ defined by (3.2.3) is orthogonal to $q_1, \ldots, q_{k-1}$ if and only if $r_{jk} = q_j^* a_k$, $j = 1, \ldots, k - 1$.

(c) Show that $\hat{q}_k \neq 0$. Deduce that $r_{kk} > 0$; hence q_k is well defined and $\| q_k \| = 1$.

(d) Verify (3.2.5). Deduce that $\mathrm{span}\{a_1, \ldots, a_k\} \subseteq \mathrm{span}\{q_1, \ldots, q_k\}$. Use a dimension argument to conclude that $\mathrm{span}\{a_1, \ldots, a_k\} = \mathrm{span}\{q_1, \ldots, q_k\}$.

(e) Show that the equations (3.2.6), $k = 1, \ldots, m$, are equivalent to the matrix equation $A = QR$ (3.2.7), where A, Q, and R are as defined in the text.

3.2.9.

(a) Substitute (3.2.4) into (3.2.3), and show that this (Gram–Schmidt) equation can be rewritten as
$$\hat{q}_k = a_k - Q_{k-1} Q_{k-1}^* a_k = (I - Q_{k-1} Q_{k-1}^*) a_k,$$
where $Q_{k-1} = \begin{bmatrix} q_1 & \cdots & q_{k-1} \end{bmatrix}$.

(b) Show that $I - Q_{k-1} Q_{k-1}^*$ is the orthoprojector of $\mathbb{F}^n$ onto $\mathrm{span}\{q_1, \ldots, q_{k-1}\}^{\perp}$.

3.3   Similarity Transformation to Upper Hessenberg Form

One way to find the eigenvalues of a matrix is to transform it to triangular form by a similarity transformation, then read the eigenvalues from the main diagonal. Schur's theorem tells us that this can always be done, but so far we do not have a practical means of doing it. Now that we have tools to introduce zeros into matrices, it is natural to ask to what extent they can help us toward this goal. As we shall see, we can get most of the way there by brute force, but the last little bit requires some finesse. This section is concerned with the brute-force part.

Let $A \in \mathbb{F}^{n \times n}$. In the previous section we saw how to reduce A to upper triangular form by applying a sequence of elimination matrices on the left. Those operations are not similarity transformations. Now we would like to carry out the same sort of procedure using similarity transformations. That means we have to apply transformations on the right as well as the left, and we will have to modify the procedure accordingly.

It is clear where the difficulty lies. Suppose we start out just as we did in the previous section, picking an elimination matrix G_1^{-1} such that

$$G_1^{-1} A = \begin{bmatrix} \alpha_1 & s^T \\ 0 & \tilde{A}_1 \end{bmatrix}.$$

Then when we complete the similarity transformation by multiplying by G_1 on the right, we find that the zeros that we so carefully created in $G_1^{-1} A$ get wiped out. There are no zeros in $G_1^{-1} A G_1$, unless some appear by accident.

We need to find a way to create zeros that we can defend when we apply the transformation on the right. The key to this is simply to be a bit less greedy. Write A in partitioned form:

$$A = \begin{bmatrix} a_{11} & b^T \\ x & \tilde{A}_1 \end{bmatrix}.$$

Let $\tilde{G}_2$ be an $n - 1 \times n - 1$ nonsingular matrix such that $\tilde{G}_2^{-1} x = \alpha_2 e_1$, and let

$$G_2 = \begin{bmatrix} 1 & 0^T \\ 0 & \tilde{G}_2 \end{bmatrix}.$$

Then

$$G_2^{-1} A = \begin{bmatrix} a_{11} & b^T \\ \alpha_2 e_1 & \tilde{G}_2^{-1} \tilde{A}_1 \end{bmatrix},$$

and moreover,

$$G_2^{-1} A G_2 = \begin{bmatrix} a_{11} & b^T \tilde{G}_2 \\ \alpha_2 e_1 & \tilde{G}_2^{-1} \tilde{A}_1 \tilde{G}_2 \end{bmatrix}.$$

Application of G_2^{-1} on the left creates $n - 2$ zeros in the first column, and application of G_2 on the right leaves them intact. This is so because the first column of G_2 is e_1. When we apply G_2 on the right, the first column of the product is unchanged. Let $A_2 = G_2^{-1} A G_2$.

For the second step we build an elimination matrix of the form

$$G_3 = \begin{bmatrix} 1 & & \\ & 1 & \\ & & \tilde{G}_3 \end{bmatrix}$$

such that $G_3^{-1} A_2$ has zeros in the second column from the fourth position downward. This transformation leaves the zeros in the first column intact, because $\tilde{G}_3^{-1} 0 = 0$. When we apply G_3 on the right to obtain $A_3 = G_3^{-1} A_2 G_3$, the zeros in the first two columns are left intact because G_3 does not alter them.

In the next step we will introduce zeros into the third column, and so on. It is now clear that we will be able to introduce many zeros by brute force, but we will not quite achieve upper triangular form, since we must leave nonzeros in the diagonal immediately below the main diagonal. The form that we do achieve is called *upper Hessenberg*, and it looks like this in the 5×5 case:

$$H = \begin{bmatrix} h_{11} & h_{12} & h_{13} & h_{14} & h_{15} \\ h_{21} & h_{22} & h_{23} & h_{24} & h_{25} \\ 0 & h_{32} & h_{33} & h_{34} & h_{35} \\ 0 & 0 & h_{43} & h_{44} & h_{45} \\ 0 & 0 & 0 & h_{54} & h_{55} \end{bmatrix}.$$

In general a matrix $H \in \mathbb{F}^{n \times n}$ is called *upper Hessenberg* if $h_{ij} = 0$ whenever $i > j + 1$.

Let us take a careful look at the kth step in the reduction of a matrix to upper Hessenberg form. After $k - 1$ steps we have $A_k = G_k^{-1} \cdots G_2^{-1} A G_2 \cdots G_k$, whose first $k - 1$ columns have been transformed to upper Hessenberg form. More precisely, A_k has the form

$$\left[\begin{array}{c|cc} H_k & \multicolumn{2}{c}{C} \\ \hline 0 & x & \tilde{A}_k \end{array} \right],$$

where H_k is a $k \times k$ upper Hessenberg matrix, and the block below H_k consists mainly of zeros. The symbol 0 denotes $k - 1$ columns of zeros, and the x denotes a single nonzero column. Let $\tilde{G}_{k+1}$ denote an $n - k \times n - k$ elimination matrix such that $\tilde{G}_{k+1}^{-1} x = \alpha_{k+1} e_1$, and let $G_{k+1} = \begin{bmatrix} I_k & 0 \\ 0 & \tilde{G}_{k+1} \end{bmatrix}$. Then letting $A_{k+1} = G_{k+1}^{-1} A_k G_{k+1}$, we have

$$A_{k+1} = \left[\begin{array}{c|c} H_k & C\tilde{G}_{k+1} \\ \hline \tilde{G}_{k+1}^{-1}0 \quad \tilde{G}_{k+1}^{-1}x & \tilde{G}_{k+1}^{-1}\tilde{A}_k\tilde{G}_{k+1} \end{array} \right] = \left[\begin{array}{c|c} H_k & \tilde{C} \\ \hline 0 \quad \alpha_{k+1}e_1 & \tilde{A}_{k+1} \end{array} \right].$$

The desired zeros have been introduced into the kth column.

After $n - 2$ steps, the matrix $H = A_{n-1}$ is upper Hessenberg. Before summarizing this result as a theorem, let us make one more observation. Let G denote the accumulated transforming matrix: $G = G_2 G_3 \cdots G_{n-1}$; $H = G^{-1} A G$. Since each of the G_k has e_1 as its first column, one easily checks that the first column of G itself is e_1. Now consider a modification of the process. Let $x \in \mathbb{F}^n$ be any nonzero vector, and let $G_1 \in \mathbb{F}^{n \times n}$ be a nonsingular matrix whose first column is proportional to x. For example, G_1 can be any elimination matrix such that $G_1^{-1} x = \alpha_1 e_1$ for some α_1. Now let $A_1 = G_1^{-1} A G_1$, and proceed with a reduction to upper Hessenberg form as described above, applying the algorithm to A_1 instead of A. The result is an upper Hessenberg matrix $H = G^{-1} A G$, where $G = G_1 G_2 G_3 \cdots G_{n-1}$. Since the first column of G_1 is proportional to x, one easily deduces (Exercise 3.3.1) that the first column of G is proportional to x. Thus we have the following result.

Theorem 3.3.1. *Let $A \in \mathbb{F}^{n \times n}$, and let $x \in \mathbb{F}^n$ be an arbitrary nonzero vector. Then there is a nonsingular matrix G whose first column is proportional to x such that the matrix $H = G^{-1} A G$ is upper Hessenberg. G can be taken to be unitary, and G and H can be constructed by a direct procedure that requires $O(n^3)$ arithmetic operations.*

Proof: We have already outlined a constructive proof. We can construct a unitary G by taking each of the G_k to be an elementary reflector, for example. In Exercise 3.3.2 it is shown that H can be obtained in about $\frac{10}{3}n^3$ flops if the G_k are taken to be elementary matrices. In many applications, assembly of G is not necessary. However, it can be done in about $\frac{4}{3}n^3$ flops if needed (Exercise 3.3.2). $\square$

The transforming matrix G is clearly far from unique. From the standpoint of stability, the best choice is to take G to be unitary. Notice that if A is Hermitian ($A = A^*$) and we take G to be unitary, then H is also Hermitian. A matrix that is both upper Hessenberg and Hermitian must be *tridiagonal*. This is a good outcome. However, if we wish to preserve some other type of structure, we might choose G differently. For example, we could make G symplectic or hyperbolic.

If A is complex, then H will be complex in general. However, it is always easy to adjust the matrix so that its subdiagonal entries are all real and even nonnegative (Exercise 3.3.5). This transformation can be of practical value. For example, if H is tridiagonal and Hermitian, and its off-diagonal entries are real, then H is real and symmetric.

Hessenberg Form and Krylov Subspaces

Let $A \in \mathbb{F}^{n \times n}$, and let $x \in \mathbb{F}^n$ be nonzero. The jth *Krylov subspace* associated with A and x, denoted $\mathcal{K}_j(A, x)$, is defined by

$$\mathcal{K}_j(A, x) = \text{span}\{x, Ax, A^2 x, \ldots, A^{j-1} x\}.$$

There is a close relationship between upper Hessenberg matrices and Krylov subspaces. An upper Hessenberg matrix H is called *properly* upper Hessenberg if all of its subdiagonal entries are nonzero, i.e., $h_{j+1,j} \neq 0$ for $j = 1, \ldots, n-1$.

Theorem 3.3.2. *Let $A \in \mathbb{F}^{n \times n}$ and $H \in \mathbb{F}^{n \times n}$, suppose $G \in \mathbb{F}^{n \times n}$ is nonsingular, and let $H = G^{-1} A G$. Let $g_1, \ldots, g_n$ denote the columns of G. Let x denote g_1 or any vector proportional to g_1.*

(a) *If H is upper Hessenberg with $h_{j+1,j} \neq 0$ for $j = 1, \ldots, m-1$ and $h_{m+1,m} = 0$, then*

$$\text{span}\{g_1, \ldots, g_j\} = \mathcal{K}_j(A, x), \qquad j = 1, \ldots, m.$$

Furthermore, $\mathcal{K}_m(A, x)$ is invariant under A.

(b) *If H is properly upper Hessenberg, then*

$$\text{span}\{g_1, \ldots, g_j\} = \mathcal{K}_j(A, x), \qquad j = 1, \ldots, n. \tag{3.3.1}$$

In this case $\mathcal{K}_n(A, x)$ has dimension n, which means that $x, Ax, \ldots, A^{n-1} x$ are linearly independent.

(c) *Conversely, H is properly upper Hessenberg if (3.3.1) holds.*

Proof: The theorem is proved by rewriting the equation $H = G^{-1} A G$ in the form $A G = G H$ and examining its consequences for the columns of G. See Exercise 3.3.6.   □

A special kind of properly Hessenberg matrix is the companion matrix. The relationship between Krylov subspaces and companion matrices is explored in Exercise 3.3.7. An interesting result about orthogonal complements of Krylov subspaces is proved in Exercise 3.3.9.

Skew-Hamiltonian Hessenberg Form

The most striking example of benefit derived from exploiting structure is that of Hermitian matrices, which we mentioned just above. Another good example is the skew-Hamiltonian case.[5] Let $B \in \mathbb{R}^{2n \times 2n}$ be a skew-Hamiltonian matrix. Then B has the form

$$B = \left[\begin{array}{cc} A & C \\ G & A^T \end{array} \right], \qquad C^T = -C, \ G^T = -G. \tag{3.3.2}$$

Notice that it is impossible to reduce B to upper Hessenberg form without destroying the skew-Hamiltonian structure. If we want to make B upper Hessenberg without destroying the skew symmetry of the G block, we have to transform that block to zero, thereby making the matrix block triangular. It turns out that we can do this. If we want to make the $(1, 1)$ block upper Hessenberg, then we will have to make the $(2, 2)$ block lower Hessenberg. We can do that as well, and we can do it all with orthogonal symplectic matrices of types (b) and (c) from Proposition 3.1.4 [208]. The orthogonality gives good stability properties, while the symplecticity preserves the skew-Hamiltonian structure.

Theorem 3.3.3. *Let* $B \in \mathbb{R}^{2n \times 2n}$ *be a skew-Hamiltonian matrix. Then there is an orthogonal, symplectic matrix* $Q \in \mathbb{R}^{2n \times 2n}$ *such that*

$$\hat{B} = Q^{-1} B Q = \left[\begin{array}{cc} H & K \\ 0 & H^T \end{array} \right],$$

where H *is upper Hessenberg (and* $K^T = -K$*). There is a backward-stable direct* $O(n^3)$ *algorithm to compute* Q *and* $\hat{B}$.

Proof: The first step of the algorithm transforms the first column of B to upper Hessenberg form. Let $\left[\begin{array}{c} x \\ y \end{array} \right]$ denote this first column. Notice that entry y_1 is already zero, as it is the $(1, 1)$ entry of the skew-symmetric matrix G. For $j = 2, \ldots, n$ choose c_j and s_j so that $c_j^2 + s_j^2 = 1$ and

$$\left[\begin{array}{cc} c_j & s_j \\ -s_j & c_j \end{array} \right] \left[\begin{array}{c} x_j \\ y_j \end{array} \right] = \left[\begin{array}{c} w_j \\ 0 \end{array} \right].$$

Then, letting $C = \mathrm{diag}\{1, c_2, \ldots, c_n\}$, $S = \mathrm{diag}\{0, s_2, \ldots, s_n\}$, and

$$w = \left[\begin{array}{cccc} x_1 & w_2 & \cdots & w_n \end{array} \right]^T,$$

we have

$$\left[\begin{array}{cc} C & S \\ -S & C \end{array} \right] \left[\begin{array}{c} x \\ y \end{array} \right] = \left[\begin{array}{c} w \\ 0 \end{array} \right].$$

The transforming matrix is orthogonal and symplectic, and it does not touch rows 1 and $n + 1$. Now let $\tilde{Q}$ be an $(n - 1) \times (n - 1)$ elementary reflector such that

$$\tilde{U} \left[\begin{array}{c} w_2 \\ w_3 \\ \vdots \\ w_n \end{array} \right] = \left[\begin{array}{c} \alpha \\ 0 \\ \vdots \\ 0 \end{array} \right],$$

---

[5] Unfortunately, the Hamiltonian case, which is of greater interest, is not so easy.

and let $U = \mathrm{diag}\{1, \tilde{U}\} \in \mathbb{R}^{n \times n}$. Then

$$\begin{bmatrix} U & 0 \\ 0 & U \end{bmatrix} \begin{bmatrix} w \\ 0 \end{bmatrix} = \begin{bmatrix} x_1 \\ \alpha \\ 0 \\ \vdots \end{bmatrix}.$$

This transformation is also orthogonal and symplectic and does not touch rows 1 or $n + 1$. Now define V by

$$V^{-1} = \begin{bmatrix} U & 0 \\ 0 & U \end{bmatrix} \begin{bmatrix} C & S \\ -S & C \end{bmatrix}.$$

If we apply V^{-1} to B on the left, then the first column of $V^{-1}B$ is upper Hessenberg. When we complete the similarity transformation by applying V on the right, columns 1 and $n + 1$ (particularly column 1) remain untouched, so $V^{-1}BV$ still has the zeros in the first column. Since the skew-Hamiltonian structure is retained, $V^{-1}BV$ actually has a few more zeros than $V^{-1}B$ does. The form of $V^{-1}BV$ is illustrated by

$$\begin{bmatrix} a & a & a & a & 0 & c & c & c \\ a & a & a & a & c & 0 & c & c \\ 0 & a & a & a & c & c & 0 & c \\ 0 & a & a & a & c & c & c & 0 \\ 0 & 0 & 0 & 0 & a & a & 0 & 0 \\ 0 & 0 & g & g & a & a & a & a \\ 0 & g & 0 & g & a & a & a & a \\ 0 & g & g & 0 & a & a & a & a \end{bmatrix}.$$

This completes the first step.

The second step is analogous to the first. It creates the desired zeros in the second column without disturbing the zeros that were created previously. The left transformation, which creates the new zeros, does not touch rows 1, 2, $n + 1$, and $n + 2$. The right transformation does not touch columns 1, 2, $n + 1$, and $n + 2$, so the zeros are preserved.

After $n - 1$ steps, the reduction is complete. $\square$

As an immediate consequence of Theorem 3.3.3, we have the following corollary.

Corollary 3.3.4. *Every eigenvalue of a skew-Hamiltonian matrix has even algebraic multiplicity. Every eigenspace has dimension at least two.*

Applying the Wintner–Murnaghan theorem (Theorem 2.2.6) to the Hessenberg matrix H, we find that there is an orthogonal $W \in \mathbb{R}^{n \times n}$ and quasi-triangular $T \in \mathbb{R}^{n \times n}$ such that $T = W^{-1}HW$. If we carry out a similarity transformation on $\hat{B}$ with the orthogonal symplectic transforming matrix $\mathrm{diag}\{W, W\}$, we obtain the following skew-Hamiltonian Wintner–Murnaghan form.

Theorem 3.3.5. *Let $B \in \mathbb{R}^{2n \times 2n}$ be a skew-Hamiltonian matrix. Then there is an orthogonal, symplectic matrix $Q \in \mathbb{R}^{2n \times 2n}$ such that*

$$\check{B} = Q^{-1}BQ = \begin{bmatrix} T & N \\ 0 & T^T \end{bmatrix},$$

where T is upper quasi-triangular (and $N^T = -N$).

In Chapter 4 we will develop an efficient, stable method (QR algorithm) for computing the Winter–Murnaghan form of an upper Hessenberg matrix. Then we will also have an efficient, backward-stable, structure-preserving method for computing the eigenvalues of a skew-Hamiltonian matrix.

If we are willing to give up orthogonality, we can push the decomposition even further. The matrix T is similar to a matrix in real Jordan canonical form (Theorem 2.4.12). That is, $T = SJS^{-1}$, where J is in real Jordan form. Applying a similarity transformation by the symplectic (but nonorthogonal) matrix $\mathrm{diag}\{S, S^{-T}\}$, we obtain the form

$$\begin{bmatrix} J & M \\ 0 & J^T \end{bmatrix}.$$

It can be shown [86] that the skew symmetric matrix M can be eliminated by a further symplectic similarity transformation, so an appropriate real Jordan form for skew-Hamiltonian matrices looks like

$$\begin{bmatrix} J & 0 \\ 0 & J^T \end{bmatrix}.$$

We mentioned above that the algebraic multiplicities of the eigenvalues of a skew-Hamiltonian matrix are all even. From this latest form we can see that the geometric multiplicities are also even. In fact, every detail of the Jordan structure appears in duplicate.

Exercises

3.3.1. Let $H, K \in \mathbb{F}^{n \times n}$, and let $z \in \mathbb{F}^n$.

    (a) Show that the first column of H is z if and only if $He_1 = z$.

    (b) Show that if the first column of H is z and the first column of K is e_1, then the first column of HK is z.

    (c) Suppose that the first column of G_1 is z and that the first columns of $G_2, \ldots, G_{n-1}$ are all e_1. Prove by induction on k that the first column of $G_1 G_2 \ldots G_k$ is z for $k = 1, \ldots, n - 1$.

3.3.2. Suppose that, in the reduction to upper Hessenberg form, each of the G_k is an elementary matrix. The effective part of G_k is $(n - k + 1) \times (n - k + 1)$. In other words, the effective part of G_{n-j} is $j + 1 \times j + 1$.

    (a) Show that the transformation $A_k \rightarrow G_{k+1}^{-1} A_k$ requires approximately $4(n - k)^2$ flops, and the transformation $G_{k+1}^{-1} A_k \rightarrow G_{k+1}^{-1} A_k G_{k+1} = A_{k+1}$ requires approximately $4(n - k)n$. Why is the second part of the transformation more expensive than the first?

    (b) Show that the total flop count for the reduction to upper Hessenberg form is about

$$\sum_{k=0}^{n-2} 4(n - k)^2 + 4(n - k)n = 4\sum_{j=2}^{n} j^2 + jn \approx 4\int_0^n j^2 + jn\,dj = \frac{10n^3}{3}.$$

(c) If we wish to assemble $G = G_1 G_2 \cdots G_{n-1}$, we can do it as follows. Start with the identity matrix. Multiply it by G_{n-1} on the left, then left multiply the result by G_{n-2}, then G_{n-3}, and so on up to G_1. For $j = 1, 2, 3, \ldots$, let $\hat{G}_{n-j} = G_{n-j} G_{n-j+1} \cdots G_{n-1}$. Show that G_{n-j} has the form

$$\begin{bmatrix} I_{n-j+1} & 0 \\ 0 & M_j \end{bmatrix},$$

where M_j is $j+1 \times j+1$. Show that the transformation $\hat{G}_{n-j+1} \to G_{n-j} \hat{G}_{n-j+1} = \hat{G}_{n-j}$ takes about $4j^2$ flops.

(d) Show that the total computation of $G = \hat{G}_1$ requires about $\frac{4n^3}{3}$ flops.

3.3.3.

(a) Review the material on reflectors in Section 3.1. Then write a MATLAB code (m-file) that takes as input a nonzero vector x and returns a reflector $Q = I - \beta u u^*$ such that $Qx = \alpha e_1$. The syntax could be something like this: `[u,beta,alpha] = reflector(x)`. How does your code compare with the code `reflector.m` from the website? [6]

(b) Using your reflector code from part (a), write a routine `[H,U] = myhess(A)` that produces unitary U and upper Hessenberg H such that $A = UHU^*$. Try your code out on some random matrices of various sizes. Compute the residual $\| A - UHU^* \|/\| A \|$ to check that your code works correctly. How does your code compare with the code `downhess.m` from the website?

3.3.4. A matrix $B \in \mathbb{R}^{n \times n}$ is called *pseudosymmetric* if there are a symmetric matrix A and a signature matrix D such that $B = AD$. Recall that a signature matrix is a diagonal matrix with entries ± 1 on the main diagonal. Suppose that the pseudosymmetric matrix $B = AD$ is reduced to upper Hessenberg form $H = G^{-1} BG$, where the transforming matrix G has the property that $\hat{D} = G^T DG$ is another signature matrix.

(a) Show that $H = \hat{A}\hat{D}$, where $\hat{A}$ is symmetric.

(b) Deduce that H is pseudosymmetric and tridiagonal.

(c) Show that $\hat{D} = G^{-1} DG^{-T}$. Show further that H can be obtained by operating on A and D directly, performing the transformations $\hat{A} = G^{-1} AG^{-T}$ and $\hat{D} = G^{-1} DG^{-T}$.

(d) Sketch an algorithm that uses $n - 2$ transformations of the type described in Theorem 3.1.5 to transform A to $\hat{A}$ and D to $\hat{D}$.

(e) The m-file `pseudotridiag.m` performs this reduction. Download this, the simple driver program `tridiago.m`, and the several m-files that they depend on, and try running it for a variety of choices of n. Recall from Theorem 3.1.5 that this algorithm can suffer breakdowns. If you run it on a large number of random matrices, you will

---

[6] www.siam.org/books/ot101

find that breakdowns are exceedingly rare. However, instability manifests itself in the residuals, which are much less satisfactory than those of downhess.m, which uses unitary transformations. The residuals get worse as n is increased.

(f) Build a 3×3 pseudosymmetric matrix that causes pseudotridiag.m to break down.

3.3.5. Let $H \in \mathbb{C}^{n \times n}$ be an upper Hessenberg matrix. Show that there is a unitary diagonal matrix D such that the matrix $\hat{H} = D^{-1}HD$ (which is still upper Hessenberg) has $\hat{h}_{j+1,j}$ real and nonnegative for $j = 1, \ldots, n - 1$.

3.3.6. Prove Theorem 3.3.2 by induction on j.

(a) Show that $\text{span}\{g_1\} = \mathcal{K}_1(A, x)$.

(b) Using the equation $AG = GH$, show that if H is upper Hessenberg, then

$$Ag_j = \sum_{i=1}^{j+1} g_i h_{ij}, \qquad j = 1, \ldots, n - 1.$$

Show that if $h_{j+1,j} \neq 0$, then

$$g_{j+1} = h_{j+1,j}^{-1}\left(Ag_j - \sum_{i=1}^{j} g_i h_{ij}\right), \qquad j = 1, \ldots, m - 1.$$

(c) Assume (inductively) that $\text{span}\{g_1, \ldots, g_j\} = \mathcal{K}_j(A, x)$. Show that under the conditions of part (b), $g_{j+1} \in \mathcal{K}_{j+1}(A, x)$. Deduce that

$$\text{span}\{g_1, \ldots, g_{j+1}\} \subseteq \mathcal{K}_{j+1}(A, x).$$

(d) Continuing from part (c), use a dimension argument to conclude that

$$\text{span}\{g_1, \ldots, g_{j+1}\} = \mathcal{K}_{j+1}(A, x).$$

This proves statement (a) of Theorem 3.3.2, by induction on j, except that we still have to show that $\mathcal{K}_m(A, x)$ is invariant. Statement (b) is essentially the case $m = n$ of statement (a), so there is nothing more to prove there.

(e) Show that if $h_{m+1,m} = 0$, then $\text{span}\{g_1, \ldots, g_m\}$ is invariant under A. Hence $\mathcal{K}_m(A, x)$ is invariant under A.

(f) Prove that if the subspace conditions (3.3.1) hold, then

$$Ag_j \in \text{span}\{g_1, \ldots, g_{j+1}\} \quad \text{and} \quad Ag_j \notin \text{span}\{g_1, \ldots, g_j\}$$

for $j = 1, \ldots, n - 1$. Using the equation $AG = GH$, deduce that H is properly upper Hessenberg. This proves statement (c) of Theorem 3.3.2.

3.3.7. Let $A \in \mathbb{F}^{n \times n}$, $x \in \mathbb{F}^n$, and suppose that $x, Ax, \ldots, A^{n-1}x$ are linearly independent. Let $G = [g_1 \ \cdots \ g_n]$ be the nonsingular matrix given by $g_j = A^{j-1}x$, $j = 1, \ldots, n$, and let $C = G^{-1}AG$. Using the equation $AG = GC$, show that C is a companion matrix,

$$
C = \begin{bmatrix}
0 & 0 & 0 & & c_0 \\
1 & 0 & 0 & & c_1 \\
0 & 1 & 0 & & c_2 \\
\vdots & \ddots & \ddots & & \vdots \\
0 & \cdots & 0 & 1 & c_{n-1}
\end{bmatrix},
$$

which is a special type of properly upper Hessenberg matrix.

3.3.8. Let $\hat{B}$ and $\check{B}$ be skew-Hamiltonian matrices in the form given by Theorems 3.3.3 and 3.3.5, respectively. Let $F \in \mathbb{R}^{n \times n}$ be the *flip matrix*: $F = [\ e_n \ \cdots \ e_2 \ e_1\ ]$, and let $Z = \mathrm{diag}\{I, F\}$.

   (a) Show that Z is orthogonal but not symplectic.

   (b) Show that $Z^{-1}\hat{B}Z$ is upper Hessenberg.

   (c) Show that $Z^{-1}\check{B}Z$ is quasi-triangular.

3.3.9. (duality and Krylov subspaces) Suppose $H = G^{-1}AG$, where H is properly upper Hessenberg. Let W be the unique matrix such that $W^* = G^{-1}$.

   (a) Let F be the flip matrix defined in Exercise 3.3.8, and let

$$
\tilde{H} = FH^*F.
$$

      Show that $\tilde{H}$ is properly upper Hessenberg (a "flipped" version of H).

   (b) Show that $A^*\tilde{W} = \tilde{W}\tilde{H}$, where $\tilde{W} = WF$.

   (c) Deduce that $\mathrm{span}\{w_n, w_{n-1}, \ldots, w_{j+1}\}$ is a Krylov subspace:

$$
\mathcal{K}_{n-j}(A^*, w_n) = \mathrm{span}\{w_n, w_{n-1}, \ldots, w_{j+1}\}, \qquad j = 0, \ldots, n-1.
$$

   (d) Show that $\mathrm{span}\{g_1, \ldots, g_j\}^\perp = \mathrm{span}\{w_n, \ldots, w_{j+1}\}$, and deduce that

$$
\mathcal{K}_j(A, g_1)^\perp = \mathcal{K}_{n-j}(A^*, w_n), \qquad j = 1, \ldots, n-1.
$$

Kressner used this relationship in his analysis of aggressive early deflation [137].

3.4 Working from Bottom to Top

In Section 3.1 we focused on elimination matrices G that have the property that, for a given vector x, $Gx = \alpha e_1$. The use of the vector e_1 was a matter of convenience. Clearly we can build a G such that $Gx = \alpha e_k$ for any k. In particular, we can build $G \in \mathbb{F}^{n \times n}$ such that $Gx = \alpha e_n$. This sort of transformation "pushes the nonzero part" of x to the bottom instead of the top. We can arrange for G to be unitary, triangular, symplectic, or whatever we need. Clearly the difference in the algorithm to make $Gx = \alpha e_n$ versus $Gx = \alpha e_1$ will be very slight.

Our focus on column vectors (ignoring row vectors) has also been no more than a matter of choice. We can obviously do eliminations in row vectors as well. Indeed, if $Gx = \alpha e_n$, then $x^T G^T = \alpha e_n^T$, so G^T is an elimination matrix for the row vector x^T.

RG Decompositions

In Section 3.2 we discussed the reduction of a matrix to upper triangular form by applying a sequence of elimination matrices that create zeros in successive columns of A. The reduction was interpreted as a GR decomposition of A.

We can equally well transform a matrix to triangular form by introducing zeros row by row instead of column by column. For simplicity we will restrict ourselves to the square case; let $A \in \mathbb{F}^{n \times n}$. Let x^T denote the bottom row of A, and let $G_1 \in \mathbb{F}^{n \times n}$ be a matrix such that $x^T G_1^{-1} = \alpha_1 e_n^T$. Then AG_1^{-1} has $n-1$ zeros in its last row:

$$AG_1^{-1} = \begin{bmatrix} \tilde{A}_1 & s \\ 0^T & \alpha_1 \end{bmatrix}.$$

Now consider the kth step. After $k-1$ steps, A has been transformed to

$$A_{k-1} = AG_1^{-1}G_2^{-1}\cdots G_{k-1}^{-1} = \begin{bmatrix} \tilde{A}_{k-1} & S_{k-1} \\ 0 & R_{k-1} \end{bmatrix},$$

where R_{k-1} is upper triangular of dimension $(k-1) \times (k-1)$. Let $\tilde{x}^T$ be the last row of $\tilde{A}_{k-1}$, and let $\tilde{G}_k$ be an elimination matrix of dimension $n-k+1$ such that $\tilde{x}^T \tilde{G}_k^{-1} = \alpha_k e_{n-k+1}^T$. Let

$$G_k = \begin{bmatrix} \tilde{G}_k & 0 \\ 0 & I_{k-1} \end{bmatrix},$$

and let $A_k = A_{k-1}G_k^{-1}$. Then

$$A_k = \begin{bmatrix} \tilde{A}_{k-1}\tilde{G}_k^{-1} & S_{k-1} \\ 0 & R_{k-1} \end{bmatrix}.$$

Since $\tilde{A}_{k-1}\tilde{G}_k^{-1}$ has $n-k$ new zeros in its bottom row, A_k has the form

$$A_k = \begin{bmatrix} \tilde{A}_k & S_k \\ 0 & R_k \end{bmatrix},$$

where R_k is $k \times k$ and upper triangular. Its top row consists of the bottom row of S_{k-1} with the scalar α_k adjoined in front. After $n-1$ steps we have reduced the matrix to upper triangular form

$$R = A_{n-1} = AG_1^{-1}\cdots G_{n-1}^{-1}.$$

Letting $G = G_{n-1}G_{n-2} \cdots G_1$, we have

$$A = RG.$$

Thus the process of transforming A to upper triangular form, a row at a time from bottom to top, can be viewed as an RG decomposition of A.

RG decompositions come in all flavors: RQ, RL, RS, RH, and so on. The arithmetic cost of an RG decomposition of an $n \times n$ matrix is exactly the same as that of a GR decomposition.

Our focus on reduction to upper triangular (to the exclusion of lower triangular) form has also been a matter of choice. We could equally well have developed GL and LG decompositions, where L is lower triangular. The algorithms are discussed in Exercise 3.4.2.

Transformation to Upper Hessenberg Form

Now that we know how to transform a matrix to triangular form by rows, from bottom to top, it must be clear that we can do the reduction to upper Hessenberg form the same way. Write A in the partition form

$$A = \begin{bmatrix} \tilde{A}_1 & b \\ x^T & a_{nn} \end{bmatrix}.$$

Let $\tilde{G}_2$ be an $(n-1) \times (n-1)$ nonsingular matrix such that $x^T \tilde{G}_2^{-1} = \alpha_2 e_{n-1}^T$, and let

$$G_2 = \begin{bmatrix} \tilde{G}_2 & 0 \\ 0^T & 1 \end{bmatrix}.$$

Then

$$AG_2^{-1} = \begin{bmatrix} \tilde{A}_1 \tilde{G}_2^{-1} & b \\ \alpha_2 e_{n-1}^T & a_{nn} \end{bmatrix}$$

and

$$G_2 A G_2^{-1} = \begin{bmatrix} \tilde{G}_2 \tilde{A}_1 \tilde{G}_2^{-1} & \tilde{G}_2 b \\ \alpha_2 e_{n-1}^T & a_{nn} \end{bmatrix}.$$

Application of G_2^{-1} on the right creates $n - 2$ zeros in the bottom row, and application of G_2 on the left leaves them intact. This is so because the bottom row of G_2 is e_n^T. When we apply G_2 on the left, the bottom row of the product is unchanged. Let $A_2 = G_2 A G_2^{-1}$. For the second step we build an elimination matrix of the form

$$G_3 = \begin{bmatrix} \tilde{G}_3 & & \\ & 1 & \\ & & 1 \end{bmatrix}$$

such that $A_2 G_3^{-1}$ has zeros in the first $n - 3$ positions of row $n - 1$. When we apply G_3 on the left to obtain $A_3 = G_3 A_2 G_3^{-1}$, the zeros in the bottom two rows are left intact because G_3 doesn't alter them.

Continuing in this manner for $n - 2$ steps, we obtain an upper Hessenberg matrix $H = A_{n-1} = G_{n-1} \cdots G_2 A G_2^{-1} \cdots G_{n-1}^{-1}$. Each of the transforming matrices G_j has e_n^T as its last row; that is, $e_n^T G_j = e_n^T$. Therefore the product $G = G_{n-1} \cdots G_2$ also has e_n^T as its last row.

We can alter the algorithm to make G have whatever last row we want it to have. Let $x \in \mathbb{F}^n$ be any nonzero vector, and let $G_1 \in \mathbb{F}^{n \times n}$ be an elimination matrix such that $x^T G_1^{-1} = \alpha_1 e_n^T$ for some nonzero α_1. Then $e_n^T G_1 = \alpha^{-1} x^T$, and so the last row of G_1 is proportional to x^T. Now let $A_1 = G_1 A G_1^{-1}$ and proceed to transform A_1 to upper Hessenberg form by the algorithm sketched above. The result is an upper Hessenberg matrix $H = GAG^{-1}$, where $G = G_{n-1} \cdots G_2 G_1$. Since the last row of G_1 is proportional to x^T, the last row of G is also proportional to x^T. Thus we have the following analogue of Theorem 3.3.1.

Theorem 3.4.1. *Let $A \in \mathbb{F}^{n \times n}$, and let $x \in \mathbb{F}^n$ be an arbitrary nonzero vector. Then there is a nonsingular matrix G whose bottom row is proportional to x^T such that the matrix $H = GAG^{-1}$ is upper Hessenberg. G can taken to be unitary, and G and H can be constructed by a direct procedure that requires $O(n^3)$ arithmetic operations.*

Obviously there are also algorithms for transforming a matrix to lower Hessenberg form. Our preference for upper Hessenberg matrices is a matter of pure prejudice. See Exercise 3.4.3.

Exercises

3.4.1.

(a) Modify your reflector code from Exercise 3.3.3 so that it builds a reflector such that $Qx = \alpha e_n$.

(b) Using your reflector code from part (a), write a routine that performs the reduction to Hessenberg form from bottom to top. Try your code out on some random matrices of various sizes. Compute a residual to check that your code works correctly. How does your code compare with the code `uphess.m` from the website?[7]

3.4.2.

(a) Sketch an algorithm to compute a decomposition $A = GL$, where L is lower triangular. Which column of A is attacked first?

(b) Sketch an algorithm to compute a decomposition $A = LG$, where L is lower triangular. Which row of A is attacked first?

(c) Show how to compute a GL or LG decomposition of A by computing a GR or RG decomposition of A^T.

3.4.3.

(a) Sketch an algorithm that performs a similarity transformation to lower Hessenberg form by introducing zeros row by row from top to bottom.

---

[7]www.siam.org/books/ot101

(b) Sketch an algorithm that performs a similarity transformation to lower Hessenberg form by introducing zeros column by column from right to left.

(c) Using the transpose operation, relate the algorithms from parts (a) and (b) to algorithms for producing upper Hessenberg matrices.

(d) Let $F = \begin{bmatrix} e_n & \cdots & e_1 \end{bmatrix}$, the flip matrix, and note that $F = F^{-1}$. Show that transforming A to upper Hessenberg form from bottom to top, as presented in this section, is the same as transforming $(FAF)^T$ to upper Hessenberg form from top to bottom, as described in Section 3.3.

Chapter 4
Iteration

In Chapter 3 we figured out how to transform an arbitrary matrix to upper Hessenberg form by the systematic, brute-force introduction of zeros. By this approach we did not succeed in getting to triangular form. Was this due to a lack of ingenuity on our part, or was there some fundamental reason for our failure? An answer to this question was provided long ago by Galois theory [109]. In principle the problem of computing the eigenvalues of an arbitrary $n \times n$ matrix is equivalent to that of computing the zeros of an arbitrary nth degree polynomial. If we had a finite prespecified procedure (that is, a *direct method*) for triangularizing a matrix, then we would have a direct procedure for computing the eigenvalues of an arbitrary matrix. This is tantamount to having a general formula for the zeros of an nth degree polynomial. One of the fundamental results of Galois theory is that no such formula exists, except for $n \leq 4$.

Thus we see that our failure to achieve triangular form was not due to a lack of ingenuity; a direct reduction to triangular form is simply impossible. As a consequence, all methods for solving eigenvalue problems are necessarily *iterative*; that is, they produce a sequence of approximants that approach the solution. We never quite get to the solution. However, if (after a number of iterations) our approximant agrees with the true solution to ten or fifteen decimal places, that is normally good enough.

In this chapter we will derive a rich variety of related iterative methods for the eigenvalue problem. Some of the methods produce a sequence of vectors that converges to an eigenvector. Others produce a sequence of subspaces that converges to an invariant subspace. Still others produce a sequence of matrices that converges to (block) upper triangular form.

The best algorithms at our disposal converge rapidly and are quite robust. However, even some of the best methods may occasionally fail, and the convergence theory is incomplete. In this chapter we will speak rather informally about convergence. Precise results and theorems will be presented in Chapter 5.

4.1 Subspace Iteration

Let $A \in \mathbb{C}^{n \times n}$. Our discussion begins with the humble power method, which can be used to find a dominant eigenvector of A, if it has one. Let $\lambda_1, \lambda_2, \ldots, \lambda_n$ denote the eigenvalues of A, ordered so that $|\lambda_1| \geq |\lambda_2| \geq \cdots \geq |\lambda_n|$. If $|\lambda_1| > |\lambda_2|$, then λ_1 is called the *dominant eigenvalue* of A, and every eigenvector of A associated with λ_1 is called a *dominant eigenvector*.

Given a nonzero starting vector $x \in \mathbb{C}^n$, the power method builds the Krylov sequence

$$x, \ Ax, \ A^2x, \ A^3x, \ldots. \tag{4.1.1}$$

If the choice of starting vector is not unlucky, this sequence, if properly rescaled, will converge to a dominant eigenvector of A, if A has one. It is not hard to see why this is so. To avoid complications, let us assume that A is semisimple. Then $\mathbb{C}^n$ has a basis $v_1, \ldots, v_n$ consisting of eigenvectors of A. Suppose $Av_j = \lambda_j v_j$ for each j, and suppose that the eigenvectors are ordered so that $|\lambda_1| > |\lambda_2| \geq \cdots \geq |\lambda_n|$. The starting vector x of (4.1.1) can be expressed as

$$x = c_1 v_1 + c_2 v_2 + \cdots + c_n v_n,$$

where the scalars c_j are unknown but almost certainly all nonzero. Then

$$Ax = c_1 \lambda_1 v_1 + c_2 \lambda_2 v_2 + \cdots + c_n \lambda_n v_n,$$

and in general

$$A^i x = c_1 \lambda_1^i v_1 + c_2 \lambda_2^i v_2 + \cdots + c_n \lambda_n^i v_n. \tag{4.1.2}$$

Since $|\lambda_1| > |\lambda_j|$ for all j, the vector $A^i x$ will point more and more in the direction of the eigenvector v_1 as i increases, assuming $c_1 \neq 0$. The rate of convergence depends on the ratio $|\lambda_2/\lambda_1|$. This argument will be made precise (and extended beyond semisimple matrices) in Section 5.1.

When we look for eigenvectors, our real objects of interest are eigenspaces. If A has a dominant eigenvalue λ_1, the associated eigenspace, which is necessarily one-dimensional, is called the *dominant eigenspace* of A. A dominant eigenvector v is simply a representative of the dominant eigenspace; any nonzero multiple of v will do just as well. Since the eigenspace is one-dimensional, v is a basis for the space. Similarly, each of the vectors in the sequence (4.1.1) can be viewed as a representative of the one-dimensional space that it spans, and (4.1.1) can be reinterpreted as a sequence of subspaces

$$\mathcal{S}, \ A\mathcal{S}, \ A^2\mathcal{S}, \ A^3\mathcal{S}, \ \ldots \tag{4.1.3}$$

(where $\mathcal{S} = \text{span}\{x\}$) that converges to the eigenspace $\text{span}\{v\}$.

More generally we can consider subspaces of dimension greater than one. Given any subspace $\mathcal{S}$ of $\mathbb{C}^n$, we can consider what happens to the sequence of subspaces (4.1.3). It turns out that if $\mathcal{S}$ is a k-dimensional space and not an unlucky choice, (4.1.3) will converge to a k-dimensional dominant invariant subspace of A, if A has one. A k-dimensional invariant subspace $\mathcal{U}$ is *dominant* if the eigenvalues associated with $\mathcal{U}$ (see Proposition 2.1.11) are $\lambda_1, \ldots, \lambda_k$, where $|\lambda_1| \geq \cdots \geq |\lambda_n|$, as before, and $|\lambda_k| > |\lambda_{k+1}|$. In the semisimple case, the dominant invariant subspace is just $\mathcal{U} = \text{span}\{v_1, \ldots, v_k\}$. The convergence rate depends on the ratio $|\lambda_{k+1}/\lambda_k|$. These claims are not surprising in light of (4.1.2); the k components that undergo the greatest magnification are the ones that survive. A rigorous proof will be given in Section 5.1. At that point we will also state what is meant by a "not unlucky" choice of initial subspace. For the remainder of this discussion we will assume that our choice was not unlucky.

This simple generalization of the power method is called *subspace iteration*. Recall from Section 2.2 that *the basic task of eigensystem computations is to find a nontrivial invariant subspace*. Subspace iteration performs this task.

For $i = 0, 1, 2, \ldots$, let $\mathcal{S}^{(i)} = A^i \mathcal{S}$. Then the sequence (4.1.3) becomes

$$\mathcal{S}^{(0)}, \ \mathcal{S}^{(1)}, \ \mathcal{S}^{(2)}, \ \mathcal{S}^{(3)}, \ \ldots.$$

Once we have $\mathcal{S}^{(i-1)}$ in hand, we can produce $\mathcal{S}^{(i)}$ by the operation $\mathcal{S}^{(i)} = A\mathcal{S}^{(i-1)}$. The question of how to carry out this operation in practice will be addressed in Section 4.2.

Shifts of Origin

The shifted matrix $A - \rho I$ has the same invariant subspaces as A has, but the eigenvalues are shifted to $\lambda_1 - \rho, \ldots, \lambda_n - \rho$. Since the convergence rate of subspace iteration depends on a ratio of eigenvalues, it may be possible to improve the convergence rate by replacing A by a shifted matrix. This is a matter of computing $\mathcal{S}^{(i)} = (A - \rho I)\mathcal{S}^{(i-1)}$ instead of $\mathcal{S}^{(i)} = A\mathcal{S}^{(i-1)}$.

As we shall see, it is usually a good idea to use shifts that are good approximations to eigenvalues of A. As we make progress in our subspace iterations, we will be able to obtain better estimates of eigenvalues, so it makes sense to replace the shift now and then by a better one. In fact, we might even change the shift at each iteration. If we do this, the ith iteration looks like $\mathcal{S}^{(i)} = (A - \rho_i I)\mathcal{S}^{(i-1)}$, where ρ_i is the shift chosen for the ith step.

If we take m steps of subspace iteration with shifts $\rho_1, \ldots, \rho_m$, we obtain

$$\mathcal{S}^{(m)} = (A - \rho_m I)(A - \rho_{m-1} I) \cdots (A - \rho_1 I)\mathcal{S},$$

which is the same as

$$\mathcal{S}^{(m)} = p(A)\mathcal{S},$$

where p is the mth degree polynomial defined by $p(z) = (z - \rho_m) \cdots (z - \rho_1)$. Thus m steps of subspace iteration with shifts are the same, in principle, as a single step of subspace iteration driven by the matrix $p(A)$. Such a step we call a *multiple step of degree m*.

The idea of aggregating steps in this manner is of both theoretical and practical importance. Let us therefore generalize our subspace iteration process so that each step is a multiple step of degree m:

$$\mathcal{S}^{(i)} = p_i(A)\mathcal{S}^{(i-1)},$$

where p_i is a polynomial of degree m. The zeros of p_i are called the *shifts* for the ith step.

To see how such a process can be made to converge rapidly, consider the following scenario. Suppose that $\rho_1, \ldots, \rho_m$ are good approximations of m distinct eigenvalues of A, and let $p(z) = (z - \rho_1)(z - \rho_2) \cdots (z - \rho_m)$. Consider the stationary subspace iteration

$$\mathcal{S}^{(i)} = p(A)\mathcal{S}^{(i-1)}. \tag{4.1.4}$$

The convergence rate depends upon a ratio of the eigenvalues of $p(A)$, which are $p(\lambda_1), \ldots, p(\lambda_n)$. Renumber the eigenvalues such that $|p(\lambda_1)| \geq \cdots \geq |p(\lambda_n)|$. If $|p(\lambda_k)| > |p(\lambda_{k+1})|$, where k is the dimension of $\mathcal{S}$, then the multiple subspace iteration (4.1.4) will converge. The rate of convergence depends on the ratio $|p(\lambda_{k+1})/p(\lambda_k)|$. The smaller this ratio is, the faster the convergence will be.

Notice that some of the eigenvalues $p(\lambda_j)$ are small. Indeed, for each λ_j that is close to one of the shifts ρ_i, $|p(\lambda_j)|$ must be small if $|\lambda_j - \rho_i|$ is sufficiently small. Thus m of the eigenvalues are small. These are $p(\lambda_{n-m+1}), \ldots, p(\lambda_n)$. Typically the other $n-m$ eigenvalues are not small, so we have $|p(\lambda_{n-m})| \gg |p(\lambda_{n-m+1})|$. Now suppose that m and k are related by $k = n - m$. Then we have $|p(\lambda_k)| \gg |p(\lambda_{k+1})|$. Thus the ratio $|p(\lambda_{k+1})/p(\lambda_k)|$ is tiny, and the subspace iteration converges rapidly. In Section 4.3 we will see how to achieve this.

Here we have considered a stationary scenario, in which the same shifts $\rho_1, \ldots, \rho_m$ are used on each multiple step. This was done solely to keep the exposition simple. If a nonstationary iteration is used, in which we change the shifts as better ones become available, the analysis becomes more difficult, but the convergence is normally much faster.

Exercises

4.1.1. Let A be a matrix with eigenvalues $\lambda_1, \ldots, \lambda_n$, and let p be a polynomial with zeros $\rho_1, \ldots, \rho_m$.

   (a) Show that $p(\lambda_1), \ldots, p(\lambda_n)$ are the eigenvalues of $p(A)$.

   (b) Show that if λ_j is very near one of the ρ_i, then $|p(\lambda_j)|$ is small.

4.1.2. (duality in subspace iteration)

   (a) Show that if $\mathcal{U}$ is invariant under A, then $\mathcal{U}^\perp$ is invariant under A^*.

   (b) Show that if ρ is a shift that is not an eigenvalue of A, then the two subspace iterations

$$\mathcal{S}, \ (A - \rho I)\mathcal{S}, \ (A - \rho I)^2 \mathcal{S}, \ldots$$

and

$$\mathcal{S}^\perp, \ (A^* - \overline{\rho} I)^{-1}\mathcal{S}^\perp, \ (A^* - \overline{\rho} I)^{-2}\mathcal{S}^\perp, \ldots$$

are equivalent, in the sense that $(A^* - \overline{\rho} I)^{-i}\mathcal{S}^\perp$ is the orthogonal complement of $(A - \rho I)^i \mathcal{S}$ for all i. The first sequence of subspaces converges to $\mathcal{U}$ if and only if the second converges to $\mathcal{U}^\perp$.

This result shows that every subspace iteration process is implicitly accompanied by an inverse iteration process (a process with negative powers).

4.2   Simultaneous Iteration and *GR* Algorithms; *QR* Algorithm

Let us now consider how subspace iteration

$$\mathcal{S}_i = p(A)\mathcal{S}_{i-1}$$

can be effected in practice. We consider stationary subspace iteration for notational simplicity; the extension to nonstationary subspace iteration is straightforward.

First of all, supposing that $\dim(\mathcal{S}_0) = k$, can we say for sure that every subspace in the sequence $(\mathcal{S}_j)$ has dimension k? No! If $\mathcal{N}(p(A)) \cap \mathcal{S}_{i-1} \neq \{0\}$, then $\dim(\mathcal{S}_i) < \dim(\mathcal{S}_{i-1})$. (See Exercise 4.2.4.) However, we will ignore this slight possibility for now and assume that all of the spaces have dimension k.

The way to get one's hands on a subspace is to pick a basis for the space, so let us suppose we have a basis $s_1^{(i-1)}, \ldots, s_k^{(i-1)}$ for $\mathcal{S}_{i-1}$. Then we can get a basis for $\mathcal{S}_i$ by applying $p(A)$ to each of these basis vectors. The vectors $p(A)s_1^{(i-1)}, \ldots, p(A)s_k^{(i-1)}$ are indeed a basis if, as we are assuming, $\mathcal{N}(p(A)) \cap \mathcal{S}_{i-1} = \{0\}$. If we are thinking in purely theoretical terms, we can take this as our basis for the next step. That is, we take $s_1^{(i)} = p(A)s_1^{(i-1)}, \ldots, s_k^{(i)} = p(A)s_k^{(i-1)}$ and then apply $p(A)$ to $s_1^{(i)}, \ldots, s_k^{(i)}$ to effect the next step.

This approach works poorly in practice because it amounts to applying the power method to each of the vectors $s_1^{(0)}, \ldots, s_k^{(0)}$ independently; we have $s_j^{(i)} = p(A)^i s_j^{(0)}$, so each of the sequences $(s_j^{(i)})_{i=0}^{\infty}$ will tend to converge to the eigenspace associated with the dominant eigenvalue of $p(A)$. This means that after several iterations all of the vectors $s_1^{(i)}, \ldots, s_k^{(i)}$ will point in nearly the same direction. Thus, although they are technically a basis for $\mathcal{S}_i$, they form a very ill-conditioned one. They represent the space poorly.

A simple remedy for this problem is to work only with orthonormal bases. Given $q_1^{(i-1)}, q_2^{(i-1)}, \ldots, q_k^{(i-1)}$, an orthonormal basis for $\mathcal{S}_{i-1}$, compute $p(A)q_1^{(i-1)}, p(A)q_2^{(i-1)}, \ldots, p(A)q_k^{(i-1)}$. This is a basis but not an orthonormal basis of $\mathcal{S}_i$, so orthonormalize it by (say) the Gram–Schmidt process to obtain an orthonormal basis $q_1^{(i)}, q_2^{(i)}, \ldots, q_k^{(i)}$.

This very satisfactory procedure is called *simultaneous iteration*. It can be expressed concisely in matrix form as follows. For each j, let $\hat{Q}_j \in \mathbb{F}^{n \times k}$ denote the matrix whose columns are $q_1^{(j)}, \ldots, q_k^{(j)}$. Let $S_i = p(A)\hat{Q}_{i-1}$. Then the columns of S_i are $p(A)q_1^{(i-1)}, \ldots, p(A)q_k^{(i-1)}$, so $\hat{Q}_i$ is obtained from S_i by orthonormalizing its columns. By Theorem 3.2.2, there is a nonsingular upper triangular matrix $R_i \in \mathbb{F}^{k \times k}$ such that $S_i = \hat{Q}_i R_i$. Thus the step from $\mathcal{S}_{i-1}$ to $\mathcal{S}_i$ is given by the two equations

$$p(A)\hat{Q}_{i-1} = S_i, \qquad S_i = \hat{Q}_i R_i.$$

In practice the condensed QR decomposition $S_i = \hat{Q}_i R_i$ can be computed by some version of Gram–Schmidt or by the elimination process using reflectors or rotators described in Section 3.2. See the discussion leading up to Theorem 3.2.1. Eliminating the symbol S_i, we make the description even more concise:

$$p(A)\hat{Q}_{i-1} = \hat{Q}_i R_i.$$

In the interest of flexibility, we now introduce a version of simultaneous iteration in which the basis vectors are not necessarily orthonormal. Briefly

$$p(A)\hat{G}_{i-1} = \hat{G}_i R_i. \tag{4.2.1}$$

Here the columns of $\hat{G}_{i-1}$ form a basis (not necessarily orthonormal) of $\mathcal{S}_{i-1}$. Thus the columns of $p(A)\hat{G}_{i-1}$ are a basis of $\mathcal{S}_i$. The equation $p(A)\hat{G}_{i-1} = \hat{G}_i R_i$ implies that the columns of $\hat{G}_i$ are also a basis for $\mathcal{S}_i$. As before, the matrix R_i is upper triangular. The process of generating $\hat{G}_i$ from $p(A)\hat{G}_{i-1}$ can be viewed as a normalization process by which one basis of $\mathcal{S}_i$ is replaced by a "better" one.

If $|p(\lambda_k)| > |p(\lambda_{k+1})|$, the subspaces $\mathcal{S}_i$ will converge to the dominant k-dimensional invariant subspace. For ease of exposition, let us continue to assume that A is semisimple, and let $v_1, \ldots, v_n$ denote linearly independent eigenvectors, corresponding to eigenvalues $\lambda_1, \ldots, \lambda_n$ of A, respectively, ordered so that $|p(\lambda_1)| \geq |p(\lambda_2)| \geq \cdots \geq |p(\lambda_n)|$. Then the dominant k-dimensional subspace is $\mathcal{U}_k = \mathrm{span}\{v_1, \ldots, v_k\}$.

Simultaneous iteration yields an important additional bonus. Let $\hat{g}_1^i, \ldots, \hat{g}_k^{(i)}$ denote the columns of $\hat{G}_i$. Then the subspace iteration can be expressed as

$$\mathrm{span}\left\{ p(A)\hat{g}_1^{(i-1)}, \ldots, p(A)\hat{g}_k^{(i-1)} \right\} = \mathrm{span}\left\{ \hat{g}_1^{(i)}, \ldots, \hat{g}_k^{(i)} \right\}.$$

However, since R_i in (4.2.1) is upper triangular, we have more generally

$$\mathrm{span}\left\{ p(A)\hat{g}_1^{(i-1)}, \ldots, p(A)\hat{g}_j^{(i-1)} \right\} = \mathrm{span}\left\{ \hat{g}_1^{(i)}, \ldots, \hat{g}_j^{(i-1)} \right\}, \qquad \text{for } j = 1, \ldots, k,$$

by Theorem 3.2.2. Thus simultaneous iteration effects nested subspace iterations on spaces of dimensions $1, \ldots, k$ all at once. Consequently we also have

$$\mathrm{span}\left\{ \hat{g}_1^{(i)}, \ldots, \hat{g}_j^{(i)} \right\} \to \mathcal{U}_j = \mathrm{span}\left\{ v_1, \ldots, v_j \right\}$$

for each $j \leq k$ for which $|p(\lambda_j)| > |p(\lambda_{j+1})|$.

Up to the 1970s, simultaneous iteration was an important practical method for computing a few dominant eigenvalues and eigenvectors of large, sparse matrices. Since then it has been supplanted by Krylov subspace methods, which we will discuss in Chapter 9. Our motive for introducing simultaneous iteration at this point is to use it as a vehicle for the introduction of GR algorithms.

GR Algorithms

Now consider simultaneous iteration (4.2.1) with $k = n$, starting from $\hat{G}_0 = I$. Then all of the matrices in (4.2.1) are $n \times n$, and the $\hat{G}_i$ are nonsingular. The equation $p(A)\hat{G}_{i-1} = \hat{G}_i R_i$ is a full $n \times n$ GR decomposition in the sense of Section 3.2. Any of the methods discussed in that section can be used to effect the GR decomposition.

This n-dimensional simultaneous iteration effects subspace iterations of dimensions $1, \ldots, n - 1$ on $\mathrm{span}\{\hat{g}_1^{(i)}, \ldots, \hat{g}_j^{(i)}\}$, $j = 1, \ldots, n - 1$. Thus we have the possibility of convergence to subspaces of dimensions $1, 2, \ldots, n - 1$ simultaneously. We need to monitor this convergence somehow. There are numerous practical ways to do this, but we will consider here only one method, one that leads naturally to the GR algorithms. Suppose that for some j, $\mathrm{span}\{\hat{g}_1^{(i)}, \ldots, \hat{g}_j^{(i)}\}$ is invariant under A. Then, if we define

$$A_i = \hat{G}_i^{-1} A \hat{G}_i, \tag{4.2.2}$$

the invariance of $\mathrm{span}\{\hat{g}_1^{(i)}, \ldots, \hat{g}_j^{(i)}\}$ will be apparent in A_i. By Proposition 2.2.2, if we partition A_i as

$$A_i = \begin{bmatrix} A_{11}^{(i)} & A_{12}^{(i)} \\ A_{21}^{(i)} & A_{22}^{(i)} \end{bmatrix}, \tag{4.2.3}$$

where $A_{11}^{(i)}$ is $j \times j$, then $A_{21}^{(i)} = 0$. If $\text{span}\{\hat{g}_1^{(i)}, \ldots, \hat{g}_j^{(i)}\}$ is not invariant but merely close to an invariant subspace, then it is reasonable to expect that $A_{21}^{(i)} \approx 0$. If $\text{span}\{\hat{g}_1^{(i)}, \ldots, \hat{g}_j^{(i)}\}$ converges to an invariant subspace as $i \to \infty$, we expect $A_{21}^{(i)} \to 0$. That is, we expect A_i to tend to block triangular form. This convergence can be taking place for many values of j at once, so A_i will converge to block triangular form with a leading $j \times j$ block for many different j at once. Thus the iterates will tend toward a form

$$
B = \begin{bmatrix}
B_{11} & B_{12} & \cdots & B_{1m} \\
0 & B_{22} & & B_{2m} \\
\vdots & \ddots & \ddots & \vdots \\
0 & \cdots & 0 & B_{mm}
\end{bmatrix},
$$

in which the main-diagonal blocks $B_{11}, \ldots, B_{mm}$ are square and mostly of relatively small size. Their eigenvalues can then be determined with relatively little extra work. In the best case, when we have $|p(\lambda_j)| > |p(\lambda_{j+1})|$ for $j = 1, \ldots, n - 1$, A_i will tend to upper triangular form; each of the B_{jj} will be 1×1, consisting of a single eigenvalue.

It is thus evident that one way to monitor convergence of the subspace iterations is to carry out the similarity transformation (4.2.2) at each step and examine A_i. This seems like a lot of work, so it is natural to ask whether there is some way to produce A_i directly from A_{i-1}, circumventing the simultaneous iteration step (4.2.1). Indeed there is! For each i, define a nonsingular matrix G_i by $G_i = \hat{G}_{i-1}^{-1}\hat{G}_i$. Then, since $\hat{G}_i = \hat{G}_{i-1}G_i$ (and $\hat{G}_0 = I$), we easily check that

$$
\hat{G}_i = G_1 G_2 \cdots G_i
$$

and

$$
A_i = G_i^{-1} A_{i-1} G_i. \tag{4.2.4}
$$

Thus we can obtain A_i by this similarity transformation if we have G_i. Premultiplying (4.2.1) by $\hat{G}_{i-1}^{-1}$, we find that

$$
\hat{G}_{i-1}^{-1} p(A) \hat{G}_{i-1} = \hat{G}_{i-1}^{-1} \hat{G}_i R_i = G_i R_i
$$

or

$$
p(A_{i-1}) = G_i R_i. \tag{4.2.5}
$$

This shows that we can obtain G_i by performing a GR decomposition of $p(A_{i-1})$. Then we get A_i by the similarity transformation (4.2.4). Summarizing this on a single line, we have

$$
p(A_{i-1}) = G_i R_i, \qquad A_i = G_i^{-1} A_{i-1} G_i. \tag{4.2.6}
$$

This is an iteration of the *generic GR algorithm driven by p*. If the degree of the polynomial p is m, we say that the *degree* of the GR iteration is m. We call it *generic* because the nature of the GR decomposition is left unspecified. As we have seen in Chapter 3, there are many different GR decompositions. For example, if we do QR decompositions in (4.2.1), then the $\hat{G}_i$ are all unitary, and so are the G_i. Thus we should do QR decompositions in (4.2.6) to produce the correct A_i. In this case we normally use the symbol Q_i in place of G_i, and we have a step of the *QR algorithm driven by p*:

$$
p(A_{i-1}) = Q_i R_i, \qquad A_i = Q_i^{-1} A_{i-1} Q_i.
$$

Similarly, if we use LR decompositions, we get an LR *algorithm*, and so on.

The equations $p(A)\hat{G}_{i-1} = \hat{G}_i R_i$ from (4.2.1) and $p(A_{i-1}) = G_i R_i$ from (4.2.5) represent the same equation in different coordinate systems: We obtained the latter from the former by premultiplying by $\hat{G}_{i-1}^{-1}$. A vector that is represented by x in the old "A" coordinate system is represented by $\hat{G}_{i-1}^{-1}x$ in the new "A_{i-1}" coordinate system.[1] This transformation maps $\hat{g}_1^{(i-1)}, \ldots, \hat{g}_n^{(i-1)}$, the columns of $\hat{G}_{i-1}$, to $e_1, \ldots, e_n$, the columns of I. The columns of $p(A_{i-1})$ are $p(A_{i-1})e_1, \ldots, p(A_{i-1})e_n$, which correspond to $p(A)\hat{g}_1^{(i-1)}, \ldots, p(A)\hat{g}_n^{(i-1)}$ in the old coordinate system. The decomposition $p(A_{i-1}) = G_i R_i$ then produces $g_1^{(i)}, \ldots, g_n^{(i)}$ (the columns of G_i), a "better" basis such that

$$\text{span}\left\{g_1^{(i)}, \ldots, g_j^{(i)}\right\} = p(A_{i-1})\text{span}\left\{e_1, \ldots, e_j\right\}, \qquad j = 1, \ldots, n-1.$$

Finally, the similarity transformation $A_i = G_i^{-1} A_{i-1} G_i$ in (4.2.6) effects a further change of coordinate system, which maps $g_1^{(i)}, \ldots, g_n^{(i)}$ to $G_i^{-1}g_1^{(i)}, \ldots, G_i^{-1}g_n^{(i)}$, which are $e_1, \ldots, e_n$.

The picture is now complete. The GR iteration (4.2.6) effects a step of simultaneous iteration on the spaces $\text{span}\left\{e_1, \ldots, e_j\right\}$, $j = 1, \ldots, n$. It then performs a change of coordinate system that maps the resulting subspaces back to $\text{span}\left\{e_1, \ldots, e_j\right\}$, $j = 1, \ldots, n$. Thus the GR algorithm is nothing but simultaneous iteration with a change of coordinate system at each step. In ordinary simultaneous iteration, the matrix stays the same (A), and the subspaces change. In the GR algorithm, the matrix changes, and the subspaces stay the same. The effect of this is that the spaces $\text{span}\left\{e_1, \ldots, e_j\right\}$ become ever closer to being invariant subspaces (under A_i), and the A_i correspondingly approach block triangular form.

The Principle of Structure Preservation

As we noted above, there are many ways to choose our G matrices, and there are also many ways to choose p. What should guide our choices in a particular case? Often the choice is determined in part by the *principle of structure preservation:* If A has any significant structure, then we should design a GR iteration that preserves that structure. Observation of this principle usually results in algorithms that are superior in speed and accuracy. For example, if A is Hermitian ($A^* = A$), we would like all of the iterates A_k to be Hermitian as well. This can be achieved by taking G to be unitary, that is, by using the QR algorithm.

As another example, suppose that A is real. This is significant structure and should be preserved. Since A may well have some complex eigenvalues, it might appear that we would be forced out of the real number system, for if we want to approximate complex eigenvalues well, we must use complex shifts. The remedy is simple: For each complex shift μ that we use, we must also use the complex conjugate $\overline{\mu}$. This guarantees that $p(A)$ stays real. Then in the decomposition $p(A) = GR$, we can take G and R to be real to guarantee that we stay within the real field. These iterations can never approach upper triangular form, but they can approach quasi-triangular form, in which each complex-conjugate pair of eigenvalues is delivered as a real 2×2 block. Thus each complex-conjugate pair of eigenvalues is obtained together.

---

[1] For example, the equation $Ax = x\lambda$ corresponds to $(\hat{G}_{i-1}^{-1}A\hat{G}_{i-1})(\hat{G}_{i-1}^{-1}x) = (\hat{G}_{i-1}^{-1}x)\lambda$, i.e., $A_{i-1}(\hat{G}_{i-1}^{-1}x) = (\hat{G}_{i-1}^{-1}x)\lambda$.

Notes and References

The earliest works on simultaneous iteration were by Bauer [22, 23] around 1957 and by Voyevodin [210]. Simultaneous iteration was also discussed in the books by Householder [119] and Wilkinson [230].

The subject of GR algorithms began with Rutishauser's work on the quotient-difference and LR algorithms [178, 179, 180, 182], starting around 1954. The QR algorithm was introduced by Francis [89] and Kublanovskaya [139]. Wilkinson's book [230] contains extensive coverage of the LR and QR algorithms, including a lot of historical information. The first study of GR algorithms in general was by Della-Dora [71]. In that work the symbol G stood for *group*. In each GR decomposition, the G factor was required to belong to some matrix group (e.g., unitary group, symplectic group, etc.). For us the "G" stands for *generic*.

Connections between simultaneous iteration and GR algorithms were made by Bauer [23], Fadeev and Fadeeva [83], Householder [119], Wilkinson [230], Buurema [54], Parlett and Poole [172], Watkins [214], and probably others.

Exercises

4.2.1. Consider a GR iteration (4.2.6).

   (a) Show that in the case $p(z) = z$ we have $A_i = R_i G_i$. Thus in this case the iteration takes the simpler form

$$A_{i-1} = G_i R_i, \qquad R_i G_i = A_i.$$

   (b) Show that in general (arbitrary p) we have $p(A_i) = R_i G_i$.

4.2.2. A primitive QR algorithm with $p(z) = z$, as in Exercise 4.2.1, is very easy to program in MATLAB: Just apply the commands

$$[Q \ R] \ = \ qr(A); \ A \ = \ R*Q \qquad (4.2.7)$$

repeatedly.

   (a) Pick a small value of n like $n = 6$ and generate a random matrix of that size (e.g., A = randn(n);). Apply several iterations of (4.2.7) to A and watch for signs of convergence. (To do the iterations easily, make use of your up arrow key.)

   (b) Make an m-file that has (4.2.7) embedded in a loop. Put in a semicolon so that A doesn't get printed out on each iteration. Also put showmatrix(A) in the loop so that a color picture of the matrix gets printed out on each iteration. Run this on larger matrices (e.g., $n = 60$) for 200 or so iterations. (Get the file showmatrix.m from the website.[2] If you need help with this, see the file primitiveqr.m.)

4.2.3. Repeat Exercise 4.2.2 with the QR decomposition replaced by an LU decomposition ([L,R] = lu(A)). This is a primitive implementation of the basic LR algorithm with pivoting.

4.2.4. Let $B \in \mathbb{F}^{n \times n}$, and let $\mathcal{S}_{i-1}$ and $\mathcal{S}_i$ be subspaces of $\mathbb{F}^n$ such that $\mathcal{S}_i = B\mathcal{S}_{i-1}$.

---

[2]www.siam.org/books/ot101

(a) Show that

$$\dim(\mathcal{S}_{i-1}) = \dim(\mathcal{S}_i) + \dim(\mathcal{N}(B) \cap \mathcal{S}_{i-1}).$$

(Let $x_1, \ldots, x_m$ be a basis for $\mathcal{N}(B) \cap \mathcal{S}_{i-1}$. Then pick $x_{m+1}, \ldots, x_p$ so that $x_1, \ldots, x_p$ together form a basis for $\mathcal{S}_{i-1}$. Then show that $Bx_{m+1}, \ldots, Bx_p$ form a basis for $\mathcal{S}_i$.)

(b) Deduce that $\dim(\mathcal{S}_i) < \dim(\mathcal{S}_{i-1})$ if and only if $\mathcal{N}(B) \cap \mathcal{S}_{i-1} \neq \{0\}$.

4.2.5. Show that if $B = G^{-1}AG$ and p is a polynomial, then $p(B) = G^{-1}p(A)G$.

4.3 Shifting to Accelerate Convergence

The GR iteration (4.2.6) must be repeated numerous times to achieve convergence. This still looks expensive. We found in Section 3.2 that it costs $O(n^3)$ flops to do a GR decomposition. The similarity transformation in (4.2.6) also costs $O(n^3)$ in general, so the cost of generic GR iteration seems to be $O(n^3)$ flops. Even if we can find all of the eigenvalues in as few as n steps (an ambitious hope), the total cost of running the algorithm will be $O(n^4)$, which is too much. Moreover, the total number of steps needed may be much more than n. If we wish to make the GR algorithm attractive, we need to do two things. (1) We must develop a strategy for accelerating convergence, so that fewer total iterations are needed. (2) We must figure out how to effect each iteration economically.

Our first step toward economy will be to incorporate a preprocessing step that reduces the matrix to upper Hessenberg form before beginning the GR iterations. We have seen in Section 3.3 how this can be done. It is an expensive process ($O(n^3)$ flops), but, as we shall see, it only needs to be done once: Once we have obtained Hessenberg form, the GR iterations preserve it. This will be demonstrated later; for now let us simply assume that the GR iterates A_i in (4.2.6) are all upper Hessenberg.

We now focus on how to choose p, which is equivalent to choosing the shifts $\rho_1, \ldots, \rho_m$. Ultimately we intend to use a different p on each iteration, but to start let us assume that our iteration is stationary. Even though we generate the iterates by (4.2.6), they still satisfy

$$A_i = \hat{G}_i^{-1} A \hat{G}_i,$$

where spaces spanned by columns of $\hat{G}_i$ are being driven toward invariant subspaces by simultaneous iterations (4.2.1). If $|p(\lambda_j)| > |p(\lambda_{j+1})|$, then the space spanned by the first j columns of $\hat{G}_i$ is converging to the dominant j-dimensional invariant subspace of $p(A)$ at a rate determined by $|p(\lambda_{j+1})/p(\lambda_j)|$. As a consequence, $A_{21}^{(i)} \to 0$, using the notation of (4.2.3).

Notice that our assumption that the matrices are upper Hessenberg implies that the form of $A_{21}^{(i)}$ is extremely simple: Its only nonzero entry is $a_{j+1,j}^{(i)}$. Thus convergence is very easy to monitor: $A_{21}^{(i)} \to 0$ if and only if $a_{j+1,j}^{(i)} \to 0$.

If for some i, $a_{j+1,j}^{(i)} = 0$ exactly, then the space spanned by the first j columns of $\hat{G}_i$ is invariant. It is exactly the invariant subspace associated with the dominant eigenvalues $p(\lambda_1), \ldots, p(\lambda_j)$. (We continue to assume that the eigenvalues are ordered so that $|p(\lambda_1)| \geq \cdots \geq |p(\lambda_n)|$.) In the semisimple case, this space is just span$\{v_1, \ldots, v_j\}$. This implies that the eigenvalues of $A_{11}^{(i)}$ are $\lambda_1, \ldots, \lambda_j$. The eigenvalues of $A_{22}^{(i)}$ must therefore be $\lambda_{j+1}, \ldots, \lambda_n$.

In reality we never have $a_{j+1,j}^{(i)} = 0$ exactly, but $a_{j+1,j}^{(i)}$ will eventually be small after we have taken enough steps. Then it is reasonable to expect that the eigenvalues of $A_{22}^{(i)}$ will be close to $\lambda_{j+1}, \ldots, \lambda_n$. Recall that good approximations to eigenvalues make good shifts. In Section 4.1 we observed that if the shifts $\rho_1, \ldots, \rho_m$ are good approximations to m distinct eigenvalues of A, then m of the eigenvalues of $p(A)$ will be small, $|p(\lambda_{n-m+1})/p(\lambda_{n-m})|$ will be small, and the subspace iteration of dimension $j = n - m$ will converge rapidly. We then expect $a_{n-m+1,n-m}^{(i)}$ to converge rapidly to zero.

This is our clue to how to choose good shifts. Consider the matrix $A_{22}^{(i)}$ from (4.2.3) in the case $j = n - m$. $A_{22}^{(i)}$ is the lower right-hand $m \times m$ submatrix of A_i. If $a_{n-m+1,n-m}^{(i)}$ is small, the eigenvalues of $A_{22}^{(i)}$ will be good approximations to $\lambda_{n-m+1}, \ldots, \lambda_n$. If we take the eigenvalues of $A_{22}^{(i)}$ as shifts to use to build a new polynomial p to use on subsequent steps, this should hasten convergence of $a_{n-m+1,n-m}^{(i)}$ to zero as $i \to \infty$. After a few more steps, of course, the eigenvalues of the new $A_{22}^{(i)}$ should be even better approximations of $\lambda_{n-m+1}, \ldots, \lambda_n$, so we should use these to form a new p to make the convergence even faster.

Some questions suggest themselves immediately. How do we choose shifts to begin with? How small does $a_{n-m+1,n-m}^{(i)}$ need to be before we dare to use the eigenvalues of $A_{22}^{(i)}$ as shifts? How often should we choose new shifts? Our answers to these questions are based on experience, not theory. Experience has shown that it is almost always okay to start using the eigenvalues of $A_{22}^{(i)}$ as shifts right from very beginning. In other words, starting from A_0, use the eigenvalues of $A_{22}^{(0)}$ as shifts, regardless of whether or not $a_{n-m+1,n-m}^{(0)}$ is small. Now, how many iterations should we take before picking a new set of shifts? Again experience has shown that it is generally a good idea to compute new shifts on every iteration. This gives the dynamic *shifted GR algorithm*:

$$
\begin{aligned}
&\text{for } i = 1, 2, 3, \ldots \\
&\left[\begin{array}{l}
\text{compute } \rho_1^{(i)}, \ldots, \rho_m^{(i)}, \text{ the eigenvalues of } A_{22}^{(i-1)} \\
\text{let } p_i(A_{i-1}) = (A_{i-1} - \rho_1^{(i)}I) \cdots (A_{i-1} - \rho_m^{(i)}I) \\
p_i(A_{i-1}) = G_i R_i \\
A_i = G_i^{-1} A_{i-1} G_i
\end{array}\right.
\end{aligned}
\tag{4.3.1}
$$

With this strategy it often happens that little progress is made in the first iterations, since the shifts are poor approximations to eigenvalues. But after a few steps, the shifts begin to approach eigenvalues, and progress becomes more rapid. In Section 5.2 we will prove a very satisfying result: Under generic conditions the shifted GR algorithm (4.3.1) is locally quadratically convergent. By this we mean that the underlying subspace iteration of dimension $n - m$ converges quadratically to an $(n - m)$-dimensional invariant subspace, and $a_{n-m+1,n-m}^{(i)}$ converges quadratically to zero as $i \to \infty$. The quadratic convergence is a not-unexpected result of the positive feedback between the improving shifts and the shrinking magnitude of $a_{n-m+1,n-m}^{(i)}$.

There are a number of details that remain to be discussed. The most important is how to perform the iterations (4.3.1) economically, but we will defer that to Section 4.5. The quadratic convergence of $a_{n-m+1,n-m}^{(i)}$ to zero implies that it will rapidly be reduced to a small enough magnitude that it can safely be regarded as zero. A commonly accepted criterion is this: Let u denote the unit roundoff of the computer. In double precision, on machines that adhere to the IEEE floating point standard [160], $u \approx 10^{-16}$. If

$$|a_{n-m+1,n-m}^{(i)}| < u \left(|a_{n-m,n-m}^{(i)}| + |a_{n-m+1,n-m+1}^{(i)}|\right), \qquad (4.3.2)$$

then set $a_{n-m+1,n-m}^{(i)}$ to zero. This criterion ensures that the error incurred by this perturbation is comparable in size to the roundoff errors that are made again and again in the course of the computation.

Once we have set $a_{n-m+1,n-m}^{(i)}$ to zero, we have

$$A_i = \left[\begin{array}{cc} A_{11}^{(i)} & A_{12}^{(i)} \\ 0 & A_{22}^{(i)} \end{array} \right],$$

and the problem can be reduced to a smaller size. The eigenvalues of the small matrix $A_{22}^{(i)}$ can easily be computed. These are m eigenvalues of A. The remaining eigenvalues are all eigenvalues of $A_{11}^{(i)}$, so we can continue the GR iterations on this submatrix instead of A_i. This process is called *deflation*. We grab new shifts from the lower $m \times m$ submatrix of $A_{11}^{(i)}$ and proceed to pursue the next batch of m eigenvalues. Thanks to quadratic convergence, we will soon be able to break off the next $m \times m$ submatrix and deflate the problem further. After approximately n/m such deflations, we will have computed all of the eigenvalues of A.

In order to carry out this program, we need an auxiliary routine that calculates eigenvalues of the $m \times m$ submatrices $A_{22}^{(i)}$. Bear in mind that m is small. If $m = 1$, we have no problem, since the eigenvalue of a 1×1 matrix is evident. If $m = 2$, we can compute the eigenvalues by a carefully implemented quadratic formula.[3] If $m > 2$, we employ an auxiliary GR routine that does steps of degree 1 or 2 to compute the eigenvalues of the $m \times m$ submatrices. For example, in LAPACK 3.0 [8] (which has been superseded) the main QR routines were normally run with $m = 6$, but there was also an auxiliary QR routine with $m = 1$ (in the complex case) or $m = 2$ (real case) to compute eigenvalues of small submatrices.[4]

Our focus has been on the entry $a_{n-m+1,n-m}^{(i)}$, which is driven rapidly to zero by subspace iterations of dimension $n - m$. Let us not forget, however, that the GR algorithm is an implementation of simultaneous iteration, and as such it effects subspace iterations of dimensions $1, 2, \ldots, n - 1$ simultaneously. As the subspace iteration of dimension j converges, it pushes $a_{j+1,j}^{(i)}$ to zero. Except for the iteration of dimension $n - m$, these subspace iterations tend to converge quite slowly. Nevertheless, they do have the effect that many of the subdiagonal entries $a_{j+1,j}^{(i)}$ drift toward zero. It behooves us, therefore, to keep tabs on these subdiagonal entries. As soon as any one of them satisfies

$$|a_{j+1,j}^{(i)}| < u \left(|a_{jj}^{(i)}| + |a_{j+1,j+1}^{(i)}|\right),$$

---

[3]Serious implementations include many details that we will not discuss here. For example, the QR algorithm code for real matrices in LAPACK [8, 188] checks whether the eigenvalues of a 2×2 matrix are real or complex. If they are real, it does an additional orthogonal similarity transformation to triangularize the matrix and thereby split out the real eigenvalues. If the eigenvalues are complex, an orthogonal similarity transform is used to transform the matrix to the form $\left[\begin{array}{cc} a & b \\ -c & a \end{array} \right]$, where $b, c > 0$, which has eigenvalues $a \pm i\sqrt{bc}$.

[4]Features of the newer LAPACK QR routines are discussed in Section 4.9.

we should set $a_{j+1,j}^{(i)}$ to zero and split the problem into two smaller subproblems. When applying the GR algorithm to large matrices (e.g., $n > 1000$), decouplings of this type occur commonly. This can result in considerable economy since, for example, the reduction of two matrices of dimension $n/2$ generally takes much less work than the reduction of a single matrix of dimension n.

Another effect of the gradual convergence is that the average quality of the shifts increases as the iterations proceed. In the beginning the shifts may be very bad approximations to eigenvalues, and it may take quite a few iterations before the the first deflation occurs. Once it has occurred, we begin to go after the second batch of m eigenvalues by computing the eigenvalues of the trailing $m \times m$ submatrix of the remaining matrix $A_{11}^{(i)}$. Since some progress toward convergence has already taken place, these new shifts may be better approximations of eigenvalues than the original shifts were. Consequently it may take fewer iterations to deflate out the second batch. Similarly the third batch may take fewer iterations than the second batch, and so on. This is not a rule; sometimes the second deflation requires more iterations than the first did. But there is a definite long-term tendency toward fewer iterations per deflation as the algorithm proceeds. By the way, experience suggests that the QR algorithm with $m = 2$ requires something like four iterations per deflation on average, when applied to random matrices. However, this number should not be taken too seriously.

Before moving on to the next topic, we must not forget to admit that the strategy for choosing shifts that we have advocated here does not always work. Exercise 4.3.5 gives an example of failure, in which the ratios $|p(\lambda_{j+1})/p(\lambda_j)|$ are all exactly 1. Because of the existence of matrices like the one in Exercise 4.3.5, the standard codes incorporate an exceptional shift strategy: Whenever ten iterations (using standard shifts as described above) have passed with no deflations having occurred, an iteration with *exceptional shifts* is taken. These ad hoc shifts can be chosen at random or by any arbitrary procedure. Their purpose is merely to break up any symmetries that might be preventing convergence.

The shift strategy that we have advocated here is known as the *generalized Rayleigh quotient shift strategy*. It is widely used, but other strategies are possible. For example, one could compute the eigenvalues of the trailing matrix of dimension $m + k$ for some $k > 0$ and somehow pick out the "most promising" m eigenvalues and use them as shifts, discarding the other k. An instance of this is the *Wilkinson shift* for the symmetric eigenvalue problem. The two eigenvalues of the lower right-hand 2×2 submatrix are computed. The one that is closer to $a_{nn}^{(i-1)}$ is used as the shift for a QR step of degree one, and the other is discarded. There is no end to the variety of shift strategies that can be dreamt up, but any good strategy will be based on what is happening in the lower right-hand corner of the matrix.

Exercises

4.3.1. It is easy to do simple experiments in MATLAB that demonstrate the rapid convergence that usually results from shifting. Create a random complex upper Hessenberg matrix. For example: n = 10; A = randn(n) + i*randn(n); A = hess(A). Then apply a sequence of shifted QR iterations of degree one to A:

```
[Q,R] = qr(A - A(n,n)*eye(n)); A = Q'*A*Q; z = abs(A(n,n-1)).
```

Iterate until $z < 10^{-16}$.

4.3.2. Repeat Exercise 4.3.1 using a real symmetric matrix. For example: `A = randn(n);` `A = A + A'; A = hess(A).` How does the rate of convergence here compare with what you saw in Exercise 4.3.1?

4.3.3. This exercise shows that you don't always get what you expect. Repeat Exercise 4.3.1 using the matrix generated by the command

```
A = toeplitz([2 1 zeros(1,n-2)]).
```

(a) Try an odd value of n, say $n = 5$. Can you explain what you saw?

(b) Try an even value of n, say $n = 4$. Look at the whole matrices. Do you have any idea what is going on here? Can you suggest a remedy?

4.3.4. Let

$$A = \begin{bmatrix} 1 & 2 & -1 & 1 \\ -1 & 1 & -1 & 1 \\ 0 & 1 & 1 & 1 \\ 0 & 0 & -1 & 1 \end{bmatrix}.$$

Here is a crude and dirty implementation of a QR iteration of degree two that you can apply to A with $n = 4$:

```
shift = eig(A(n-1:n,n-1:n));
[Q R] = qr(real((A-shift(1)*eye(n))*(A-shift(2)*eye(n))));
A = Q'*A*Q
```

Do several iterations. Look at the matrices using `format short e`. What evidence for convergence do you see?

4.3.5. Let A denote the $n \times n$ circulant shift matrix

$$A = \begin{bmatrix} e_2 & e_3 & \cdots & e_n & e_1 \end{bmatrix}.$$

The 4×4 version of A is

$$\begin{bmatrix} & & & 1 \\ 1 & & & \\ & 1 & & \\ & & 1 & \end{bmatrix}.$$

(a) Show that A is unitary.

(b) Show that if we choose shifts from the lower right-hand $m \times m$ submatrix of A with $m < n$, all the shifts are zero. Consequently $p(A) = A^m$, which is also unitary.

(c) Show that in the QR decomposition $p(A) = QR$ (which is unique by Theorem 1.3.8), $Q = p(A)$, and $R = I$. Deduce that the QR iteration $A_1 = Q^{-1}AQ$ makes no progress; that is, $A_1 = A$.

(d) Show that the eigenvalues of A are the nth roots of unity $\lambda_k = e^{2k\pi i/n}$, and deduce that $|p(\lambda_{k+1})/p(\lambda_k)| = 1$ for $k = 1, \ldots, n-1$.

The zero shifts are poor approximations to all of the eigenvalues and equally far from all of them, so no eigenvalue is favored over the others. An absolute deadlock results. One iteration with random shifts will suffice to break up the deadlock and start progress toward convergence.

4.4 The Differential Quotient-Difference Algorithm

Most practical GR algorithms are implemented implicitly as bulge-chasing algorithms, as will be described in the Section 4.5. An interesting exception is the differential quotient-difference (dqd) algorithm, which we now describe. This excellent high-accuracy algorithm has become one of LAPACK's main tools for computing eigenvalues of symmetric tridiagonal matrices. This section can be skipped without loss of continuity, as it is independent of our main development. Moreover, an independent derivation of the dqd algorithm as a bulge-chasing algorithm will be presented in Section 8.5.

We begin with the Cholesky variant of the LR algorithm. Let T be a Hermitian, positive-definite matrix, and let B denote its Cholesky factor (Theorem 1.4.1). One iteration (with zero shift) is as follows:

$$T = B^*B, \qquad BB^* = T_1. \tag{4.4.1}$$

That is, compute the Cholesky decomposition of T, then reverse the factors and multiply them back together to get the next iterate T_1. Since $T_1 = B^{-*}TB^* = BTB^{-1}$, this is a GR iteration in which the roles of G and R are played by B^* and B, respectively. The iteration preserves the symmetry. It also preserves bandedness: If T is banded with bandwidth $2s + 1$, then so is T_1 (Exercise 4.4.1). Of course, the factor B is also banded. Consequently the operation count for Cholesky LR steps on narrow-band matrices is much less than for full matrices. In the most extreme case, T is tridiagonal. One can always transform a positive definite matrix to real, symmetric, tridiagonal form by the method of Section 3.3, using unitary transformations. Then the tridiagonal form is retained under the Cholesky LR iterations, and each iteration requires only $O(n)$ computations. It is easy to incorporate shifts; however, each shift must be smaller than the smallest eigenvalue (among those that have not yet been deflated), since otherwise the shifted matrix will not be positive definite (Exercise 2.3.9). Exercise 4.4.2 shows that two (zero-shift) Cholesky LR steps are equal to one (zero-shift) QR iteration.

Another way of preserving the symmetric positive definite structure begins by deliberately breaking the symmetry. Suppose that T is real and tridiagonal, say

$$T = \begin{bmatrix} t_1 & s_1 & & & \\ s_1 & t_2 & \ddots & & \\ & \ddots & \ddots & s_{n-1} \\ & & s_{n-1} & t_n \end{bmatrix} \tag{4.4.2}$$

with all $s_j > 0$. Since T is positive definite, all of the t_j are positive as well. It is easy to show (Exercise 4.4.3) that there is a diagonal matrix $D = \mathrm{diag}\{d_1, \ldots, d_n\}$ with all $d_i > 0$

such that

$$
\tilde{T} = D^{-1}TD = \begin{bmatrix} t_1 & 1 & & & \\ s_1^2 & t_2 & \ddots & & \\ & \ddots & \ddots & 1 & \\ & & s_{n-1}^2 & t_n \end{bmatrix}. \tag{4.4.3}
$$

Conversely, any matrix of the form (4.4.3) can be transformed to the symmetric form (4.4.2) by the inverse diagonal similarity transformation. Thus the map ϕ defined by $T \mapsto \tilde{T}$ gives a one-to-one correspondence between matrices of the forms (4.4.2) and (4.4.3).

$\tilde{T}$ has a unique decomposition $\tilde{T} = LR$, where L is unit lower triangular and R is upper triangular (Theorem 1.3.4). It is easy to show (Exercise 4.4.4) that L and R are both bidiagonal and that, moreover, the superdiagonal of R consists entirely of ones:

$$
\tilde{T} = LR = \begin{bmatrix} 1 & & & \\ l_1 & 1 & & \\ & \ddots & \ddots & \\ & & l_{n-1} & 1 \end{bmatrix} \begin{bmatrix} r_1 & 1 & & \\ & r_2 & \ddots & \\ & & \ddots & 1 \\ & & & r_n \end{bmatrix}. \tag{4.4.4}
$$

The entries l_j and r_j are all positive.

Since $\tilde{T}$ is so closely related to T, we expect there to be a close relationship between the LR decomposition of $\tilde{T}$ and the Cholesky decomposition $T = B^*B$. Let

$$
B = \begin{bmatrix} \alpha_1 & \beta_1 & & \\ & \alpha_2 & \ddots & \\ & & \ddots & \beta_{n-1} \\ & & & \alpha_n \end{bmatrix}
$$

with all α_j and β_j positive. We can certainly obtain L and R from B and vice versa via T and $\tilde{T}$, but it ought to be possible to take a more direct route. Indeed it is; Exercise 4.4.5 shows that $l_j = \beta_j^2$ and $r_j = \alpha_j^2$ for all j.

Related to the Cholesky LR algorithm we have the standard LR algorithm with zero shift:

$$
\tilde{T} = LR, \qquad RL = \tilde{T}_1. \tag{4.4.5}
$$

Since $\tilde{T}_1 = L^{-1}TL$, this is a GR iteration with $G = L$. It is easy to show that the special form of $\tilde{T}$ is preserved by this iteration. That is, $\tilde{T}_1$ is tridiagonal with 1's on the superdiagonal and positive entries on the main diagonal and subdiagonal. Furthermore, the effect of the LR step is easily seen to be essentially the same as that of the Cholesky LR step (Exercise 4.4.6).

A sequence of LR iterations produces a sequence of iterates $\tilde{T}, \tilde{T}_1, \tilde{T}_2, \tilde{T}_3, \dots$. Each iterate is obtained from the previous one by computing an LR decomposition, reversing the factors, and multiplying them back together. The quotient-difference (qd) algorithm is obtained from the LR algorithm by shifting our attention from the iterates to the decompositions. Given the LR factors of, say, T_j, we compute the LR factors of T_{j+1} directly, without computing the intermediate result T_{j+1}. Before we explain how to do this, perhaps we should say a word about why it might be worthwhile. It turns out that the factors encode the eigenvalues more accurately than the tridiagonal matrices themselves do. If we work exclusively with the factors, we have a chance to come up with a more accurate algorithm. This issue will be discussed in Section 8.5.

To see how to jump from one set of factors to the next, consider the first step. Say we have L and R from (4.4.5). We know that $RL = \tilde{T}_1$. Then to do the next step, we need the decomposition $\tilde{T}_1 = \hat{L}\hat{R}$. The challenge is to get directly from L and R to $\hat{L}$ and $\hat{R}$ without ever computing $\tilde{T}_1$. If the algorithm is to be effective, it must allow for shifts, so let us introduce a shift into the step. Instead of computing the decomposition of $\tilde{T}$, we will decompose the shifted matrix $\tilde{T}_1 - \rho I$. Our basic equations are therefore

$$RL = \tilde{T}_1, \qquad \tilde{T}_1 - \rho I = \hat{L}\hat{R}.$$

Eliminating T_1 from these equations, we obtain

$$RL - \rho I = \hat{L}\hat{R}. \tag{4.4.6}$$

We have L and R, which means we have positive numbers $l_1, \dots, l_{n-1}$ and $r_1, \dots, r_n$, as shown in (4.4.4). Since $\hat{L}$ and $\hat{R}$ have the same general form as L and R, computing them amounts to computing $2n - 1$ numbers $\hat{l}_1, \dots, \hat{l}_{n-1}$ and $\hat{r}_1, \dots, \hat{r}_n$. It is very easy to see how to do this. Just compute both sides of (4.4.6) and equate them to get

$$r_j + l_j - \rho = \hat{r}_j + \hat{l}_{j-1}, \qquad j = 1, \dots, n, \tag{4.4.7}$$

with the convention that $\hat{l}_0 = 0 = l_n$ and

$$r_{j+1}l_j = \hat{l}_j\hat{r}_j, \qquad j = 1, \dots, n - 1, \tag{4.4.8}$$

then reorganize these equations to get the algorithm

$$
\begin{aligned}
&\hat{l}_0 \leftarrow 0 \\
&\text{for } j = 1, \dots, n - 1 \\
&\quad \left[\begin{array}{l} \hat{r}_j \leftarrow r_j + l_j - \rho - \hat{l}_{j-1} \\ \hat{l}_j \leftarrow l_j r_{j+1}/\hat{r}_j \end{array} \right. \\
&\hat{r}_n \leftarrow r_n - \rho - \hat{l}_{n-1}.
\end{aligned}
\tag{4.4.9}
$$

This is one iteration of the *shifted quotient-difference* algorithm (qds). If we apply it repeatedly, we perform a sequence of shifted LR iterations without ever forming the tridiagonal matrices $\tilde{T}_j$. As l_{n-1} tends to zero, r_n converges to an eigenvalue.

This is not our final version of the algorithm, but before we go on, we should make some observations about the shifts. First of all, we are handling them differently than we did earlier in this section. At each step we subtract off a shift, but we never restore it. Since $RL - \rho I = \hat{L}\hat{R}$, the eigenvalues of $\hat{L}\hat{R}$ are all ρ less than those of RL. In the next iteration we will work with $\hat{R}\hat{L}$, which has the same eigenvalues as $\hat{L}\hat{R}$ since those two matrices are similar ($\hat{R}\hat{L} = \hat{L}^{-1}(\hat{L}\hat{R})\hat{L}$). If we wanted to get back to the original eigenvalues, we should work with $\hat{R}\hat{L} + \rho I$. We do not do that, as it is pointless. Instead we just keep track of the accumulated shift $\hat{\rho}$. We start with $\hat{\rho} = 0$, and at each iteration we add the current shift to $\hat{\rho}$ to get a new $\hat{\rho}$. Then, whenever an eigenvalue emerges, we have to add $\hat{\rho}$ to it to get a correct eigenvalue of the original matrix.

The other point about the shifts is that they must not be too large. We know that (at each iteration) the r_j and l_j quantities are all positive because of the connection with the Cholesky variant of the LR algorithm. This depends on positive definiteness. If we take the shift too big, the shifted matrix $\tilde{T} - \rho I$ will no longer be (similar to) a positive definite matrix, and negative r_j and l_j quantities will appear. As we shall see below, this is something we would like to avoid. In order to maintain positive definiteness, we must always choose shifts that are smaller than the smallest eigenvalue. That way, the shifted matrix will always have positive eigenvalues. Our objective in shifting is to get close to an eigenvalue, so obviously we must try to make ρ as close to the smallest eigenvalue as possible without overstepping it. If we do accidentally exceed it, negative quantities will appear, and we will be forced to restart the iteration with a smaller value of ρ. Once we have deflated an eigenvalue, we can safely go after the next one. At any stage, we just need to make sure the shift is no larger than the smallest eigenvalue that has not yet been removed by deflation.

Algorithm (4.4.9) is not as accurate as it could be because it contains some unnecessary subtractions. To obtain a superior algorithm, introduce the auxiliary quantities $\tilde{r}_j = r_j - \rho - \hat{l}_{j-1}$, $j = 1, \ldots, n$, where $\hat{l}_0 = 0$. Then (4.4.9) can be rewritten (Exercise 4.4.7) as

$$
\begin{aligned}
&\tilde{r}_1 \leftarrow r_1 - \rho \\
&\text{for } j = 1, \ldots, n-1 \\
&\quad \left[
\begin{aligned}
&\hat{r}_j \leftarrow \tilde{r}_j + l_j \\
&q \leftarrow r_{j+1}/\hat{r}_j \\
&\hat{l}_j \leftarrow l_j q \\
&\tilde{r}_{j+1} \leftarrow \tilde{r}_j q - \rho
\end{aligned}
\right. \\
&\hat{r}_n \leftarrow \tilde{r}_n.
\end{aligned}
\tag{4.4.10}
$$

This is *dqds*, the *differential* version of the shifted qd algorithm, and it has proved to be remarkably stable. Consider first the zero-shift case. Examining (4.4.10), we see that when $\rho = 0$, the algorithm involves no subtractions. There is only one addition in the loop, and the two quantities that are added are both positive. Thus there is no possibility of cancellation, and all quantities in (4.4.10) are computed to high relative accuracy. If the process is repeated to convergence, the resulting eigenvalue will be computed to high relative accuracy, even if it is very tiny. This finding is supported by a simple MATLAB experiment in Exercise 4.4.9.

If the shift is nonzero but small enough to preserve positive definiteness, then all quantities remain positive. The only cancellation occurs when the shift is subtracted. However, whatever is subtracted is no greater than what will be restored when the accumulated shift is added to the computed eigenvalue in the end. It follows that every eigenvalue is computed to high relative accuracy.

The algorithm can also be used in the nonpositive case (i.e., shift larger than the smallest eigenvalue), and it can even be applied to nonsymmetrizable tridiagonal matrices ((4.4.3) with some of the s_j^2 replaced by negative numbers), but in either case, negative values of r_j and l_j enter the computation, and the accuracy guarantee is lost. This does not imply that the algorithm will necessarily give bad results. However, if at some stage one of the $\hat{r}_j$ should be exactly zero, the algorithm will break down because of a division by zero.

The qd algorithm is of great historical interest. Introduced by Rutishauser in the 1950s [178, 179, 180], it was the first GR algorithm. That makes Rutishauser the grandfather of this subject. See the original work to understand why the algorithm is called what it is. (However, it is certainly clear from (4.4.9) that quotients and differences are involved.) The differential formulation of the algorithm first appeared in Rutishauser's book [182], which was published posthumously. It is interesting that this version of the algorithm is so successful precisely because it gets rid of the differences. They are hidden in the auxiliary quantities $\tilde{r}_j$, which are obtained by division and multiplication, not by subtraction (except for the shift). The dqd algorithm was largely forgotten until the 1990s, when Fernando and Parlett [87] and Parlett [170] pointed out its excellent characteristics and advocated its use.

In Section 8.5 we will present an independent derivation of the dqd algorithm as a bulge-chasing algorithm. An advantage of that approach is that the auxiliary quantities $\tilde{r}_j$ arise naturally, whereas here they were introduced without motivation. An advantage of the approach given here is that it is easy to see how to introduce shifts, whereas in the bulge-chasing derivation it is not clear how to shift.

Exercises

4.4.1. Consider a Cholesky LR step (4.4.1).

(a) Show that $T_1 = BTB^{-1}$.

(b) A matrix A is called s-Hessenberg if $a_{ij} = 0$ whenever $i - j > s$. (Thus, for example, an upper Hessenberg matrix is 1-Hessenberg.) Show that if A is s-Hessenberg and C is upper triangular, then both AC and CA are s-Hessenberg.

(c) Show that a Hermitian matrix T is s-Hessenberg if and only if it is banded with band width $2s + 1$.

(d) Now, returning to the Cholesky LR step, suppose that T has band width $2s + 1$. Use the equation from part (a) to show that T_1 is s-Hessenberg, hence banded with band width $2s + 1$.

(e) Show that T_1 is tridiagonal if T is.

4.4.2. Let T be positive definite, and let T_2 be the matrix obtained after two Cholesky LR iterations:

$$T = B_1^* B_1, \qquad B_1 B_1^* = T_1,$$
$$T_1 = B_2^* B_2, \qquad B_2 B_2^* = T_2.$$

Let A be the matrix obtained from T after one QR iteration with zero shift:

$$T = QR, \qquad RQ = Q^{-1}TQ = A.$$

(a) Show that $T_2 = (B_2 B_1)T(B_2 B_1)^{-1}$.

(b) Show that $A = RTR^{-1}$.

(c) Show that $T^2 = (B_2 B_1)^*(B_2 B_1)$.

(d) Show that $T^2 = T^*T = R^*R$.

(e) Invoke uniqueness of the Cholesky decomposition to deduce that $R = B_2 B_1$, and hence $T_2 = A$. Thus two (zero-shift) Cholesky LR iterations equal one (zero-shift) QR iteration.

4.4.3. Let T be a real, symmetric, tridiagonal matrix as in (4.4.2) with $s_j > 0$ for all j. Show that there is a diagonal matrix $D = \text{diag}\{d_1, \ldots, d_n\}$ with all $d_j > 0$ such that $\tilde{T} = D^{-1} T D$ has the form

$$
\begin{bmatrix}
\hat{t}_1 & 1 & & \\
\hat{s}_1 & \hat{t}_2 & \ddots & \\
& \ddots & \ddots & 1 \\
& & \hat{s}_{n-1} & \hat{t}_n
\end{bmatrix}.
$$

Show that D is not unique; however, once d_1 (or any one of the d_i) has been chosen arbitrarily, then all d_j are uniquely determined by the recurrence $d_{j+1} = d_j / s_j$. Show furthermore that $\hat{t}_j = t_j$ for $j = 1, \ldots, n$, and $\hat{s}_j = s_j^2$ for $j = 1, \ldots, n-1$. Thus $\tilde{T}$ has the form shown in (4.4.3).

4.4.4. Show that there exist unique positive $l_1, \ldots, l_{n-1}$ and $r_1, \ldots, r_n$ such that the equation $\tilde{T} = LR$ given by (4.4.4) holds. Write down an algorithm that computes the l_j and r_j in $O(n)$ flops. (Positivity of all of the r_j can be inferred from the positive definiteness of T, or it can be deferred to Exercise 4.4.5.)

4.4.5. Let T and $\tilde{T} = D^{-1} T D$ be as in (4.4.2) and (4.4.3), respectively, and let $T = B^T B$ be the Cholesky decomposition and $\tilde{T} = LR$ the LR decomposition, as in (4.4.4).

(a) Use the uniqueness of the LR decomposition to show that there is a unique diagonal matrix $C = \text{diag}\{c_1, \ldots, c_n\}$ with positive main-diagonal entries such that $L = D^{-1} B^T C$ and $R = C^{-1} B D$. Show that $c_j = d_j / \alpha_j$, $j = 1, \ldots, n$.

(b) Show that the entries in the LR factorization satisfy $l_j = \beta_j^2$ and $r_j = \alpha_j^2$.

4.4.6. Let T be symmetric, positive definite, and tridiagonal with positive off-diagonal entries, as in (4.4.2), and suppose $\tilde{T} = \phi(T)$, where ϕ is the one-to-one map $T \mapsto \tilde{T}$ with $\tilde{T}$ given by (4.4.3). Let T_1 and $\tilde{T}_1$ be given by one Cholesky LR step (4.4.1) and one standard LR step (4.4.5), respectively.

(a) Show that $\tilde{T}_1$ is tridiagonal with 1's on the superdiagonal and positive entries on the main and subdiagonals.

(b) Show that $\tilde{T}_1 = \phi(T_1)$. Thus the Cholesky LR and standard LR steps have the same effect.

4.4.7.

(a) Compute the left- and right-hand sides of (4.4.6), and infer that (4.4.6) implies (4.4.7) and (4.4.8) if we make the conventions $\hat{l}_0 = 0$ and $l_n = 0$.

(b) Derive algorithm (4.4.9) from (4.4.7) and (4.4.8).

(c) Let $\tilde{r}_j = r_j - \rho - \hat{l}_{j-1}$, $j = 1, \ldots, n$, where $\hat{l}_0 = 0$. Show that in algorithm (4.4.9), $\tilde{r}_j = \hat{r}_j - l_j$. Introducing $q_j = r_{j+1}/\hat{r}_j$, show that $\tilde{r}_{j+1} = (\hat{r}_j - l_j)q_j - \rho = \tilde{r}_j q - \rho$. Deduce that (4.4.10) is equivalent to (4.4.9).

4.4.8. Show that if T is tridiagonal and $T = LR$, where L and R are of the form (4.4.4) and $l_{n-1} = 0$, then r_n is an eigenvalue of T. Show that the remaining eigenvalues of T are the eigenvalues of $L_{n-1} R_{n-1}$, where L_{n-1} and R_{n-1} are obtained from L and R, respectively, by deleting the nth row and column.

4.4.9. Let

$$T = LR = \begin{bmatrix} 1 & & \\ x & 1 & \\ & x & 1 \end{bmatrix} \begin{bmatrix} 1 & 1 & \\ & 1 & 1 \\ & & 1 \end{bmatrix},$$

where $x = 123456789$.

(a) Use MATLAB to compute T explicitly. Then use MATLAB's `eig` command to compute the eigenvalues of T. This treats T as an ordinary unsymmetric matrix and uses the unsymmetric QR algorithm. There are two large eigenvalues around 123456789 and one tiny one. The large ones are accurate.

(b) Show that $\det T = 1$. This is the product of the eigenvalues. Compute the product of your three computed eigenvalues and notice that it is nowhere near 1. Thus the computed tiny eigenvalue is completely inaccurate. To get an accurate value of the tiny eigenvalue λ_3, solve the equation $\lambda_1 \lambda_2 \lambda_3 = 1$: $\lambda_3 = 1/(\lambda_1 \lambda_2)$. Since the large eigenvalues are accurate, this will give the correct value of the tiny eigenvalue to about 15 decimal places. Use `format long e` to view all digits of λ_3.

(c) Write a MATLAB m-file that implements an iteration of the dqds algorithm (4.4.10). Apply the dqds algorithm with $\rho = 0$ to the factors of T. Notice that after one iteration $\hat{r}_3$ agrees with the correct value of the tiny eigenvalue to about eight decimal places. Perform a second iteration to get the correct value to about 15 decimal places.

4.5 Economical *GR* Iterations by Bulge-Chasing

Continuing our main line of development from Section 4.3, we come to the question of how to implement a step of the GR algorithm efficiently. Since we are now going to focus on a single iteration, let us drop the subscripts from (4.2.6) and consider a GR iteration driven by p, mapping a matrix A to $\hat{A}$:

$$p(A) = GR, \qquad \hat{A} = G^{-1}AG. \tag{4.5.1}$$

This will allow us to reintroduce subscripts for a different purpose. We have already mentioned that we are going to reduce the matrix to upper Hessenberg form before beginning, so let us assume that A is upper Hessenberg. First off, we wish to show that if A is upper Hessenberg, then so is $\hat{A}$.

Theorem 4.5.1. *Let A, G, R, and $\hat{A}$ be related by the generic GR iteration (4.5.1), and suppose that $p(A)$ is nonsingular. Then $\hat{A} = RAR^{-1}$. If A is upper Hessenberg, then $\hat{A}$ is upper Hessenberg. If A is properly upper Hessenberg, then $\hat{A}$ is properly upper Hessenberg.*

The easy proof is worked out in Exercise 4.5.3. Theorem 4.5.1 is not strictly true in the case when $p(A)$ is singular. However, in the singular case, we have greater flexibility in how we carry out the GR decomposition. We will see shortly that, even when $p(A)$ is singular, it is always possible to build G in such a way that the Hessenberg form is preserved.

To get an idea of how the use of Hessenberg matrices can make the algorithm economical, consider a GR iteration of degree one. In this case $p(A) = A - \rho I$, where ρ is a single shift, so $p(A)$ is an upper Hessenberg matrix. Now consider the cost of computing the decomposition $p(A) = GR$ using the construction described in Section 3.2. The matrix G is a product $G = G_1 G_2 \cdots G_{n-1}$, where each G_i^{-1} is an elimination matrix that transforms the ith column of $G_{i-1}^{-1} \cdots G_1^{-1} p(A)$ to upper triangular form. Since $p(A)$ is upper Hessenberg to begin with, each of these eliminations creates only a single zero. This can be done by a matrix that is effectively 2×2, acting only on rows i and $i + 1$. For example, G_i can be a plane rotator. The cost of generating such a transformation is $O(1)$, and the cost of applying it to a matrix is $O(n)$. Since the total number of transforms is also $O(n)$, the total cost of the GR decomposition is $O(n^2)$. A more precise count is made in Exercise 4.5.4. Recalling from Section 3.2 that GR decompositions cost $O(n^3)$ flops in general, we see that Hessenberg GR decompositions are relatively cheap.

Now consider the similarity transformation $\hat{A} = G^{-1}AG$. In performing the GR decomposition, we do not explicitly build the matrix G; we keep it in the factored form $G_1 G_2 \cdots G_{n-1}$, and we perform the similarity transformation by stages. Let $A_0 = A$, $A_1 = G_1^{-1} A_0 G_1$ and, in general, $A_i = G_i^{-1} A_{i-1} G_i$. Then $A_{n-1} = \hat{A}$. Notice that we do not even need to save up the G_i: As soon as each G_i is produced, it can be used to effect the similarity transformation $A_i = G_i^{-1} A_{i-1} G_i$ and then be discarded. Each of these transformations is inexpensive. The left multiplication by G_i^{-1} affects two rows of A_{i-1} and can be done in $O(n)$ flops. Similarly, the right multiplication by G_i affects two columns and can be done in $O(n)$ flops. Since the total number of transformations is $O(n)$, the total work for the transformation is $O(n^2)$ flops.

We conclude that the total cost of a GR iteration of degree one on an upper Hessenberg matrix is $O(n^2)$. We know from Section 3.3 that the cost of reducing a matrix to upper Hessenberg form is $O(n^3)$ flops. If we are willing to do this expensive step once, then the subsequent GR iterations are relatively inexpensive.

So far we have been discussing GR iterations of degree one. For iterations of higher degree the situation is more complicated. Suppose that $p(A)$ has degree two, i.e., $p(A) = (A - \rho_1 I)(A - \rho_2 I)$. Then the very computation of $p(A)$ is expensive. Even if A is upper Hessenberg, it costs $O(n^3)$ flops to multiply the matrices together. It thus appears at first glance that it will be impossible to effect multiple GR steps efficiently. Fortunately this turns out not to be so. There is a way of implementing multiple GR steps known as bulge-chasing that avoids the explicit computation of $p(A)$. All that's needed is the first column of $p(A)$, and this can be computed cheaply.

Chasing the Bulge

We now describe a *generic bulge-chasing algorithm* and demonstrate that it effects a multiple *GR* iteration on an upper Hessenberg matrix without explicitly computing $p(A)$. Since it effects a *GR* step implicitly, the generic bulge-chasing algorithm is also called an *implicit GR step* or *implicitly shifted GR step*. By contrast, the iteration (4.5.1), if executed in the straightforward way, is called an *explicit GR step*.

We require that A be properly upper Hessenberg, which means that it is upper Hessenberg and $a_{j+1,j} \neq 0$ for $j = 1, \ldots, n-1$. There is no loss of generality in this assumption. If A is upper Hessenberg but not properly upper Hessenberg, then some $a_{j+1,j}$ is zero. But then A is reducible:

$$A = \left[\begin{array}{cc} A_{11} & A_{12} \\ 0 & A_{22} \end{array} \right],$$

where A_{11} is $j \times j$. Both A_{11} and A_{22} are upper Hessenberg. The eigenvalues of A can be found by finding those of A_{11} and A_{22} separately. If either of these is not properly upper Hessenberg, it can be reduced further. Thus it suffices to consider properly upper Hessenberg matrices. As we remarked earlier, reductions of this type are highly desirable.

Suppose we wish to effect a *GR* iteration (4.5.1), where p has degree m, say $p(A) = (A - \rho_1 I)(A - \rho_2 I) \cdots (A - \rho_m I)$. In practice $m \ll n$; we allow the case $m = 1$. The Hessenberg form is not inherited by $p(A)$ (unless $m = 1$); however, $p(A)$ does have a Hessenberg-like property, which we call *m-Hessenberg*. A matrix B is *m-Hessenberg* if $b_{ij} = 0$ whenever $i > j + m$. This means that B can have nonzero entries in the first m diagonals below the main diagonal, but all entries beyond the mth subdiagonal must be zero. An ordinary upper Hessenberg matrix is 1-Hessenberg in this terminology. The matrix B is *properly m-Hessenberg* if it is m-Hessenberg and $a_{j+m,j} \neq 0$ for $j = 1, \ldots, n-m$. It is easy to show that $p(A)$ is properly m-Hessenberg (Exercise 4.5.5).

The generic bulge-chasing algorithm begins by computing a vector that is proportional to $p(A)e_1$, the first column of $p(A)$. Since $p(A)$ is m-Hessenberg, only the first $m+1$ entries of $p(A)e_1$ are nonzero. The computation is quite inexpensive and can be accomplished by the following procedure:

$$\begin{array}{l} x \leftarrow e_1 \\ \text{for } k = 1, \ldots, m \\ \left[\; x \leftarrow \alpha_k (A - \rho_k I)x, \right. \end{array} \qquad (4.5.2)$$

where α_k is any convenient scaling factor. The total amount of work in this step is $O(m^3)$ (Exercise 4.5.6), which is negligible if $m \ll n$.

Let $x = \alpha p(A)e_1$ be the vector produced by (4.5.2). Construct a matrix $G_1 \in \mathbb{F}^{n \times n}$ whose first column is proportional to x. Since x consists mostly of zeros, we can write $x = \left[\begin{array}{c} y \\ 0 \end{array} \right]$, where $y \in \mathbb{F}^{m+1}$, and G_1 can be chosen to have the form $G_1 = \text{diag}\left\{ \tilde{G}_1, I \right\}$, where $\tilde{G}_1 \in \mathbb{F}^{(m+1) \times (m+1)}$ is a matrix whose first column is proportional to y. Notice that $\tilde{G}_1 e_1 = \beta y$ if and only if $\tilde{G}_1^{-1} y = \beta^{-1} e_1$, so $\tilde{G}_1$ is an elimination matrix and can be constructed by the methods of Section 3.1.

Once we have G_1, let

$$A_1 = G_1^{-1} A G_1.$$

The transformation $A \rightarrow G_1^{-1}A$ acts on the first $m + 1$ rows of A, disturbing the upper Hessenberg form. The subsequent transformation $G_1^{-1}A \rightarrow G_1^{-1}AG_i$ acts on the first $m + 1$ columns and does a bit more damage, filling in row $m + 2$. More precisely, consider the partition

$$A = \begin{bmatrix} A_{11} & A_{12} \\ A_{21} & A_{22} \end{bmatrix}, \tag{4.5.3}$$

where A_{11} is $(m + 1) \times (m + 1)$. A_{11} and A_{22} are upper Hessenberg, and A_{21} has only one nonzero entry in its upper right-hand corner. Then, since $G_1 = \text{diag}\{\tilde{G}_1, I\}$,

$$A_1 = \begin{bmatrix} \tilde{G}_1^{-1}A_{11}\tilde{G}_1 & \tilde{G}_1^{-1}A_{12} \\ A_{21}\tilde{G}_1 & A_{22} \end{bmatrix}.$$

The Hessenberg form is preserved in the $(2, 2)$ block, which is the bulk of the matrix. It is completely destroyed in the $(1, 1)$ block. In the $(2, 1)$ block, $A_{21}\tilde{G}_1$ has nonzero entries only in the first row, since the one nonzero entry of A_{21} is in its first row. Thus A_1 has the form

$$A_1 = \quad \tag{4.5.4}$$

where all entries outside of the outlined region are zero. A_1 is Hessenberg with a bulge.

The rest of the generic bulge-chasing algorithm consists of reducing A_1 to upper Hessenberg form by the method described in Section 3.3. Specifically, the bulge-chasing algorithm ends with an upper Hessenberg matrix

$$\hat{A} = G_{n-1}^{-1} \cdots G_2^{-1}A_1G_2 \cdots G_{n-1}.$$

Here the notation is consistent with that in Section 3.3; for each i, the transformation G_i^{-1} is responsible for returning column $i - 1$ to upper Hessenberg form. Since A_1 is already quite close to Hessenberg form, each of these transformations is considerably less expensive than it would be in general. For example, consider G_2. Since only the first $m + 2$ entries of the first column of A_1 are nonzero, G_2 can be taken to be of the form $G_2 = \text{diag}\{1, \tilde{G}_2, I\}$, where $\tilde{G}_2^{-1}$ is an $(m + 1) \times (m + 1)$ elimination matrix that creates m zeros in the first column of $G_2^{-1}A_1$, returning that column to upper Hessenberg form. One column has been shaved from the bulge. To complete the similarity transformation, we must multiply by G_2 on the right. From the form of G_2 we see that this transformation acts on columns 2 through $m + 2$. Before the transformation, the bulge extends down to row $m + 2$. Since $a_{m+3,m+2} \neq 0$, the right multiplication by G_2 adds one additional row to the bulge. Since everything in columns 2 through $m + 2$ below row $m + 3$ is zero, the zeros below row $m + 3$ stay zero. Letting $A_2 = G_2^{-1}A_1G_2$, we see that the tip of the bulge of A_2 is at position $(m + 3, 2)$. The bulge in A_2 is of the same size as the bulge in A_1, but it has been moved one row down and one column to the right. Now it is clear that G_3 will have the form $G_3 = \text{diag}\{1, 1, \tilde{G}_3, I\}$, where $\tilde{G}_3^{-1}$ is an $(m + 1) \times (m + 1)$ elimination matrix that returns the second column of

$A_3 = G_3^{-1} A_2 G_3$ to upper Hessenberg form. The effect of this transformation is completely analogous to the effect of the transformation from A_1 to A_2: the bulge is moved over one column and down one row.

Now it is clear why we call this a bulge-chasing algorithm. At each step the bulge is pushed down and to the right. In midchase we have

$$A_k = \begin{bmatrix} & & \\ & & \\ & & \end{bmatrix},$$

(4.5.5)

where the tip of the bulge is at position $(m + k + 1, k)$. In the end, the bulge is pushed off the bottom of the matrix. The last few transformations have relatively less work to do.

This completes the description of the generic bulge-chasing algorithm. We still have to show that it effects an iteration of the generic *GR* algorithm (4.5.1), but first let us make a couple of observations. An easy flop count (Exercise 4.5.8) shows that each iteration of the bulge-chasing algorithm requires only $O(n^2 m)$ work (assuming $m \ll n$), so it is relatively economical. Exercise 4.5.9 estimates the total work to compute all of the eigenvalues of an $n \times n$ matrix by repeated applications of the bulge-chasing algorithm and concludes that it is $O(n^3)$.

Note also that the generic bulge-chasing algorithm always produces an upper Hessenberg matrix. We proved previously that the *GR* algorithm, applied to a Hessenberg matrix, preserves the Hessenberg form, provided that $p(A)$ is nonsingular. Now we see that if we effect the algorithm by bulge-chasing, we always preserve the Hessenberg form, regardless of whether or not $p(A)$ is singular.

Now let us get on with the task of showing that the generic bulge-chasing algorithm does indeed effect an iteration of the generic *GR* algorithm. First of all, the bulge-chasing algorithm performs a similarity transformation $\hat{A} = G^{-1} A G$, where $G = G_1 G_2 \cdots G_{n-1}$. It is easy to check that the first column of G is the same as the first column of G_1, which was chosen to be proportional to the first column of $p(A)$ (Exercise 3.3.1). Thus there is a $\gamma \neq 0$ such that

$$p(A)e_1 = \gamma G e_1. \tag{4.5.6}$$

The transformation $\hat{A} = G^{-1} A G$ is an instance of Theorem 3.3.1.

We will make essential use of Krylov matrices. Given $B \in \mathbb{F}^{n \times n}$ and $z \in \mathbb{F}^n$, the *Krylov matrix* associated with B and z, denoted $\kappa(B, z)$, is the $n \times n$ matrix whose columns are the Krylov sequence:

$$\kappa(B, z) = \begin{bmatrix} z & Bz & B^2 z & \cdots & B^{n-1} z \end{bmatrix}.$$

The leading columns of Krylov matrices span Krylov subspaces, which we introduced in Section 3.3. The following lemmas about Krylov matrices are easily verified.

Lemma 4.5.2. *Suppose* $G\hat{A} = AG$. *Then for any* $z \in \mathbb{F}^n$,

$$G\kappa(\hat{A}, z) = \kappa(A, Gz).$$

Lemma 4.5.3. *For any $A \in \mathbb{F}^{n \times n}$ and any $z \in \mathbb{F}^n$,*

$$p(A)\kappa(A, z) = \kappa(A, p(A)z).$$

Lemma 4.5.4. *Let B be an upper Hessenberg matrix. Then*

(a) $\kappa(B, e_1)$ *is upper triangular;*

(b) $\kappa(B, e_1)$ *is nonsingular if and only if B is properly upper Hessenberg.*

The desired result is an easy consequence of the lemmas.

Theorem 4.5.5. *The generic bulge-chasing algorithm, applied to a properly upper Hessenberg matrix, effects an iteration of the generic GR algorithm.*

Proof: The algorithm begins with a properly upper Hessenberg matrix A and ends with an upper Hessenberg $\hat{A} = G^{-1}AG$, where the transforming matrix G satisfies (4.5.6). Applying Lemma 4.5.2, equation (4.5.6), and then Lemma 4.5.3, we obtain

$$G\kappa(\hat{A}, e_1) = \kappa(A, Ge_1) = \gamma^{-1}\kappa(A, p(A)e_1) = \gamma^{-1}p(A)\kappa(A, e_1). \quad (4.5.7)$$

By Lemma 4.5.4, $\kappa(\hat{A}, e_1)$ and $\kappa(A, e_1)$ are upper triangular, and the latter is nonsingular. Rearranging (4.5.7), we obtain

$$p(A) = G(\gamma\kappa(\hat{A}, e_1)\kappa(A, e_1)^{-1}) = GR,$$

where $R = \gamma\kappa(\hat{A}, e_1)\kappa(A, e_1)^{-1}$ is upper triangular. Thus we have

$$\hat{A} = G^{-1}AG, \qquad \text{where} \qquad p(A) = GR.$$

We conclude that the generic bulge-chasing algorithm effects an iteration of the generic GR algorithm. $\square$

Specific versions of the GR algorithm are obtained by making specific choices of elimination matrices. If we implement the bulge-chasing algorithm with each G_i taken to be unitary, say a reflector, then the resulting transformation G will be unitary, and we rename it Q. We have $\hat{A} = Q^{-1}AQ$, where $p(A) = QR$. This is the essentially unique QR decomposition of $p(A)$ (Theorem 1.3.8), and the bulge-chasing algorithm effects an iteration of the QR algorithm in this case. This is an *implicitly shifted QR iteration*. If, on the other hand, we implement the bulge-chase with lower triangular Gauss transforms, the resulting G will be unit lower triangular, and we rename it L. We have $\hat{A} = L^{-1}AL$, with $p(A) = LR$. This is the unique LR decomposition of $p(A)$ (Theorem 1.3.4), and the bulge-chase effects an iteration of the LR algorithm. Not every LR iteration succeeds. At any point, the desired Gauss transform may fail to exist due to a zero pivot. (Not every $p(A)$ *has* an LR decomposition.) Worse yet, the LR iteration may succeed but be unstable, due to small pivots and large multipliers. A more robust algorithm can be obtained by using Gauss transforms with partial pivoting. This results in the LR algorithm with partial pivoting. Stability is not guaranteed; experience suggests that this algorithm yields good eigenvalues but can produce suspect eigenvectors. Other types of elimination matrices result in other types of GR iterations. Furthermore, there is nothing in the theory that rules out the mixing of transformations of different types, e.g., Gauss transforms and reflectors [108].

Notes and References

The first bulge-chasing algorithm was the double-shift QR algorithm of Francis [89]. The implicit LR algorithm was discussed by Wilkinson [230]. A QR/LR hybrid algorithm that mixes reflectors and Gauss transforms in the bulge-chase was introduced by Haag and Watkins [108]. Bai and Demmel [13] were the first to try QR iterations of high degree. If the degree is made too large, trouble is caused by shift blurring [217]. See Section 7.1.

Exercises

4.5.1.

    (a) Download the file `downchase.m` from the website.[5] Examine the file to find out what it does. Download the file `qrgo.m`, which is a simple driver program for `downchase.m`, and check it out.

    (b) Try out the codes with various choices of matrix size n and shift degree m. Don't take n too large, unless you turn off the plot feature or are extremely patient. (The choice $n = 50$ and $m = 10$ makes a nice picture.) Large values of m make prettier pictures, but if you make m too big, you may experience a convergence failure due to shift blurring. (Section 7.1).

    (c) If you want to see the same things in reverse, try out the codes `upchase.m` and `rqgo.m`.

4.5.2.

    (a) Download the files `hrchase.m` and `hrgo.m`, and figure out what they do.

    (b) Try out the codes with various choices of matrix size n and shift degree $2 \leq m \leq 6$. The choice $n = 50$ and $m = 6$ makes a nice picture. If you take m much bigger than 6, you will experience lack of convergence due to shift blurring (Section 7.1). This algorithm makes interesting pictures, but the lack of stability shows itself in the large residuals.

4.5.3. This exercise works out the proof of Theorem 4.5.1. Suppose that the relationships (4.5.1) hold.

    (a) Show that $Ap(A) = p(A)A$.

    (b) Show that $G^{-1}p(A) = R$ and, if $p(A)$ is nonsingular, then $G = p(A)R^{-1}$. Deduce that $\hat{A} = RAR^{-1}$.

    (c) Show that if B is upper triangular and C is upper Hessenberg, then both BC and CB are upper Hessenberg.

    (d) Deduce that $\hat{A}$ is upper Hessenberg if A is.

---

[5] www.siam.org/books/ot101

(e) Where does this argument break down if $p(A)$ is singular?

(f) Show that if B is upper triangular and nonsingular and C is properly upper Hessenberg, then both BC and CB are properly upper Hessenberg.

(g) Deduce that $\hat{A}$ is properly upper Hessenberg if A is.

4.5.4. Consider a decomposition $A - \rho I = GR$, where A is upper Hessenberg, by the procedure of Section 3.2. Then $G = G_1 \cdots G_{n-1}$ and $R = G_{n-1}^{-1} \cdots G_1^{-1}(A - \rho I)$. The ith matrix G_i^{-1} is responsible for transforming the ith column to upper triangular form.

(a) Show that G_i^{-1} needs only to set the $(i+1, i)$ entry to zero, so it can be taken to be effectively a 2×2 matrix: $G_i = \operatorname{diag}\left\{ I_{i-1}, \tilde{G}_i, I_{n-i-1} \right\}$.

(b) The cost of a computation $y = \tilde{G}_i^{-1} x$, where $x, y \in \mathbb{F}^2$, depends on the details of $\tilde{G}_i$, but it can never be more than a few flops. Let's say the cost is C. Show that the cost of premultiplying $G_{i-1}^{-1} \cdots G_1^{-1}(A - \rho I)$ by G_i^{-1} is $C(n - i + 1)$.

(c) Assuming that the cost of applying all of the G_i is the same, show that the total cost of the GR decomposition is about $Cn^2/2$.

(d) Show that the matrix $A_1 = G_1^{-1} A G_1$ is upper Hessenberg, except for a "bulge" in the $(3, 1)$ position.

(e) Letting $A_i = G_i^{-1} A_{i-1} G_i$, one can check that A_i is upper Hessenberg, except for a bulge in the $(i+2, i)$ position. Assuming this form, show that the transformation $A_i = G_i^{-1} A_{i-1} G_i$ costs $C(n+3)$.

(f) Deduce that the total cost of the similarity transformation $\hat{A} = G^{-1} A G$ is about Cn^2. Thus the total cost of a GR iteration of degree one is about $1.5Cn^2$.

4.5.5.

(a) Show that if B is k-Hessenberg and C is j-Hessenberg, then BC is $(k+j)$-Hessenberg.

(b) Show that if B is properly k-Hessenberg and C is properly j-Hessenberg, then BC is properly $(k + j)$-Hessenberg.

(c) Show that if A is properly upper Hessenberg and $p(A) = (A - \rho_1 I) \cdots (A - \rho_m I)$ is a polynomial of degree m, then $p(A)$ is properly m-Hessenberg.

4.5.6. Show that after j times through the loop in (4.5.2), the vector x has (at most) $j + 1$ nonzero entries. Show that the amount of arithmetic in the kth step of the loop is $O(k^2)$. Show that the total amount of arithmetic in the algorithm (4.5.2) is $O(m^3)$.

4.5.7. This exercise confirms some of the properties of the generic bulge-chasing algorithm.

(a) Let $\begin{bmatrix} y \\ 0 \end{bmatrix} = x = \alpha p(A) e_1$, where $\alpha \neq 0$, A is properly upper Hessenberg, and p has degree m. Show that $y_{m+1} \neq 0$.

(b) Show that in (4.5.3), $A_{21} = e_1 a_{m+2,m+1} e_{m+1}^T$.

(c) Show that $A_{21} \tilde{G}_1 = e_1 z^T$, where z^T is proportional to the last row of $\tilde{G}_1$. Thus only the first row of $A_{21} \tilde{G}_1$ is nonzero.

(d) Show that the (1, 1) entry of $A_{21} \tilde{G}_1$ is nonzero. This is the tip of the bulge in A_1.

(e) Verify that $A_2 = G_2^{-1} A_1 G_2$ is upper Hessenberg, except for a bulge extending to the $(m+3, 2)$ entry. That is, $a_{ij}^{(2)} = 0$ if $i > j + 1$ and either $j < 2$ or $i > m + 3$.

(f) Verify that the entry in the $(m+3, 2)$ position of A_2 is nonzero.

(g) Letting $A_k = G_k^{-1} A_{k-1} G_k$, prove by induction that for $k = 1, \dots, n - m - 1$, A_k is upper Hessenberg, except for a bulge extending to position $(m + k + 1, k)$. That is, $a_{ij}^{(k)} = 0$ if $i > j + 1$ and either $j < k$ or $i > m + k + 1$. Prove by induction that $a_{m+k+1,k}^{(k)}$ is nonzero.

4.5.8. Suppose that each step $A_k = G_k^{-1} A_{k-1} G_k$ of the generic bulge-chasing algorithm is effected by an elementary transformation. In other words, each G_k is an elementary matrix. At each step the effective part $\tilde{G}_k$ is $(m+1) \times (m+1)$. (In the last few steps, the effective part is even smaller. However, we will ignore this detail, which has a negligible effect on the operation count.)

(a) Show that each transformation $A_k = G_k^{-1} A_{k-1} G_k$ requires about $2(n+m+3)(m+1)$ multiplications and as many additions. Assuming $m \ll n$, this number is about $2n(m+1)$.

(b) Show that the total cost of the generic bulge-chasing algorithm is about $4n^2(m+1)$ flops.

4.5.9. Let $A \in \mathbb{F}^{n \times n}$ be an upper Hessenberg matrix. In this exercise we make a rough estimate of the total cost of finding all of the eigenvalues of A by the generic GR algorithm. We have already estimated (Section 3.3) that the cost of getting the matrix into upper Hessenberg form is $O(n^3)$. Now we compute the additional costs. The GR algorithm is iterative, so in principle it requires an infinite amount of work. The flop count becomes finite in practice only because we are willing to set $a_{j+1,j}^{(i)}$ to zero once it becomes sufficiently tiny. Assume that six iterations are needed for each deflation, and that each deflation breaks off an $m \times m$ chunk, delivering exactly m eigenvalues. Assume the cost of an mth degree bulge-chase is about $4n^2 m$ flops, as shown in Exercise 4.5.8.

(a) Show that under the stated assumptions, about $24n^2 m$ flops will be needed for the first batch of eigenvalues, then $24(n - m)^2 m$ flops will be needed for the next batch, $24(n - 2m)^2 m$ for the third batch, and so on. About how many batches are there in all?

(b) By approximating a sum by an integral, show that the total cost of all of the GR iterations is about $8n^3$ flops.

(c) The count in part (b) ignores the auxiliary computations of eigenvalues of $m \times m$ sub-matrices needed for the shift computations and the final computation of eigenvalues of each deflated $m \times m$ block. Assuming the cost of each such auxiliary computation to be $O(m^3)$, show that the total cost of all such computations is $O(nm^2)$, which is insignificant if $m \ll n$.

Because of counts like this, the cost of computing eigenvalues by implicit GR (bulge-chasing) algorithms is generally considered to be $O(n^3)$. This count assumes that only the eigenvalues are computed. If eigenvectors are wanted as well, additional work must be done, but the cost is still $O(n^3)$. Computation of eigenvectors will be discussed in Section 4.8.

4.5.10. Suppose that the cost of computing the eigenvalues of an $n \times n$ proper upper Hessenberg matrix A is Cn^3, where C is a constant (perhaps $C = 8$, as suggested by the previous exercise). Show that if $a_{j+1,j} = 0$, where $j \approx n/2$, then the cost of finding the eigenvalues is reduced by a factor of 4. This shows that the time savings resulting from finding zeros on the subdiagonal can be substantial.

4.5.11. Suppose $G\hat{A} = AG$.

(a) Show that $G\hat{A}^k = A^k G$ for $k = 0, 1, \ldots, n - 1$.

(b) Prove Lemma 4.5.2.

4.5.12. Prove Lemma 4.5.3, using the fact that $p(A)A = Ap(A)$. (Indeed, Lemma 4.5.3 is redundant; it is a special case of Lemma 4.5.2.)

4.5.13. Prove Lemma 4.5.4.

4.6   The Singular Case

We have seen that if $p(A)$ is nonsingular, then the upper Hessenberg form is preserved under a GR iteration. This result is not strictly true if $p(A)$ is singular, which might give the reader the false impression that the singular case is an undesirable nuisance case that we should avoid. In fact just the opposite is true; it is the most desirable case. We always wish to choose shifts that are as close to eigenvalues of A as possible. When we happen to hit an eigenvalue exactly, then we are in the singular case. As we shall see, a GR iteration with a singular $p(A)$ leads to a deflation in one iteration. This is not at all unexpected. We have seen that the convergence rates depend upon the ratios $|p(\lambda_{j+1})/p(\lambda_j)|$. In the singular case, at least one eigenvalue of $p(A)$ is zero. Suppose that j eigenvalues are zero: $p(\lambda_{n-j+1}) = p(\lambda_{n-j+2}) = \cdots = p(\lambda_n) = 0$. Then $|p(\lambda_{n-j+1})/p(\lambda_{n-j})| = 0$, predicting extremely rapid convergence. Convergence in one iteration is as rapid as you can get.

Our scenario is the same as in the previous section: We consider a single GR iteration of degree m, mapping the properly upper Hessenberg matrix A to $\hat{A}$:

$$p(A) = GR, \qquad \hat{A} = G^{-1}AG. \tag{4.6.1}$$

The matrix $p(A) = (A - \rho_1 I)(A - \rho_2 I) \cdots (A - \rho_m I)$ is singular if and only if one or more of the factors $A - \rho_j I$ is singular, and this is true in turn if and only if one or more of the shifts ρ_j is an eigenvalue of A. A shift that is an eigenvalue of A is called an *exact shift*. The following theorem tells what happens in the singular case.

Theorem 4.6.1. *Consider a GR iteration (4.6.1), where A is properly upper Hessenberg and $p(A)$ is singular. Suppose that j (a positive integer) is the number of shifts that are* *exact, and denote these shifts by* $\rho_1, \ldots, \rho_j$.[6] *Then*

$$\hat{A} = \begin{bmatrix} \hat{A}_{11} & \hat{A}_{12} \\ 0 & \hat{A}_{22} \end{bmatrix},$$

where $\hat{A}_{22}$ *is* $j \times j$. *The eigenvalues of* $\hat{A}_{22}$ *are* $\rho_1, \ldots, \rho_j$. $\hat{A}_{11}$ *is properly upper Hessenberg.*

The matrix $\hat{A}_{22}$ is not necessarily upper Hessenberg in general. However, if the GR iteration is implemented by bulge-chasing as described in Section 4.5, it certainly will be. Theorem 4.6.1 shows that we can deflate the j eigenvalues $\rho_1, \ldots, \rho_j$ after a single iteration and continue the iterations with the properly upper Hessenberg submatrix $\hat{A}_{11}$.

This section is devoted to proving Theorem 4.6.1. Before we begin, we should admit that this result is mainly of theoretical interest. It holds in exact arithmetic but not in floating point. Theorem 4.6.1 says that in exact arithmetic we have $\hat{a}_{n-j+1,n-j} = 0$ exactly. In floating point arithmetic we do not get an exact zero, but we expect to get $\hat{a}_{n-j+1,n-j} \approx 0$. It turns out in practice that $\hat{a}_{n-j+1,n-j}$ is usually small but not close enough to zero to permit a deflation using the criterion (4.3.2). Typically two or more iterations using exact shifts will be needed before we can safely deflate.

Another point worth mentioning is that in practice we almost never have exact shifts. We may have some ρ_i that agree with eigenvalues to sixteen decimal places, but they are almost never exactly equal to eigenvalues of A. Thus the singular case is one to which we aspire but that we seldom actually attain.

Now let's get to work. A matrix is called *derogatory* if it has an eigenvalue that has more than one Jordan block (Theorem 2.4.11) associated with it. Conversely, it is called *nonderogatory* if each eigenvalue has only one Jordan block associated with it. For example, the identity matrix $\begin{bmatrix} 1 & 0 \\ 0 & 1 \end{bmatrix}$ is derogatory, because it has two Jordan blocks, [1] and [1], associated with the eigenvalue 1. However, the defective matrix $\begin{bmatrix} 1 & 1 \\ 0 & 1 \end{bmatrix}$ is nonderogatory, because it has (i.e., it *is*) a single Jordan block associated with the sole eigenvalue 1.

Lemma 4.6.2. *Let* $A \in \mathbb{F}^{n \times n}$ *be properly upper Hessenberg. Then, for every* $\rho \in \mathbb{C}$, rank$(A - \rho I) \geq n - 1$. *Consequently* A *is nonderogatory.*

The proof is worked out in Exercise 4.6.2.

Lemma 4.6.3. *Let* $A \in \mathbb{F}^{n \times n}$ *be properly upper Hessenberg, and suppose* $p(A) = (A - \rho_1 I)(A - \rho_2 I) \cdots (A - \rho_m I)$. *Suppose that* j *of the shifts* $\rho_1, \ldots, \rho_m$ *are eigenvalues*[7] *of* A. *Then* rank$(p(A)) = n - j$. *Furthermore, the first* $n - j$ *columns of* $p(A)$ *are linearly independent.*

The rank of $p(A)$ is easily established by use of the Jordan canonical form of A (Exercise 4.6.3). Once we know the rank of A, it is easy to show, in almost all cases, that the leading $n - j$ columns of $p(A)$ are linearly independent (Exercise 4.6.4), but we must

---

[6]Usually the j exact shifts will be distinct, but we do not strictly require this. If λ is a multiple eigenvalue of A of algebraic multiplicity k, then we allow as many as k of the shifts to equal λ. If more than k of the shifts are equal to λ, we count only k of them as exact shifts.

[7]In counting the number of perfect shifts, we apply the rules laid out in the footnote to Theorem 4.6.1.

take a roundabout route to get the result in full generality (Exercise 4.6.5). Once we have Lemma 4.6.3, the next theorem follows easily (Exercise 4.6.6).

Theorem 4.6.4. *Under the hypotheses of Lemma 4.6.3, let $p(A) = GR$, where G is nonsingular and R is upper triangular. Then*

$$R = \begin{bmatrix} R_{11} & R_{12} \\ 0 & 0 \end{bmatrix},$$

where R_{11} is $(n - j) \times (n - j)$ and nonsingular.

One more lemma will bring us to the desired conclusion.

Lemma 4.6.5. *Let A, $\hat{A}$, and R be as in (4.6.1). Then*

$$\hat{A}R = RA. \tag{4.6.2}$$

Lemma 4.6.5 generalizes the statement in Theorem 4.5.1 that $\hat{A} = RAR^{-1}$. The restatement $\hat{A}R = RA$ is valid, regardless of whether or not R has an inverse. The straightforward proof is worked out in Exercise 4.6.7. In our current scenario, R does not have an inverse. Consider a partitioned version of (4.6.2):

$$\begin{bmatrix} \hat{A}_{11} & \hat{A}_{12} \\ \hat{A}_{21} & \hat{A}_{22} \end{bmatrix} \begin{bmatrix} R_{11} & R_{12} \\ 0 & 0 \end{bmatrix} = \begin{bmatrix} R_{11} & R_{12} \\ 0 & 0 \end{bmatrix} \begin{bmatrix} A_{11} & A_{12} \\ A_{21} & A_{22} \end{bmatrix}. \tag{4.6.3}$$

Equating the $(2, 1)$ blocks, we find that

$$\hat{A}_{21} R_{11} = 0.$$

Since R_{11} is nonsingular, this implies that $\hat{A}_{21} = 0$, as claimed in Theorem 4.6.1. Equating the $(1, 1)$ blocks in (4.6.3), and taking the nonsingularity of R_{11} into account, we obtain

$$\hat{A}_{11} = R_{11} A_{11} R_{11}^{-1} + R_{12} A_{21} R_{11}^{-1},$$

from which we can deduce that $\hat{A}_{11}$ is properly upper Hessenberg (Exercise 4.6.8).

The proof of Theorem 4.6.1 will be complete once we show that the eigenvalues of $\hat{A}_{22}$ are $\rho_1, \dots, \rho_j$. This result ought to seem "obvious" by now. There are several ways to prove it; however, the details are messy in the defective case. Two styles of proof are laid out in Exercises 4.6.9 and 4.6.10. In Exercise 4.6.9 we restrict our attention to the semisimple case for clarity, but in Exercise 4.6.10 we obtain the general result.

Exercises

4.6.1. The codes `qrexact.m` and `downchase2.m` demonstrate the use of exact shifts in the QR algorithm. Download them from the website[8] and try them out with various values of n and m.

4.6.2. This exercise proves Lemma 4.6.2. Suppose that $A \in \mathbb{F}^{n \times n}$ is properly upper Hessenberg.

---

[8]www.siam.org/books/ot101

(a) Show that for any $\rho \in \mathbb{C}$, the first $n - 1$ columns of $A - \rho I$ are linearly independent. Deduce that $\text{rank}(A - \rho I) \geq n - 1$.

(b) Show that if two matrices are similar, they have the same rank.

(c) Show that if $B \in \mathbb{F}^{n \times n}$ is a derogatory matrix, and ρ is an eigenvalue of B that has two or more Jordan blocks associated with it, then $\text{rank}(B - \rho I) \leq n - 2$.

(d) Deduce that A is nonderogatory.

4.6.3. This exercise works out the first part of the proof of Lemma 4.6.3. Let $A \in \mathbb{F}^{n \times n}$ be properly upper Hessenberg, and let $p(A) = (A - \rho_1 I) \cdots (A - \rho_m I)$. Suppose that j of the shifts are eigenvalues of A, using the rules laid out in the footnote to Theorem 4.6.1. Let the Jordan canonical form of A be $J = \text{diag}\{J_1, \ldots, J_k\}$, where each J_i is a Jordan block corresponding to an eigenvalue μ_i. Since A is nonderogatory, the μ_i are distinct.

(a) Show that if the eigenvalue μ_i appears j_i times in the list of exact shifts, then the nullity of $p(J_i)$ is j_i. (By rule, j_i does not exceed the dimension of the block J_i.)

(b) Show that the nullity of $p(J)$ is $j = \sum j_i$.

(c) Deduce that the rank of $p(J)$, and hence also of $p(A)$, is $n - j$.

4.6.4. This exercise nearly completes the proof of Lemma 4.6.3. Let $A \in \mathbb{F}^{n \times n}$ be properly upper Hessenberg, and let $p(A) = (A - \rho_1 I) \cdots (A - \rho_m I)$. Let $\rho_1, \ldots, \rho_j$ denote the shifts that are equal to eigenvalues of A, by the rules laid out in the footnote to Theorem 4.6.1, and suppose that $\rho_{j+1}, \ldots, \rho_m$ are not eigenvalues of A. Let $p_1(A) = (A - \rho_1 I) \cdots (A - \rho_j I)$ and $p_2(A) = (A - \rho_{j+1} I) \cdots (A - \rho_m I)$. Then since the factors of $p(A)$ commute, $p(A) = p_2(A) p_1(A)$. We wish to show that the first $n - j$ columns of $p(A)$ are linearly independent.

(a) Recalling that $p(A)$ is properly m-Hessenberg (Exercise 4.5.5), show that the first $n - m$ columns of $p(A)$ are linearly independent. This takes care of the case $j = m$, in which all m of the shifts are perfect.

(b) Show that $p_1(A)$ is properly j-Hessenberg. Deduce that the first $n - j$ columns of $p_1(A)$ are linearly independent.

(c) Show that $p_2(A)$ is nonsingular. Deduce that the first $n - j$ columns of $p(A)$ are linearly independent.

The only cases not covered by this exercise are those for which the number of shifts equal to a single eigenvalue exceeds the algebraic multiplicity of the eigenvalue. For example, if ρ is an eigenvalue with multiplicity 3 and it appears five times in the list of shifts, we can only count it as a perfect shift three times, according to the rules. Then the list $\rho_{j+1}, \ldots, \rho_m$ contains some eigenvalues, $p_2(A)$ is singular, and the argument given here breaks down.

4.6.5. This exercise completes the proof of Lemma 4.6.3. The task is the same as in Exercise 4.6.4, but the argument given here is valid in all cases. Under the assumptions in Lemma 4.6.3 and Exercise 4.6.4, let $H \in \mathbb{F}^{n \times n}$ and $G \in \mathbb{F}^{n \times n}$ be matrices such that H

is upper Hessenberg, G is nonsingular, the first column of G is proportional to the first column of $p(A)$, and $H = G^{-1}AG$. The existence of these matrices is guaranteed by Theorem 3.3.1. These matrices can be generated by the algorithm outlined in Section 3.3, which reduces to the bulge-chasing algorithm described in Section 4.5 in this special case.

(a) Show that $p(A) = GR$, where $R = \gamma \kappa(H, e_1)\kappa(A, e_1)^{-1}$ for some $\gamma \neq 0$. (This is a review of the proof of Theorem 4.5.5.) Our assumptions imply that rank$(p(A)) = n - j$. We want to show that the first $n - j$ columns of $p(A)$ are linearly independent. We will accomplish this by exploiting the properties of $\kappa(H, e_1)$.

(b) For any $B \in \mathbb{F}^{n \times n}$ and $x \in \mathbb{F}^n$, show that if $x, Bx, \ldots, B^{i-1}x$ are linearly independent, but $x, Bx, \ldots, B^i x$ are linearly dependent, then $B^k x \in \text{span}\{x, Bx, \ldots, B^{i-1}x\}$ for all $k \geq i$.

(c) Show that if rank$(\kappa(B, x)) = i$, then the first i columns of $\kappa(B, x)$ are linearly independent.

(d) Show that rank$(\kappa(H, e_1)) = n - j$, same as $p(A)$.

(e) Recall from Lemma 4.5.4 that $\kappa(H, e_1)$ is upper triangular. Show that

$$\kappa(H, e_1) = \begin{bmatrix} K_{11} & K_{12} \\ 0 & 0 \end{bmatrix},$$

where K_{11} is $(n - j) \times (n - j)$, upper triangular, and nonsingular.

(f) Deduce that

$$R = \begin{bmatrix} R_{11} & R_{12} \\ 0 & 0 \end{bmatrix},$$

where R_{11} is $(n - j) \times (n - j)$, upper triangular, and nonsingular. Thus the first $n - j$ columns of R are linearly independent.

(g) From the equation $p(A) = GR$, deduce that the first $n - j$ columns of $p(A)$ are linearly independent.

4.6.6. Suppose $p(A) = GR$, where G is nonsingular and R is upper triangular. Suppose rank$(p(A)) = n - j$, and that the first $n - j$ columns of $p(A)$ are linearly independent. Show rank$(R) = n - j$, and that the first $n - j$ columns of R are linearly independent. Deduce that the form of R is as claimed in Theorem 4.6.4.

4.6.7. This exercise proves Lemma 4.6.5. Show that if (4.6.1) holds, then $G^{-1}p(A) = R$ and $\hat{A}G^{-1} = G^{-1}A$. Multiply the latter equation on the right by $p(A)$, and deduce that $\hat{A}R = RA$.

4.6.8. This exercise proves parts of Theorem 4.6.1.

(a) Using (4.6.3), show that $\hat{A}_{21} R_{11} = 0$, and deduce that $\hat{A}_{21} = 0$.

(b) Using (4.6.3), show that

$$\hat{A}_{11} = R_{11} A_{11} R_{11}^{-1} + R_{12} A_{21} R_{11}^{-1}.$$

(c) Prove that $R_{11}A_{11}R_{11}^{-1}$ is properly upper Hessenberg (cf. Exercise 4.5.3). Recalling that A_{21} has only one nonzero entry and taking into account the form of R_{11}^{-1}, show that $R_{12}A_{21}R_{11}^{-1}$ has nonzeros only in its last column. Deduce that $\hat{A}_{11}$ is properly upper Hessenberg.

4.6.9. This exercise completes the proof of Theorem 4.6.1 in the semisimple case. Assume that A is properly upper Hessenberg and semisimple.

(a) Show that A has n distinct eigenvalues. (Remember, A is nonderogatory.) The perfect shifts ρ_1, ρ_2, ..., ρ_j are therefore distinct. As always, we order the eigenvalues $\lambda_1, \ldots, \lambda_n$ of A so that $|p(\lambda_1)| \geq \cdots \geq |p(\lambda_n)|$. Then $p(\lambda_{n-j}) \neq 0$, $p(\lambda_{n-j+1}) = \cdots = p(\lambda_n) = 0$, and $\lambda_{n-j+1}, \ldots, \lambda_n$ are equal to $\rho_1, \ldots, \rho_j$.

(b) Suppose $p(A) = GR$, and partition G as $G = \begin{bmatrix} G_1 & G_2 \end{bmatrix}$, where G_1 has $n - j$ columns. Taking into account the form of R given by Theorem 4.6.4, show that $\mathcal{R}(p(A)) = \mathcal{R}(G_1)$.

(c) Show that $\mathcal{R}(p(A)) = \mathrm{span}\{v_1, \ldots, v_{n-j}\}$, where $v_1, \ldots, v_{n-j}$ are eigenvectors of A associated with $\lambda_1, \ldots, \lambda_{n-j}$, respectively.

(d) Show that $AG_1 = G_1\hat{A}_{11}$. Deduce that the eigenvalues of $\hat{A}_{11}$ are $\lambda_1, \ldots, \lambda_{n-j}$.

(e) Deduce that the eigenvalues of $\hat{A}_{22}$ are $\rho_1, \ldots, \rho_j$.

4.6.10. This exercise completes the proof of Theorem 4.6.1.

(a) Show that $p(\hat{A}) = G^{-1}p(A)G = RG$.

(b) Taking into account the form of R given by Theorem 4.6.4 and the form of $p(\hat{A})$, show that $p(\hat{A}_{22}) = 0$.

(c) Recalling that A is nonderogatory, deduce that $\hat{A}$ and $\hat{A}_{22}$ are nonderogatory.

(d) Let $\hat{A}_{22} = SJS^{-1}$ denote the Jordan decomposition of $\hat{A}_{22}$. Show that $p(J) = 0$.

(e) Show that the condition $p(J) = 0$ implies that each eigenvalue of $\hat{A}_{22}$ is one of the zeros of p. Here we count multiplicity. For example, if λ_i is an eigenvalue of $\hat{A}_{22}$ of multiplicity 3, corresponding to a 3×3 Jordan block, then λ_i must appear at least 3 times as a zero of p. Thus all of the eigenvalues of $\hat{A}_{22}$ are zeros of p.

(f) Deduce that the eigenvalues of $\hat{A}_{22}$ are $\rho_1, \ldots, \rho_j$.

4.7 A Closer Look at Bulge-Chasing

In Section 4.5 we developed the generic bulge-chasing algorithm and demonstrated that it effects an iteration of the generic GR algorithm. In this section we present an alternate derivation in the spirit of [155] that gives a more complete picture of the relationship between the explicit and implicit versions of the GR algorithm.

As in the previous two sections, consider a single GR iteration of degree m, mapping A to $\hat{A}$:

$$p(A) = GR, \qquad \hat{A} = G^{-1}AG. \tag{4.7.1}$$

A is assumed to be properly upper Hessenberg. We begin by taking a detailed look at what would be involved if we were to execute the step (4.7.1) explicitly. First we would have to compute the matrix $p(A)$. Then we would do a GR decomposition. This is effected by reducing $p(A)$ to upper triangular form as described in Section 3.2. We have

$$R = G_{n-1}^{-1} \cdots G_2^{-1} G_1^{-1} p(A), \tag{4.7.2}$$

where G_k^{-1} is an elimination matrix responsible for transforming the kth column to upper triangular form. Then $p(A) = GR$, where $G = G_1 G_2 \cdots G_{n-1}$. Since A is properly upper Hessenberg and p has degree m, $p(A)$ is properly m-Hessenberg (Exercise 4.5.5). Therefore most of the elimination matrices G_k^{-1} have less work to do than they otherwise would. Since each column of $p(A)$ has at most m nonzero entries below the main diagonal, the effective part of each G_k^{-1} is $(m+1) \times (m+1)$ or smaller. Specifically, for $k = 1, \ldots, n-m$, $G_k^{-1} = \mathrm{diag}\{I_{k-1}, \tilde{G}_k^{-1}, I_{n-k-m}\}$, where $\tilde{G}_k^{-1}$ is an $(m+1) \times (m+1)$ elimination matrix. For $k = n-m+1, \ldots, n-1$, $G_k^{-1} = \mathrm{diag}\{I_{k-1}, \tilde{G}_k^{-1}\}$, where $\tilde{G}_k^{-1}$ is $(n-k+1) \times (n-k+1)$. The G_k are generated one after the other, so we can perform the similarity transformation $\hat{A} = G^{-1}AG$ by stages: Starting with $A_0 = A$, we generate $A_1, A_2, A_3, \ldots$ by

$$A_k = G_k^{-1} A_{k-1} G_k, \qquad k = 1, 2, \ldots, n-1, \tag{4.7.3}$$

and end with $A_{n-1} = \hat{A}$.

This algorithm requires that we form the matrix $p(A)$, so that we can do the GR decomposition. As we have remarked previously, computing $p(A)$ is expensive if $m > 1$, and we really cannot afford to do it. We thus pose the following question: How can we generate the transforming matrices G_k without computing $p(A)$? If we can generate the G_k, then we can perform the similarity transformations (4.7.3) and thereby execute the GR iteration.

We begin with G_1. The job of G_1^{-1} is to transform the first column of $p(A)$ to upper triangular form. This means that we can determine G_1 if we know just the first column of $p(A)$ or a vector x proportional to the first column. In Section 4.5 we noted that such an x can be computed by (4.5.2), and this computation is inexpensive if $m \ll n$ (Exercise 4.5.6). Once we have x, we choose G_1 so that $G_1^{-1}x = \alpha e_1$ for some $\alpha \neq 0$. Because only the first $m+1$ entries of x are nonzero, we can take G_1 to have the form $G_1 = \mathrm{diag}\{\tilde{G}_1, I\}$, where $\tilde{G}_1 \in \mathbb{F}^{(m+1) \times (m+1)}$. In fact this G_1 is the same as the G_1 of Section 4.5. We observed in Section 4.5 that the transformation $A_1 = G_1^{-1}AG_1$ creates a bulge (4.5.4) with its tip at position $(m+2, 1)$.

Now how do we generate G_2? If we were doing the GR step explicitly, we would have computed

$$R_1 = G_1^{-1} p(A) = \begin{bmatrix} r_{11}^{(1)} & \cdots \\ 0 & R_{22}^{(1)} \end{bmatrix}.$$

The first column of R_1 is in triangular form, and $R_{22}^{(1)}$ is a proper m-Hessenberg matrix. The task of G_2^{-1} is to transform the first column of $R_{22}^{(1)}$ to upper triangular form. The nonzero entries of that column comprise a vector

$$y = \begin{bmatrix} r_{22}^{(1)} \\ \vdots \\ r_{m+2,2}^{(1)} \end{bmatrix} \in \mathbb{F}^{m+1}. \tag{4.7.4}$$

Thus $G_2 = \mathrm{diag}\left\{1, \tilde{G}_2, I_{n-m-2}\right\}$, where $\tilde{G}_2^{-1} y = \beta e_1$ for some $\beta \neq 0$.

Since we are not doing the GR step explicitly, we do not have R_1; we have A_1. How do we generate G_2, given A_1? Well, it turns out that the vector y of (4.7.4) is proportional to the vector

$$z = \begin{bmatrix} a_{21}^{(1)} \\ \vdots \\ a_{m+2,1}^{(1)} \end{bmatrix} \tag{4.7.5}$$

from the first column of the bulge in A_1. (We defer the proof for now.) Thus the first column of A_1 contains the information we need to compute G_2: We build $\tilde{G}_2$ so that $\tilde{G}_2^{-1} z = \gamma e_1$. Then we get $\tilde{G}_2^{-1} y = \beta e_1$ as a consequence. This shows that G_2 is exactly the same as the G_2 from Section 4.5. Thus the transformation $A_1 \to G_2^{-1} A_1$ will return the first column of A_1 to upper Hessenberg form, and the transformation $G_2^{-1} A_1 \to G_2^{-1} A_1 G_2 = A_2$ extends the bulge downward by one column.

The reader can easily guess what comes next. The transforming matrix G_3, which transforms the third column of $R_2 = G_2^{-1} G_1^{-1} p(A)$ to upper triangular form in the explicit algorithm, turns out to be exactly the transformation that transforms the second column of A_2 to upper Hessenberg form. Thus the transformation from A_2 to $A_3 = G_3^{-1} A_2 G_3$ pushes the bulge over one column and down one row. This pattern holds up essentially for the rest of the GR iteration.

To make precise conclusions, we need to distinguish whether $p(A)$ is nonsingular or singular. If $p(A)$ is nonsingular, the transformations $G_1, \ldots, G_{n-1}$ produced in the bulge-chasing algorithm of Section 4.5 are identical to the $G_1, \ldots, G_{n-1}$ produced in the GR decomposition (4.7.2). If $p(A)$ is singular, the identicality of the transformations holds for $G_1, \ldots, G_{n-j}$, where $n - j$ is the rank of $p(A)$, but after that there need not be agreement. This is a consequence of extra freedom in the GR decomposition in the singular case and will be discussed below.

Proving Proportionality

We still have to show that the vector y in (4.7.4) is proportional to the vector z in (4.7.5), and that similar proportionalities hold up throughout the algorithm. For $k = 1, \ldots, n - 1$, let $\hat{G}_k = G_1 G_2 \cdots G_k$, and let

$$R_k = G_k^{-1} \cdots G_2^{-1} G_1^{-1} p(A) = \hat{G}_k^{-1} p(A). \tag{4.7.6}$$

Then

$$R_k = \begin{bmatrix} R_{11}^{(k)} & R_{12}^{(k)} \\ 0 & R_{22}^{(k)} \end{bmatrix},$$

where $R_{11}^{(k)} \in \mathbb{F}^{k \times k}$ is upper triangular. Let y denote the first column of $R_{22}^{(k)}$. This vector has at most $m + 1$ nonzero entries. If $k < n - m - 1$, some of the bottom entries of y will be zeros, but for the coming argument we do not need to think explicitly about those zeros. The transformation G_{k+1}, which produces R_{k+1} by $R_{k+1} = G_{k+1}^{-1} R_k$, will have the form diag$\{I_k, \tilde{G}_{k+1}\}$, where $\tilde{G}_{k+1}^{-1} y = \beta_{k+1} e_1$ for some $\beta_{k+1} \neq 0$.

We have established that A_1 has the form (4.5.4), i.e., almost Hessenberg, with a bulge extending to position $(m + 2, 1)$. Now assume inductively that A_k has the form (4.5.5), almost Hessenberg, with a bulge extending to position $(m + k + 1, k)$. More precisely (taking into account that the bulge is eventually shoved off the end of the matrix) we should say that the bulge extends to position $(\tilde{p}, k)$, where $\tilde{p} = \min\{m + k + 1, n\}$. Still more precisely, we can say that we are assuming $a_{ij}^{(k)} = 0$ if $i > j + 1$ and either $j < k$ or $i > m + k + 1$. Write A_k in partitioned form:

$$A_k = \begin{bmatrix} A_{11}^{(k)} & A_{12}^{(k)} \\ A_{21}^{(k)} & A_{22}^{(k)} \end{bmatrix},$$

where $A_{11}^{(k)} \in \mathbb{F}^{k \times k}$. Our assumption about the shape of A_k implies that $A_{11}^{(k)}$ is upper Hessenberg, and $A_{21}^{(k)}$ has nonzero entries only in its last column. Let z denote the last column of $A_{21}^{(k)}$. This is the first column of the bulge. If $k < n - m - 1$, some of the bottom entries of z will be zeros. We need not pay explicit attention to these. We would like to show that z is proportional to y, the first column of $R_{22}^{(i)}$. If this is so, then the G_{k+1}^{-1} that transforms R_k to R_{k+1} also serves to chase the bulge. The desired result is a consequence of the following lemma.

Lemma 4.7.1. *For $k = 1, \ldots, n - 1$,*

$$A_k R_k = R_k A.$$

This is a variation on Lemma 4.6.5. The easy proof is worked out in Exercise 4.7.1. If we now consider the equation $A_k R_k = R_k A$ in partitioned form,

$$\begin{bmatrix} A_{11}^{(k)} & A_{12}^{(k)} \\ A_{21}^{(k)} & A_{22}^{(k)} \end{bmatrix} \begin{bmatrix} R_{11}^{(k)} & R_{12}^{(k)} \\ 0 & R_{22}^{(k)} \end{bmatrix} = \begin{bmatrix} R_{11}^{(k)} & R_{12}^{(k)} \\ 0 & R_{22}^{(k)} \end{bmatrix} \begin{bmatrix} A_{11} & A_{12} \\ A_{21} & A_{22} \end{bmatrix}, \qquad (4.7.7)$$

where the $(1, 1)$ blocks are all $k \times k$, and equate the $(2, 1)$ blocks of the products, we find that

$$A_{21}^{(k)} R_{11}^{(k)} = R_{22}^{(k)} A_{21}. \qquad (4.7.8)$$

Recall that A_{21} has only one nonzero entry, $a_{k+1,k}$, in the upper right-hand corner. Therefore only the last column of $R_{22}^{(k)} A_{21}$ is nonzero. This column is easily seen to be $y a_{k+1,k}$, where y denotes, as before, the first column of $R_{22}^{(k)}$. Recall that only the last column of $A_{21}^{(k)}$ is nonzero. That column, which we called z, contains the first column of the bulge in A_k. We easily check that the final column of $A_{21}^{(k)} R_{11}^{(k)}$ is $z r_{kk}^{(k)}$. Therefore, by (4.7.8),

$$z r_{kk}^{(k)} = y a_{k+1,k}. \qquad (4.7.9)$$

This proves that y and z are proportional, provided that the scalars in (4.7.9) are nonzero. Since A is assumed properly upper Hessenberg, $a_{k+1,k}$ is certainly nonzero. As for $r_{kk}^{(k)}$, let us consider first the nonsingular case.

If $p(A)$ is nonsingular, then each of the matrices R_k is nonsingular, so all of the $r_{kk}^{(k)}$ are nonzero. Consequently, y and z are proportional in (4.7.9). This is what we wanted to show. It implies that the transformation $A_k \rightarrow G_{k+1}^{-1} A_k G_{k+1} = A_{k+1}$ clears out the kth column of the bulge, moving the bulge down and to the right. This completes the induction on the form of the bulge in the matrices A_k.

To say that each G_k serves to chase the bulge is the same as to say that the G_k of this section are the same as the G_k of Section 4.5. This is true for $k = 1, \ldots, n - 1$, that is, for all k, assuming that $p(A)$ is nonsingular.

Let us now check how well these results hold up when $p(A)$ is singular. In this case we will eventually have a zero $r_{kk}^{(k)}$, and the correspondence will break down. Indeed, it is easy to check that the main-diagonal entries in the final matrix R are given by $r_{kk} = r_{kk}^{(k)}$, so we can look at R to determine when the breakdown will occur. If $\text{rank}(p(A)) = n - j$, indicating j perfect shifts, then the form of R is dictated by Theorem 4.6.4, which shows that $r_{kk} \neq 0$ for $k = 1, \ldots, n - j$, and $r_{kk} = 0$ for $k = n - j + 1, \ldots, n$. Thus we conclude that the G_k of this section are the same as the G_k that chase the bulge for $k = 1, \ldots, n - j$, but after that the correspondence breaks down.

Now let us take a closer look at the GR decomposition as represented in (4.7.2). Theorem 4.6.4 shows that the bottom j rows of R are zero. This implies that the GR decomposition is complete after $n - j$ steps; the transformations $G_{n-j+1}, \ldots, G_{n-1}$ are extraneous. They can be chosen arbitrarily, subject to the condition that they act only on the last j rows. If we specify as above that G_k should have the form $G_k = \text{diag}\{I_{k-1}, \tilde{G}_k\}$, then each $\tilde{G}_k$ can be chosen arbitrarily for $k = n - j + 1, \ldots, n - 1$. The easiest choice is to take $\tilde{G}_k = I, k = n - j + 1, \ldots, n - 1$. Another choice is choose each $\tilde{G}_k$ so that the bulge in A_{k-1} is pushed forward. If we make this choice, then the G_k in (4.7.2) will coincide with the G_k of the bulge chase from start to finish.

Exercises

4.7.1. This exercise proves Lemma 4.7.1.

(a) Show that
$$p(A) = \hat{G}_k R_k, \qquad A_k = \hat{G}_k^{-1} A \hat{G}_k.$$

This is a *partial GR step*, since R_k is "partially" upper triangular.

(b) Show that $A_k \hat{G}_k^{-1} = \hat{G}_k^{-1} A$. Multiply this equation on the right by $p(A)$, and deduce that $A_k R_k = R_k A$.

Notice that this proof is identical to the proof of Lemma 4.6.5.

4.7.2. Referring to (4.7.8), verify that

(a) the last column of $R_{22}^{(k)} A_{21}$ is $y a_{k+1,k}$, and all other columns are zero;

(b) the last column of $A_{21}^{(k)} R_{11}^{(k)}$ is $z r_{kk}^{(k)}$, and all other columns are zero;

(c) in the final matrix R of (4.7.2) we have $r_{kk} = r_{kk}^{(k)}$ for $k = 1, \ldots, n - 1$. Deduce that $r_{kk}^{(k)} \neq 0$ for $k = 1, \ldots, n - 1$, if $p(A)$ is nonsingular.

4.7.3. In this section we used an induction argument to demonstrate that each A_k has a bulge extending to the $(\tilde{p}, k)$ entry, where $\tilde{p} = \min\{n, m + k + 1\}$. However, the induction argument was not strictly needed, since the shape of A_k can also be deduced by other means.

(a) Show that $\hat{G}_k = \text{diag}\{\check{G}_k, I\}$, where $\check{G}_k$ is $(m+k) \times (m+k)$. Deduce that the entries of $A_k = \hat{G}_k^{-1} A \hat{G}_k$ satisfy $a_{ij}^{(k)} = 0$ if $i > j + 1$ and $i > m + k + 1$.

(b) Assuming $r_{jj}^{(k)} \neq 0$ for $j = 1, \ldots, k$, deduce from (4.7.7) that

$$A_{11}^{(k)} = R_{11}^{(k)} A_{11} R_{11}^{(k)^{-1}} + R_{12}^{(k)} A_{21} R_{11}^{(k)^{-1}}.$$

(c) Show that $R_{11}^{(k)} A_{11} R_{11}^{(k)^{-1}}$ is upper Hessenberg (cf. Exercise 4.5.3). Considering the form of A_{21}, show that the only nonzero elements in $R_{12}^{(k)} A_{21} R_{11}^{(k)^{-1}}$ are in the last column. Deduce that $A_{11}^{(k)}$ is upper Hessenberg.

(d) Continuing to assume that $r_{jj}^{(k)} \neq 0$ for $j = 1, \ldots, k$, deduce from (4.7.8) that $A_{21}^{(k)} = R_{22}^{(k)} A_{21} R_{11}^{(k)^{-1}}$. Show that the only nonzero entries in $A_{21}^{(k)}$ are in the last column.

(e) From parts (c) and (d) deduce that the entries of A_k satisfy $a_{ij}^{(k)} = 0$ if $i > j + 1$ and $j < k$.

Combining the results of parts (a) and (e), we see that $a_{ij}^{(k)} = 0$ if $i > j + 1$ and either $i > m + k + 1$ or $j < k$. This implies that A_k is upper Hessenberg, except for a bulge extending to entry $(\tilde{p}, k)$, where $\tilde{p} = \min\{n, m + k + 1\}$.

4.8   Computing Eigenvectors and Invariant Subspaces

So far we have focused on the computation of eigenvalues by transforming the matrix to (block) triangular form. If eigenvectors or invariant subspaces are wanted, more work must be done.

We start with a matrix A whose eigenvectors are wanted. We transform A to upper Hessenberg form. Then we perform a large number of GR iterations to transform the Hessenberg matrix to block upper triangular form. The entire process can be summarized as a single similarity transformation

$$T = S^{-1} A S. \tag{4.8.1}$$

In the complex case, T is upper triangular, but in the real case, in which we do everything in real arithmetic, the final T is quasi-triangular.

If only eigenvalues are wanted, we can pay little attention to the transforming matrices that contribute to S in (4.8.1). We produce them; we use them; then we throw them away. However, if we want more than just eigenvalues, we must build the matrix S. This is easy in both principle and practice. Assuming that the work was begun with a reduction to upper Hessenberg form, the transforming matrix G for that step can be assembled as described in

Exercise 3.3.2. Then, for each similarity transformation G_i that we apply as part of a GR iteration, we update G by the transformation $G \leftarrow GG_i$. In the end, the array that initially held G contains the overall transformation matrix S. This is easy to implement, but it more than doubles the flop count. Nevertheless, we are still in the realm of $O(n^3)$.

The way we deal with the upper Hessenberg iterates A_i also depends upon whether or not we want more than just eigenvalues. It happens repeatedly that we can reduce the matrix:

$$A_i = \begin{bmatrix} A_{11}^{(i)} & A_{12}^{(i)} \\ 0 & A_{22}^{(i)} \end{bmatrix},$$

If just eigenvalues are wanted, all subsequent operations can be restricted to $A_{11}^{(i)}$ and $A_{22}^{(i)}$; the block $A_{12}^{(i)}$ can be ignored from this point on. However, if we want more than eigenvalues, we need to compute the correct quasi-triangular matrix T in (4.8.1). This means that we need to continue updating $A_{12}^{(i)}$ after the reduction. Each update $A_{11} \leftarrow G_1^{-1} A_{11} G_1$ or $A_{22} \leftarrow G_2^{-1} A_{22} G_2$ must be accompanied by an update $A_{12} \leftarrow G_1^{-1} A_{12}$ or $A_{12} \leftarrow A_{12} G_2$.

In Exercise 4.8.1 we estimate that the total work to produce the quasi-triangular form is about $30n^3$ flops. This compares with about $12n^3$ if only the eigenvalue computation had been done. These numbers are only very crude estimates; the actual count will vary from one matrix to the next.

A modest reduction of the total work can be realized by using the *ultimate shift strategy*: We invest $\frac{14}{3}n^3$ flops to get the Hessenberg form and an additional $8n^3$ to get the eigenvalues of A without accumulating the similarity transformation. We then return to the Hessenberg form, having saved it, and repeat the GR iterations, using the computed eigenvalues as shifts. This time we accumulate the similarity transformation. Since we are now using the *ultimate* shifts, many fewer iterations will be needed, so much less arithmetic will be done. In principle, each bulge-chase of degree m will result in the deflation of m eigenvalues (Theorem 4.6.1). In practice, because of roundoff errors, something like two bulge-chases will be needed per m eigenvalues deflated. This is still far less than the four or six that would normally be needed. The calculations in Exercise 4.8.2 suggest that the ultimate shift strategy will result in a net savings of about $8n^3$ flops. Again, these flop counts should not be taken too seriously, but they do suggest that the ultimate shift strategy will yield a modest savings.

In principle the order in which the ultimate shifts are applied is immaterial. In practice, however, better results will be obtained if the shifts are applied in the order in which they emerged in the original eigenvalue computation. For examples of really bad results obtained by applying ultimate shifts in the wrong order, see [173].

Any intelligent implementation of the ultimate shift strategy will take into account that the strategy could occasionally fail due to roundoff errors. If that should happen, the algorithm should revert to a standard shift strategy.

Computing Eigenvectors

Once we have the quasi-triangular form $T = S^{-1}AS$, we still have to do more work to get whatever it is we want, whether it be eigenvectors or invariant subspaces. Eigenvectors can be obtained rather cheaply at this point. Exercise 4.8.3 shows that eigenvectors of quasi-triangular matrices can be computed easily. If x is an eigenvector of T, associated with the eigenvalue λ, then Sx is a corresponding eigenvector of A. Each eigenvector can be computed at a cost of $3n^2$ flops or less, so a complete set of eigenvectors can be obtained in fewer than $3n^3$ flops.

In Exercise 4.8.3 we assume that the eigenvalues of T are distinct. This is what is seen in practice; even if A had some multiple eigenvalues to begin with, roundoff errors would break each multiple eigenvalue into a tight cluster of simple eigenvalues. We have to reckon, though, that the eigenvectors associated with such eigenvalues are ill conditioned for sure. A safer object to compute would be the invariant subspace associated with the cluster of eigenvalues.

Computing Invariant Subspaces

Once we have the triangularization

$$T = S^{-1}AS,$$

we already have a number of invariant subspaces. Letting $s_1, \ldots s_n$ denote the columns of S, we know that, for each j, $\mathrm{span}\{s_1, \ldots, s_j\}$ is invariant under A. If $\lambda_1, \ldots, \lambda_j$ are the first j main-diagonal entries of T, and they are distinct from the other eigenvalues of A, then $\mathrm{span}\{s_1, \ldots, s_j\}$ is the invariant subspace of A associated with the eigenvalues $\lambda_1, \ldots, \lambda_j$. This is the situation if T is upper triangular. If T is merely quasi-triangular, then $\mathrm{span}\{s_1, \ldots, s_j\}$ will be invariant if and only if it doesn't "split" a 2×2 block. That is, there should not be a 2×2 block in T in rows and columns j and $j + 1$.

The invariant subspaces that we get automatically from a triangularization may not be the ones we want. Suppose we want the invariant subspace associated with j eigenvalues $\mu_1, \ldots, \mu_j$ that are not the leading j main-diagonal entries of T. If we can effect a similarity transformation that reorders the main diagonal entries, that is, that gives us a new upper triangular matrix $\hat{T} = \hat{S}^{-1}A\hat{S}$ whose leading main-diagonal entries are $\mu_1, \ldots, \mu_j$ (in any order), then $\mathrm{span}\{\hat{s}_1, \ldots, \hat{s}_j\}$ is the desired invariant subspace.

We will be able to achieve our objective if we can develop a way of "interchanging" eigenvalues. More generally, we can think about interchanging blocks. Given a block triangular matrix

$$\begin{bmatrix} B_{11} & B_{12} & \cdots & B_{1k} \\ 0 & B_{22} & & B_{2k} \\ \vdots & & \ddots & \vdots \\ 0 & & & B_{kk} \end{bmatrix},$$

can we perform a similarity transformation that produces a new block triangular matrix in which the positions of blocks B_{jj} and $B_{j+1,j+1}$ have been reversed? Actually we do not need to recreate the blocks exactly. We just need for the new (j, j) and $(j + 1, j + 1)$ block to be similar to the old $B_{j+1,j+1}$ and B_{jj}, respectively, so that the positions of the eigenvalues are exchanged. If we can do this, we will be able to rearrange the blocks in any desired order.

Obviously it suffices to consider the case $k = 2$. If we can swap the blocks in

$$\begin{bmatrix} B_{11} & B_{12} \\ 0 & B_{22} \end{bmatrix},$$

we can do anything. Let's begin by swapping in the crudest imaginable way. Applying a similarity transformation by

$$\begin{bmatrix} 0 & I \\ I & 0 \end{bmatrix},$$

we obtain

$$\begin{bmatrix} B_{22} & 0 \\ B_{12} & B_{11} \end{bmatrix},$$

which has the diagonal blocks in the desired places but is no longer block upper triangular. Now we can apply the block-diagonalization method introduced in Section 2.4 to annihilate the block B_{12}. Recall from (2.4.4) that this requires solving the Sylvester equation

$$X B_{22} - B_{11} X = B_{12}.$$

Recall from Theorem 2.4.2 (and Exercise 2.4.5) that this has a unique solution if and only if B_{11} and B_{22} have no eigenvalues in common. Assuming this to be the case, then, as in (2.4.3), we have

$$\begin{bmatrix} I & 0 \\ -X & I \end{bmatrix} \begin{bmatrix} B_{22} & 0 \\ B_{12} & B_{11} \end{bmatrix} \begin{bmatrix} I & 0 \\ X & I \end{bmatrix} = \begin{bmatrix} B_{22} & 0 \\ 0 & B_{11} \end{bmatrix}. \tag{4.8.2}$$

This transformation succeeds in annihilating B_{12} because the space spanned by the columns of $\begin{bmatrix} I \\ X \end{bmatrix}$ is invariant under $\begin{bmatrix} B_{22} & 0 \\ B_{12} & B_{11} \end{bmatrix}$ (Proposition 2.2.2); it is the invariant subspace associated with the eigenvalues of B_{22}. In practice we prefer not to perform the transformation (4.8.2) literally, since the transforming matrix can be ill conditioned (when $\| X \|$ is large). Instead we can perform a similarity transformation by a related matrix whose leading columns span the same invariant subspace. Suppose that we perform a QR decomposition

$$\begin{bmatrix} I & 0 \\ X & I \end{bmatrix} = QR = \begin{bmatrix} Q_{11} & Q_{12} \\ Q_{21} & Q_{22} \end{bmatrix} \begin{bmatrix} R_{11} & R_{12} \\ 0 & R_{22} \end{bmatrix}.$$

Then Q is unitary, and we can use it to perform the exchange, since the leading columns $\begin{bmatrix} Q_{11} \\ Q_{21} \end{bmatrix}$ span the same invariant subspace as the columns of $\begin{bmatrix} I \\ X \end{bmatrix}$. Therefore

$$\begin{bmatrix} Q_{11}^* & Q_{21}^* \\ Q_{12}^* & Q_{22}^* \end{bmatrix} \begin{bmatrix} B_{22} & 0 \\ B_{12} & B_{11} \end{bmatrix} \begin{bmatrix} Q_{11} & Q_{12} \\ Q_{21} & Q_{22} \end{bmatrix} = \begin{bmatrix} C_{11} & C_{12} \\ 0 & C_{22} \end{bmatrix}, \tag{4.8.3}$$

where C_{11} and C_{22} have the same eigenvalues as B_{22} and B_{11}, respectively. Thus the blocks have been swapped.

Equation (4.8.3) is true in exact arithmetic. However, when the computation is done in floating point, the block C_{21} will not be exactly zero. If its norm lies under some specified threshold, it can be set to zero. If not, the block swap must be rejected on the grounds that the eigenvalues of B_{11} are not sufficiently separated from those of B_{22}. This is an indication that some or all of the eigenvalues in B_{11} and B_{22} should be treated as a single cluster, and no attempt should be made to separate them. If we accept this verdict, then there is never any reason to swap these two blocks.

In order to carry out the block-swap procedure, we need to have a way to solve the Sylvester equations that arise. But notice that in this application the Sylvester equations are really small. The blocks we are swapping are never bigger than 2×2. If we want to swap two 2×2 blocks, corresponding to two pairs of complex eigenvalues, the Sylvester system consists of four equations in four unknowns, which we can easily solve without any special techniques. If we want to swap a 1×1 and a 2×2 block, we have to solve two equations in two unknowns, and to swap two 1×1 blocks we have to solve one equation in one unknown. For more details consult Bai and Demmel [14]. An alternate procedure was described by Bojanczyk and Van Dooren [43].

Now returning to the big picture, recall that our objective was to compute the invariant subspace associated with selected eigenvalues $\mu_1, \ldots, \mu_j$. Once we have swapped the blocks containing these eigenvalues to the top, updating the transforming matrix S at every step, we have achieved our objective. The leading j columns of the updated S span the desired invariant subspace.

Once we have the invariant subspace, we might also like to estimate its condition number. This can be done accurately if all of the transformation matrices were unitary, that is, if we used the QR algorithm. At the end of the computation, we have the matrix in triangular form

$$\left[\begin{array}{cc} A_{11} & A_{12} \\ 0 & A_{22} \end{array} \right],$$

with the block A_{11} corresponding to the invariant subspace. From Theorem 2.7.13 we know that the condition number is $\| A \|/\operatorname{sep}(A_{11}, A_{22})$. This is a condition number for $\mathcal{E}_k$; if the transforming matrix S is unitary, it is also a condition number for $\mathcal{S}$. The challenge is to estimate $\operatorname{sep}(A_{11}, A_{22})$, which is the reciprocal of the norm of φ^{-1}, the inverse Sylvester operator. Recall that $\varphi^{-1}(Y) = X$ if and only if

$$X A_{11} - A_{22} X = Y.$$

Once again, we have an incentive to solve Sylvester equations, but now the Sylvester system may be large, so we would like to solve it as efficiently as possible. A method for doing this, the Bartels–Stewart algorithm [21], is outlined in Exercise 4.8.5.

If we can evaluate $\varphi^{-1}(Y)$ for a few values of Y, we can use a condition number estimator of Hager type [114] to get a good estimate of $\| \varphi^{-1} \|$ and of the condition number of the invariant subspace. Routines that perform these computations are provided in LAPACK [8]. See also [16].

The Normal Case

Everything associated with eigenvector computation is greatly simplified in the normal case. The principle of structure preservation dictates that if A is normal ($AA^* = A^*A$), then we should try to preserve the normality. We do this by applying only unitary transformations to A. Thus the reduction to Hessenberg form is effected by unitary transformations, and then the QR algorithm is applied. In the end we arrive at the diagonal matrix guaranteed by Spectral Theorem 2.3.5: We have $A = U \Lambda U^{-1}$, where U is unitary and Λ is diagonal. The problem of eigenvector computation is now trivial; each column of U is an eigenvector of A. Block swapping is also trivial; for each pair of eigenvalues you swap, just interchange the corresponding columns of U. In fact block swapping is not really necessary. To obtain a basis for the invariant subspace associated with specified eigenvalues $\mu_1, \ldots, \mu_j$, just pick out the corresponding columns of U.

Condition number computation is also easy. The eigenvalues are always well conditioned, and the condition numbers of invariant subspaces are determined by gaps between eigenvalues.

Exercises

4.8.1. Let $A \in \mathbb{F}^{n \times n}$. In this exercise we make a crude estimate of the total cost of computing the quasi-triangular form $T = S^{-1}AS$ (4.8.1) by the reduction to upper Hessenberg form followed by the generic GR algorithm. We have already estimated (Section 3.3) that the cost of the transformation to upper Hessenberg form is $\frac{14}{3}n^3$ flops, including the cost of assembling the transforming matrix. Now we compute the additional costs. This is an update of Exercise 4.5.9. Assume that six iterations are needed for each deflation, and that each deflation breaks off an $m \times m$ chunk, delivering exactly m eigenvalues. Assume that the cost of an mth degree bulge-chase is about $4n^2m$ flops, as shown in Exercise 4.5.8.

(a) Show that, under the stated assumptions, about $24n^2m$ flops will be needed for the first batch of eigenvalues, then $24n(n-m)m$ flops will be needed for the next batch, $24n(n-2m)m$ for the third batch, and so on, assuming that after we have made a deflation
$$\left[\begin{array}{cc} A_{11} & A_{12} \\ 0 & A_{22} \end{array} \right],$$
we must continue to update the $(1, 2)$ block of the matrix.

(b) By approximating a sum by an integral, show that the total flop count from part (a) is about $12n^3$. (This compares with $8n^3$ if the updates of the $(1, 2)$ block are not needed (Exercise 4.5.9).)

(c) Now we consider the cost of accumulating the transforming matrix S. Assume that the reduction to Hessenberg form has been done, and we have the transforming matrix G from that reduction. Each time we apply an elementary matrix G_i to A, we must perform an update $G \leftarrow GG_i$. Show that the cost of each such update is about $4nm$ flops. Show that under the assumptions we have made, the total number of updates is about $n^2/(2m)$. Thus the total work to produce S from G is about $12n^3$ flops.

(d) Deduce that the total operation count for the reduction to quasi-triangular form is about $\frac{86}{3}n^3$ flops, which we can round to about $30n^3$. (This compares with about $12n^3$ if only eigenvalues are wanted.)

4.8.2. Suppose we run the GR algorithm with ultimate shifts, and assume that two iterations are needed for each deflation (instead of the six that were assumed in Exercise 4.8.1), and that each deflation breaks off an $m \times m$ chunk, delivering exactly m eigenvalues. Show that the total cost of the ultimate shift phase of the computation is $8n^3$ flops (instead of $24n^3$). Taking into account that this method requires some extra work to get the eigenvalues to use as shifts, deduce that the net savings for the ultimate shift strategy is about $8n^3$ flops.

(Obviously this estimate is very crude, since it depends on the iteration numbers 2 and 6, which are speculative.)

4.8.3. In this exercise we investigate the computation of eigenvectors of a quasi-triangular matrix T. We will assume that the eigenvalues of T are distinct, which is what occurs in practice. If λ is an eigenvalue of T, then it is either an entry on the main diagonal (a 1×1 block) or an eigenvalue of a 2×2 block on the main diagonal. In either event, we can partition T as

$$\begin{bmatrix} T_{11} & T_{12} & T_{13} \\ & T_{22} & T_{23} \\ & & T_{33} \end{bmatrix},$$

where T_{22} is either 1×1 or 2×2, and λ is an eigenvalue of T_{22}. Let x denote an eigenvector of T associated with λ. Partition x conformably with the partition of T.

(a) Show that

$$\begin{bmatrix} T_{11} - \lambda I & T_{12} & T_{13} \\ & T_{22} - \lambda I & T_{23} \\ & & T_{33} - \lambda I \end{bmatrix} \begin{bmatrix} x_1 \\ x_2 \\ x_3 \end{bmatrix} = \begin{bmatrix} 0 \\ 0 \\ 0 \end{bmatrix}.$$

(b) Show that $x_3 = 0$.

(c) Show that x_2 is an eigenvector of T_{22} associated with the eigenvalue λ. What does this mean in the case T_{22} is 1×1? Show that in either case, x_2 can be computed for $O(1)$ work.

(d) Show that x_1 satisfies $(T_{11} - \lambda I)x_1 = -T_{12}x_2$ and that x_1 is uniquely determined once x_2 has been chosen. Show that if T_{11} is $j \times j$, x_1 can be computed by back substitution with about j^2 flops.

(e) Show that if $A = STS^{-1}$, then $v = Sx$ is an eigenvector of A associated with eigenvalue λ. Show that the total cost of computing v (given S and T) is not more than $3n^2$ flops.

(f) If A is real and λ is complex, explain how to get an eigenvector associated with $\bar{\lambda}$ for free, once you have an eigenvector associated with λ.

4.8.4. Suppose that $T \in \mathbb{C}^{n \times n}$ is upper triangular and has λ as an eigenvalue with algebraic multiplicity two.

(a) Show how to compute an eigenvector of T associated with λ.

(b) Show that there may or may not be a second linearly independent eigenvector associated with λ. State precise conditions under which a second linearly independent eigenvector exists.

(c) If the above-diagonal entries of T are "random," is it more likely that the dimension of the eigenspace is one or two?

4.8.5. This exercise outlines the Bartels–Stewart algorithm [21] for solving Sylvester equations. A variant is given in Exercise 4.8.6.

(a) Suppose we wish to solve the Sylvester equation $AX - XB = C$. Apply Schur's theorem to A and B, and deduce that the equation can be transformed to a related equation $T\hat{X} - \hat{X}S = \hat{C}$, where T and S are upper triangular. (If A, B, and C are real, we can stay within the real number system by making S and T quasi-triangular.)

(b) Now suppose we wish to solve a Sylvester equation $TX - XS = C$, where T and S are upper triangular and have no eigenvalues in common. Let $x_1, \ldots, x_k$ denote the columns of X, and likewise for C. Show that x_1 is the solution of $(T - s_{11}I)x_1 = c_1$. This triangular system can be solved by back substitution. Write down a triangular system that can be solved for x_2, given x_1.

(c) Write down an algorithm that solves for $x_1, x_2, \ldots, x_k$ in succession by solving a sequence of upper triangular systems. (Each is a small Sylvester equation in its own right.) Show that each of the systems to be solved is nonsingular. This is the basic Bartels–Stewart algorithm.

(d) If T is merely block triangular, the algorithm from part (c) remains applicable, except that a block variant of back substitution must be used. Show that each step of each back solve amounts to solving a small Sylvester equation.

(e) Now suppose that both T and S are block triangular. Show that the computation of X can be achieved by solving a sequence of small Sylvester equations.

(f) Write down an algorithm that solves $TX - XS = C$ when both T and S are quasi-triangular. This is essentially the algorithm that was published by Bartels and Stewart in [21].

(g) Show that (any version of) the Bartels–Stewart algorithm requires $O(m^3 + k^3)$ flops to solve a system of mk equations. Which is more expensive, the Schur decomposition phase or the back substitution phase? How does the flop count compare to that for dense Gaussian elimination (LU decomposition)?

4.8.6. This exercise outlines a variation on the Bartels–Stewart algorithm. We restrict ourselves to the complex case for simplicity.

(a) Suppose we wish to solve the Sylvester equation $AX - XB = C$. Apply Schur's theorem to A^T and B, and deduce that the equation can be transformed to a related equation $T\hat{X} - \hat{X}S^T = \hat{C}$, where T and S are *lower* triangular.

(b) Now suppose we wish to solve a Sylvester equation $TX - XS^T = C$, where T and S are lower triangular and have no eigenvalues in common. Use the Kronecker product (Exercises 2.4.3 and 2.4.4) to rewrite this equation as

$$(I \otimes T - S \otimes I)\mathrm{vec}(X) = \mathrm{vec}(C).$$

(c) Show that the coefficient matrix $I \otimes T - S \otimes I$ is lower triangular. This system can be solved efficiently by forward substitution.

(d) How does this algorithm compare with the (complex version of) the Bartels–Stewart algorithm?

4.9   Practical Bulge-Chasing Algorithms

General Matrices

The standard codes for general matrices usually begin with a preprocessing step that *balances* the matrix [159, 174, 232]. This procedure isolates certain obvious invariant subspaces, if there are any. Then it does a diagonal similarity transformation that attempts to minimize the norm of the matrix. This move substantially improves the accuracy of the computed eigenvalues in some cases [15, § 7.2].[9]

Once the balancing has been done, the standard codes use exclusively unitary transformations in the interest of numerical stability. Let us consider a complex matrix A first. The matrix is reduced to upper Hessenberg form by unitary transformations. If the transformation is done in such a way that all of the subdiagonal entries of the resulting upper Hessenberg matrix are real (always possible), then some economy is gained on the subsequent QR iterations.

LAPACK version 3.0 [8] does this, then it applies one of two complex QR codes, depending on the size of the matrix.[10] The "small-matrix" code is used whenever the matrix is 50×50 or smaller. This is an implicit single-shift ($m = 1$) QR algorithm using Wilkinson shifts. Recall that this means that the eigenvalue of

$$
\begin{bmatrix}
a_{n-1,n-1}^{(k-1)} & a_{n-1,n}^{(k-1)} \\
a_{n,n-1}^{(k-1)} & a_{n,n}^{(k-1)}
\end{bmatrix}
\tag{4.9.1}
$$

that is closer to $a_{n,n}^{(k-1)}$ is used as the shift for the kth step. Since this strategy occasionally fails to yield convergence (Exercise 4.3.5), an exceptional shift is used whenever 10 or 20 iterations have passed since the most recent deflation.

If the matrix is over 50×50, the "large-matrix" QR code is used. This is a multishift QR algorithm with degree $m = 6$. The eigenvalues of the lower right-hand $m \times m$ submatrix are used as shifts. If 10 or 20 iterations have passed with no deflations, exceptional shifts are used. The small-matrix code is used to compute the shifts. It is also used whenever a matrix of size 50 or less is isolated via deflations.

The authors of the large-matrix code [13] would have liked to use a much larger value of m than 6 but were stopped by an unexpected phenomenon. If m is taken much bigger than 6, roundoff errors prevent the algorithm from converging (or at least drastically slow it down) when it ought to converge quadratically. This phenomenon is explained in terms of *shift blurring* in Section 7.1.

The LAPACK 3.0 codes for real matrices are about the same as for complex matrices, except that the small-matrix algorithm does double shifts ($m = 2$) to stay in real arithmetic. The two shifts for the kth iteration are eigenvalues of (4.9.1). Since the matrix is real, these are either both real or a complex-conjugate pair. If they are complex, the two complex-conjugates are used as shifts. If they are real, the one that is closer to $a_{nn}^{(k-1)}$ (a Wilkinson

---

[9]Balancing can also sometimes degrade the accuracy. For example, it seems to be a bad idea to balance a matrix that has already been reduced to upper Hessenberg form [225].

[10]As this book was nearing completion, LAPACK version 3.1 was released, rendering this material out of date. I decided to leave it in for two reasons: (1) It gives historical perspective on the development of the QR algorithm, and (2) it provides a smooth lead-in to the discussion of the newer codes.

shift) is used twice [68]. In either case the driving polynomial

$$(A_{k-1} - \rho_1 I)(A_{k-1} - \rho_2 I)$$

is real, so the iteration stays within the real number system. As in the complex case, exceptional shifts are used occasionally. This code differs hardly at all from the algorithm originally proposed by Francis [89].

The large-matrix code for the real case is a multishift QR code with degree $m = 6$, just as in the complex case. Since all of the shifts are either real or complex-conjugate pairs, the iterations are done entirely in real arithmetic.

High Performance QR Codes

An important and beneficial side effect of the development of the multishift QR algorithm [13] is that it got people to think about new shift strategies. If one wants to perform a QR iteration of degree 40, say, then one needs 40 shifts. Had it not been for this, one might never have thought to compute the eigenvalues of the lower right-hand 40×40 submatrix and use them as shifts (generalized Rayleigh quotient shift strategy). Important experiments of Dubrulle [77] showed that this shift strategy works very well even if we intend to take QR steps of lower degree. If we have 40 shifts, we can use them to perform one QR iteration of degree 40 or, for example, 20 iterations of degree two. In other words, we can chase one large bulge or 20 small bulges. In theory the two different procedures should yield exactly the same result (Exercise 4.9.1). In practice they do not, because they have vastly different roundoff errors. Because of shift blurring, the high-degree QR iteration is completely ineffective. In contrast, the method that performs 20 steps of degree two works very well. The quadratic convergence demonstrated in [227] (and in Chapter 5) is realized in practice.

Parallel QR algorithm

If one is going to compute a large number of shifts in advance and use them to create and chase a large number of small bulges, one sees an opportunity for parallelism. If the matrix is very large and distributed over a number of processors, then one can start a bulge and begin chasing it. Then, once it has been passed into the territory of the next processor, the first processor can start a new bulge, and so on. In this way we can achieve parallelism by chasing many bulges in pipeline fashion. A successful parallel QR code built on this principle is described in [112]. This code is included in ScaLAPACK [39], a parallel version of LAPACK.

The idea of chasing many bulges in pipeline fashion, as just described, is quite old, but it did not catch on right away because, for a long time, nobody thought to change the shifting strategy. It was assumed that if one were going to chase bulges of degree two, then one should use the eigenvalues of the lower right-hand 2×2 submatrix as shifts. The entries of this submatrix are among the last to be computed in a QR iteration, as the bulge is chased from top to bottom. If one wishes to start a new iteration before the current bulge has reached the bottom of the matrix, one is forced to use old shifts, because the new ones are not available yet. If one wants to keep a steady stream of, say, four bulges running in

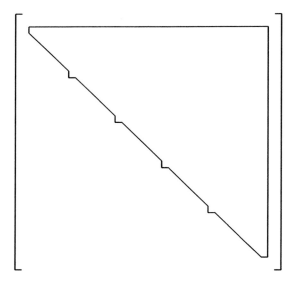

Figure 4.1. *Pipelined QR steps.*

the pipeline, as in Figure 4.1, one is forced to use shifts that are three iterations out of date. Van de Geijn [205] analyzed the effect of using such shifts, which he called deferred shifts, on the rate of convergence. He showed that if deferred shifts are used, the convergence rate ceases to be quadratic but remains superlinear. Precisely, he showed that if the shifts are deferred by h iterations, the convergence rate is τ, where τ is the unique positive root of $t^{h+1} - t^h - 1 = 0$. One easily shows that τ lies strictly between 1 and 2, which means that the convergence is superlinear but subquadratic. When $h = 1$, which means that the shifts are one iteration out of date, $\tau = (1 + \sqrt{5})/2 \approx 1.62$, the golden ratio. This is the same as the convergence rate of the secant method for solving nonlinear equations. As h is increased, τ decreases steadily. When $h = 3$, $\tau \approx 1.38$, and when $h = 20$, $\tau \approx 1.11$. Thus we see that if we pipeline large numbers of bulges using the standard shifting strategy, the convergence rate will be degraded severely.

In contrast to these results, if we use the generalized Rayleigh quotient shifting strategy with, say, $m = 40$, we can chase up to 20 bulges in the pipeline without losing quadratic convergence. Suddenly the idea of chasing bulges in pipeline fashion seems much more appealing than it did before.

Efficient use of cache

A related idea was considered by Braman, Byers, and Mathias [44] and Lang [144]. They use the generalized Rayleigh quotient shift strategy and chase a large number of bulges in pipeline fashion, but instead of spreading the bulges out for parallelism, they bunch them together for efficient cache use. Standard QR iterations of low degree (e.g., $m = 2$) are unable to make efficient use of cache because they operate on only a few rows and columns

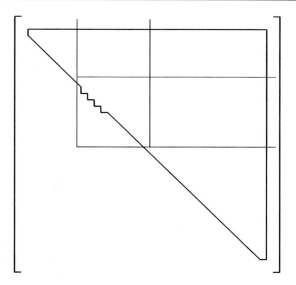

Figure 4.2. *Tightly packed bulges.*

at a time. We know from Exercise 3.2.5 that we can can use the cache efficiently if we can arrange the arithmetic so that it consists of matrix-matrix multiplications in which none of the matrix dimensions is small. The code described in [44] does just this. Consider a multiple QR iteration in which a large number of small bulges ($m = 2$) are chased, one right after the other, as depicted in Figure 4.2. The arithmetic generated by each bulge-chase is confined to the rows and columns currently occupied by the bulge. In Figure 4.2 we have depicted a small window in which the bulges lie. If we chase the bulges from one end of the window to the other, all of the arithmetic takes place in that window and the window to the right and the window above. In other words, if we partition the matrix as

$$\begin{bmatrix} A_{11} & A_{12} & A_{13} \\ A_{21} & A_{22} & A_{23} \\ 0 & A_{32} & A_{33} \end{bmatrix}, \tag{4.9.2}$$

where A_{22} is the window in which the bulges lie, then all of the arithmetic takes place in A_{22}, A_{23}, and A_{12}. Let $t \times t$ denote the window size; we envision a scenario in which $1 \ll t \ll n$. Let $Q \in \mathbb{F}^{t \times t}$ denote the product of all the transformations that are used to chase the bulges from the top to the bottom of the window. Then this portion of the bulge-chase is a similarity transformation on (4.9.2) by a matrix of the form diag$\{I, Q, I\}$. Thus (4.9.2) is transformed to

$$\begin{bmatrix} A_{11} & A_{12}Q & A_{13} \\ A_{21} & Q^{-1}A_{22}Q & Q^{-1}A_{23} \\ 0 & A_{32} & A_{33} \end{bmatrix}.$$

We note that A_{21} and A_{32} each have only one nonzero entry, and that is untouched by Q, i.e., $Q^{-1}A_{21} = A_{21}$ and $A_{32}Q = A_{32}$. The QR code described in [44] performs the bulge-chasing operations on A_{22} in the standard way (with the entire submatrix A_{22} in cache) but defers the operations on A_{12} and A_{23}. As the bulges are chased from one end of the window to the other, the $t \times t$ matrix Q is assembled. Once Q is available, the transformations $A_{12} \rightarrow A_{12}Q$ and $A_{23} \rightarrow Q^{-1}A_{23}$ are applied as conventional matrix-matrix multiplies. If t is large enough, then these operations can make efficient use of cache. Q is $t \times t$, A_{12} is $i \times t$, where i is large (except at the very beginning of the bulge-chase), and A_{23} is $t \times j$, where j is large (except at the very end of the bulge-chase). Since the vast majority of the arithmetic takes place within A_{12} and A_{23} (assuming $1 \ll t \ll n$), the performance is enhanced significantly. To complete the description we simply note that once the bulges have reached the bottom of the window, that window is dissolved and a new window is chosen, such that the bulges are now at the top of the new window and ready to be pushed further. The bulges are pushed from one window to the next, until they drop off the bottom of the matrix. Notice that the windows are not disjoint; each pair of successive windows has a significant overlap.

A price is paid for the improved cache use. This rearrangement of the arithmetic results in a significant increase in the number of floating point operations that must be performed. The exact penalty depends on the window size and other details. The computations in Exercise 4.9.2 indicate that the optimal window size is about twice the length of the chain of bulges. This does not imply a constraint on window size. We can achieve whatever window size we want for optimal cache use, then adjust the length of the chain of bulges so that it is half the window size. With this choice, the flop count is roughly doubled. Amazingly, even with this steep flop-count penalty, the cache-efficient method is much faster than conventional methods on large problems. For more details see [44]. LAPACK version 3.1 contains such a code, which has supplanted the multishift code ($m = 6$) used in LAPACK 3.0.

Kressner [134] has experimented with different sizes of bulges in the bulge chain. A bulge of degree m occupies $m + 1$ rows and $m + 1$ columns, so larger bulges allow more compact bulge chains. For example, if we have 24 shifts, we can pack them into twelve bulges of degree 2 or six bulges of degree 4 or four bulges of degree 6, resulting in chains of length $12(2 + 1) = 36$, $6(4 + 1) = 30$, or $4(6 + 1) = 28$, respectively. With the reduction in bulge chain length there is a corresponding decrease in the flop count. Of course we cannot take m too large without risking serious problems with shift blurring. Kressner's experiments suggest that the choices $m = 4$ and $m = 6$ do provide a modest improvement over the choice $m = 2$.

We have described two methods that chase small bulges in pipeline fashion. One spreads the bulges out for parallelism, while the other bunches them together for efficient cache use. These approaches are not incompatible. If the matrix is large enough, we can profitably chase several tightly bunched chains of bulges simultaneously to obtain both parallelism and efficient cache use.

Aggressive early deflation

The standard method of deflation is to examine the subdiagonal entries $a_{j+1,j}$, and if any of them is sufficiently small, i.e.,

$$|a_{j+1,j}| \le u(|a_{jj}| + |a_{j+1,j+1}|),$$

then set it to zero. Braman, Byers, and Mathias [45] observed that this strategy is less effective than one might hope when used in the context of chasing many small bulges, as described just above. They noticed that when large numbers of shifts are computed for use in chasing many small bulges, it often happens that many of the shifts are already equal to eigenvalues to working precision. This is especially so after a few iterations have passed, and it is so even though none of the subdiagonal entries is small enough to allow a deflation. Surely it will be possible to deflate those eigenvalues in another iteration (i.e., multiple bulge-chase) or two, but wouldn't it be better if we could somehow deflate the converged eigenvalues immediately and go after some eigenvalues that have not already converged? Braman, Byers, and Mathias [45] developed a new strategy called *aggressive early deflation* that does just this. This supplements the standard deflation strategy; it does not replace it.

Suppose we are going to compute some large number s of shifts ($1 \ll s \ll n$) for use in the next sequence of bulge-chases. Pick a number $k > s$, for example, $k \approx 1.5s$. Let A denote the current iterate, and consider the partition.

$$A = \left[\begin{array}{cc} A_{11} & A_{12} \\ A_{21} & A_{22} \end{array} \right],$$

where A_{22} is $k \times k$. Since A is upper Hessenberg, A_{21} has only one nonzero entry, $a_{n-k+1,n-k}$. Use a small-matrix QR code to compute the Schur decomposition $A_{22} = QTQ^{-1}$ (Theorem 2.2.4). In the real case we compute the Wintner–Murnaghan form (Theorem 2.2.6). This gives us the k eigenvalues of A_{22}. We will use s of them as shifts, but we also hope that some of them are eigenvalues of A that we will be able to deflate.

Perform a similarity transformation by diag$\{I, Q\}$ on A to obtain

$$\left[\begin{array}{cc} A_{11} & A_{12}Q \\ Q^{-1}A_{21} & T \end{array} \right].$$

Since A_{21} has only one nonzero entry, and that in the last column, $Q^{-1}A_{21}$ has nonzeros only in its last column. Call that column x. The transformed matrix is no longer upper Hessenberg; it is "Hessenberg + spike," the spike being the vector x. The T part is upper triangular in the complex case and quasi-triangular in the real case. For simplicity we will discuss the triangular case, the extension to the quasi-triangular case being routine. For illustration, the bottom portion of the matrix looks as follows in the case $k = 5$:

$$\left[\begin{array}{ccccccc} h & h & h & h & h & h & h \\ h & h & h & h & h & h & h \\ x & t & t & t & t & t \\ x & & t & t & t & t \\ x & & & t & t & t \\ x & & & & t & t \\ x & & & & & t \end{array} \right].$$

Of course, we have much larger values of k in mind. From the illustration we see that if, say, the bottom three components of x are zero, then the matrix is block upper triangular with a 3×3 block in the lower right-hand corner, and we can deflate three eigenvalues immediately. More generally, if the bottom t entries $x_{k-t+1}, \ldots, x_k$ of the "spike" vector x are zero, we can deflate t eigenvalues. In fact, the bottom entries of x will never be exactly zero, but some of them may be near enough to zero that we are willing to set them to zero and do a deflation. From the construction of x we see immediately that $\| x \| = | a_{n-k+1,n-k} |$. After some iterations it will often be the case that $| a_{n-k+1,n-k} |$ is small. Each entry of x is even smaller. Moreover, it often happens that some of the bottom entries of x are much smaller than $| a_{n-k+1,n-k} |$. If they are small enough, we can do some deflations.

The *aggressive early deflation* procedure is roughly as follows. Check x_k. If it is small enough, set it to zero. Then check x_{k-1}, and so on. Keep going until you hit an x_{k-m} that is not small enough to be set to zero. In Section 4.8 we developed a procedure for interchanging eigenvalues in a triangular or quasi-triangular matrix. Use this procedure to move the entry $t_{k-m,k-m}$ to the top of T, and make the corresponding updates to x. If the new x_{k-m} is small enough, set it to zero. Otherwise move $t_{k-m,k-m}$ to the top of T, updating x, and try again. Continue until every eigenvalue of T has been inspected for deflatability. Then, if all of the entries $x_k, \ldots, x_{k-t+1}$ have been set to zero, we can deflate t eigenvalues. Of the eigenvalues of T that are not deflatable, we use the "bottommost" s as shifts for the next QR iteration. Before the iteration can take place, the matrix must be returned to upper Hessenberg form. This is inexpensive because the part of the matrix above the spike is still upper Hessenberg form. Thus the reduction can begin in the "spike" column and consists of the final $k - t - 1$ steps of an ordinary reduction to upper Hessenberg form.

If the deflation procedure was sufficiently successful, the undeflatable part of T will be smaller than $s \times s$, and there will not be enough shifts left for the next multiple QR step. In this case, simply repeat the procedure, perhaps multiple times, until enough shifts become available. In extreme cases it can happen that the entire matrix is processed by aggressive early deflations [45].

The procedure that we have outlined here is slightly different from the one described in [45]. For one thing, we have left out some details. One important question is this: How small does an entry of x have to be before we can safely set it to zero? For a detailed discussion of this question, see [45]. Here is one possibility. Set x_j to zero if $|x_j| < u \max\{| a_{n-k+1,n-k} |, | t_{jj} | \}$.

Kressner [137] has shown that aggressive early deflation seeks converged Ritz values of a certain Krylov subspace. Convergence theorems and numerical results in [137] suggest that aggressive early deflation works well even in conjunction with standard QR iterations (without pipelined bulges) and is generally superior to the standard deflation strategy.

On large matrices aggressive early deflation can dramatically reduce computing time. It is included in the new large-matrix QR code in LAPACK 3.1.[11]

Hermitian Matrices

If A is Hermitian ($A = A^*$), we should preserve the Hermitian structure by applying only unitary similarity transformations. When we reduce A to upper Hessenberg form, the resulting Hessenberg matrix T is Hermitian and therefore tridiagonal. In the reduction we can even arrange that the off-diagonal entries of T are real (Exercise 4.9.5). Since the main-diagonal entries of T are automatically real, we conclude that T is real and symmetric.

---

[11]The small-matrix QR code in LAPACK 3.1 is the same as the one in LAPACK 3.0, except that it has been made more robust against overflow and underflow, and a more conservative deflation criterion has been installed [4].

The standard implementations of the QR algorithm do iterations of degree one ($m = 1$), although there is no reason why a higher degree (e.g., $m = 2$) could not be used. Thus we have

$$T_{m-1} - \rho_{m-1} I = Q_m R_m, \qquad Q_m^T T_{m-1} Q_m = T_m.$$

When steps of degree one are done implicitly, the bulge that is chased consists of a single entry, which has a symmetric reflection above the superdiagonal. The initial transformation, which acts on rows and columns 1 and 2 creates a bulge in positions $(3, 1)$ and $(1, 3)$, so that the matrix has the form

$$\begin{bmatrix} t & t & b & & \\ t & t & t & & \\ b & t & t & t & \\ & & t & t & t \\ & & & t & t \end{bmatrix}.$$

The second transformation pushes the bulge down and over one position:

$$\begin{bmatrix} t & t & & & \\ t & t & t & b & \\ & t & t & t & \\ & b & t & t & t \\ & & & t & t \end{bmatrix},$$

and so on. Symmetry is preserved and should be exploited at every step. A variety of good implementations of this iteration have been developed [169].

Because of the tridiagonal form, applying each transformation in the bulge-chase takes only $O(1)$ work. Since $n - 1$ transformations are used in each bulge-chase, the flop count for a single symmetric, tridiagonal QR iteration is $O(n)$, an order of magnitude less than the cost of a nonsymmetric QR iteration.

The simplest choice of shift is the *Rayleigh quotient shift* $\rho = t_{nn}$. This strategy generally works very well. Theorem 5.2.9 will show that its local convergence rate is cubic. However, it is not globally convergent (Exercise 4.9.6). A better strategy is the *Wilkinson shift:* Compute the eigenvalues of the lower right-hand 2×2 submatrix

$$\begin{bmatrix} t_{n-1,n-1} & t_{n-1,n} \\ t_{n,n-1} & t_{n,n} \end{bmatrix}$$

and use as the shift the one that is closer to $t_{n,n}$. This strategy has been shown to converge globally, and it generally converges at least cubically [169].

If the transforming matrix is updated at each similarity transformation, then in the end its columns are n orthonormal eigenvectors of A. The extra cost of the updates is significant though: $O(n^2)$ flops per QR iteration. Thus the cost of the updates for a QR iteration is much greater than the cost of the QR iteration itself. Lang [145] has shown how to speed up this phase of the computation.

There are other interesting methods for solving the symmetric eigenvalue problem. These include Jacobi's method from 1846 [120, 75], the bisection method, Cuppen's divide-and-conquer method [61, 76], and the dqd algorithm [182, 87, 170], which was discussed in Section 4.4 and will come up again in Section 8.5. A good method for computing eigenvectors is the multiple relatively robust representation (MRRR) method [171]. These methods are summarized briefly in [221, § 6.6]. Some of them are so good with respect to speed and/or accuracy that they threaten to squeeze the QR algorithm completely out of the arena of symmetric eigenvalue problems.

The Positive-Definite Case

If T is positive-definite, then T has a Cholesky decomposition $T = B^T B$, in which B is upper bidiagonal. If we compute the SVD of B, then we get the spectral decomposition of T immediately. From the standpoint of accuracy and stability, this SVD route to the eigensystem is superior. We will have more to say on this when we discuss the SVD in Chapter 8.

Other Matrix Structures

Special versions of the QR algorithm have been developed for real orthogonal and complex unitary matrices. Some of these are discussed in Section 8.10. A version of the QR algorithm for real skew symmetric matrices is discussed in Section 8.6 in connection with the SVD. There is no need for a special method for complex skew-Hermitian matrices, since a skew-Hermitian multiplied by i makes a Hermitian matrix.

Pseudosymmetric matrices

Recall from Section 3.3 that a matrix $A \in \mathbb{R}^{n \times n}$ is *pseudosymmetric* if there is a signature matrix D and a symmetric matrix T such that $A = TD$. A pseudosymmetric matrix that is upper Hessenberg must be tridiagonal and have the form

$$
A = \begin{bmatrix}
a_1 & b_1 & & & \\
c_1 & a_2 & \ddots & & \\
& \ddots & \ddots & b_{n-1} \\
& & c_{n-1} & a_n
\end{bmatrix},
\tag{4.9.3}
$$

where $c_j = \pm b_j$ for $j = 1, \ldots, n-1$. Almost any matrix can be transformed to pseudosymmetric tridiagonal form by a similarity transformation [230], although there are stability questions. The unsymmetric Lanczos algorithm [143] generates a pseudosymmetric tridiagonal matrix whose eigenvalues must then be computed. (See Section 9.6.)

The special GR algorithm that preserves pseudosymmetry is called the HR algorithm [47, 50]. Suppose that $A = TD$ is pseudosymmetric. An iteration of the HR algorithm has the form

$$
p(A) = HR, \qquad \hat{A} = H^{-1}AH,
$$

where H is chosen so that $H^T D H = \hat{D}$, where $\hat{D}$ is another signature matrix. This is the HR decomposition of Theorem 1.3.12. In Exercise 4.9.7 it is shown that this algorithm preserves pseudosymmetric form: $\hat{A} = \hat{T}\hat{D}$, where $\hat{T}$ is symmetric. From the proof of Theorem 3.1.5 one sees that H can be built as a product of orthogonal and hyperbolic transformations. When the algorithm is implemented implicitly, as a bulge-chasing algorithm, the eliminations are effected by transformations like those described in Theorem 3.1.5. The symbol H stands for *hyperbolic* in recognition of the key role played by hyperbolic transformations here. We will describe this algorithm in greater detail in Sections 6.4 and 8.9.

The eliminations described in Theorem 3.1.5 cannot always be carried out. Consequently the HR algorithm can suffer breakdowns and near-breakdowns, resulting in instability. Therefore the HR algorithm cannot be recommended for general use, and it certainly cannot be applied successfully to very large matrices. However, it is useful as a helper algorithm to perform the restarts in several variants of the unsymmetric Lanczos process, which will be discussed in Chapter 9. In this role it is usually implemented as a bulge-chasing algorithm of degree $m = 2$ to accommodate complex-conjugate shifts in real arithmetic.

Hamiltonian matrices

The Hamiltonian eigenvalue problem arises in the quadratic regulator problem of optimal control theory [53, 141, 146] and other contexts. By Exercise 2.2.17, the Hamiltonian structure is preserved by symplectic similarity transformations, so the structure-preserving algorithm for Hamiltonian matrices is the SR algorithm

$$p(H) = SR, \qquad \hat{H} = S^{-1}HS,$$

where S is symplectic and R is $\hat{P}$-triangular (Theorem 1.3.11). For the sake of stability, one would prefer to make S both symplectic and orthogonal, but this is hard to achieve. First we will discuss algorithms that do not insist on orthogonality.

Bunse–Gerstner and Mehrmann [53] showed that any Hamiltonian matrix can be transformed by a symplectic similarity transformation into the extremely condensed form

$$H = \begin{bmatrix} E & T \\ D & -E \end{bmatrix}, \tag{4.9.4}$$

where D and E are diagonal and T is tridiagonal and symmetric. This is a disguised upper Hessenberg matrix; if one applies the permutation similarity transformation $\hat{P}H\hat{P}^T$ (cf. Theorem 1.3.11 and accompanying discussion), shuffling the rows and columns, one obtains a (very sparse) upper Hessenberg matrix.

In [53] an implicit SR algorithm was developed for the condensed form (4.9.4). Once one applies the shuffling permutation, this algorithm becomes an instance of the generic bulge-chasing algorithm of Section 4.5. The algorithm is of degree four with shifts ρ, $\bar{\rho}$, $-\rho$, $-\bar{\rho}$ because the eigenvalues appear in quadruples λ, $\bar{\lambda}$, $-\lambda$, $-\bar{\lambda}$.[12] Like the HR algorithm, the Hamiltonian SR algorithm can suffer breakdowns and instability and is not recommended for general use.

In the reduction to the condensed form (4.9.4), there is some slack in the choice of the entries of D, E, and T. If one wishes, one can take $E = 0$ and D a signature matrix. The SR algorithm on this extra condensed form is equivalent to the HR algorithm, as we will

---

[12]This is the form of the shifts if they are neither real nor purely imaginary. If some of the shifts are real or purely imaginary, then other configurations are possible, but in any event the four shifts are the eigenvalues of a 4×4 Hamiltonian matrix.

show in Section 8.9. When we need to do SR iterations to restart the Hamiltonian Lanczos method in Chapter 9, we will effect SR iterations via HR iterations, which are simpler.

Now we consider algorithms that utilize transformations that are both symplectic and orthogonal. The orthogonality guarantees backward-stability. There is one special situation for which such an algorithm has been known for many years. If the Hamiltonian matrix

$$H = \left[\begin{array}{cc} A & N \\ K & -A^T \end{array} \right]$$

arises from a control system with a single output, then K has rank one.[13] In this case, the matrix can be transformed by an orthogonal symplectic similarity transformation to a form in which A is upper Hessenberg (hence $-A^T$ is lower Hessenberg), and K has a single nonzero entry in position (n, n). This is called Hamiltonian Hessenberg form. Notice that if one reverses the last n rows and columns, the matrix becomes upper Hessenberg in the standard sense. Byers [55] developed an implicit QR algorithm for this form that uses exclusively orthogonal symplectic transformations. Thus this is both a QR and an SR algorithm. It is backward-stable, and it preserves Hamiltonian structure.

For general Hamiltonian matrices the problem is more difficult, but good progress has been made. Chapter 4 of [134] has the most complete discussion of these methods, many of which have been included in a package called HAPACK [33]. For the latest information, see [34, 57, 226]. We will discuss some of these methods in Section 8.8.

Skew-Hamiltonian matrices

The related skew-Hamiltonian eigenvalue problem is much easier. From Theorem 3.3.3 we know that a skew-Hamiltonian matrix can be reduced by an orthogonal symplectic similarity transformation (requiring $O(n^3)$ flops) to skew-Hamiltonian Hessenberg form

$$\left[\begin{array}{cc} H & K \\ 0 & H^T \end{array} \right].$$

Using the QR algorithm, we can find an orthogonal Q and quasi-triangular T such that $H = QTQ^{-1}$. This also requires $O(n^3)$ flops. Applying a similarity transform by the orthogonal symplectic matrix diag$\{Q, Q\}$, we arrive at the skew-Hamiltonian Schur form

$$\left[\begin{array}{cc} T & M \\ 0 & T^T \end{array} \right]. \tag{4.9.5}$$

This displays the eigenvalues; invariant subspaces can also be inferred.

Symplectic matrices

The symplectic eigenvalue problem arises in discrete-time quadratic regulator problems [115, 141, 152, 167]. Since the symplectic matrices form a group, the symplectic form is preserved by symplectic similarity transformations. Thus the structure-preserving GR

---

[13] In the case of a single input, N has rank one, and we can work with the transposed matrix.

algorithm for the symplectic eigenvalue problem is the SR algorithm. The details of applying SR algorithms to symplectic matrices have been discussed in [17, 18, 29, 32, 85, 88].

Banse [17] and Banse and Bunse-Gerstner [18] showed that every symplectic matrix can be reduced by a symplectic similarity transformation to a condensed *butterfly form*

$$B = \begin{bmatrix} E & T_1 \\ D & T_2 \end{bmatrix}, \tag{4.9.6}$$

where E and D are diagonal and T_1 and T_2 are tridiagonal. The submatrices are not uniquely determined; there is some slack in the choice of parameters. If one is after the simplest possible butterfly form, then one can choose $E = 0$, which immediately forces $T_2 = -D^{-1}$ by the symplectic property of B. The main-diagonal entries of D can be rescaled by arbitrary positive factors, so we can rescale it to make it a signature matrix. Suppose we do that. Then the symplectic property of B also implies that the matrix defined by $T = DT_2 = D^{-1}T_2$ is symmetric. With these choices, the symplectic butterfly matrix takes the form

$$B = \begin{bmatrix} 0 & -D \\ D & DT \end{bmatrix}, \tag{4.9.7}$$

where D is a signature matrix (i.e., diagonal with main-diagonal entries ± 1) and T is symmetric and tridiagonal. Notice that the characteristics of D and T are the same as those possessed by the same symbols in the discussion of Hamiltonian matrices above. The symplectic SR algorithm on this condensed butterfly form is equivalent to the HR algorithm, as we will show in Section 8.9. When we need to do SR iterations to restart the symplectic Lanczos method in Chapter 9, we will effect the symplectic SR iterations via much simpler HR iterations.

Exercises

4.9.1. Consider two consecutive QR iterations of degrees m and k, driven by p and q, respectively:

$$p(A_0) = Q_1 R_1, \qquad A_1 = Q_1^{-1} A_0 Q_1,$$
$$q(A_1) = Q_2 R_2, \qquad A_2 = Q_2^{-1} A_1 Q_2.$$

(a) Obtain an expression for $q(A_0)$ in terms of Q_1 and A_1, and deduce that

$$q(A_0)p(A_0) = QR, \qquad A_2 = Q^{-1} A_0 Q,$$

where $Q = Q_1 Q_2$ and $R = R_2 R_1$. This implies that the two QR iterations of degree m and k are equivalent to a single iteration of degree $m + k$, driven by the polynomial qp.

(b) Show that if s shifts are chosen and used to do either one QR iteration of degree s or several QR iterations of lower degree, the outcome is the same in principal.

4.9.2. In this exercise we estimate the optimal size for the window in (4.2), based on flop counts. Suppose we are chasing k bulges of degree m. (m is small, e.g., 2 or 4.) Each bulge occupies

$m + 1$ rows and columns, so the entire chain of bulges occupies $(m + 1)k$ rows and columns. Let us define $l = (m + 1)k$. This is the length of the bulge chain. To push the bulge chain from the top to the bottom of a $t \times t$ window, we must push each bulge forward by s rows/columns, where $s = t - l$. The number of flops that take place in the window A_{22} is the same, regardless of which method we use, and it is small compared to what happens in A_{12} and A_{23}. Remember that we are assuming $1 \ll t \ll n$.

(a) The size of the submatrices A_{12} and A_{23} depends on where we are in the bulge-chase. Show that no matter where we are, the number of rows in A_{12} plus the number of columns in A_{23} is $n - t$.

(b) Show that if we update A_{12} and A_{23} in the conventional manner, applying the reflectors individually, the total flop count for the transformations $A_{12} \to A_{12}Q$ and $A_{23} \to Q^{-1}A_{23}$ is about $4(m + 1)ks(n - t) = 4ls(n - t)$.

(c) Show that if we update A_{12} and A_{23} in the cache-efficient manner, performing the transformations $A_{12} \to A_{12}Q$ and $A_{23} \to Q^{-1}A_{23}$ explicitly, the total flop count is about $2t^2(n - t) = 2(l + s)^2(n - t)$.

(d) Let $f(s)$ denote the ratio of flop counts from parts (c) and (b). Clearly $f(s) = (l + s)^2/(2ls)$. If we minimize f, then we minimize the flop-count penalty. Use calculus to show that f is minimized when $s = l$, i.e., when $t = 2l$. This means that for optimal flop count we should take the window to be about twice as long as the bulge chain. Show that $f(l) = 2$, which means that the flop count for the cache-efficient method is about twice that for the conventional method.

4.9.3. Show that the "spike" vector x in aggressive early deflation satisfies $\| x \| = | a_{n-k+1,n-k} |$.

4.9.4. How does the aggressive early deflation procedure need to be modified in the case when T is quasi-triangular instead of triangular?

4.9.5. Let $T \in \mathbb{C}^{n \times n}$ be a Hermitian tridiagonal matrix.

(a) Show that the main-diagonal entries of T are all real.

(b) Show that there is a unitary diagonal matrix D such that all of the subdiagonal entries of $\hat{T} = D^*TD$ are real and nonnegative. Hence $\hat{T}$ is real and symmetric.

4.9.6. Consider the symmetric tridiagonal matrix $T = \begin{bmatrix} 0 & 1 \\ 1 & 0 \end{bmatrix}$. Show that an iteration of the QR algorithm on T with Rayleigh quotient shift $\rho = 0$ goes nowhere. (Notice that Q is orthogonal. What are the eigenvalues of Q? Where does the shift lie, relative to the two eigenvalues?)

4.9.7. This exercise demonstrates some basic facts about pseudosymmetric matrices and the HR algorithm.

(a) Show that a tridiagonal matrix A is pseudosymmetric if and only if it has the form (4.9.3), where $c_j = \pm b_j$ for $j = 1, \ldots, n - 1$.

(b) Show that the decomposition $A = TD$ of a pseudosymmetric tridiagonal matrix is not unique and can always be organized so that the off-diagonal entries of T are nonnegative.

(c) Suppose that $A = TD$ is pseudosymmetric, $\hat{A} = H^{-1}AH$, and $H^T DH = \hat{D}$. Let $\hat{T} = \hat{A}\hat{D}^{-1} = \hat{A}\hat{D}$, so that $\hat{A} = \hat{T}\hat{D}$. Show that $\hat{T} = H^{-1}TH^{-T}$, and deduce that $\hat{T}$ is symmetric. Thus $\hat{A} = \hat{T}\hat{D}$ is pseudosymmetric. The structure is preserved by iterations of the HR algorithm

4.9.8. Show that if H is a Hamiltonian matrix of the form (4.9.4) with D and E diagonal and T tridiagonal, and $\hat{P}$ is the perfect-shuffle permutation (1.3.2), then $\hat{P}H\hat{P}^T$ is upper Hessenberg.

Chapter 5
Convergence

In the early sections of Chapter 4 we made numerous vague claims about the convergence of subspace iteration and GR algorithms. In this chapter we will make those claims precise and prove them. We will draw heavily from [227].

5.1 Convergence of Subspace Iteration

Every variant of subspace iteration generates a sequence of subspaces $(\mathcal{S}_i)$. The task is to prove that $\mathcal{S}_i \to \mathcal{U}$, where $\mathcal{U}$ is a subspace that is invariant under A. To measure convergence, we will use the metric $d(\cdot, \cdot)$ on subspaces introduced in Section 2.6. $\mathcal{S}_i \to \mathcal{U}$ means $d(\mathcal{S}_i, \mathcal{U}) \to 0$. We will make use of several claims from Section 2.6, including Propositions 2.6.14 and 2.6.16.

We will begin by stating a basic theorem about stationary subspace iteration with semisimple matrices. Then we will generalize it to general matrices and nonstationary subspace iterations. We will discuss the theorems, then we will prove them.

Theorem 5.1.1. *Let $A \in \mathbb{C}^{n \times n}$ be semisimple, and let p be a polynomial of degree $< n$. Let $\lambda_1, \ldots, \lambda_n$ denote the eigenvalues of A, ordered so that $|p(\lambda_1)| \geq |p(\lambda_2)| \geq \cdots \geq |p(\lambda_n)|$. Suppose that k is an integer satisfying $1 \leq k < n$ for which $|p(\lambda_k)| > |p(\lambda_{k+1})|$, and let $\rho = |p(\lambda_{k+1})|/|p(\lambda_k)| < 1$. Let $\mathcal{U}$ and $\mathcal{V}$ be the invariant subspaces of A associated with $\lambda_1, \ldots, \lambda_k$ and $\lambda_{k+1}, \ldots, \lambda_n$, respectively. Consider the stationary subspace iteration*

$$\mathcal{S}_i = p(A)\mathcal{S}_{i-1}, \qquad i = 1, 2, 3, \ldots,$$

where $\mathcal{S}_0 = \mathcal{S}$ is a k-dimensional subspace satisfying $\mathcal{S} \cap \mathcal{V} = \{0\}$. Then there is a constant C such that

$$d(\mathcal{S}_i, \mathcal{U}) \leq C\rho^i, \quad i = 1, 2, 3, \ldots.$$

In typical applications, p is a polynomial of modest degree. We could consider more general functions than polynomials, but there is no gain in generality in doing so. For every function f for which $f(A)$ is defined, there is a unique polynomial p of degree $< n$ such that $f(A) = p(A)$ [142, Ch. 9].

The theorem states that the sequence of subspaces converges to the dominant k-dimensional invariant subspace $\mathcal{U}$ linearly and with the contraction number $\rho = |p(\lambda_{k+1})|/|p(\lambda_k)|$, as we claimed in Chapter 4.

213

The condition $\mathcal{S} \cap \mathcal{V} = \{0\}$ is satisfied by almost any $\mathcal{S}$ chosen at random. This is so because $\dim(\mathcal{S}) + \dim(\mathcal{V}) = k + (n - k) = n$. Since the sum of the dimensions does not exceed n, the subspaces are not forced to have a nontrivial intersection, and this implies that they almost certainly will not. This fact is most easily understood by visualizing low-dimensional cases, e.g., $n = 2$ and $k = 1$ or $n = 3$ and $k = 1$ or 2. If we happen to choose a starting subspace $\mathcal{S}$ that satisfies $\mathcal{S} \cap \mathcal{V} \neq \{0\}$, then we have been really unlucky. In Section 4.1, when we spoke of not being unlucky in our choice of $\mathcal{S}$, we meant that our $\mathcal{S}$ satisfies $\mathcal{S} \cap \mathcal{V} = \{0\}$. This condition guarantees that the dimensions of the subspaces does not decrease in the course of the iterations: $\dim(\mathcal{S}_i) = k$ for all i (Exercise 5.1.1).

Theorem 5.1.2. *Let all of the hypotheses be as in Theorem 5.1.1, except that A need not be semisimple. Then for every $\hat{\rho}$ satisfying $\rho < \hat{\rho} < 1$ there is a constant $\hat{C}$ such that*

$$d(\mathcal{S}_i, \mathcal{U}) \leq \hat{C}\hat{\rho}^i, \quad i = 1, 2, 3, \ldots.$$

Again we have linear convergence to the dominant invariant subspace, but we have had to give up a little bit on the contraction number. We can make $\hat{\rho}$ as close to ρ as we please, but the closer $\hat{\rho}$ is to ρ, the larger the constant $\hat{C}$ is.

Now we consider nonstationary subspace iteration.

Theorem 5.1.3. *Let $A \in \mathbb{C}^{n \times n}$ be any matrix, semisimple or not, and let p be a polynomial of degree m, where $m < n$. Let $\lambda_1, \ldots, \lambda_n$ denote the eigenvalues of A, ordered so that $|p(\lambda_1)| \geq |p(\lambda_2)| \geq \cdots \geq |p(\lambda_n)|$. Suppose that k is an integer satisfying $1 \leq k < n$ for which $|p(\lambda_k)| > |p(\lambda_{k+1})|$, and let $\rho = |p(\lambda_{k+1})|/|p(\lambda_k)| < 1$. Let $\mathcal{U}$ and $\mathcal{V}$ be the invariant subspaces of A associated with $\lambda_1, \ldots, \lambda_k$ and $\lambda_{k+1}, \ldots, \lambda_n$, respectively. Let (p_i) be a sequence of polynomials such that $\lim_{i \to \infty} p_i = p$, and for all i, $p_i(\lambda_j) \neq 0$ for $j = 1, \ldots, k$. Consider the nonstationary subspace iteration*

$$\mathcal{S}_i = p_i(A)\mathcal{S}_{i-1}, \quad i = 1, 2, 3, \ldots,$$

where $\mathcal{S}_0 = \mathcal{S}$ is a k-dimensional subspace satisfying $\mathcal{S} \cap \mathcal{V} = \{0\}$. Then for every $\hat{\rho}$ satisfying $\rho < \hat{\rho} < 1$, there is a constant $\hat{C}$ such that

$$d(\mathcal{S}_i, \mathcal{U}) \leq \hat{C}\hat{\rho}^i, \quad i = 1, 2, 3, \ldots.$$

The technical condition $p_i(\lambda_j) \neq 0$, $j = 1, \ldots, k$, guarantees that $\dim(\mathcal{S}_i) = \dim(\mathcal{S}_{i-1})$ for all i (Exercise 5.1.2), and we shall see that it is crucial to the proof of the theorem. Clearly this condition is nearly always satisfied. Moreover, nothing bad happens if it is violated. In the context of GR iterations, if $p_i(\lambda_j) = 0$, then λ_j is an exact shift for the ith iteration and will be deflated on that iteration, as we saw in Section 4.6.

The conclusion of Theorem 5.1.3 is the same as that of Theorem 5.1.2. The difference is that for the scenario in Theorem 5.1.3, we can do the subspace iterations without knowing p in advance. The objective is to choose the p_i by some dynamic strategy in such a way that they converge to some good polynomial p. The best sort of p is one whose zeros are eigenvalues of A. If at least one zero of p is an eigenvalue of A, then $p(\lambda_n) = 0$. Normally there will be some k such that $p(\lambda_{k+1}) = \cdots = p(\lambda_n) = 0$ and $p(\lambda_k) \neq 0$.[1] For this value

---

[1] The only exception to this is the case when $p(\lambda_j) = 0$ for all j, but we are never that lucky.

of k we have $\rho = |\,p(\lambda_{k+1})\,|/|\,p(\lambda_k)\,| = 0$, so if we do subspace iterations of this dimension, we have that for every $\hat{\rho} > 0$ there is a $\hat{C}$ such that $d(\mathcal{S}_i, \mathcal{U}) \le \hat{C}\hat{\rho}^i$. This means that the convergence is superlinear. We state this important result as a corollary.

Corollary 5.1.4. *Under the conditions of Theorem* 5.1.3, *if some of the zeros of the limit polynomial* p *are eigenvalues of* A, *and* $p(\lambda_{k+1}) = \cdots = p(\lambda_n) = 0$, *and* $p(\lambda_k) \ne 0$ *for some* k, *then subspace iterations of dimension* k *converge superlinearly.*

This is the main point: A nonstationary iteration allows the possibility of superlinear convergence. Looking at Corollary 5.1.4, you might ask how we know in advance which value of k to use to get superlinear convergence. Remember, though, that GR algorithms automatically execute subspace iterations of all dimensions simultaneously, so if subspace iterations of any dimension k are converging superlinearly, the GR algorithm will benefit from it. As we shall see, we are often able to achieve superlinear convergence in the context of GR algorithms. Usually the convergence is quadratic, and sometimes it is even cubic.

Proof of Convergence of Subspace Iteration

The proofs of all of our convergence theorems begin by changing to a coordinate system in which the matrix has a simple form. We will use the Jordan canonical form (Theorem 2.4.11), although there are other forms that would work equally well.[2] Let J denote the Jordan form of A. Assume that the Jordan blocks are ordered in such a way that

$$J = \begin{bmatrix} J_1 & \\ & J_2 \end{bmatrix},$$

where J_1 is $k \times k$ and contains the blocks corresponding to eigenvalues $\lambda_1, \ldots, \lambda_k$. Then J_2 is $(n - k) \times (n - k)$ and contains the blocks corresponding to $\lambda_{k+1}, \ldots, \lambda_n$. In the semisimple case, $J_1 = \operatorname{diag}\{\lambda_1, \ldots, \lambda_k\}$ and $J_2 = \operatorname{diag}\{\lambda_{k+1}, \ldots, \lambda_n\}$.

We have $A = VJV^{-1}$ for some nonsingular matrix V. Make a partition $V = \begin{bmatrix} V_1 & V_2 \end{bmatrix}$, where $V_1 \in \mathbb{C}^{n \times k}$. Then the equation $AV = VJ$ implies $AV_1 = V_1 J_1$ and $AV_2 = V_2 J_2$. Thus $\mathcal{R}(V_1)$ and $\mathcal{R}(V_2)$ are the invariant subspaces of A associated with $\lambda_1, \ldots, \lambda_k$ and $\lambda_{k+1}, \ldots, \lambda_n$, respectively. In the notation of our convergence theorems, $\mathcal{R}(V_1) = \mathcal{U}$ and $\mathcal{R}(V_2) = \mathcal{V}$.

Now we make our change of coordinate system. For each i let $\tilde{\mathcal{S}}_i = V^{-1}\mathcal{S}_i$, $\tilde{\mathcal{U}} = V^{-1}\mathcal{U}$, and $\tilde{\mathcal{V}} = V^{-1}\mathcal{V}$. Notice that $\tilde{\mathcal{U}} = V^{-1}\mathcal{R}(V_1) = \mathcal{R}(V^{-1}V_1) = \operatorname{span}\{e_1, \ldots, e_k\} = \mathcal{E}_k$, and similarly $\tilde{\mathcal{V}} = \operatorname{span}\{e_{k+1}, \ldots, e_n\} = \mathcal{E}_k^\perp$. These spaces are invariant under J. The subspace iteration

$$p_i(A)\mathcal{S}_{i-1} = \mathcal{S}_i$$

is equivalent to

$$p_i(J)\tilde{\mathcal{S}}_{i-1} = \tilde{\mathcal{S}}_i.$$

We will study this iteration and show that $d(\tilde{\mathcal{S}}_i, \tilde{\mathcal{U}}) \to 0$ at the claimed rate. Then we will be able to deduce immediately that $d(\mathcal{S}_i, \mathcal{U}) \to 0$ at the same rate, thanks to Proposition 2.6.14, which implies that

$$d(\mathcal{S}_i, \mathcal{U}) \le \kappa(V)\, d(\tilde{\mathcal{S}}_i, \tilde{\mathcal{U}}). \tag{5.1.1}$$

---

[2]For example, we could use the block diagonal form guaranteed by Proposition 2.4.4, which is much easier to derive than the Jordan form.

For each i, let $\hat{p}_i(x)$ be the product of all polynomials that have been applied in the first i subspace iterations. In the nonstationary case we have $\hat{p}_i(x) = p_i(x)p_{i-1}(x) \cdots p_1(x)$, while in the stationary case we have $\hat{p}_i(x) = p(x) \cdots p(x) = p(x)^i$. Clearly

$$\tilde{S}_i = \hat{p}_i(J)\tilde{S}_0. \tag{5.1.2}$$

We will introduce simple matrix representations of the subspaces. Let $S = \begin{bmatrix} S_1 \\ S_2 \end{bmatrix} \in \mathbb{C}^{n \times k}$ be a full-rank matrix such that $\mathcal{R}(S) = \tilde{S} = \tilde{S}_0$, partitioned so that S_1 is $k \times k$. It is not hard to show (Exercise 5.1.3) that the hypothesis $\mathcal{S} \cap \mathcal{V} = \{0\}$ implies that S_1 is nonsingular. Letting $X = S_2 S_1^{-1}$, we have immediately that

$$S = \begin{bmatrix} I \\ X \end{bmatrix} S_1,$$

so $\begin{bmatrix} I \\ X \end{bmatrix}$ represents $\tilde{S}$ as well, in the sense that its columns span $\tilde{S}$. By Proposition 2.6.16, $\| X \| = \tan \theta_k$, where θ_k is the largest principal angle between $\tilde{S}$ and $\tilde{\mathcal{U}} = \mathcal{E}_k$.

Since we wish to analyze the iteration (5.1.2), we note that

$$\hat{p}_i(J) \begin{bmatrix} I \\ X \end{bmatrix} = \begin{bmatrix} \hat{p}_i(J_1) \\ \hat{p}_i(J_2)X \end{bmatrix} = \begin{bmatrix} I \\ \hat{p}_i(J_2)X\hat{p}_i(J_1)^{-1} \end{bmatrix} \hat{p}_i(J_1), \tag{5.1.3}$$

assuming that $\hat{p}_i(J_1)$ is nonsingular. But this is an easy consequence of the hypotheses of the theorems (Exercise 5.1.4). Let

$$S_i = \begin{bmatrix} I \\ \hat{p}_i(J_2)X\hat{p}_i(J_1)^{-1} \end{bmatrix}.$$

Then, by (5.1.2) and (5.1.3), $\tilde{S}_i = \mathcal{R}(S_i)$. Thus, by Proposition 2.6.16,

$$\| \hat{p}_i(J_2)X\hat{p}_i(J_1)^{-1} \| = \tan \theta_k^{(i)},$$

where $\theta_k^{(i)}$ is the largest principal angle between $\tilde{S}_i$ and $\tilde{\mathcal{U}}$. Consequently

$$d(\tilde{S}_i, \tilde{\mathcal{U}}) = \sin \theta_k^{(i)} \le \tan \theta_k^{(i)} \le \| X \| \, \| \hat{p}_i(J_2) \| \, \| \hat{p}_i(J_1)^{-1} \|.$$

Combining this with (5.1.1), we have

$$d(\mathcal{S}_i, \mathcal{U}) \le \kappa(V) \, \| X \| \, \| \hat{p}_i(J_2) \| \, \| \hat{p}_i(J_1)^{-1} \|. \tag{5.1.4}$$

This is our main bound, which we will use to prove all of our convergence theorems.[3] From this point on, we consider the three theorems separately.

---

[3]Kressner [137] has proved variants of this bound that demonstrate the convergence rates for individual eigenvectors in connection with aggressive early deflation.

The Stationary Case

We begin with the semisimple, stationary case. Now J_1 and J_2 are diagonal, $\hat{p}_i(J_2) = p(J_2)^i = \text{diag}\{p(\lambda_{k+1})^i, \ldots, p(\lambda_n)^i\}$, and $\|\hat{p}_i(J_2)\| = |p(\lambda_{k+1})|^i$. Similarly, $\hat{p}_i(J_1)^{-1} = p(J_1)^{-i} = \text{diag}\{p(\lambda_1)^{-i}, \ldots, p(\lambda_k)^{-i}\}$, and $\|\hat{p}_i(J_1)^{-1}\| = |p(\lambda_k)|^{-i}$. Substituting the values of the norms into (5.1.4), we find that

$$d(\mathcal{S}_i, \mathcal{U}) \leq \kappa(V) \|X\| \frac{|p(\lambda_{k+1})|^i}{|p(\lambda_k)|^i} = C\rho^i,$$

where $C = \kappa(V)\|X\|$. This proves Theorem 5.1.1.

The size of the constant $C = \kappa(V)\|X\|$ is of practical interest. The condition number $\kappa(V)$ is a measure of the distortion introduced when we changed coordinate systems. If A is normal, $\kappa(V) = 1$; otherwise $\kappa(V) > 1$. As A approaches a defective matrix, $\kappa(V) \to \infty$. The norm of X is a measure of how well the transformed initial subspace $\tilde{S}$ approximates the transformed invariant subspace $\tilde{\mathcal{U}} = \mathcal{E}_k$. A good approximation is rewarded by a small constant $\|X\|$ and conversely.

Now consider the nonsemisimple case, which requires a bit more work. Now J_1 and J_2 are not necessarily diagonal, but they are upper triangular, so $p(J_1)$ and $p(J_2)$ are also upper triangular. Write $p(J_2) = D + N$, where D is diagonal and N is strictly upper triangular. Clearly $D = \text{diag}\{p(\lambda_{k+1}), \ldots, p(\lambda_n)\}$, and $\|D\| = |p(\lambda_{k+1})|$. Since we are still talking about stationary iterations,

$$\hat{p}_i(J_2) = p(J_2)^i = (D + N)^i. \tag{5.1.5}$$

We will use this characterization to get an upper bound on $\|\hat{p}_i(J_2)\|$ for use in (5.1.4). If we expand the binomial in (5.1.5), we obtain a sum of 2^i terms, each of which is a product of i factors, some diagonal (D) and some strictly upper triangular (N). Since the matrices are of order $n - k$, any product that contains $n - k$ or more strictly upper triangular factors will be zero (Exercise 5.1.6). For each $h < n - k$, each term having h strictly upper triangular factors must have $i - h$ diagonal factors. Thus the norm of each such term is bounded above by $\|N\|^h \|D\|^{i-h} = \|N\|^h |p(\lambda_{k+1})|^{i-h}$. For each $h < n - k$ there are $\binom{i}{h}$ terms having exactly h strictly upper triangular factors, and so

$$\|\hat{p}_i(J_2)\| = \|(D + N)^i\| \leq \sum_{h=0}^{n-k-1} \binom{i}{h} \|N\|^h |p(\lambda_{k+1})|^{i-h}$$

$$= |p(\lambda_{k+1})|^i \sum_{h=0}^{n-k-1} \binom{i}{h} \|N\|^h |p(\lambda_{k+1})|^{-h}. \tag{5.1.6}$$

Since $\binom{i}{h}$ is a polynomial in i of degree h, the sum in (5.1.6) is just a polynomial in i. Consequently

$$\|\hat{p}_i(J_2)\| \leq \pi_2(i)|p(\lambda_{k+1})|^i, \tag{5.1.7}$$

where π_2 is a polynomial of degree at most $n - k - 1$.

The same sort of argument can be applied to obtain a bound on $\|\hat{p}_i(J_1)^{-1}\|$. The matrix $p(J_1)$ is upper triangular, and so is $p(J_1)^{-1}$. It can be expressed as a sum $p(J_1)^{-1} = E + M$, where $E = \text{diag}\{p(\lambda_1)^{-1}, \ldots, p(\lambda_k)^{-1}\}$ and M is strictly upper triangular. Arguing as in the previous paragraph (Exercise 5.1.8), we can show that

$$\| \hat{p}_i(J_1)^{-1} \| \leq \pi_1(i) |p(\lambda_k)|^{-i}, \tag{5.1.8}$$

where π_1 is a polynomial of degree at most $k - 1$.

Substituting the inequalities (5.1.7) and (5.1.8) into (5.1.4), we obtain

$$d(\mathcal{S}_i, \mathcal{U}) \leq \pi(i)\rho^i,$$

where $\pi = \kappa(V) \| X \| \pi_1 \pi_2$ is a polynomial. Given any $\hat{\rho}$ satisfying $\rho < \hat{\rho} < 1$, we have $d(\mathcal{S}_i, \mathcal{U}) \leq \pi(i)(\rho/\hat{\rho})^i \hat{\rho}^i$. Since $\rho/\hat{\rho} < 1$ and π is a mere polynomial, $\pi(i)(\rho/\hat{\rho})^i \to 0$ as $i \to \infty$. In particular, this factor is bounded: There is a $\hat{C}$ such that $\pi(i)(\rho/\hat{\rho})^i \leq \hat{C}$ for all i. Therefore $d(\mathcal{S}_i, \mathcal{U}) \leq \hat{C}\hat{\rho}^i$ for all i. This proves Theorem 5.1.2.

The Nonstationary Case

Finally we prove Theorem 5.1.3. The proof is similar to the proof in the stationary case, but now we must deal not with just one polynomial but with a whole sequence of polynomials $p_i \to p$. We have not said what we mean by $p_i \to p$, but as it happens, all reasonable notions of convergence in this context turn out to be equivalent. More precisely, any two norms on a finite-dimensional space induce the same topology [202, § II.3]. So by $p_i \to p$ we mean convergence with respect to the unique norm topology on the space of polynomials of degree $< m$. This implies that the coefficients of p_i converge to the coefficients of p, from which it follows that $p_i(M) \to p(M)$ for all matrices $M \in \mathbb{C}^{n \times n}$.

Let $\nu = |p(\lambda_{k+1})|$ and $\delta = |p(\lambda_k)|$, so that $\rho = \nu/\delta$. Given $\hat{\rho}$ with $\rho < \hat{\rho} < 1$, choose $\tilde{\rho}$ so that $\rho < \tilde{\rho} < \hat{\rho}$. There is a unique ϵ such that $0 < \epsilon < \delta$ and $\tilde{\rho} = (\nu + \epsilon)/(\delta - \epsilon)$. Let $\tilde{\nu} = \nu + \epsilon$ and $\tilde{\delta} = \delta - \epsilon$, so that $\tilde{\rho} = \tilde{\nu}/\tilde{\delta}$. Since $p_i \to p$, we have $p_i(\lambda_j) \to p(\lambda_j)$ for all j, so $\max_{k+1 \leq j \leq n} |p_i(\lambda_j)| \to \max_{k+1 \leq j \leq n} |p(\lambda_j)| = \nu$. Moreover, $\min_{1 \leq j \leq k} |p_i(\lambda_j)| \to \min_{1 \leq j \leq k} |p(\lambda_j)| = \delta$. Therefore there is an i_0 such that, for all $i > i_0$, $\max_{k+1 \leq j \leq n} |p_i(\lambda_j)| \leq \tilde{\nu}$ and $\min_{1 \leq j \leq k} |p_i(\lambda_j)| \geq \tilde{\delta}$. Let $C_2 = \| \hat{p}_{i_0}(J_2) \|$ and $C_1 = \| \hat{p}_{i_0}(J_1)^{-1} \|$. Then

$$\| \hat{p}_i(J_2) \| \leq C_2 \| \prod_{j=i_0+1}^{i} p_j(J_2) \|$$

and

$$\| \hat{p}_i(J_1)^{-1} \| \leq C_1 \| \prod_{j=i_0+1}^{i} p_j(J_1)^{-1} \|.$$

The matrix $p(J_2)$ is upper triangular, so it can be expressed as a sum $p(J_2) = D + N$, where $D = \text{diag}\{p(\lambda_{k+1}), \ldots, p(\lambda_n)\}$, and N is strictly upper triangular. Each of the matrices $p_j(J_2)$ has an analogous representation $p_j(J_2) = D_j + N_j$. Since $p_j \to p$ as $j \to \infty$, we have also $p_j(J_2) \to p(J_2)$, $D_j \to D$, and $N_j \to N$. Thus there is a constant K such that $\| N_j \| \leq K$ for all j. Also, by definition of i_0, $\| D_j \| \leq \tilde{\nu}$ for all $j > i_0$. Now

$$\prod_{j=i_0+1}^{i} p_j(J_2) = \prod_{j=i_0+1}^{i} (D_j + N_j),$$

which can be expanded as a sum of 2^{i-i_0} terms each of which is a product of $i - i_0$ factors, some diagonal and some strictly upper triangular. Each term that has $n - k$ or more strictly upper triangular factors must be zero (Exercise 5.1.6). For each $h < n - k$, each term having h strictly upper triangular factors must also have $i - i_0 - h$ diagonal factors. The norm of each of the strictly upper triangular factors is bounded above by K, the diagonal factors by $\tilde{v}$, so the norm of the entire term is bounded above by $K^h \tilde{v}^{i-i_0-h}$. For each $h < m$ there are $\binom{i-i_0}{h}$ terms having exactly h strictly upper triangular factors, so

$$\| \hat{p}_i(J_2) \| \leq \tilde{v}^i C_2 \sum_{h=0}^{n-k-1} \binom{i-i_0}{h} K^h \tilde{v}^{-i_0-h}.$$

The sum is a polynomial in i of degree at most $n - k - 1$, so

$$\| \hat{p}_i(J_2) \| \leq \pi_2(i) \tilde{v}^i, \tag{5.1.9}$$

where π_2 is a polynomial.

Now we apply a similar argument to get a bound on $\| \hat{p}_i(J_1)^{-1} \|$. The matrix $p(J_1)^{-1}$ is upper triangular and can be expressed as a sum $p(J_1)^{-1} = E + M$, where $E = \text{diag}\{p(\lambda_1)^{-1}, \ldots, p(\lambda_k)^{-1}\}$ and M is strictly upper triangular. Each $p_j(J_1)^{-1}$ has an analogous representation $p_j(J_1)^{-1} = E_j + M_j$. Since $p_j(J_1) \to p(J_1)$, we also have $p_j(J_1)^{-1} \to p(J_1)^{-1}$, $E_j \to E$, and $M_j \to M$. Thus there exists K' such that $\| M_j \| \leq K'$ for all j. Furthermore, for $j > i_0$, $\| E_j \| \leq \tilde{\delta}^{-1}$. Thus, reasoning as in the previous paragraph, we conclude that

$$\| \hat{p}_i(J_1)^{-1} \| \leq \pi_1(i) \tilde{\delta}^{-i}, \tag{5.1.10}$$

where π_1 is a polynomial of degree at most $k - 1$.

Substituting the inequalities (5.1.9) and (5.1.10) into (5.1.4), we obtain

$$d(\mathcal{S}_i, \mathcal{U}) \leq \pi(i) \tilde{\rho}^i,$$

where π is a polynomial. Consequently $d(\mathcal{S}_i, \mathcal{U}) \leq \pi(i)(\tilde{\rho}/\hat{\rho})^i \hat{\rho}^i$. Since $\tilde{\rho}/\hat{\rho} < 1$ and π is a polynomial, the factor $\pi(i)(\tilde{\rho}/\hat{\rho})^i$ tends to zero as $i \to \infty$. Thus there is a $\hat{C}$ such that $\pi(i)(\tilde{\rho}/\hat{\rho})^i \leq \hat{C}$ for all i, and therefore $d(\mathcal{S}_i, \mathcal{U}) \leq \hat{C}\hat{\rho}^i$ for all i. This proves Theorem 5.1.3.

Exercises

5.1.1. In this exercise we show that in Theorems 5.1.1 and 5.1.2, all of the iterates $\mathcal{S}_i$ have dimension k.

(a) Show that $\dim(\mathcal{S}_1) < \dim(\mathcal{S}_0)$ if and only if $\mathcal{S} = \mathcal{S}_0$ contains a nonzero vector x such that $p(A)x = 0$. Notice that any such x is an eigenvector of $p(A)$.

(b) Show that any such x must be a member of $\mathcal{V}$. Deduce that the condition $\mathcal{S} \cap \mathcal{V} = \{0\}$ forces $\dim(\mathcal{S}_1) = \dim(\mathcal{S}_0)$.

(c) Show that $\dim(\mathcal{S}_i) < \dim(\mathcal{S}_0)$ if and only if $\mathcal{S}$ contains a nonzero vector x such that $p(A)^i x = 0$. Such an x is an eigenvector of $p(A)^i$ but not necessarily of $p(A)$.

(d) Use the Jordan canonical form of A to show that a nonzero vector x as in part (c) must belong to $\mathcal{V}$. Hence the condition $\mathcal{S} \cap \mathcal{V} = \{0\}$ forces $\dim(\mathcal{S}_i) = k$ for all i.

(e) Show that $\mathcal{S}_i \cap \mathcal{V} = \{0\}$ for all i.

5.1.2. Let all hypotheses be as in Theorem 5.1.3.

(a) Using the Jordan form of A, show that the condition $p_i(\lambda_j) \neq 0$ for $j = 1, \ldots, k$ and all i, together with $\mathcal{S} \cap \mathcal{V} = \{0\}$, implies that $\mathcal{S}_i \cap \mathcal{V} = \{0\}$ for all i.

(b) Show that $\dim(\mathcal{S}_i) < \dim(\mathcal{S}_{i-1})$ if and only if there is a nonzero $x \in \mathcal{S}_{i-1}$ such that $p_i(A)x = 0$. Show that the conditions $\mathcal{S}_{i-1} \cap \mathcal{V} = \{0\}$ and $p_i(\lambda_j) \neq 0$ for $j = 1, \ldots, k$, guarantee that this cannot happen. Thus $\dim(\mathcal{S}_i) = k$ for all i.

5.1.3.

(a) Show that the condition $\mathcal{S} \cap \mathcal{V} = \{0\}$ implies that $\tilde{\mathcal{S}} \cap \tilde{\mathcal{V}} = \{0\}$.

(b) Show that $\tilde{\mathcal{V}} = \mathcal{R}(E'_{n-k})$, where $E'_{n-k} = \begin{bmatrix} 0 \\ I_{n-k} \end{bmatrix} \in \mathbb{C}^{n \times (n-k)}$.

(c) If $\tilde{\mathcal{S}} = \mathcal{R}(S)$, where $S = \begin{bmatrix} S_1 \\ S_2 \end{bmatrix}$ is a full-rank matrix and $S_1 \in \mathbb{C}^{k \times k}$, show that if S_1 is singular, then there is a nonzero vector $x \in \tilde{\mathcal{S}} \cap \tilde{\mathcal{V}}$.

(d) Deduce that the condition $\mathcal{S} \cap \mathcal{V} = \{0\}$ implies that S_1 is nonsingular.

5.1.4.

(a) Show that, under the hypotheses of Theorem 5.1.3, all of the eigenvalues of $\hat{p}_i(J_1)$ are nonzero. Therefore $\hat{p}_i(J_1)$ is nonsingular.

(b) Show that, under the hypotheses of Theorems 5.1.1 and 5.1.2, all of the eigenvalues of $\hat{p}_i(J_1)$ are nonzero.

(c) Show that the nonsingularity of $\hat{p}_i(J)$ implies (again) that $\dim(\mathcal{S}_i) = k$.

5.1.5. Check that under the hypotheses of Theorem 5.1.1 (semisimple, stationary case) the following equations hold:

(a) $\hat{p}_i(J_2) = p(J_2)^i = \operatorname{diag}\{p(\lambda_{k+1})^i, \ldots, p(\lambda_n)^i\}$.

(b) $\|\hat{p}_i(J_2)\| = |p(\lambda_{k+1})|^i$.

(c) $\hat{p}_i(J_1)^{-1} = p(J_1)^{-i} = \operatorname{diag}\{p(\lambda_1)^{-i}, \ldots, p(\lambda_k)^{-i}\}$.

(d) $\|\hat{p}_i(J_1)^{-1}\| = |p(\lambda_k)|^{-i}$.

(e) Combine these facts with inequality (5.1.4) to deduce that Theorem 5.1.1 holds with $C = \kappa(V) \|X\|$.

5.1.6. Let $T = T_1 T_2 \cdots T_j$ be a product of $j \times j$ upper triangular matrices.

(a) Show that if any one of the factors is strictly upper triangular, then T is strictly upper triangular.

(b) Show that if m of the factors are strictly upper triangular, then $t_{rs} = 0$ if $r > s - m$.

(c) Deduce that $T = 0$ if j or more of the factors are strictly upper triangular.

5.1.7. Confirm that $\binom{i}{h}$ is a polynomial in i of degree h, and the sum in (5.1.6) is a polynomial of degree at most $n - k - 1$.

5.1.8. In this exercise you will prove inequality (5.1.8).

(a) Show that $\hat{p}_i(J_1)^{-1} = p(J_1)^{-i} = (E + M)^i$, where

$$E = \operatorname{diag}\{p(\lambda_1)^{-1}, \ldots, p(\lambda_k)^{-1}\}$$

and M is strictly upper triangular. Show that $\| E \| = |p(\lambda_k)|^{-1}$.

(b) Show that $\| \hat{p}_i(J_1)^{-1} \| \le |p(\lambda_k)|^{-i} \sum_{h=0}^{k-1} \binom{i}{h} \| M \|^h |p(\lambda_k)|^h$.

(c) Show that the sum in part (b) is a polynomial in i of degree at most $k - 1$, thus confirming (5.1.8).

5.1.9. Obtain tighter bounds on the degrees of the polynomials π_1 and π_2 in terms of the size of the Jordan blocks in J.

5.2 Convergence of *GR* Algorithms

Given $A = A_0 \in \mathbb{C}^{n \times n}$, consider *GR* iterations

$$p_i(A_{i-1}) = G_i R_i, \quad A_i = G_i^{-1} A_{i-1} G_i, \quad i = 1, 2, 3, \ldots. \tag{5.2.1}$$

Each G_i is nonsingular, and each R_i is upper triangular. The A_i are not necessarily upper Hessenberg. We will assume that the p_i are polynomials of degree m that converge to some limiting polynomial p. In the simplest situation, $p_i = p$ for all i, but our real objective is to develop a shifting strategy such that the p_i converge to a p whose zeros are eigenvalues of A.

In Section 4.2 we introduced the generic *GR* algorithm and showed that it is simultaneous iteration with a change of coordinate system at each step. Since simultaneous iteration is nested subspace iteration, we can use our convergence theorems for subspace iteration to prove convergence theorems for *GR* iterations. In Section 4.2 we established the connection between *GR* iterations and subspace iteration on a step-by-step basis. For the purposes of convergence theory, one must make the connection on a global or cumulative basis. To this end, define the accumulated transforming matrices

$$\hat{G}_i = G_1 G_2 \cdots G_i, \qquad \hat{R}_i = R_i \cdots R_2 R_1.$$

Then from (5.2.1) we have immediately

$$A_i = \hat{G}_i^{-1} A \hat{G}_i \tag{5.2.2}$$

for all i. The accumulated transforming matrices also satisfy the following fundamental identity.

Theorem 5.2.1. *Let $\hat{p}_i = p_i \cdots p_2 p_1$, as before. Then*

$$\hat{p}_i(A) = \hat{G}_i \hat{R}_i \tag{5.2.3}$$

for all i.

The easy induction proof is worked out in Exercise 5.2.2. Equation (5.2.3) is the key to our analysis.[4] Obviously $\hat{R}_i$ is upper triangular, so (5.2.3) shows that the accumulated transformations give a generic GR decomposition of $\hat{p}_i(A)$. Assuming (for simplicity) that $\hat{p}_i(A)$ is nonsingular, (5.2.3) shows (using Theorem 3.2.2) that, for any k, the space spanned by the first k columns of $\hat{G}_i$ is just $\hat{p}_i(A)\mathcal{E}_k$, where $\mathcal{E}_k = \mathrm{span}\{e_1, \ldots, e_k\}$. Being the result of i steps of subspace iteration starting from $\mathcal{E}_k$, this space is (hopefully) close to an invariant subspace, in which case $A_i = \hat{G}_i^{-1} A \hat{G}_i$ is (hopefully) close to block triangular form (cf. Proposition 2.2.2).

Notice that the sequence of subspaces $\hat{p}_i(A)\mathcal{E}_k$ is determined by the choice of polynomials p_i and is independent of what type of GR iterations (e.g., QR, LR, HR, etc.) are being done. The choice of GR decomposition affects the sequence of matrix iterates (A_i) through the cumulative transforming matrices $\hat{G}_i$. We cannot guarantee convergence unless these are reasonably well behaved. Precise conditions for convergence are given in Theorem 5.2.3, which depends heavily upon the following preliminary result.

Proposition 5.2.2. *Let $A \in \mathbb{C}^{n \times n}$, and let $\mathcal{U}$ be a k-dimensional subspace of $\mathbb{C}^n$ that is invariant under A. Let $\mathcal{S}$ be a k-dimensional subspace of $\mathbb{C}^n$ that approximates $\mathcal{U}$, and let $G \in \mathbb{C}^{n \times n}$ be a nonsingular matrix whose first k columns span $\mathcal{S}$. Let $B = G^{-1}AG$, and consider the partitioned form*

$$B = \begin{bmatrix} B_{11} & B_{12} \\ B_{21} & B_{22} \end{bmatrix},$$

where $B_{11} \in \mathbb{C}^{k \times k}$. Then

$$\| B_{21} \| \le 2\sqrt{2} \, \kappa(G) \, \| A \| \, d(\mathcal{S}, \mathcal{U}).$$

The proof of Proposition 5.2.2 is worked out in Exercise 5.2.3. This result is a generalization of Proposition 2.2.2. If the leading k columns of G span the invariant subspace $\mathcal{U}$, then $\mathcal{S} = \mathcal{U}$, $d(\mathcal{S}, \mathcal{U}) = 0$, and $B_{21} = 0$. Proposition 5.2.2 shows that if the leading k columns of G nearly span an invariant subspace, then B_{21} is nearly zero.

Theorem 5.2.3. *Let $A \in \mathbb{C}^{n \times n}$, and consider the GR iteration (5.2.1), starting with $A_0 = A$, where the mth degree polynomials p_i converge to some mth degree polynomial p. Let $\lambda_1, \ldots, \lambda_n$ denote the eigenvalues of A, ordered so that $|p(\lambda_1)| \ge \cdots \ge |p(\lambda_n)|$. Assume $p_i(\lambda_j) \ne 0$ for $j = 1, \ldots, k$ and all i. Suppose that for some k satisfying $1 \le k < n$,*

---

[4] Indeed, every convergence analysis of which I am aware makes use of an equation like (5.2.3).

$|p(\lambda_k)| > |p(\lambda_{k+1})|$. *Let* $\mathcal{U}$ *and* $\mathcal{V}$ *denote the invariant subspaces of A associated with* $\lambda_1, \ldots, \lambda_k$ *and* $\lambda_{k+1}, \ldots, \lambda_n$, *respectively, and suppose* $\mathcal{E}_k \cap \mathcal{V} = \{0\}$. *Partition the GR iterates* A_i *as*

$$A_i = \begin{bmatrix} A_{11}^{(i)} & A_{12}^{(i)} \\ A_{21}^{(i)} & A_{22}^{(i)} \end{bmatrix}.$$

Let $\rho = |p(\lambda_{k+1})|/|p(\lambda_k)| < 1$. *Then for every* $\hat{\rho}$ *satisfying* $\rho < \hat{\rho} < 1$ *there is a constant* $\hat{K}$ *such that*

$$\| A_{21}^{(i)} \| \leq \hat{K}\kappa(\hat{G}_i)\hat{\rho}^i, \qquad i = 1, 2, 3, \ldots,$$

where $\hat{G}_i$ *is the cumulative transforming matrix:* $\hat{G}_i = G_1 G_2 \cdots G_i$.

This theorem follows immediately from Theorem 5.1.3, Proposition 5.2.2, and (5.2.3). The details are worked out in Exercise 5.2.4.

Theorem 5.2.3 guarantees convergence only if the condition numbers $\kappa(\hat{G}_i)$ are not too badly behaved. If $\kappa(\hat{G}_i)\hat{\rho}^i \to 0$, then $\| A_{21}^{(i)} \| \to 0$, and the iterates A_i converge to block triangular form. This can happen even if $\kappa(\hat{G}_i) \to \infty$, provided that the rate of growth is less than the rate at which $\hat{\rho} \to 0$. If the condition numbers remain bounded, we have convergence for sure.

Corollary 5.2.4. *Under the conditions of Theorem 5.2.3, suppose that there is a constant* $\hat{\kappa}$ *such that* $\kappa(\hat{G}_i) \leq \hat{\kappa}$ *for all i. Then for every* $\hat{\rho}$ *satisfying* $\rho < \hat{\rho} < 1$ *there is a constant* $\hat{M}$ *such that*

$$\| A_{21}^{(i)} \| \leq \hat{M}\hat{\rho}^i, \qquad i = 1, 2, 3, \ldots.$$

Thus the iterates A_i *converge to block triangular form. The convergence is linear if* $\rho > 0$ *and superlinear if* $\rho = 0$.

In general it is difficult to verify that the condition numbers $\kappa(\hat{G}_i)$ remain bounded. The one case in which we can be sure of it is the *QR* algorithm, for which $\kappa(\hat{G}_i) = 1$ for all i.

Corollary 5.2.5. *Under the conditions of Theorem 5.2.3, the QR algorithm converges to block triangular form: For every* $\hat{\rho}$ *satisfying* $\rho < \hat{\rho} < 1$ *there is a constant* $\hat{K}$ *such that*

$$\| A_{21}^{(i)} \| \leq \hat{K}\hat{\rho}^i, \qquad i = 1, 2, 3, \ldots.$$

The convergence is linear if $\rho > 0$ *and superlinear if* $\rho = 0$.

Convergence of the matrices to block triangular form implies convergence of the eigenvalues.

Theorem 5.2.6. *Under the conditions of Theorem 5.2.3, if* $A_{21} \to 0$ *as* $i \to \infty$, *then the eigenvalues of* $A_{11}^{(i)}$ *converge to* $\lambda_1, \ldots, \lambda_k$, *and those of* $A_{22}^{(i)}$ *converge to* $\lambda_{k+1}, \ldots, \lambda_n$.

This theorem ought to seem obvious, but it does require proof. A proof is worked out in Exercises 5.2.5 and 5.2.6.

If the conditions of Corollary 5.2.4 hold for one value of k, then they typically hold for many values of k at once, and the sequence (A_i) converges to a correspondingly finer block triangular form. In the best possible case, in which the hypotheses are satisfied for all k, the limiting form is upper triangular.

In practice the p_i are polynomials of low degree $m \ll n$. When the GR algorithm functions as intended (and this is usually the case) the p_i converge to a limiting polynomial whose zeros are all eigenvalues of A. Thus, if we take $k = n - m$, then $\rho = 0$, and the iterates will effectively converge after just a few iterations to the block triangular form

$$\begin{bmatrix} A_{11} & A_{12} \\ 0 & A_{22} \end{bmatrix},$$

with $A_{22} \in \mathbb{C}^{m \times m}$. Since m is small, it is a simple matter to compute the eigenvalues of A_{22}. The rest of the eigenvalues of A are eigenvalues of A_{11}, so subsequent GR iterations can be applied to the reduced matrix A_{11}. Now we are just repeating observations that were made in greater detail in Section 4.3.

One of the hypotheses of Theorem 5.2.3 is the subspace condition $\mathcal{E}_k \cap \mathcal{U} = \{0\}$. As we observed earlier in this section, this type of condition almost never fails to hold in practice. It is satisfying, nevertheless, that if we apply the GR algorithm to a properly upper Hessenberg matrix, the condition is guaranteed to hold for all k [168].

Proposition 5.2.7. *Let A be a properly upper Hessenberg matrix, and let k be an integer satisfying $1 \le k < n$. Let $\mathcal{V}$ be any subspace of dimension $n - k$ that is invariant under A. Then $\mathcal{E}_k \cap \mathcal{V} = \{0\}$.*

A proof is worked out in Exercise 5.2.7.

Quadratic and Cubic Convergence

We now address the question of how to choose the p_i. In Section 4.3 we advocated use of the generalized Rayleigh quotient shift strategy. We remarked there that this strategy is not globally convergent, so it cannot be used without (occasional) supplementary exceptional shifts. However, it generally works quite well, and the exceptional shifts are almost never needed. Here we will show that, under generic conditions, the asymptotic convergence rate of the generalized Rayleigh quotient shift strategy is quadratic. By "generic conditions" we mean that A has distinct eigenvalues. Only exceptional matrices have repeated eigenvalues. Furthermore, we are *not* proving global convergence. We will show that *if* the iterations converge, then the convergence rate is (usually) quadratic.

Let us recall the generalized Rayleigh quotient shift strategy. After $i - 1$ iterations we have

$$A_{i-1} = \begin{bmatrix} A_{11}^{(i-1)} & A_{12}^{(i-1)} \\ A_{21}^{(i-1)} & A_{22}^{(i-1)} \end{bmatrix}.$$

Suppose the partition is such that $A_{22}^{(i-1)}$ is $m \times m$. The *generalized Rayleigh quotient shift strategy* takes the m shifts for the ith iteration to be the eigenvalues of $A_{22}^{(i-1)}$. Thus $p_i(x) = (x - \rho_1) \cdots (x - \rho_m)$, where $\rho_1, \ldots, \rho_m$ are the eigenvalues of $A_{22}^{(i-1)}$. Another way to say this is that p_i is the characteristic polynomial of $A_{22}^{(i-1)}$.

Theorem 5.2.8. *Let $A \in \mathbb{C}^{n \times n}$ have distinct eigenvalues. Suppose that the conditions of Theorem 5.2.3 are satisfied with $k = n - m$. Suppose that there is a constant $\hat{\kappa}$ such that $\kappa(\hat{G}_i) \le \hat{\kappa}$ for all i, so that the A_i converge to block triangular form in the sense described in Theorem 5.2.3, with $k = n - m$. Then the convergence is quadratic.*

Let $\mathcal{S} = \mathcal{S}_0 = \mathcal{E}_k$ and $\mathcal{S}_i = p_i(\mathcal{S}_{i-1})$, as in the proof of Theorem 5.2.3. The subspace iterates $\mathcal{S}_i$ converge to the invariant subspace $\mathcal{U}$, and from that we deduced the convergence of the GR iterates. In the scenario of Theorem 5.2.3 we have $d(\mathcal{S}, \mathcal{U}) \to 0$ linearly (at least). If we use Rayleigh quotient shifts, we can show that there is a constant M such that, for all sufficiently small ϵ, if $d(\mathcal{S}_i, \mathcal{U}) = \epsilon$, then $d(\mathcal{S}_{i+1}, \mathcal{U}) \leq M\epsilon^2$. This suffices to establish quadratic convergence. The idea of the proof is simple. If $d(\mathcal{S}_i, \mathcal{U})$ is very small, then A_i is very close to block triangular form, so the shifts are excellent approximations to the eigenvalues $\lambda_{k+1}, \ldots, \lambda_n$, the contraction number for the ith step is small, and $d(\mathcal{S}_{i+1}, \mathcal{U})$ is much smaller than $d(\mathcal{S}_i, \mathcal{U})$. The details are worked out in Exercise 5.2.11.

For certain classes of matrices possessing special structure, the generalized Rayleigh quotient shift strategy yields cubic convergence.

Theorem 5.2.9. *Under the hypotheses of Theorem 5.2.8, suppose that each of the iterates*

$$A_i = \left[\begin{array}{cc} A_{11}^{(i)} & A_{12}^{(i)} \\ A_{21}^{(i)} & A_{22}^{(i)} \end{array} \right]$$

satisfies $\| A_{12}^{(i)} \|_F = \| A_{21}^{(i)} \|_F$, *where* $\| \cdot \|_F$ *denotes the Frobenius norm (1.2.3). Then the iterates converge cubically if they converge.*

The proof is sketched in Exercises 5.2.12 and 5.2.13. We now give several specific examples where cubic convergence occurs.

Example 5.2.10. *If A is normal and we apply the QR algorithm to A, then all A_i are normal. Hence* $\| A_{12}^{(i)} \|_F = \| A_{21}^{(i)} \|_F$ *by Proposition 2.3.2. Thus the QR algorithm with the generalized Rayleigh quotient shift strategy applied to normal matrices converges cubically when it converges. Recall that Hermitian, skew-Hermitian, and unitary matrices are all normal.*

Example 5.2.11. *If A is tridiagonal and pseudosymmetric (4.9.3) and we apply the HR algorithm to A, then all A_i are tridiagonal and pseudosymmetric (Exercise 4.9.7). This implies trivially that* $\| A_{12}^{(i)} \|_F = \| A_{21}^{(i)} \|_F$. *Thus the HR algorithm with the generalized Rayleigh quotient shift strategy applied to pseudosymmetric tridiagonal matrices converges cubically when it converges.*

Example 5.2.12. *If A is Hamiltonian and we apply the SR algorithm to A, then all A_i are Hamiltonian. The SR algorithm does not exactly fit the scenario of this chapter, because the R-matrix in the SR decomposition (Theorem 1.3.11) is $\hat{P}$-triangular, not triangular. However, if we apply the shuffling transformation $M \to \hat{P} M \hat{P}^T$ (see (1.3.2)) to every matrix in sight, the resulting shuffled SR algorithm does fit, and Theorem 5.2.9 can be applied. It is not hard to check that the shuffled matrices $\hat{A}_i = \hat{P} A_i \hat{P}^{-1}$ satisfy* $\| \hat{A}_{12}^{(i)} \|_F = \| \hat{A}_{21}^{(i)} \|_F$ *for even values of k. Since the eigenvalues of Hamiltonian matrices come in pairs and are extracted in pairs, only the even values of k matter. Thus the SR algorithm with the generalized Rayleigh quotient shift strategy applied to Hamiltonian matrices converges cubically when it converges.*

Difficult Cases

All of our convergence theorems rely on there being some separation of the eigenvalues: $|p(\lambda_k)| > |p(\lambda_{k+1})|$. This is (or seems) impossible to achieve if the matrix has a single eigenvalue of algebraic multiplicity n. More generally, if a matrix has many eigenvalues, but one of them has high multiplicity j, then the matrix can be decoupled to an extent by the GR algorithm, but in the end there will be a $j \times j$ block containing the multiple eigenvalue that cannot be reduced further, at least according to the theorems we have proved in this section. Thus it is natural to ask what happens when the GR algorithm is applied to a matrix with a single eigenvalue of high algebraic multiplicity.

Let $A \in \mathbb{C}^{n \times n}$ be a properly upper Hessenberg matrix that has a single eigenvalue λ of algebraic multiplicity n. Then the Jordan canonical form of A must consist of a single Jordan block (Lemma 4.6.2). This is the case in theory. In practice we will never see such a configuration. The first roundoff error will break the single eigenvalue into a cluster of eigenvalues. The cluster will not be all that tight, because a Jordan block explodes when disturbed. (See Example 2.7.2 and Exercises 2.7.1 and 2.7.2.) Once we find a shift that is in or near the cluster, the relative tightness of the cluster is even less. Convergence is slow at first, but eventually the eigenvalue separation is good enough to get us into the regime of rapid convergence. Of course, the computed eigenvalues in this cluster will be ill conditioned and will not be computed very accurately.

Exercises

5.2.1.

(a) Rework Exercise 4.3.1. What evidence for quadratic convergence do you see?

(b) Rework Exercise 4.3.2. What evidence for cubic convergence do you see?

(c) Rework Exercise 4.3.4. What evidence for quadratic convergence do you see?

5.2.2. This exercise proves Theorem 5.2.1 by induction on i.

(a) Show that the case $i = 1$ of (5.2.3) follows immediately from the case $i = 1$ of (5.2.1).

(b) Now assume $i = j > 1$ and suppose that (5.2.3) holds for $i = j - 1$. Show that $\hat{p}_j = p_j \hat{p}_{j-1}$, $\hat{G}_j = \hat{G}_{j-1} G_j$, and $\hat{R}_j = R_j \hat{R}_{j-1}$. Then show that $\hat{p}_j(A) = p_j(A)\hat{G}_{j-1}\hat{R}_{j-1}$.

(c) Use (5.2.2) followed by (5.2.1) to show that $p_j(A) = \hat{G}_{j-1} p(A_{j-1})\hat{G}_{j-1}^{-1} = \hat{G}_j R_j \hat{G}_{j-1}^{-1}$.

(d) Combine the results of parts (b) and (c) to conclude that $\hat{p}_j(A) = \hat{G}_j \hat{R}_j$. This completes the proof of Theorem 5.2.1.

5.2.3. This exercise proves Proposition 5.2.2. Consider a decomposition $G = QR$, where Q is unitary and R is upper triangular. Make the partitions $Q = \begin{bmatrix} Q_1 & Q_2 \end{bmatrix}$, where $Q_1 \in \mathbb{C}^{n \times k}$, and

$$R = \begin{bmatrix} R_{11} & R_{12} \\ 0 & R_{22} \end{bmatrix},$$

where $R_{11} \in \mathbb{C}^{k \times k}$.

(a) Show that $\mathcal{S} = \mathcal{R}(Q_1)$ and $\mathcal{S}^\perp = \mathcal{R}(Q_2)$.

(b) Show that $B_{21} = R_{22}^{-1} Q_2^* A Q_1 R_{11}$.

(c) Show that $\| R_{22}^{-1} \| \le \| G^{-1} \|$ and $\| R_{11} \| \le \| G \|$. Deduce that

$$\| B_{21} \| \le \kappa(G) \| Q_2^* A Q_1 \|.$$

(d) By Proposition 2.6.13 there exist $U_1 \in \mathbb{C}^{n \times k}$ and $U_2 \in \mathbb{C}^{n \times (n-k)}$ with orthonormal columns such that $\mathcal{R}(U_1) = \mathcal{U}$, $\mathcal{R}(U_2) = \mathcal{U}^\perp$,

$$\| Q_1 - U_1 \| \le \sqrt{2} d(\mathcal{S}, \mathcal{U}), \quad \text{and} \quad \| Q_2 - U_2 \| \le \sqrt{2} d(\mathcal{S}^\perp, \mathcal{U}^\perp).$$

Corollary 2.6.10 implies that $d(\mathcal{S}^\perp, \mathcal{U}^\perp) = d(\mathcal{S}, \mathcal{U})$. Show that $U_2^* A U_1 = 0$, then show that

$$Q_2^* A Q_1 = Q_2^* A Q_1 - U_2^* A U_1 = (Q_2 - U_2)^* A Q_1 + U_2^* A (Q_1 - U_1).$$

(e) Deduce that
$$\| Q_2^* A Q_1 \| \le 2\sqrt{2} \| A \| d(\mathcal{S}, \mathcal{U}).$$

Combine this with part (c) to complete the proof of Proposition 5.2.2.

5.2.4. This exercise proves Theorem 5.2.3. Let $\mathcal{S} = \mathcal{E}_k$ and $\mathcal{S}_i = \hat{p}_i(A)\mathcal{S}$ for all i.

(a) Check that all of the hypotheses of Theorem 5.1.3 are satisfied, and deduce that there is a $\hat{C}$ such that $d(\mathcal{S}_i, \mathcal{U}) \le \hat{C} \hat{\rho}^i$ for all i.

(b) Consider the partition $\hat{G}_i = [\ G_1^{(i)} \quad G_2^{(i)}\ ]$, where $G_1^{(i)} \in \mathbb{C}^{n \times k}$. Use (5.2.3) to show that $\mathcal{S}_i = \mathcal{R}(G_1^{(i)})$ for all i. Notice that this is so, regardless of whether or not $\hat{p}_i(A)$ is singular, since $\mathcal{S}_i$ and $\mathcal{R}(G_1^{(i)})$ both have dimension k.

(c) Apply Proposition 5.2.2 with the roles of G and B played by $\hat{G}_i$ and A_i, respectively, to conclude that $\| A_{21}^{(i)} \| \le \hat{K} \kappa(\hat{G}_i) \hat{\rho}^i$, where $\hat{K} = 2\sqrt{2} \| A \| \hat{C}$.

5.2.5. This and the next exercise prove Theorem 5.2.6. In this exercise we consider the "QR" case, in which all of the $\hat{G}_i$ are unitary. We will use the same notation as in Theorem 5.2.3, except that will write $\hat{Q}_i$ instead of $\hat{G}_i$ to remind us that the transformations are all unitary. Make the partition $\hat{Q}_i = [\ \hat{Q}_1^{(i)} \quad \hat{Q}_2^{(i)}\ ]$, where $\hat{Q}_1^{(i)} \in \mathbb{C}^{n \times k}$. We know from (5.2.3) that $\mathcal{R}(\hat{Q}_1^{(i)}) = \mathcal{S}_i = \hat{p}_i(A)\mathcal{S}$.

(a) Let $U \in \mathbb{C}^{n \times k}$ be a matrix with orthonormal columns such that $\mathcal{R}(U) = \mathcal{U}$, and let $C = U^* A U$. Show that the eigenvalues of C are $\lambda_1, \ldots, \lambda_k$.

(b) By Proposition 2.6.13 we know that for every i there is a matrix $S_i \in \mathbb{C}^{n \times k}$ such that $\mathcal{R}(S_i) = \mathcal{S}_i$ and $\| S_i - U \| \le \sqrt{2} d(\mathcal{S}_i, \mathcal{U})$. By Proposition 1.5.11 there is a unitary matrix $X_i \in \mathbb{C}^{k \times k}$ such that $S_i = \hat{Q}_1^{(i)} X_i$. Let $B_i = S_i^* A S_i \in \mathbb{C}^{k \times k}$. Show that B_i is unitarily similar to $A_{11}^{(i)}$: $B_i = X_i^* A_{11}^{(i)} X_i$. Thus B_i has the same eigenvalues as $A_{11}^{(i)}$.

(c) Show that
$$B_i - C = (S_i - U)^* A S_i + U^* A(S_i - U).$$

(d) Deduce that $\| B_i - C \| \le 2\sqrt{2} \, \| A \| \, d(\mathcal{S}_i, \mathcal{U}) \le 2\sqrt{2} \, \| A \| \| \hat{C} \hat{\rho}^i$.

(e) Deduce that $B_i \to C$ and, using Theorem 2.7.1, conclude that the eigenvalues of $A_{11}^{(i)}$ converge to $\lambda_1, \ldots, \lambda_k$.

(f) Construct a similar argument to show that the eigenvalues of $A_{22}^{(i)}$ converge to $\lambda_{k+1}, \ldots,$ λ_n. You might start as follows: Let $\check{U} \in \mathbb{C}^{n \times (n-k)}$ be a matrix with orthonormal columns such that $\mathcal{R}(\check{U}) = \mathcal{U}^\perp$, and let $\check{C} = \check{U}^* A \check{U}$. Show that the eigenvalues of $\check{C}$ are $\lambda_{k+1}, \ldots, \lambda_n$. For every i, $\mathcal{R}(\hat{Q}_2^{(i)}) = \mathcal{S}_i^\perp$, $\mathcal{S}_i^\perp \to \mathcal{U}^\perp$ because $d(\mathcal{S}_i^\perp, \mathcal{U}^\perp) = d(\mathcal{S}_i, \mathcal{U})$, and so on.

Example 2.7.2 and the matrices in Exercises 2.7.1 and 2.7.2 show that the rate of convergence of the eigenvalues can be arbitrarily slow. However, in most cases the eigenvalues converge just as fast as the subspace iterations do. This is the case, for example, if $\lambda_1, \ldots, \lambda_k$ are distinct, since then the eigenvalues vary Lipschitz continuously with respect to the matrix entries. This is the case in the proof of quadratic convergence (Exercise 5.2.11). Sometimes the eigenvalues converge much faster than the subspace iterations do, as the proof of cubic convergence (Exercise 5.2.13) shows.

5.2.6. This exercise proves Theorem 5.2.6. For each i consider a QR decomposition $\hat{G}_i = \hat{Q}_i T_i$, which we can write in partitioned form as

$$\begin{bmatrix} \hat{G}_1^{(i)} & \hat{G}_2^{(i)} \end{bmatrix} = \begin{bmatrix} \hat{Q}_1^{(i)} & \hat{Q}_2^{(i)} \end{bmatrix} \begin{bmatrix} T_{11}^{(i)} & T_{12}^{(i)} \\ 0 & T_{22}^{(i)} \end{bmatrix}.$$

Let $U_i = T_i^{-1}$, and let $F_i = \hat{Q}_i^* A \hat{Q}_i$. We also partition these matrices conformably with the others.

(a) Show that
$$A_{11}^{(i)} = U_{11}^{(i)} F_{11}^{(i)} T_{11}^{(i)} + U_{12}^{(i)} F_{21}^{(i)} T_{11}^{(i)}.$$

Now the plan is roughly this: We can show that the last term goes to zero, because $F_{21}^{(i)} \to 0$, so $A_{11}^{(i)}$ is nearly similar to $F_{11}^{(i)}$, whose eigenvalues converge to $\lambda_1, \ldots, \lambda_k$ by Exercise 5.2.5. We have to proceed carefully.

(b) Let U, C, S_i, and B_i be defined as in Exercise 5.2.5. Note that $\mathcal{S}_i = \mathcal{R}(S_i) = \mathcal{R}(\hat{G}_1^{(i)}) = \mathcal{R}(\hat{Q}_1^{(i)})$. Define X_i by $S_i = \hat{Q}_1^{(i)} X_i$, as before. Let $W_i = X_i^* T_{11}^{(i)} A_{11}^{(i)} U_{11}^{(i)} X_i$. Check that W_i is similar to $A_{11}^{(i)}$, so that W_i has the same eigenvalues as $A_{11}^{(i)}$.

(c) Show that
$$W_i = B_i + X_i^* T_{11}^{(i)} U_{12}^{(i)} F_{21}^{(i)} X_i.$$

We will use this equation to show that W_i is close to B_i (which is close to C).

(d) Show that $\| T_{11}^{(i)} \| \le \| \hat{G}_i \|$, $\| U_{12}^{(i)} \| \le \| \hat{G}_i^{-1} \|$, and $\| T_{11}^{(i)} \| \| U_{12}^{(i)} \| \le \kappa(\hat{G}_i) \le \hat{\kappa}$. Use Proposition 5.2.2 to show that $\| F_{21}^{(i)} \| \le 2\sqrt{2} \, \| A \| \, d(\mathcal{S}_i, \mathcal{U})$. Deduce that
$$\| W_i - B_i \| \le 2\sqrt{2} \, \hat{\kappa} \, \| A \| \, d(\mathcal{S}_i, \mathcal{U}).$$

(e) Using part (d) of Exercise 5.2.5, conclude that

$$\| W_i - C \| \le 2\sqrt{2}(\hat{\kappa} + 1)\| A \| d(\mathcal{S}_i, \mathcal{U}) \le M\hat{\rho}^i,$$

where M is a constant. Deduce that the eigenvalues of $A_{11}^{(i)}$ converge to $\lambda_1, \ldots, \lambda_k$.

(f) Show that

$$A_{22}^{(i)} = U_{22}^{(i)} F_{21}^{(i)} T_{12}^{(i)} + U_{22}^{(i)} F_{22}^{(i)} T_{22}^{(i)}.$$

Devise a proof that the eigenvalues of $A_{22}^{(i)}$ converge to $\lambda_{k+1}, \ldots, \lambda_n$.

5.2.7. This exercise proves Proposition 5.2.7. Suppose that A is properly upper Hessenberg. Thus $a_{ij} = 0$ for $i > j + 1$, and $a_{ij} \ne 0$ for $i = j + 1$.

(a) Let x be a nonzero vector in $\mathcal{E}_k = \text{span}\{e_1, \ldots, e_k\}$. There is at least one component x_j such that $x_j \ne 0$. Show that the largest j for which $x_j \ne 0$ satisfies $j \le k$. What can you say about Ax? A^2x?

(b) Prove that $x, Ax, A^2x, \ldots, A^{n-k}x$ are linearly independent.

(c) Prove that any subspace that is invariant under A and contains x must have dimension at least $n - k + 1$. Deduce that the space $\mathcal{V}$ cannot contain x.

(d) Deduce that $\mathcal{E}_k \cap \mathcal{V} = \{0\}$.

5.2.8. In this exercise we explore an interesting case of nonconvergence of a *GR* algorithm. Let

$$A = \begin{bmatrix} a & b \\ b & 0 \end{bmatrix},$$

where $0 < a < b$. Consider stationary *GR* iterations of degree 1 with shift 0 at each step. This means that $p_i(A_{i-1}) = p(A_{i-1}) = A_{i-1}$ at each step.

(a) Show that all of the conditions of Theorem 5.2.3 with $k = 1$ are satisfied. In particular, $|p(\lambda_1)| > |p(\lambda_2)|$.

(b) Consider a *GR* algorithm in which each *GR* decomposition is an *LR* decomposition with partial pivoting (Theorem 1.3.5). Thus $A_0 = G_1 R_1$, where $G_1 = PL$, P is a permutation matrix, and L is unit lower triangular with $|l_{21}| \le 1$. Show that the condition $0 < a < b$ forces $P = \begin{bmatrix} 0 & 1 \\ 1 & 0 \end{bmatrix}$, $L^{-1} = \begin{bmatrix} 1 & 0 \\ -a/b & 1 \end{bmatrix}$, and $L = \begin{bmatrix} 1 & 0 \\ a/b & 1 \end{bmatrix}$. (Think of doing Gaussian elimination with partial pivoting to transform A_0 to upper triangular form, as in Section 3.2.)

(c) Show that $A_1 = G_1^{-1} A_0 G_1 = A_0$. Show that $A_0 = A_1 = A_2 = A_3 = \cdots$. Thus the *GR* iteration is stationary; it fails to converge.

(d) Explain why this *GR* iteration can fail to converge, despite satisfaction of all the hypotheses of Theorem 5.2.3.

This is an interesting case in which the decay of ρ^i is exactly matched by the growth of $\kappa(\hat{G}_i)$. Two more examples are given in the next two exercises. These are rare cases; usually the LR algorithm with partial pivoting works fairly well.

5.2.9. Study the LR algorithm with partial pivoting for the matrix

$$A = \begin{bmatrix} a & c \\ b & 0 \end{bmatrix},$$

where $0 < a < \min\{b, c\}$. This shows that symmetry of the matrix was not essential to the outcome of Exercise 5.2.8.

5.2.10. Study the LR algorithm with partial pivoting for the matrix

$$A = \begin{bmatrix} a & b & c \\ d & 0 & 0 \\ 0 & e & 0 \end{bmatrix},$$

where c, d, and e are large relative to a and b.

5.2.11. This exercise proves Theorem 5.2.8, quadratic convergence of GR algorithms. Let $\mathcal{U}$ be the invariant subspace of A associated with eigenvalues $\lambda_1, \ldots, \lambda_k$. Let $\mathcal{S}_0 = \mathcal{E}_k$, and let $\mathcal{S}_{i+1} = p_{i+1}(A)\mathcal{S}_i$, $i = 0, 1, 2, 3, \ldots$. We will show that there is a constant M such that, for all sufficiently small $\epsilon > 0$, if $d(\mathcal{S}_i, \mathcal{U}) = \epsilon$, then $d(\mathcal{S}_{i+1}, \mathcal{U}) \leq M\epsilon^2$. This guarantees quadratic convergence.

(a) Let $d(\mathcal{S}_i, \mathcal{U}) = \epsilon$. (For now we are vague about how small ϵ needs to be.) Use Proposition 5.2.2 to show that $\| A_{21}^{(i)} \| \leq M_1 \epsilon$, where $M_1 = 2\sqrt{2}\hat{\kappa} \| A \|$.

(b) Since A has distinct eigenvalues, A is semisimple (Section 2.1), and so it has the form $A = V \Lambda V^{-1}$, where $\Lambda = \text{diag}\{\lambda_1, \ldots, \lambda_n\}$. Thus $A_i = \hat{G}_i^{-1} V \Lambda V^{-1} \hat{G}_i$. Apply the Bauer–Fike theorem (Theorem 2.7.5) to A_i to show that the eigenvalues of

$$\begin{bmatrix} A_{11}^{(i)} & A_{12}^{(i)} \\ 0 & A_{22}^{(i)} \end{bmatrix} \tag{5.2.4}$$

are close to the eigenvalues of A_i. Specifically, for every eigenvalue ρ of (5.2.4) there is an eigenvalue λ of A_i such that $|\lambda - \rho| \leq M_2 \epsilon$, where $M_2 = \hat{\kappa}\kappa(V)M_1$.

(c) The polynomial used for the $(i+1)$th step is $p_{i+1}(x) = (x - \rho_1)(x - \rho_2) \cdots (x - \rho_m)$, where $\rho_1, \ldots, \rho_m$ are the eigenvalues of $A_{22}^{(i)}$. Since these are eigenvalues of (5.2.4), each is within $M_2\epsilon$ of an eigenvalue of A. Show that if ϵ is sufficiently small, no two of the ρ_i are within $M_2\epsilon$ of the same eigenvalue of A. Show that

$$|\lambda_{k+j} - \rho_j| \leq M_2\epsilon, \quad j = 1, \ldots, m,$$

if the ρ_i are numbered appropriately. Show that if $M_2\epsilon \leq 1$, then for $j = k+1, \ldots, n$, $|p_{i+1}(\lambda_j)| \leq M_3\epsilon$, where $M_3 = (2\| A \| + 1)^{m-1} M_2$.

(d) Let $\Lambda_1 = \text{diag}\{\lambda_1, \ldots, \lambda_k\}$ and $\Lambda_2 = \text{diag}\{\lambda_{k+1}, \ldots, \lambda_n\}$, so that $\Lambda = \text{diag}\{\Lambda_1, \Lambda_2\}$. Show that $\| p_{i+1}(\Lambda_2) \| \leq M_3\epsilon$.

(e) Let γ denote half the minimum distance between $\{\lambda_1, \ldots, \lambda_k\}$ and $\{\lambda_{k+1}, \ldots, \lambda_n\}$. Show that if $M_2\epsilon \leq \gamma$, then for $j = 1, \ldots, k$, $|p_{i+1}(\lambda_j)| \geq \gamma^m$. Deduce that $p_{i+1}(\Lambda_1)$ is nonsingular, and $\| p_{i+1}(\Lambda_1)^{-1} \| \leq 1/\gamma^m$.

(f) Now we work in a convenient coordinate system. Let $\tilde{\mathcal{S}}_i = V^{-1}\mathcal{S}_i, \tilde{\mathcal{U}} = V^{-1}\mathcal{U} = \mathcal{E}_k$, and $\tilde{\mathcal{V}} = V^{-1}\mathcal{V} = \mathcal{E}_k^\perp$. Show that $\tilde{\mathcal{S}}_{i+1} = p_{i+1}(\Lambda)\tilde{\mathcal{S}}_i$.

(g) Let $\beta = d(\tilde{\mathcal{S}}_i, \tilde{\mathcal{U}})$. By Proposition 2.6.14, $\beta \leq \kappa(V)\epsilon$, so we can guarantee that $\beta \leq \sqrt{3}/2$ by making ϵ small enough. Deduce that $\tilde{\mathcal{S}}_i \cap \tilde{\mathcal{V}} = \{0\}$ (Proposition 2.6.15), and consequently there is a matrix of the form $Y = \begin{bmatrix} I \\ X \end{bmatrix} \in \mathbb{C}^{n \times k}$ such that $\tilde{\mathcal{S}}_i = \mathcal{R}(Y)$. Show that
$$d(\tilde{\mathcal{S}}_{i+1}, \tilde{\mathcal{U}}) \leq \frac{\beta}{\sqrt{1 - \beta^2}} \| p_{i+1}(\Lambda_2) \| \| p_{i+1}(\Lambda_1)^{-1} \|.$$

This can be done by a variant of the argument in Section 5.1 from approximately (5.1.2) to (5.1.4).

(h) Deduce that $d(\mathcal{S}_{i+1}, \mathcal{U}) \leq M\epsilon^2$, where $M = 2\kappa(V)^2 M_3/\gamma^m$.

5.2.12. Our proof of cubic convergence of GR iterations will make use of the following lemma [227], a variation on the Bauer–Fike theorem: Let $B = \text{diag}\{B_1, B_2\}$ be a block diagonal matrix whose blocks are semisimple: $C_1^{-1}B_1C_1 = D_1 = \text{diag}\{\mu_1, \ldots, \mu_k\}$, and $C_2^{-1}B_2C_2 = D_2 = \text{diag}\{\mu_{k+1}, \ldots, \mu_n\}$. Let
$$E = \begin{bmatrix} 0 & E_{12} \\ E_{21} & 0 \end{bmatrix},$$
where $E_{21} \in \mathbb{C}^{(n-k)\times k}$. If λ is an eigenvalue of $B + E$, then
$$\min_{1 \leq j \leq k} |\mu_j - \lambda| \min_{k+1 \leq j \leq n} |\mu_j - \lambda| \leq \kappa(C_1)\kappa(C_2) \| E_{12} \| \| E_{21} \|. \tag{5.2.5}$$

Prove this lemma as follows. Let x be an eigenvector of $B+E$ associated with the eigenvalue λ, and let $C = \text{diag}\{C_1, C_2\}, z = C^{-1}x$, and $z = \begin{bmatrix} z_1 \\ z_2 \end{bmatrix}$, where $z_1 \in \mathbb{C}^k$.

(a) Show that $C^{-1}(B + E)Cz = \lambda z$. Then deduce that
$$(\lambda I - D_1)z_1 = C_1^{-1}E_{12}C_2z_2 \quad \text{and} \quad (\lambda I - D_2)z_2 = C_2^{-1}E_{21}C_1z_1.$$

(b) Show that (5.2.5) is trivially true if $\lambda I - D_1$ or $\lambda I - D_2$ is singular. Now assume that these matrices are nonsingular, and show that $z_1 \neq 0, z_2 \neq 0$, and
$$z_1 = (\lambda I - D_1)^{-1}C_1^{-1}E_{12}C_2(\lambda I - D_2)^{-1}C_2^{-1}E_{21}C_1z_1.$$

(c) Take norms of both sides to show that
$$1 \leq \| (\lambda I - D_1)^{-1} \| \| (\lambda I - D_2)^{-1} \| \kappa(C_1)\kappa(C_2) \| E_{12} \| \| E_{21} \|.$$

Deduce (5.2.5).

5.2.13. This exercise proves Theorem 5.2.9, cubic convergence of certain GR iterations. The proof is the same as the proof of Theorem 5.2.8 (Exercise 5.2.11), except that in this special case the shifts can be shown to differ from eigenvalues of A_i by $O(\epsilon^2)$ instead of $O(\epsilon)$. We assume $d(\mathcal{S}_i, \mathcal{U}) = \epsilon$, as before.

(a) Show that the hypothesis $\| A_{12} \|_F = \| A_{21} \|_F$ implies that $\| A_{12}^{(i)} \| \leq \sqrt{k} \| A_{21}^{(i)} \|$ (cf. Exercise 1.2.10).

(b) Use the Bauer–Fike theorem to show that there is a constant M_2, independent of i, such that the eigenvalues of $B_i = \begin{bmatrix} A_{11}^{(i)} & 0 \\ 0 & A_{22}^{(i)} \end{bmatrix}$ differ from those of A_i by less than $M_2\epsilon$. Show that if ϵ is small enough, no two eigenvalues of B_i are within $M_2\epsilon$ of the same eigenvalue of A_i.

(c) Let $\mu_1, \ldots, \mu_k$ and $\mu_{k+1}, \ldots, \mu_n$ denote the eigenvalues of $A_{11}^{(i)}$ and $A_{22}^{(i)}$, respectively, ordered so that $|\lambda_j - \mu_j| \leq M_2\epsilon$ for $j = 1, \ldots, n$. Let γ be as defined in part (e) of Exercise 5.2.11, and suppose $M_2\epsilon \leq \gamma$. Use the lemma from Exercise 5.2.12 with $B = B_i$ and $E = A_i - B_i$ to show that for $j = k + 1, \ldots, n$,

$$|\mu_j - \lambda_j| \leq \gamma^{-1} \kappa(C_1) \kappa(C_2) \| A_{21}^{(i)} \| \| A_{12}^{(i)} \|,$$

where $C = \text{diag}\{C_1, C_2\}$ is a matrix that diagonalizes B_i. Certainly such a C exists. Furthermore, Theorems 3 and 5 of [190] guarantee that $\kappa(C_1)\kappa(C_2)$ can be bounded above independently of i.

(d) Deduce that there is a constant M_5 such that

$$|\mu_j - \lambda_j| \leq M_5\epsilon^2, \qquad j = k + 1, \ldots, n.$$

(e) Now proceed as in Exercise 5.2.11 to show that there is a constant M such that $d(\mathcal{S}_{i+1}, \mathcal{U}) \leq M\epsilon^3$.

Chapter 6

The Generalized Eigenvalue Problem

Many eigenvalue problems that arise in applications are most naturally formulated as *generalized* eigenvalue problems

$$Av = \lambda Bv,$$

where A and B are $n \times n$ matrices. In this chapter we will discuss how the standard and generalized eigenvalue problems are similar and how they are different. We will introduce GZ algorithms, generalizations of GR algorithms, for solving the generalized eigenvalue problem, and we will show how GZ algorithms can be implemented by bulge-chasing.

6.1   Introduction

Consider an ordered pair (A, B) of matrices in $\mathbb{C}^{n \times n}$. A nonzero vector $v \in \mathbb{C}^n$ is called an *eigenvector* of the pair (A, B) if there exist $\mu, \nu \in \mathbb{C}$, not both zero, such that

$$\mu Av = \nu Bv. \tag{6.1.1}$$

The scalars μ and ν are not uniquely determined by v, but their ratio is (except in the *singular* case $Av = Bv = 0$, which we will mention again below). If $\mu \neq 0$, (6.1.1) is equivalent to

$$Av = \lambda Bv, \tag{6.1.2}$$

where $\lambda = \nu / \mu$. The scalar λ is then called an *eigenvalue* of the pair (A, B) associated with the eigenvector v. If (6.1.1) holds with $\mu = 0$ and $\nu \neq 0$, then (A, B) is said to have an *infinite eigenvalue*; the eigenvalue of (A, B) associated with v is ∞.

This is the generalized eigenvalue problem. In the case $B = I$ it reduces to the standard eigenvalue problem. The following proposition records some fairly obvious facts.

Proposition 6.1.1. *Let $A, B \in \mathbb{C}^{n \times n}$, and let $\lambda \in \mathbb{C}$ be nonzero.*

   (a) *λ is an eigenvalue of (A, B) if and only if $1/\lambda$ is an eigenvalue of (B, A).*

   (b) *∞ is an eigenvalue of (A, B) if and only if 0 is an eigenvalue of (B, A).*

   (c) *∞ is an eigenvalue of (A, B) if and only if B is a singular matrix.*

(d) *If B is nonsingular, the eigenvalues of (A, B) are exactly the eigenvalues of AB^{-1} and $B^{-1}A$. If v is an eigenvector of (A, B) with associated eigenvalue λ, then v is an eigenvector of $B^{-1}A$ with eigenvalue λ, and Bv is an eigenvector of AB^{-1} with eigenvalue λ.*

The expression $A - \lambda B$ with indeterminate λ is commonly called a *matrix pencil.* The terms "matrix pair" and "matrix pencil" are used more or less interchangeably. For example, if v is an eigenvector of the pair (A, B), we also say that v is an eigenvector of the matrix pencil $A - \lambda B$.

Clearly $\lambda \in \mathbb{C}$ is an eigenvalue of (A, B) if and only if the matrix $A - \lambda B$ is singular, and this is in turn true if and only if

$$\det(\lambda B - A) = 0. \tag{6.1.3}$$

This is the *characteristic equation* of the pair (A, B). From the definition of the determinant it follows easily that the function $\det(\lambda B - A)$ is a polynomial in λ of degree n or less. We call it the *characteristic polynomial* of the pair (A, B).

It can happen that the characteristic polynomial is identically zero. For example, if there is a $v \neq 0$ such that $Av = Bv = 0$, then $(\lambda B - A)v = 0$ for all λ, and $\det(\lambda B - A)$ is identically zero. Every λ is an eigenvalue. The pair (A, B) is called a *singular pair* if its characteristic polynomial is identically zero. Otherwise it is called a *regular pair*. We will focus mainly on regular pairs. An example of a singular pair that does not have a common null vector is given in Exercise 6.1.2. For more on singular pairs see [92, 73, 206, 231].

If B is nonsingular, then the pair (A, B) is certainly regular. $\det(\lambda B - A) = \det(B) \det(\lambda I - B^{-1}A)$, so the characteristic polynomial is the same as that of the matrix $B^{-1}A$, except for a constant factor. The pair (A, B) has n finite eigenvalues, and they are the same as the eigenvalues of $B^{-1}A$. The pair (A, B) is also guaranteed regular if A is nonsingular. An example of a regular pencil for which neither A nor B is nonsingular is given in Exercise 6.1.4.

Two pairs (A, B) and $(\tilde{A}, \tilde{B})$ are *strictly equivalent* if there exist nonsingular matrices U and V such that $A = U\tilde{A}V$ and $B = U\tilde{B}V$. This is the same as to say that $A - \lambda B = U(\tilde{A} - \lambda\tilde{B})V$ for all $\lambda \in \mathbb{C}$. Two pairs that are strictly equivalent have the same eigenvalues, and their eigenvectors satisfy simple relationships (Exercise 6.1.5).

Two pairs are strictly equivalent if and only if they have the same *Kronecker canonical form.* You can read about this generalization of the Jordan canonical form in [92, Ch. XII]. For stable computational techniques, see [206] and [73].

Two pairs (A, B) and $(\tilde{A}, \tilde{B})$ are *strictly unitarily equivalent* if there exist *unitary* matrices U and V such that $A = U\tilde{A}V$ and $B = U\tilde{B}V$.

Theorem 6.1.2 (generalized Schur theorem). *Let A, $B \in \mathbb{C}^{n \times n}$. Then there exist unitary Q, $Z \in \mathbb{C}^{n \times n}$ and upper triangular T, $S \in \mathbb{C}^{n \times n}$ such that $A = QTZ^*$ and $B = QSZ^*$. In other words,*

$$A - \lambda B = Q(T - \lambda S)Z^*.$$

Thus every pair is strictly unitarily equivalent to an upper triangular pair. A proof is worked out in Exercise 6.1.6.

The eigenvalues of an upper triangular pair (S, T) are evident. The characteristic equation is clearly

$$\det(\lambda T - S) = \prod_{j=1}^{n}(\lambda t_{jj} - s_{jj}) = 0.$$

If $t_{jj} = 0$ and $s_{jj} = 0$ for some j, the pair (S, T) is singular; otherwise it is regular (Exercise 6.1.7). In the regular case, each factor for which $t_{jj} \neq 0$ yields a finite eigenvalue $\lambda_j = s_{jj}/t_{jj}$. If there is a factor with $t_{jj} = 0$ (and hence $s_{jj} \neq 0$), then ∞ is an eigenvalue. We define the algebraic multiplicity of the infinite eigenvalue to be the number of factors $\lambda t_{jj} - s_{jj}$ for which $t_{jj} = 0$. With this convention, each regular pair has exactly n eigenvalues, some of which may be infinite.

The proof of Theorem 6.1.2 outlined in Exercise 6.1.6 is nonconstructive; it gives no clue how to compute the triangular pair (S, T) without knowing some eigenvectors in advance. Practical computational procedures will be developed in Sections 6.2 and 6.3.

We have tipped our hand about how we intend to attack the generalized eigenvalue problem. Another method that suggests itself when B is nonsingular is to form either AB^{-1} or $B^{-1}A$ and solve the standard eigenvalue problem for that matrix. As it turns out, this approach is usually inadvisable, as accuracy can be lost through the inversion of B and through formation of the product.

Example 6.1.3. *Here is a small MATLAB example in which a superior outcome is obtained by keeping A and B separate.*

```
format long e
randn('state',123);
[M,R] = qr(randn(2));
[N,R] = qr(randn(2));
A = [ 2 1; 0 1e-8];   B = [ 1 1; 0 1e-8];
A = M*A*N; B = M*B*N;
Smart_Result = eig(A,B)
Other_Result = eig(A*inv(B))
```

This code produces a 2×2 matrix pair (A, B) for which the eigenvalues are $\lambda_1 = 2$ and $\lambda_2 = 1$. B is ill conditioned, as is A. The eigenvalues are computed two different ways. The command eig(A,B) *uses the QZ algorithm (Section 6.3), which works with A and B separately to compute the eigenvalues. The command* eig(A*inv(B)) *obviously computes AB^{-1} and then finds the eigenvalues. Here are the results:*

```
Smart_Result =

    1.999999999475219e+00
    1.000000000557143e+00


Other_Result =

    1.997208963148296e+00
    1.002791036386043e+00
```

We see that the QZ computation got the eigenvalues correct to about ten decimal places. The fact that it did not do better can be attributed to the ill conditioning of the eigenvalues. As we see, the computation that formed AB^{-1} explicitly did much worse.

Exercises

6.1.1. Prove Proposition 6.1.1.

6.1.2. Let

$$A = \begin{bmatrix} 1 & 0 & 0 \\ 0 & 0 & 0 \\ 0 & 0 & 1 \end{bmatrix}$$

and

$$B = \begin{bmatrix} 0 & 1 & 0 \\ 0 & 0 & 1 \\ 0 & 0 & 0 \end{bmatrix}.$$

(a) Show that $\det(\lambda B - A)$ is identically zero. Thus (A, B) is a singular pair.

(b) Show that there is no nonzero v such that $Av = Bv = 0$.

(c) For each $\lambda \in \mathbb{C}$, find a nonzero v (depending on λ) such that $(\lambda B - A)v = 0$.

6.1.3. Show that if A is nonsingular, then the pair (A, B) is regular.

6.1.4. Let $A = \begin{bmatrix} 1 & 0 \\ 0 & 0 \end{bmatrix}$ and $B = \begin{bmatrix} 0 & 0 \\ 0 & 1 \end{bmatrix}$. Compute the characteristic polynomial of (A, B). Show that (A, B) is a regular pair with eigenvalues 0 and ∞.

6.1.5. Let (A, B) and $(\tilde{A}, \tilde{B})$ be equivalent pairs. Let U and V be nonsingular matrices such that $A - \lambda B = U(\tilde{A} - \lambda \tilde{B})V$.

(a) Show that (A, B) and $(\tilde{A}, \tilde{B})$ have the same characteristic polynomial to within a constant. Thus they have the same eigenvalues.

(b) Show that x is an eigenvector of (A, B) associated with eigenvalue λ if and only if Vx is an eigenvector of $(\tilde{A}, \tilde{B})$ associated with λ.

(c) A nonzero vector y^T is called a *left eigenvector* of (A, B) associated with the eigenvalue λ if $y^T A = \lambda y^T B$. Show that y^T is a left eigenvector of (A, B) if and only if $y^T U$ is a left eigenvector of $(\tilde{A}, \tilde{B})$.

6.1.6. This exercise proves the generalized Schur theorem (Theorem 6.1.2) by induction on n. Clearly the result holds when $n = 1$. Now we prove that it holds for matrices of size $n \times n$, assuming that it holds for $(n - 1) \times (n - 1)$ matrices.

(a) The fundamental task is to produce two vectors $x, y \in \mathbb{C}^n$ such that $\| x \| = 1$, $\| y \| = 1$, $Ax = \mu y$, $Bx = \nu y$, for some $\mu, \nu \in \mathbb{C}$. First assume that B is nonsingular, and let x be an eigenvector of $B^{-1}A$: $B^{-1}Ax = \lambda x$, with $\| x \| = 1$. Now take y to be an appropriate multiple of Bx, and show that this x and y do the job for some μ and ν satisfying $\lambda = \nu/\mu$.

(b) Now suppose that B is singular. Then there is a vector x with $\| x \| = 1$ and $Bx = 0$. Consider two cases: $Ax \neq 0$ (infinite eigenvalue) and $Ax = 0$ (singular pencil). Show that in each case there is a vector y satisfying $\| y \| = 1$ and scalars μ and ν such that $Ax = \mu y$ and $Bx = \nu y$.

(c) Let $Z_1 \in \mathbb{C}^{n \times n}$ and $Q_1 \in \mathbb{C}^{n \times n}$ be unitary matrices with x and y as first column, respectively. Show that $Q_1^* A Z_1$ and $Q_1^* B Z_1$ are both block triangular:

$$
Q_1^*(A - \lambda B)Z_1 =
\begin{bmatrix}
\mu & * & \cdots & * \\
0 & & & \\
\vdots & & \hat{A} & \\
0 & & &
\end{bmatrix}
- \lambda
\begin{bmatrix}
\nu & * & \cdots & * \\
0 & & & \\
\vdots & & \hat{B} & \\
0 & & &
\end{bmatrix}.
$$

Here you make essential use of the fact that Q_1 is unitary.

(d) Finish the proof of the theorem by induction on n.

6.1.7. Let (S, T) be a pair in which both S and T are upper triangular.

(a) Show that $\det(\lambda T - S) = \prod_{j=1}^{n}(\lambda t_{jj} - s_{jj})$.

(b) Show that the pair is regular if and only if there is no j for which $s_{jj} = t_{jj} = 0$.

(c) Suppose that the pair (S, T) is regular and has ∞ as an eigenvalue. Show that the algebraic multiplicity of the eigenvalue ∞ (the number of j for which $t_{jj} = 0$) is equal to the algebraic multiplicity of 0 as an eigenvalue of (T, S).

(d) Give an example of a pair (S, T) that is regular but has no finite eigenvalues.

6.2 Reduction to Hessenberg-Triangular Form

According to Theorem 6.1.2, every pair (A, B) is strictly unitarily equivalent to an upper triangular pair (S, T). If we know (S, T), then we know the eigenvalues of (A, B). We now begin to tackle the question of how to compute (S, T). The solution of the standard eigenvalue problem usually begins by reducing the matrix to a condensed form, usually upper Hessenberg form. An analogous procedure works for the generalized eigenvalue problem: We can transform (A, B) to (H, T), where H is upper Hessenberg and T is upper triangular. We call this *Hessenberg-triangular* form.

Theorem 6.2.1. *Let A, $B \in \mathbb{C}^{n \times n}$. Then the pair (A, B) is strictly equivalent to a pair (H, T) that can be computed by a direct algorithm in $O(n^3)$ flops, where H is upper Hessenberg and T is upper triangular. That is, $A = UHV$ and $B = UTV$, where U and V are nonsingular, and U, V, H, and T can be computed in $O(n^3)$ flops. U and V can be taken to be unitary.*

Proof: We will outline a class of algorithms that transform (A, B) to the Hessenberg-triangular form. Each of the matrices U and V will be constructed as a product of elimination matrices. If unitary transformations are desired, each elimination matrix should be taken to be unitary.

The first step is to transform B immediately to triangular form. Let $B = GR$ be any decomposition of B for which G is nonsingular and R is upper triangular. (If you insist on a unitary transformation, do a QR decomposition.) As we know, this can be done in $O(n^3)$ flops. Then the pair $(G^{-1}A, R)$ is strictly equivalent to (A, B). The computation of $G^{-1}A$ also costs $O(n^3)$ flops. G will become a factor in the transforming matrix U.

In light of the previous paragraph, let us now consider a pair (A, B) in which B is already upper triangular. The following algorithm transforms A to upper Hessenberg form while defending the upper triangular form of B. In what follows, the symbols A and B will refer to the transformed matrices that are on their way to becoming H and T. We start with an elimination matrix (e.g., a rotator) that acts on rows $n - 1$ and n, which transforms $a_{n,1}$ to 0. This is a transformation on the left; it will become a part of U. It must also be applied to B, and when it is, the upper triangular form is disturbed: $b_{n,n-1}$ becomes nonzero. We return this entry to zero by applying an elimination matrix to columns $n - 1$ and n. This is a right transformation; it will become a part of V. This transformation must also be applied to A. It affects columns $n - 1$ and n and does not disturb the zero that was introduced at position $(n, 1)$.

The next step is to apply an elimination matrix acting on rows $n - 2$ and $n - 1$ to transform $a_{n-1,1}$ to zero. When this transformation is applied to B, it disturbs the triangular form, creating a nonzero in position $b_{n-1,n-2}$. This can be eliminated by a transformation on the right that acts on columns $n - 2$ and $n - 1$. When this transformation is applied to A, it does not disturb the zeros that were introduced previously.

Continuing in this manner, we transform the entries $a_{n-2,1}, \ldots, a_{31}$ to 0 one by one. This is as far as we can go in column one. If we try to transform a_{21} to 0, we will end up destroying all the zeros that we previously created. (We cannot violate Galois theory.) We move on to the second column and transform $a_{n,2}, a_{n-1,2}, \ldots, a_{42}$ to 0. Then we move on to the third column, the fourth column, and so on. In the end, A will have been transformed to upper Hessenberg form, and B will still be upper triangular. It is easy to check that the total computational effort is $O(n^3)$ (Exercise 6.2.1).   □

Second Reduction Algorithm

The proof of Theorem 6.2.1 outlines a direct method for transforming a matrix pair to Hessenberg-triangular form. We now develop a second procedure that is less efficient in general but is useful in special cases. We will use it for bulge-chasing in Section 6.3.

Unlike the first method, the second method does not begin by transforming B to upper triangular form. Instead it transforms A and B to Hessenberg and triangular form together. Transformations on the right (resp., left) are used to create zeros in B (resp., A). The first step creates the desired zeros in the first columns of A and B, beginning with B. Needed is a transformation $BH^* = \tilde{B}$ such that $\tilde{b}_{21} = \cdots = \tilde{b}_{n1} = 0$. This is a different sort of elimination than those we studied in Section 3.1, but fortunately it is not hard to accomplish, at least in theory. In fact, it can be seen as an elimination on the first column of B^{-1} in the sense of Section 3.1 (Exercise 6.2.2). Fortunately the elimination can be done even if

B^{-1} does not exist. The conditions $\tilde{b}_{21} = \cdots = \tilde{b}_{n1} = 0$ just mean that the first row of H must be orthogonal to the second through nth rows of B, so we need to build an H with this property. A stable way to do this is to compute a decomposition $B = RQ$, where R is upper triangular and Q is unitary. This can be done by a process analogous to that of computing a QR decomposition (Exercise 6.2.3). Then the equation $BQ^* = R$ (first column) shows that the first row of Q is orthogonal to rows 2 through n of B. Now we just need to take H to be a matrix whose first row is proportional to the first row of Q. We could take H to be Q itself, but in general Q is quite an expensive transformation.[1] We prefer to build a cheap transformation to do the same task. This is just a matter of building an H^* with a specified first column. We have seen in Section 3.1 how to do this economically. Once we have H, we can make the updates $B \leftarrow BH^*$, $A \leftarrow AH^*$, creating the desired zeros in B.

The second part of the first step creates zeros in the first column of A in positions $(3, 1), \ldots, (n, 1)$. This is a routine elimination that can be done as described in Section 3.1. The elimination matrix has the form $G^{-1} = \mathrm{diag}\{1, \tilde{G}^{-1}\}$. The transformation $A \leftarrow G^{-1}A$, $B \leftarrow G^{-1}B$ creates the desired zeros in the first column of A without disturbing the zeros that were previously created in B. This completes the first step, which can be summarized as follows.

Second reduction to Hessenberg-triangular form (first step)

$$
\begin{bmatrix}
\text{Compute decomposition } B = RQ. \\
\text{Compute elimination matrix } H \text{ with } H^*e_1 = \alpha Q^*e_1. \\
A \leftarrow AH^* \\
B \leftarrow BH^* \quad \text{(creating zeros in } B) \\
\text{Let } x = [a_{21}\ a_{31}\ \cdots\ a_{n1}]^T \\
\text{Compute } \tilde{G} \text{ such that } \tilde{G}e_1 = x \\
\text{let } G = \mathrm{diag}\{1, \tilde{G}\} \\
A \leftarrow G^{-1}A \quad \text{(creating zeros in } A) \\
B \leftarrow G^{-1}B
\end{bmatrix}
\tag{6.2.1}
$$

The second step introduces zeros in the second columns of A and B. It is identical to the first step, except that its action is restricted to the submatrix obtained by ignoring the first row and column of each matrix. One easily checks that the operations of the second step do not disturb the zeros that were created in the first step. The third step operates on even smaller submatrices, and so on. After $n-1$ steps, the pair has been transformed to Hessenberg-triangular form. The final step creates a single zero in B and no zeros in A.

This algorithm is uneconomical in general. If it is implemented in the obvious way, there is an RQ decomposition costing $O(n^3)$ work at each step. Thus the total work will be $O(n^4)$.[2]

---

[1]The first reduction process that we described began with a QR decomposition of B, but it could just as well have begun with an RQ decomposition. Now, in our second reduction process, we are proposing to start with an RQ decomposition of B. Since we are going to all that expense, why don't we just use the RQ decomposition to triangularize B and then proceed as in the first reduction? Good question! The answer is that we are just developing an approach here that is not economical in general but will be useful later when we talk about bulge-chasing.

[2]There is a way to bring the flop count down to $O(n^3)$. A new RQ decomposition does not need to be computed from scratch at each step. Instead it can be obtained by an $O(n^2)$ updating procedure from the RQ decomposition from the previous step [134]. The resulting algorithm seems still to be somewhat more expensive than the first reduction algorithm that we described.

Exercises

6.2.1. This exercise checks out some of the details of the reduction to Hessenberg-triangular form (Theorem 6.2.1).

   (a) Check that the algorithm sketched in the proof of Theorem 6.2.1 succeeds in transforming A to upper Hessenberg form. In particular, none of the right transformations, which are required to preserve the triangular form of B, destroys any of the zeros that have been previously introduced into A.

   (b) Show that the total number of elimination matrices that are used (not counting the preliminary reduction of B to triangular form) is approximately n^2.

   (c) Show that each of the transformations requires $O(n)$ (or less) computational effort. Thus the total computational effort to produce H and T is $O(n^3)$.

   (d) Describe how the transforming matrices U and V can be accumulated. Show that the total work to construct U and V is $O(n^3)$.

6.2.2.

   (a) Suppose that $T \in \mathbb{C}^{n \times n}$ is nonsingular and block upper triangular:

$$T = \begin{bmatrix} T_{11} & T_{12} \\ 0 & T_{22} \end{bmatrix},$$

where T_{11} is square. Show that T^{-1} has the same block triangular structure as T. You will probably find it useful to apply this fact to the matrix $\tilde{B}$ in part (c) of this exercise.

   (b) Give formulas for the blocks of T^{-1}.

   (c) Suppose that $B \in \mathbb{C}^{n \times n}$ is nonsingular, let Z be a nonsingular matrix, and let $\tilde{B} = BZ$. Show that $\tilde{b}_{21} = \cdots = \tilde{b}_{n1} = 0$ if and only if Z^{-1} is an elimination matrix for the first column of B^{-1}.

6.2.3. Outline an algorithm that computes a decomposition $B = RQ$, where R is upper triangular and Q is unitary. Use unitary elimination matrices to transform B to upper triangular form. Start with one that produces the desired zeros in the bottom row of R. Show that this decomposition takes the same amount of work as a QR decomposition.

6.2.4. Let B, $H \in \mathbb{C}^{n \times n}$, and suppose $BH^* = \tilde{B}$. Let $b_1, \ldots, b_n$ and $h_1, \ldots, h_n$ denote the *rows* of B and G, respectively.

   (a) Show that the conditions $\tilde{b}_{21} = \cdots = \tilde{b}_{n1} = 0$ just mean that h_1 is orthogonal to $b_2, \ldots, b_n$.

   (b) Suppose $B = RQ$, where R is upper triangular and Q is unitary with *rows* $q_1, \ldots, q_n$. Show that q_1 is orthogonal to $b_2, \ldots, b_n$.

6.2.5. Check that the second and subsequent steps of the algorithm outlined in (6.2.1) do not disturb the zeros that were created in the earlier steps.

6.3 *GZ* Algorithms

Consider a pair (A, B) for which B is nonsingular. In this section we describe a family of iterative methods called GZ algorithms that transform A and B to triangular form. For our initial description it is crucial that B be nonsingular, but later on we will see that nothing bad happens if B is singular. We draw substantially from [228].

Before we begin our iterations of the GZ algorithm, we transform the pair to some initial form

$$A_0 = G_0^{-1} A Z_0, \qquad B_0 = G_0^{-1} B Z_0.$$

For example, (A_0, B_0) could be in Hessenberg-triangular form. Later on we will assume that (A_0, B_0) does have this form, but for now we allow them to have any form. For example, we could take $G_0 = Z_0 = I$.

The explicit form of the generic GZ algorithm can be described as follows. The ith iteration transforms (A_{i-1}, B_{i-1}) to (A_i, B_i) by a strict equivalence transformation determined by A_{i-1}, B_{i-1}, a *shift polynomial* p_i, and two GR decompositions. Specifically, let G_i and Z_i be nonsingular matrices, and let R_i and S_i be upper triangular matrices such that

$$p_i(A_{i-1}B_{i-1}^{-1}) = G_i R_i \quad \text{and} \quad p_i(B_{i-1}^{-1}A_{i-1}) = Z_i S_i. \tag{6.3.1}$$

Then let

$$A_i = G_i^{-1} A_{i-1} Z_i \quad \text{and} \quad B_i = G_i^{-1} B_{i-1} Z_i. \tag{6.3.2}$$

In the special case $B_{i-1} = I$, $Z_i = G_i$, $S_i = R_i$, this algorithm reduces to the GR algorithm (4.3.1).

An easy computation shows that

$$A_i B_i^{-1} = G_i^{-1}(A_{i-1}B_{i-1}^{-1})G_i.$$

Since G_i was obtained from the decomposition $p_i(A_{i-1}B_{i-1}^{-1}) = G_i R_i$, we see that the transformation $A_{i-1}B_{i-1}^{-1} \to A_i B_i^{-1}$ is an iteration of the GR algorithm (4.3.1). At the same time we have

$$B_i^{-1}A_i = Z_i^{-1}(B_{i-1}^{-1}A_{i-1})Z_i,$$

where $p_i(B_{i-1}^{-1}A_{i-1}) = Z_i S_i$, so the transformation $B_{i-1}^{-1}A_{i-1} \to B_i^{-1}A_i$ is also a GR iteration.

Since the GZ algorithm automatically effects GR iterations on AB^{-1} and $B^{-1}A$, we expect the sequences $(A_i B_i^{-1})$ and $(B_i^{-1}A_i)$ to converge to (block) triangular form. If the shifts are chosen well, we expect quadratic convergence. In [228] it is shown that under suitable conditions the sequences (A_i) and (B_i) separately converge to (block) triangular form, exposing the eigenvalues. With a suitable choice of shifts, quadratic convergence can be obtained. The developments of Chapter 8 show that, at least in the Hessenberg-triangular case, convergence of the GZ algorithm follows immediately from the convergence of the GR algorithm.

The generic GZ algorithm is actually a large class of algorithms. Specific instances are obtained by specifying the exact form of the GR decompositions and how the p_i are to be chosen. The choice of p_i is generally the same as it is for GR iterations: The zeros of p_i, which are called the *shifts* for the ith iteration, should be estimates of eigenvalues. The degree of p_i is called the *degree* of the ith GZ iteration. If the G_i and Z_i are always taken to be unitary, implying that the decompositions in (6.3.1) are (more or less unique) QR decompositions, the algorithm is called the QZ algorithm [156]. If the G_i and Z_i are lower triangular, it is an LZ algorithm. If they are products of lower triangular and permutation matrices, it is an LZ algorithm with pivoting [129].

The Generic Bulge-Chasing Algorithm

If we execute the generic GZ algorithm as shown in (6.3.1) and (6.3.2), each step will be very expensive. We have to invert B_{i-1}, multiply it by A_{i-1} in both orders, and compute $p_i(A_{i-1}B_{i-1}^{-1})$ and $p_i(B_{i-1}^{-1}A_{i-1})$. These are all expensive operations, as are the two GR decompositions that follow. It is imperative that we find a more efficient way to do these iterations. We do not want to do any of the operations explicitly; in particular, we want to avoid inverting B_{i-1}. In the end we hope to have an algorithm that works well even if B_{i-1} does not have an inverse. To this end we introduce the generic bulge-chasing algorithm.

Since we are now going to consider just a single iteration, we drop the subscripts (so that we can reintroduce subscripts for a different purpose) and consider a single iteration that starts with a pair (A, B) and ends with a strictly equivalent pair $(\hat{A}, \hat{B})$ via

$$p(AB^{-1}) = GR, \qquad p(B^{-1}A) = ZS \qquad (6.3.3)$$

and

$$\hat{A} = G^{-1}AZ, \qquad \hat{B} = G^{-1}BZ. \qquad (6.3.4)$$

At this point we will assume that A is upper Hessenberg and B is upper triangular. The iteration will preserve the Hessenberg-triangular form. We may assume further that A is a proper upper Hessenberg matrix, since otherwise the eigenvalue problem can be decoupled to produce two or more smaller eigenvalue problems.

We will begin by describing the generic bulge-chasing algorithm. Then we will show that it effects an iteration of the generic GZ algorithm. (Thus the generic bulge-chasing algorithm is also known as an *implicit* GZ algorithm.) To get the algorithm started, we just need a vector proportional to the first column of $p(AB^{-1})$. This can be computed inexpensively if, as we shall assume, the degree of p (call it m) satisfies $m \ll n$. Since A is upper Hessenberg, so is AB^{-1}. Thus $p(AB^{-1})$ is m-Hessenberg, which implies that only the first $m + 1$ entries of the first column of $p(AB^{-1})$ can be nonzero. Let $\rho_1, \ldots, \rho_m$ denote the zeros of p (the *shifts*). Then the first column of $p(AB^{-1})$ is

$$p(AB^{-1})e_1 = (AB^{-1} - \rho_m I)(AB^{-1} - \rho_{m-1}I) \cdots (AB^{-1} - \rho_1 I)e_1,$$

so a vector x proportional to the first column of $p(AB^{-1})$ can be computed by the following algorithm:

$$\begin{array}{l} x \leftarrow e_1 \\ \text{for } k = 1, \ldots, m \\ \left[\begin{array}{l} y \leftarrow B^{-1}x \\ x \leftarrow \alpha_k(Ay - \rho_k x) \end{array} \right. \end{array} \qquad (6.3.5)$$

Here α_k can be any convenient scaling factor. The step $y \leftarrow B^{-1}x$ looks like it might be expensive, but fortunately it is not. For example, consider what happens on the first time through the loop. Initially we have $x = e_1$, so $B^{-1}x = b_{11}^{-1}e_1$, because B is upper triangular. The operation $x \leftarrow \alpha_1(Ay - \rho_1 x)$ results in a new x whose first two components are (potentially) nonzero, all other components being zero. On the second time through the loop, the operation $y \leftarrow B^{-1}x$ results in a y whose first two components are the only ones that can be nonzero because B^{-1} is upper triangular. The operation $x \leftarrow \alpha_2(Ay - \rho_2 x)$ then results in an x that can be nonzero only in the first three components. Each time through the loop adds one more nonzero to x. On the kth time through the loop, the current x has zeros everywhere except in the first k positions. The operation $y \leftarrow B^{-1}x$ results in a y that is also zero everywhere except in the first k positions. y can be computed in $O(k^2)$ flops by back substitution (Exercise 6.3.1). The operation $x \leftarrow \alpha_k(Ay - \rho_k x)$ also costs $O(k^2)$ flops and results in an x that is zero except in the first $k + 1$ components. After m steps, the final x is nonzero only in the first $m + 1$ components. The total cost of producing x is $O(m^3)$, which is small if $m \ll n$.

Once x has been produced, we build an elimination matrix G_1 such that $G_1 e_1 = \beta_1 x$. Because of the form of x, G_1 can take the form

$$G_1 = \begin{bmatrix} \tilde{G}_1 & \\ & I_{n-m-1} \end{bmatrix}.$$

Let

$$A_{1/2} = G_1^{-1}A \quad \text{and} \quad B_{1/2} = G_1^{-1}B.$$

The action of G_1^{-1} affects the first $m + 1$ rows of A and B, disturbing the Hessenberg-triangular form. For example, in the case $m = 2$, $A_{1/2}$ and $B_{1/2}$ have the forms

$$\begin{bmatrix} a & a & a & a & a \\ a & a & a & a & a \\ + & a & a & a & a \\ & & a & a & a \\ & & & a & a \end{bmatrix} \quad \text{and} \quad \begin{bmatrix} b & b & b & b & b \\ + & b & b & b & b \\ + & + & b & b & b \\ & & & b & b \\ & & & & b \end{bmatrix},$$

respectively. The plus symbols indicate bulges that have crept into the matrices.

The rest of the iteration consists of returning the pair to Hessenberg-triangular form using the second algorithm discussed in Section 6.2, whose first step is sketched in (6.2.1). As we shall see, this amounts to a bulge-chase. The first task is to return the first column of $B_{1/2}$ to upper triangular form. From (6.2.1) we recall that this requires an RQ decomposition of $B_{1/2}$, which is normally an expensive process. In this case it is not, because $B_{1/2}$ is nearly triangular to begin with. We can write $B_{1/2}$ in the partitioned form

$$B_{1/2} = \begin{bmatrix} B_{11} & B_{12} \\ & B_{22} \end{bmatrix},$$

where B_{11} is $(m + 1) \times (m + 1)$ and B_{22} is upper triangular. If $B_{11} = R_{11}Q_{11}$ is an RQ decomposition of B_{11}, then an RQ decomposition of $B_{1/2}$ is given by

$$B_{1/2} = \begin{bmatrix} B_{11} & B_{12} \\ & B_{22} \end{bmatrix} = \begin{bmatrix} R_{11} & B_{12} \\ & B_{22} \end{bmatrix} \begin{bmatrix} Q_{11} & \\ & I \end{bmatrix}.$$

Thus the decomposition costs only $O(m^3)$ flops, which is negligible if m is small. Let $\tilde{Z}_1 \in \mathbb{C}^{(m+1)\times(m+1)}$ be an elimination matrix such that the first row of $\tilde{Z}_1^*$ is proportional to the first row of Q_{11}, let $Z_1 = \text{diag}\{\tilde{Z}_1, I_{n-m-1}\}$, and let

$$A_1 = A_{1/2}Z_1, \qquad B_1 = B_{1/2}Z_1.$$

This completes the first part of (6.2.1). The first column of B_1 has zeros below the main diagonal. The transforming matrix Z_1 acts on the first $m+1$ columns of $A_{1/2}$, with the result that one row is added to the bottom of the bulge. In the case $m = 2$, A_1 and B_1 have the forms

$$
\begin{bmatrix}
a & a & a & a & a \\
a & a & a & a & a \\
+ & a & a & a & a \\
+ & + & a & a & a \\
  &   &   & a & a
\end{bmatrix}
\quad \text{and} \quad
\begin{bmatrix}
b & b & b & b & b \\
  & b & b & b & b \\
  & + & b & b & b \\
  &   &   & b & b \\
  &   &   &   & b
\end{bmatrix},
$$

respectively. The bulge in B has been shrunk at the expense of expanding the bulge in A. The second half of (6.2.1) transforms the first column of A_1 to Hessenberg form. This is done by a transformation

$$A_{3/2} = G_2^{-1}A_1, \qquad B_{3/2} = G_2^{-1}B_1,$$

where $G_2 = \text{diag}\{1, \tilde{G}_2, I_{n-m-2}\}$. The transformation $B_{3/2} = G_2^{-1}B_1$ acts on rows 2 through $m+2$ of B, adding a row to the bulge. In the case $m = 2$, $A_{3/2}$ and $B_{3/2}$ have the forms

$$
\begin{bmatrix}
a & a & a & a & a \\
a & a & a & a & a \\
  & a & a & a & a \\
  & + & a & a & a \\
  &   &   & a & a
\end{bmatrix}
\quad \text{and} \quad
\begin{bmatrix}
b & b & b & b & b \\
  & b & b & b & b \\
  & + & b & b & b \\
  & + & + & b & b \\
  &   &   &   & b
\end{bmatrix},
$$

respectively. The bulge in A has been shrunk and the bulge in B expanded. Both bulges have now been moved down one row and over one column. The pattern should now be clear. The next transformation, Z_2, clears out the second column of $B_{3/2}$. This requires another RQ decomposition, but this is again inexpensive, because the part of $B_{3/2}$ that is not already upper triangular is only $(m+1) \times (m+1)$. This acts on columns 2 through $m+2$ of $A_{3/2}$, adding a row to the bottom of the bulge. Next, a transformation G_3^{-1} clears out the second column of A_2, adding a row to the bottom of the bulge in B_2. Continuing in this manner, we chase the bulges downward and eventually off of the bottom of the matrices, returning the pair to Hessenberg-triangular form, at which point we have completed an iteration of the generic bulge-chasing algorithm. This takes a total of $n-1$ transformations on left and right. The result is

$$\hat{A} = A_{n-1} = G_{n-1}^{-1}\cdots G_1^{-1}AZ_1\cdots Z_{n-1},$$

$$\hat{B} = B_{n-1} = G_{n-1}^{-1}\cdots G_1^{-1}BZ_1\cdots Z_{n-1}.$$

In the case $B = I$ and $Z_i = G_i$ for all i, this algorithm reduces to the generic bulge-chasing algorithm for the standard eigenvalue problem described in Section 4.5.

Letting

$$G = G_1G_2\cdots G_{n-1} \quad \text{and} \quad Z = Z_1Z_2\cdots Z_{n-1}, \tag{6.3.6}$$

we have

$$\hat{A} = G^{-1}AZ \quad \text{and} \quad \hat{B} = G^{-1}BZ,$$

as in (6.3.4). To show that the generic bulge-chasing algorithm effects a GZ iteration, we must show that the matrices G and Z of (6.3.6) satisfy equations of the form (6.3.3). To this end we begin by noting that the form of the matrices $G_2, \ldots, G_{n-1}$ implies that the first column of G is the same as the first column of G_1 (Exercise 6.3.2). Thus the first column of G is proportional to the first column of $p(AB^{-1})$; that is, $Ge_1 = \alpha p(AB^{-1})e_1$ for some $\alpha \neq 0$. The following proposition then gives us the result.

Proposition 6.3.1. *Suppose the matrix pair (A, B) is properly Hessenberg-triangular; that is, A is properly upper Hessenberg and B is upper triangular and nonsingular. Suppose further that*

$$\hat{A} = G^{-1}AZ \quad \text{and} \quad \hat{B} = G^{-1}BZ,$$

where G and Z are nonsingular, and that the pair $(\hat{A}, \hat{B})$ is also Hessenberg-triangular. Finally, suppose $Ge_1 = \alpha p(AB^{-1})e_1$ for some $\alpha \neq 0$. Then there exist upper triangular matrices R and S such that

$$p(AB^{-1}) = GR \quad \text{and} \quad p(B^{-1}A) = ZS.$$

Proof: Applying Lemmas 4.5.3 and 4.5.2 with A replaced by AB^{-1}, and recalling that $G(\hat{A}\hat{B}^{-1}) = (AB^{-1})G$, we find that

$$p(AB^{-1})\kappa(AB^{-1}, e_1) = \kappa(AB^{-1}, p(AB^{-1})e_1) \tag{6.3.7}$$
$$= \alpha^{-1}\kappa(AB^{-1}, Ge_1) = \alpha^{-1}G\kappa(\hat{A}\hat{B}^{-1}, e_1).$$

By Lemma 4.5.4, $\kappa(AB^{-1}, e_1)$ and $\kappa(\hat{A}\hat{B}^{-1}, e_1)$ are both upper triangular, and the former is nonsingular. Letting $R = \alpha^{-1}\kappa(\hat{A}\hat{B}^{-1}, e_1)\kappa(AB^{-1}, e_1)^{-1}$, we see that R is upper triangular and, by (6.3.8),

$$p(AB^{-1}) = GR.$$

To get the equation $p(B^{-1}A) = ZS$, we start by figuring out what the first column of Z is. Noting that $Z = B^{-1}G\hat{B}$, we have

$$Ze_1 = B^{-1}G\hat{B}e_1 = \hat{b}_{11}B^{-1}Ge_1 = \hat{b}_{11}\alpha B^{-1}p(AB^{-1})e_1$$
$$= \hat{b}_{11}\alpha p(B^{-1}A)B^{-1}e_1 = \hat{b}_{11}\alpha b_{11}^{-1}p(B^{-1}A)e_1.$$

Thus $Ze_1 = \beta p(B^{-1}A)e_1$, where $\beta = \hat{b}_{11}\alpha b_{11}^{-1}$: The first column of Z is proportional to the first column of $p(B^{-1}A)$. Now we can use Lemmas 4.5.3, 4.5.2, and 4.5.4, just as in the previous paragraph, to show that $p(B^{-1}A) = ZS$, where $Z = \beta^{-1}\kappa(\hat{B}^{-1}\hat{A}, e_1)\kappa(B^{-1}A, e_1)^{-1}$ is upper triangular. $\square$

We have shown that the generic bulge-chasing algorithm satisfies (6.3.4), where G and Z are from GR decompositions (6.3.3). Therefore an iteration of the generic bulge-chasing algorithm effects an iteration of the generic GZ algorithm. It is not hard to show (Exercise 6.3.3) that the cost of a generic bulge-chasing iteration is $O(n^2m)$ if m is small, so this is much less expensive than if we did the generic GZ step explicitly.

A more thorough exposition of the connection between explicit and implicit GZ algorithms is given in [228]. There it is shown, by arguments like those in Section 4.7, that the transformations used in explicit and implicit (bulge-chasing) GZ iterations on a Hessenberg-triangular pair are essentially identical.

Repeatedly applying the generic bulge-chasing algorithm with appropriate shifts, we normally get rapid convergence, which is reflected in convergence of the subdiagonal entries $a_{j+1,j}$ to zero. Monitoring these entries and setting each one to zero once it has become sufficiently small, we repeatedly deflate and decouple the problem until all of the eigenvalues have been found.

If the generic bulge-chasing algorithm is effected using only unitary elimination matrices, it becomes an implicit QZ algorithm, as exemplified by Moler and Stewart's original QZ algorithm [156]. If only Gaussian elimination transformations with permutations are used, it becomes the LZ algorithm with partial pivoting, as exemplified by Kaufman [129]. There is nothing to rule out hybrids that mix unitary and Gauss transforms. For example, in [130], Kaufman proposed an algorithm that uses unitary elimination matrices for the "G" transformations and Gaussian elimination matrices with pivoting for the "Z" transformations.

In Section 4.9, we discussed a number of practical details for the implementation of GR algorithms for the standard eigenvalue problem. Most of these apply to GZ algorithms for the generalized eigenvalue problem as well. Ward [212] showed how to balance the generalized eigenvalue problem. Once balancing has been done, the standard software for general regular pairs uses only unitary transformations in the interest of numerical stability. Shift blurring is a problem if m is made too large. However, large numbers of small bulges can be chased in pipeline fashion to achieve improved performance due to better cache use, resulting in significant performance improvements on large matrices. Performance can also be improved by aggressive early deflation [123]. These innovations will be incorporated in the large-matrix QZ code in a future release of LAPACK.

Exercises

6.3.1. The operation $y \leftarrow B^{-1}x$ in (6.3.5) is effected by solving the system $By = x$. Show that if all but the first k entries of x are zero, then the same will be true of y, y can be computed in $O(k^2)$ flops by back substitution, and only the upper left-hand $k \times k$ submatrix of B participates in the computation. An easy way to work this problem is to partition the equation $By = x$ appropriately.

6.3.2. This exercise addresses some of the details of the generic bulge-chasing algorithm.

(a) Describe the kth step of the generic bulge-chasing algorithm, transforming A_k and B_k to $A_{k+1} = G_k^{-1} A_k Z_k$ and $B_{k+1} = G_k^{-1} B_k Z_k$. What can you say about the form of each of A_k, B_k, G_k, Z_k, A_{k+1}, and B_{k+1}?

(b) Show that the first column of $G = G_1 \cdots G_{n-1}$ is the same as the first column of G_1.

(c) Check the details in the argument that shows that $p(AB^{-1}) = GR$, where R is upper triangular.

(d) Show that $B^{-1}p(AB^{-1}) = p(B^{-1}A)B^{-1}$ for any polynomial p. Using the upper triangularity of B and $\hat{B}$, show that $p(B^{-1}A)e_1 = \delta Z e_1$, where $\delta = \gamma b_{11} \hat{b}_{11}^{-1}$.

(e) Use Lemmas 4.5.3 and 4.5.2, with A replaced by $B^{-1}A$, to show that

$$p(B^{-1}A)\kappa(B^{-1}A, e_1) = \delta Z\kappa(\hat{B}^{-1}\hat{A}, e_1).$$

Then use Lemma 4.5.4 to show that the matrix

$$S = \delta\kappa(\hat{B}^{-1}\hat{A}, e_1)\kappa(B^{-1}A, e_1)^{-1}$$

exists and is upper triangular. Then deduce that $p(B^{-1}A) = ZS$.

6.3.3. Show that the cost of an iteration of the generic bulge-chasing algorithm is $O(n^2m)$ if $m \le \sqrt{n}$ by the following steps:

   (a) Show that the cost of computing the first column of G_1 is $O(m^3) = O(nm)$.

   (b) Show that the cost of setting up and applying each G_k^{-1} transformation is $O(nm)$.

   (c) Show that the cost of setting up and applying each Z_k transformation is $O(nm)$. Do not overlook the cost of the RQ decomposition that is used in the setup phase.

   (d) Show that the entire flop count for an iteration of the generic bulge-chasing algorithm is $O(n^2m)$.

   (e) At what points did you use the assumption $m \le \sqrt{n}$?

6.3.4. An alternate way of implementing the implicit GZ algorithm can be built on the principle of defending the upper triangular form of B vigorously. In this variant, the "A" matrix carries a bulge of the same size as in the variant described in the text, but a bulge is never allowed to build up in the "B" matrix.

   (a) Show that each of the transformations G_k (including G_1) can be constructed as a product $G_k = G_{k1}G_{k2}\cdots G_{km}$ of m transformations, each of which operates on two adjacent rows, such that (for $k > 1$) each G_{ki}^{-1} creates one new zero in the "A" bulge. A_{k1}^{-1} zeros the bottom element, A_{k2}^{-1} zeros the next one, and so on upwards.

   (b) Z_k is created and applied simultaneously with G_k. Supposing that the "B" matrix is upper triangular, show that each G_{ki}^{-1} creates a nonzero in the subdiagonal of B, thus disturbing the upper triangular form. Show that there is a Z_{ki} that acts on two adjacent columns to return B to triangular form. Show that Z_{ki} adds one new entry to the bottom of the bulge in A. Let $Z_k = Z_{k1}\cdots Z_{km}$. This is a bit different from the Z_k of the algorithm described in the text. Let $A_k = G_k^{-1}A_{k-1}Z_k$ and $B_k = G_k^{-1}B_{k-1}Z_k$.

6.4 The *HZ* Algorithm

In Section 4.9 we introduced and briefly discussed the HR algorithm for pseudosymmetric matrices. This can also be formulated as an HZ algorithm for a certain generalized eigenvalue problem. Recall that a matrix $A \in \mathbb{R}^{n \times n}$ that is tridiagonal and pseudosymmetric can be expressed as a product $A = TD$, where T is tridiagonal and symmetric and D is a signature matrix. That is, D is a diagonal matrix whose main diagonal entries are ± 1. Since $D = D^{-1}$, we see that the eigenvalue problem for $A = TD$ is the same as the generalized eigenvalue problem for the pair (T, D).

The Explicit *HZ* Algorithm

Recall that an HR iteration on a pseudosymmetric matrix $A = TD$ is a special kind of GR iteration of the form

$$p(A) = HR, \qquad \hat{A} = H^{-1}AH,$$

where H satisfies $H^T DH = \hat{D}$, with $\hat{D}$ another signature matrix. One easily checks (Exercise 4.9.7) that $\hat{A} = \hat{T}\hat{D}$, where $\hat{T}$ is symmetric and tridiagonal.

Let us now see that we can formulate this as an HZ algorithm, a special structure-preserving GZ algorithm, on the pair (T, D). Referring back to (6.3.1) or (6.3.3), we see that an HZ iteration on (T, D) should begin with a pair of HR decompositions

$$p(TD^{-1}) = HR, \qquad p(D^{-1}T) = \tilde{H}\tilde{R}.$$

These are not independent; each can be derived from the other. Suppose that we have a decomposition $p(TD^{-1}) = HR$, where R is upper triangular and $H^T DH = \hat{D}$ for some signature matrix $\hat{D}$. Then $p(D^{-1}T) = D^{-1}p(TD^{-1})D = D^{-1}HRD = (DH\hat{D})(\hat{D}RD)$. Let $\tilde{H} = DH\hat{D}$ and $\tilde{R} = \hat{D}RD$. Then clearly $\tilde{R}$ is upper triangular, and $\tilde{H}$ satisfies $\tilde{H}^T D\tilde{H} = \hat{D}$. (See Exercise 6.4.1 for this and other details.) Thus the decomposition $p(D^{-1}T) = \tilde{H}\tilde{R}$ is an HR decomposition. It is easy to check that $\tilde{H} = H^{-T}$, so the decomposition can be written as $p(D^{-1}T) = H^{-T}\tilde{R}$. Thus we can take the HR version of (6.3.3) as

$$p(TD^{-1}) = HR, \qquad p(D^{-1}T) = H^{-T}\tilde{R}. \tag{6.4.1}$$

Then the HR version of (6.3.4) is

$$\hat{T} = H^{-1}TH^{-T}, \qquad \hat{D} = H^{-1}DH^{-T}. \tag{6.4.2}$$

If we invert the second equation here, we recognize it as a variant of the equation $\hat{D} = H^T DH$. Also note that the first equation guarantees that $\hat{T}$ is symmetric. Equations (6.4.1) and (6.4.2) together constitute an iteration of the HZ algorithm. Since $\hat{T}\hat{D}^{-1} = H^{-1}(TD^{-1})H$, the HZ iteration effects an HR iteration on $A = TD^{-1}$.

The Implicit *HZ* Algorithm

Now let us consider how the HZ iteration can be implemented implicitly, as a bulge-chasing algorithm. For simplicity consider first an iteration of degree $m = 1$. The iteration begins by computing a vector x as in (6.3.5). In this special case x is proportional to $(T - \rho D)e_1$, where ρ is the single shift. The first transformation matrix H_1 will satisfy $H_1 e_1 = \beta_1 x$ for some convenient nonzero scalar β_1. Only the first two entries of x are nonzero, so the first transformation acts only on rows 1 and 2. The step

$$T_{1/2} = H_1^{-1}T, \qquad D_{1/2} = H_1^{-1}D$$

creates a bulge in $D_{1/2}$. From the description of the generic bulge-chasing algorithm we know that the next transformation (Z_1) is supposed to remove the bulge from $D_{1/2}$. On the other hand, (6.4.2) suggests that Z_1 should be H_1^{-T}, giving

$$T_1 = H_1^{-1}TH_1^{-T}, \qquad D_1 = H_1^{-1}DH_1^{-T}.$$

At this point the bulge should be removed from D_1. Moreover, in the end we want $\hat{D}$ to be a signature matrix, so it is only reasonable that we should require D_1 to be a signature matrix. Indeed, at each step of the bulge-chase we should defend the signature matrix form of the "D" matrix. Then, in the end, $\hat{D}$ will be a signature matrix.

This determines the form of H_1 (and all subsequent transformation matrices). If d_{11} and d_{22} have the same sign, then H_1 should be an orthogonal matrix, for example, a rotator. Then $D_1 = D$ is automatically satisfied. If d_{11} and d_{22} have opposite sign, H_1 should be a hyperbolic transformation of type 1 (3.1.3) or type 2. In the former case we have $D_1 = D$; in the latter case D_1 is the signature matrix obtained from D by swapping the $(1, 1)$ and $(2, 2)$ entries.

After the first step, T_1 is symmetric but not tridiagonal. It has a bulge in position $(3, 1)$ and a symmetric bulge at $(1, 3)$. The bulge at $(3, 1)$ is removed by a transformation H_2^{-1}, applied on the left, that acts on rows 2 and 3. The form of H_2 is determined by the signs of the $(2, 2)$ and $(3, 3)$ entries of D_1. If they have the same sign, H_2 should be orthogonal; if they have opposite sign, H_2 should be hyperbolic. Either way, $D_2 = H_2^{-1} D_1 H_2^{-T}$ is a signature matrix. $T_2 = H_2^{-1} T_1 H_2^{-T}$ is a symmetric, nearly tridiagonal, matrix with a bulge in position $(4, 2)$. The next transformation is set up in an entirely analogous manner and pushes the bulge from $(4, 2)$ to $(5, 3)$, and so on. After $n - 1$ transformations, the bulge-chase is complete. An HZ iteration has been completed.

An implicit HZ iteration of higher degree can be arranged similarly. At each step, more than one entry has to be eliminated. This can be done by a transformation of the type guaranteed by Theorem 3.1.5. The method of constructing such transformations is outlined in the proof of the theorem. Their use guarantees that the "D" matrix remains a signature matrix, and an HZ iteration is effected.

We should not neglect to mention that the HZ bulge-chasing process can sometimes break down. If we wish to use a hyperbolic transformation on the vector $\begin{bmatrix} \alpha \\ \beta \end{bmatrix}$ to zero out β, the desired transformation does not exist if $\alpha^2 = \beta^2$.

There is one situation when we know we will never have a breakdown. If all the entries of D have the same sign, $D = \pm I$, all of the transformations will be orthogonal. Then the HZ algorithm reduces to the QR algorithm on the symmetric matrix T.

Exercises

6.4.1.

(a) Suppose $\hat{D} = H^T D H$, where D and $\hat{D}$ are signature matrices. Let $\tilde{H} = H^{-T}$. Show that $\tilde{H} = D H \hat{D}$ and $\tilde{H}^T D \tilde{H} = \hat{D}$.

(b) Show that $p(D^{-1}T) = D^{-1} p(TD^{-1}) D$ for every polynomial p.

6.4.2. Write an algorithm that does an HZ iteration of degree m.

6.5   Infinite Eigenvalues

Consider once again a pair (A, B) with no special structure. So far we have made the assumption that B is nonsingular, which implies that all eigenvalues of (A, B) are finite.

Now let us consider briefly what happens if B is singular. If we do not know beforehand whether B is singular or not, it will usually be evident once (A, B) is in Hessenberg-triangular form. Singularity will be signaled by one or more zeros on the main diagonal of the triangular matrix B.[3] This signals either an infinite eigenvalue or a singular pair.

Pushing the Zeros to the Top

We begin by showing that if B is singular, then the reduction to Hessenberg-triangular form tends to push the main-diagonal zeros to the top of B [220]. Consider the reduction to Hessenberg-triangular form outlined in the proof of Theorem 6.2.1. This begins by reducing B to upper-triangular form, at which point there will already be zeros on the main diagonal of B. We will show that the rest of the algorithm, which reduces A to upper Hessenberg form, tends to push the zeros to the top of B.

After the initial transformation of B to triangular form, b_{11} remains untouched by the algorithm. If it happens to be zero, it stays zero. Now suppose $b_{kk} = 0$ for some $k > 1$, and let us consider what happens to that zero subsequently.

Recall that the algorithm applies elimination matrices (e.g., rotators) to adjacent pairs of rows to transform $a_{n1}, a_{n-1,1}, \ldots, a_{31}$ successively to zero. Each of these transformations is applied to the B matrix as well and is immediately followed by a transformation on two adjacent columns to return B to triangular form. Once these transformations have been done, the first column of A is in Hessenberg form. A similar process transforms the second column of A to Hessenberg form, and so on.

If $b_{kk} = 0$, then it will remain zero until one of the elimination matrices touches the kth row or column. The first one to do so acts on rows k and $k + 1$. This one leaves b_{kk} at zero, since it recombines the zeros in positions (k, k) and $(k + 1, k)$ to create new zeros in those positions. The next transformation acts on rows $k - 1$ and k, and this one normally does make b_{kk} nonzero. Let us focus on the 2×2 submatrix consisting of rows and columns $k - 1$ and k. Before the transformation it has the form

$$\begin{bmatrix} b_{k-1,k-1} & b_{k-1,k} \\ 0 & 0 \end{bmatrix}. \tag{6.5.1}$$

Its rank is obviously one. When the transformation is applied, it disturbs both of the zeros. The submatrix now looks like

$$\begin{bmatrix} \tilde{b}_{k-1,k-1} & \tilde{b}_{k-1,k} \\ \tilde{b}_{k,k-1} & \tilde{b}_{k,k} \end{bmatrix}, \tag{6.5.2}$$

but its rank is still one. The next step in the reduction is to apply an elimination matrix to columns $k-1$ and k that annihilates $\tilde{b}_{k,k-1}$, returning B to upper triangular form. Application of this elimination matrix transforms the submatrix to

$$\begin{bmatrix} 0 & \hat{b}_{k-1,k} \\ 0 & \hat{b}_{kk} \end{bmatrix}. \tag{6.5.3}$$

---

[3] In practice there may not be any exact zeros, but any number that is on the order of the unit roundoff, relative to other matrix entries, can be considered to be a zero. However, we must caution that it can happen that a triangular matrix is nearly singular without any of its main-diagonal entries being tiny.

Since the rank is still one, $\hat{b}_{k-1,k-1}$ is forced to be zero. The zero has been moved from position (k, k) to position $(k - 1, k - 1)$.

The next pair of transformations acts on rows and columns $k - 2$ and $k - 1$ and pushes the zero up to position $(k - 2, k - 2)$ by the same process. The zero thus normally continues upward until it either arrives at position $(2, 2)$ or collides with another zero. The only way the upward movement can be stopped is if a trivial elimination transformation is applied on the left at some point. This happens when and only when the entry of A that is to be eliminated is already zero to begin with. In this event the zero in B stops and remains where it is until the next upward wave of rotators (corresponding to elimination of the next column of A) passes through.

Collision of Two Zeros

Suppose that B has two or more zeros on the main diagonal. Note that the number of zeros on the main diagonal is not necessarily equal to the number of infinite eigenvalues of the pair (A, B). We will see examples below in which B is strictly upper triangular, having nothing but zeros on its main-diagonal, and yet the pair (A, B) has only one infinite eigenvalue. During the reduction to Hessenberg-triangular form, the number of zeros on the main diagonal of B need not remain constant. Consider what happens when a zero that is moving upward on the main diagonal of B runs into another zero. Then we have the configuration

$$\begin{bmatrix} 0 & b_{k-1,k} \\ 0 & 0 \end{bmatrix}. \tag{6.5.4}$$

The left transformation of rows $k - 1$ and k then gives

$$\begin{bmatrix} 0 & \tilde{b}_{k-1,k} \\ 0 & \tilde{b}_{k,k} \end{bmatrix}, \tag{6.5.5}$$

in which $\tilde{b}_{kk}$ is normally nonzero. Once it becomes nonzero, it stays nonzero. The next transformation acts on columns $k - 1$ and k to annihilate $\tilde{b}_{k,k-1}$, but this transformation is trivial because $\tilde{b}_{k,k-1}$ is already zero. We conclude that in a collision of two zeros, the lower zero is normally destroyed. This can be prevented by two things: (i) $b_{k-1,k}$ could also be zero, and (ii) the transformation on rows $k - 1$ and k could be trivial. The conclusion is that when two zeros collide, the upper one survives but the lower one is normally (but not necessarily) destroyed.

The number of zeros on the main diagonal of B can decrease, but it cannot increase. If the matrix

$$\begin{bmatrix} b_{k-1,k-1} & b_{k-1,k} \\ 0 & b_{kk} \end{bmatrix}$$

has rank two before the transformations on rows and columns $k - 1$ and k, then it must still have rank two afterward. Thus zeros cannot spontaneously appear. The same is true during the GZ iterations, as we will see. It follows that the number of zeros on the main diagonal of B can never be less than the number of infinite eigenvalues, since the generalized Schur form at which we arrive in the end must have one zero on the main diagonal for each infinite eigenvalue.

Preliminary Removal of Infinite Eigenvalues

Once the pair (A, B) is in Hessenberg-triangular form, it is easy to remove the infinite eigenvalues at the outset if one wishes to. If the Hessenberg-triangular form was reached by the algorithm described above, the zeros on the main diagonal of B will tend to be near the top. Suppose first that $b_{11} = 0$. Then a rotator or other elimination matrix that acts on rows 1 and 2 can be applied (on the left) to transform a_{21} to 0. When this same transformation is applied to B, it leaves $b_{11} = 0$ and does not disturb the upper triangular form. The resulting pair has a form exemplified by

$$
\begin{bmatrix}
a_{11} & a_{12} & a_{13} & a_{14} \\
0 & a_{22} & a_{23} & a_{24} \\
0 & a_{32} & a_{33} & a_{34} \\
0 & 0 & a_{43} & a_{44}
\end{bmatrix},
\qquad
\begin{bmatrix}
0 & b_{12} & b_{13} & b_{14} \\
0 & b_{22} & b_{23} & b_{24} \\
0 & 0 & b_{33} & b_{34} \\
0 & 0 & 0 & b_{44}
\end{bmatrix}.
\qquad (6.5.6)
$$

Now we deflate an infinite eigenvalue by deleting the first row and column.[4] Now suppose $b_{11} \neq 0$ but $b_{22} = 0$. Then the deflation begins with a rotator acting on columns 1 and 2, transforming b_{11} to zero. This leaves b_{22} at zero. When the same transformation is applied to A, it creates a bulge at a_{31}. Now a transformation on the left, acting on rows 2 and 3, can eliminate a_{31}, returning A to Hessenberg form. When the same transformation is applied to B, it does not disturb the triangular form of B because $b_{22} = 0$. Now a left transformation on rows 1 and 2 transforms a_{21} to zero. When this transformation is applied to B, it destroys the zero at b_{22} but leaves $b_{11} = 0$. Now we are in the situation exemplified by (6.5.6), and we can deflate an infinite eigenvalue (or discover that the pair is singular). This process can be generalized to remove a zero from any position on the main diagonal of B (Exercise 6.5.1). A sequence of rotators moves the zero to the (1, 1) position, from which it can be deflated. An analogous algorithm can be used to drive a zero to the bottom of B for deflation at the bottom (Exercise 6.5.1). That variant is obviously more economical for zeros that are near the bottom to begin with.

Nonremoval of the Infinite Eigenvalues

We show next that it is not strictly necessary to remove the infinite eigenvalues beforehand. However experiments of Kågström and Kressner [123] suggest that the following material should be viewed as theoretical rather than practical. If infinite eigenvalues are not extracted immediately, they might never be detected, as they will be masked by subsequent roundoff errors.

If the Hessenberg-triangular pair was not obtained by a reduction process, the zeros can be anywhere in B. We have seen that it is possible to remove the infinite eigenvalues in advance. We now wish to show that there is no immediate need to do so, unless the zeros are very near the top or the bottom of B. The reason for removing zeros from the bottom is that we do not want them to get in the way of the shift computation. Every shift strategy gets its information from the lower right corner of the matrix. A common strategy is as

---

[4]If $a_{11} = 0$, the pair is singular. In that case one can, optionally, determine the fine structure (Kronecker canonical form) of the pair by the staircase algorithm [206, 73], which we choose not to discuss.

follows. If m shifts are needed, consider the pair (A, B) in the partitioned form

$$\begin{bmatrix} A_{11} & A_{12} \\ A_{21} & A_{22} \end{bmatrix}, \quad \begin{bmatrix} B_{11} & B_{12} \\ 0 & B_{22} \end{bmatrix},$$

where A_{22} and B_{22} are $m \times m$. Take the m shifts to be the eigenvalues of the pair (A_{22}, B_{22}). Since we would like all of the shifts to be finite numbers, we cannot tolerate zeros on the main diagonal of B_{22}. Therefore, if there are any such zeros, they should be pushed to the bottom and deflated out by the algorithm of part (b) of Exercise 6.5.1.

The reason for removing zeros from the top of B is that they will interfere with the computation of $p(AB^{-1})e_1$ in (6.3.5). Exercise 6.3.1 makes it clear that successful completion of (6.3.5) does not actually depend on the existence of B^{-1}. The step $y \leftarrow B^{-1}x$ is effected by solving $By = x$ by back substitution. Since only the first few entries of x are nonzero, the process involves only the upper left-hand corner of B. To be precise, one easily checks that (6.3.5) requires only that $b_{11}, \ldots, b_{mm}$ be nonzero. If any of those entries is zero, it should be pushed to the top by the algorithm of part (a) of Exercise 6.5.1 and deflated before the bulge-chase is begun.

We now consider what happens to zeros on the main diagonal of B in the course of a bulge-chase. This was studied in the QZ case (with $m = 1$ or 2) by Ward [211] and much later but in greater generality in [220], which was written in ignorance of [211]. Since the GZ iteration consists of an initial transformation followed by a reduction to Hessenberg-triangular form, we expect the GZ iteration to have the effect of moving the zeros upward. However, the bulge-chase is a very special reduction to Hessenberg-triangular form in which most of the entries to be zeroed out are already zero to begin with. Thus most of the transformations in the reduction are trivial and therefore skipped. Thus we expect the GZ iteration to have a much weaker upward-moving effect on the zeros on the main diagonal of B.

First consider the case $m = 1$. The bulge is 1×1, and it is chased back and forth between B and A. Let's suppose $b_{kk} = 0$ and see what happens as the bulge passes through that part of the matrix. When the bulge is at $a_{k,k-2}$, a transformation acting on rows $k - 1$ and k is used to set that entry to zero. Just before that transformation is applied to B, the submatrix consisting of rows and columns $k - 1$ and k has the form

$$\begin{bmatrix} b_{k-1,k-1} & b_{k-1,k} \\ 0 & 0 \end{bmatrix},$$

since $b_{kk} = 0$. This matrix definitely has rank one. After the transformation, which hits rows $k - 1$ and k, the submatrix looks like

$$\begin{bmatrix} \tilde{b}_{k-1,k-1} & \tilde{b}_{k-1,k} \\ \tilde{b}_{k,k-1} & \tilde{b}_{kk} \end{bmatrix}.$$

The entry $\tilde{b}_{k,k-1}$, which is the new bulge, is definitely nonzero by Exercise 6.5.3. Furthermore, the submatrix has rank one, since the transformation did not alter the rank. The next transformation acts on columns $k - 1$ and k, annihilating the bulge. The submatrix looks like

$$\begin{bmatrix} \hat{b}_{k-1,k-1} & \hat{b}_{k-1,k} \\ 0 & \hat{b}_{kk} \end{bmatrix},$$

it still has rank one, and $\hat{b}_{kk}$ is definitely nonzero by Exercise 6.5.3. Therefore $\hat{b}_{k-1,k-1}$ must be zero. The bulge-chase is now past this point, so this entry will not undergo any further changes. At the end of the bulge-chase we will have $\hat{b}_{k-1,k-1} = 0$. We conclude that the bulge-chase moves each zero on the main diagonal of B up by one position. After $k-1$ bulge-chases, a zero that was originally in position (k, k) will have moved up to position $(1, 1)$, and an infinite eigenvalue can be deflated. A bulge-chase neither creates nor destroys zeros. However, a zero can be destroyed in the deflation process (Exercise 6.5.4).

Since a GZ iteration of degree m is equivalent to m GZ iterations of degree 1, we might hope that each bulge-chase of degree m moves a zero up m positions. It turns out that that is what normally happens. Let us suppose $b_{kk} = 0$ and that a bulge of degree m is about to reach the kth row of B. Consider the submatrix consisting of rows and columns $k - m$ through k. This is the part of B that contains the bulge, and just before the step that returns column $k - m - 1$ of A to Hessenberg form it has the form

$$\check{B} = \left[\begin{array}{ccc|c} \check{b}_{k-m,k-m} & \cdots & \check{b}_{k-m,k-1} & \check{b}_{k-m,k} \\ \vdots & & \vdots & \vdots \\ \check{b}_{k-1,k-m} & \cdots & \check{b}_{k-1,k-1} & \check{b}_{k-1,k} \\ \hline 0 & \cdots & 0 & 0 \end{array}\right],$$

because $b_{kk} = 0$, and the bulge-chase has not yet touched the kth row. Let $\mathcal{S}$ denote the space spanned by the rows of $\check{B}$, and let r denote its dimension. Then clearly $r = \mathrm{rank}(\check{B}) \leq m$.

The operation that returns row $k - m - 1$ of A to Hessenberg form acts on rows $k - m$ through k. When it is applied to B, it recombines the rows of $\check{B}$, filling in the zeros in the bottom row. The result, $\tilde{B} = G^{-1}\check{B}$, has the form

$$\tilde{B} = \left[\begin{array}{ccc} \tilde{b}_{k-m,k-m} & \cdots & \tilde{b}_{k-m,k-1} & \tilde{b}_{k-m,k} \\ \vdots & & \vdots & \vdots \\ \tilde{b}_{k-1,k-m} & \cdots & \tilde{b}_{k-1,k-1} & \tilde{b}_{k-1,k} \\ \tilde{b}_{k,k-m} & \cdots & \tilde{b}_{k,k-1} & \tilde{b}_{kk} \end{array}\right].$$

Although the zeros have been filled in, the rank is still r. The rows of $\tilde{B}$ span $\mathcal{S}$. The next step applies a right transformation to columns $k - m$ through k to return the $(k - m)$th column of B (the first column of $\tilde{B}$) to triangular form. The result, $\hat{B} = \tilde{B}H^*$,

$$\hat{B} = \left[\begin{array}{c|ccc} \hat{b}_{k-m,k-m} & \hat{b}_{k-m,k-m+1} & \cdots & \hat{b}_{k-m,k} \\ \hline 0 & \hat{b}_{k-m+1,k-m+1} & \cdots & \hat{b}_{k-m+1,k} \\ \vdots & \vdots & & \vdots \\ 0 & \hat{b}_{k,k-m+1} & \cdots & \hat{b}_{kk} \end{array}\right].$$

The zeros appear because rows 2 through $m + 1$ of $\tilde{B}$ are orthogonal to the first row of the transforming matrix H. Recall that the rows of $\tilde{B}$ are linearly dependent. It is possible to throw out one row of $\tilde{B}$, and the remaining m rows will still span $\mathcal{S}$. Sometimes it matters which row is thrown out, but in most cases (almost always) it does not. If the vectors are in generic position, you can throw out any row, and the remaining rows will still span $\mathcal{S}$.

We would like to throw out the first row and keep the others. We hope that rows 2 through $m + 1$ span $\mathcal{S}$. Whether or not this is so depends upon the transforming matrix G^{-1}. Since this matrix has no special relationship with $\check{B}$ or $\tilde{B}$, we expect that rows 2 through $m + 1$ of $\tilde{B}$ will span $\mathcal{S}$.

Let us now suppose that rows 2 through $m + 1$ of $\tilde{B}$ span $\mathcal{S}$. The first row of H is orthogonal to all of these rows, so it is orthogonal to $\mathcal{S}$. But the first row of $\tilde{B}$ also lies in $\mathcal{S}$, so the first row of H must be orthogonal to it. This implies that $\hat{b}_{k-m,k-m} = 0$. That entry will not be touched by any subsequent operations of the bulge-chase, so at the end of the iteration we have $\hat{b}_{k-m,k-m} = 0$. The zero has moved up m positions. In conclusion, we expect each bulge-chase of degree m to move the zero up m positions. After a number of bulge-chases, the zero will have been moved up to one of the top m positions on the main diagonal of B, at which point it must be removed by the algorithm of part (a) of Exercise 6.5.1.

This argument proves nothing, because we cannot say for sure that our assumption that rows 2 through $m + 1$ of $\tilde{B}$ span $\mathcal{S}$ is correct. Therefore we did some experiments to see what happens in practice. Our experiments confirmed that the bulge-chase does normally move the zero up m positions.[5]

We have now seen how zeros on the main diagonal of B move during a GZ iteration. Before we are satisfied, we must ask one additional question. Do the zeros impede the convergence of the GZ iterations in any way? Numerical experiments [220] show that they do not; the good quadratic convergence is maintained. This is so because the infinite eigenvalues do not interfere with the process by which shifts are transmitted through the matrix during the bulge-chase. This topic is discussed thoroughly in Chapter 7.

Exercises

6.5.1. Let (A, B) be a Hessenberg-triangular pair with $b_{ii} = 0$. This exercise develops two algorithms for removing the zero.

    (a) Describe an algorithm that pushes the zero up to the upper left-hand corner of B for deflation there. The first step acts on columns $i - 1$ and i to transform $b_{i-i,i-1}$ to zero. This creates a bulge in A. Then a transformation from the left is used to return A to Hessenberg form. The next transformation sets $b_{i-2,i-2}$ to zero, and so on.

    (b) Describe an algorithm that pushes the zero downward to the bottom right-hand corner of B for deflation there. The first transformation acts on rows i and $i + 1$ of B and transforms $b_{i+1,i+1}$ to zero.

6.5.2. Let (A, B) be a pair in Hessenberg-triangular form, and suppose that B has k zeros on its main diagonal.

    (a) Show that k is an upper bound on the number of infinite eigenvalues of (A, B) (algebraic multiplicity).

---

[5]The experiments were QZ iterations of degrees 2, 3, 4, and 5. We did sometimes find that after a large number of iterations, the zero would not have moved up as far as it should have. In other words, after k iterations of degree m, the zero might have moved up only $km - 1$ or $km - 2$ positions, instead of the expected km. We suspect that this was caused by roundoff errors rather than any rank-deficiency problems. We believe it would not have happened if the arithmetic had been exact.

(b) Give a simple example that shows that the number of infinite eigenvalues of (A, B) can be less than k.

6.5.3. Consider a bulge-chase of degree $m = 1$ on a pair (A, B) for which A is properly upper Hessenberg and B is upper triangular with $b_{11} \neq 0$.

(a) The vector x that determines the initial transformation is generated by (6.3.5) with $m = 1$. This has only two nonzero entries. Show that the nonzero part of x is proportional to $\begin{bmatrix} a_{11} - \rho \\ a_{21} \end{bmatrix}$. Deduce that $x_2 \neq 0$. Show that after the initial transformation, the bulge at position b_{21} is definitely nonzero, given that $b_{11} \neq 0$.

(b) Show that the transformation that transforms the bulge at b_{21} to zero creates a bulge at a_{31} that is definitely nonzero. Also, show that the new entry at position b_{22} is definitely nonzero (even if the original b_{22} was zero).

(c) Prove by induction that every bulge that appears in the bulge-chase with $m = 1$ is certainly nonzero. (In other words, the bulge cannot suddenly become zero by accident.)

6.5.4. Suppose that (A, B) is a Hessenberg-triangular pair, A is properly upper Hessenberg, and $b_{11} = b_{22} = 0$. Show that in the process of deflating an infinite eigenvalue corresponding to $b_{11} = 0$, the entry in position $(2, 2)$ becomes nonzero, unless b_{12} is zero as well. Thus two adjacent zeros can correspond to only one infinite eigenvalue.

6.6 Deflating Subspaces

We observed early on that the concept of an invariant subspace is central to the theory and practice of the standard eigenvalue problem. For the generalized eigenvalue problem the analogous concept is that of a *deflating pair* of subspaces. This is an important idea, and we would have introduced it sooner if we had been taking a more theoretical approach. In most of this chapter we have restricted our attention to pairs (A, B) for which B is nonsingular, in which case the concept of a deflating pair reduces to that of invariant subspaces (of AB^{-1} and $B^{-1}A$). Material developed in this section will be used in Section 7.3.

Let (A, B) be a regular pair, and let $(\mathcal{S}, \mathcal{U})$ be a pair of subspaces of $\mathbb{C}^n$ of the same dimension k. Then $(\mathcal{S}, \mathcal{U})$ is called a *deflating pair* for (A, B) if $A\mathcal{S} \subseteq \mathcal{U}$ and $B\mathcal{S} \subseteq \mathcal{U}$. When B is nonsingular, the situation is as follows (Exercise 6.6.1).

Proposition 6.6.1. *Suppose that B is nonsingular.*

(a) *If $(\mathcal{S}, \mathcal{U})$ is a deflating pair of (A, B), then $\mathcal{U} = B\mathcal{S}$, $\mathcal{S} = B^{-1}\mathcal{U}$, $\mathcal{S}$ is invariant under $B^{-1}A$, and $\mathcal{U}$ is invariant under AB^{-1}.*

(b) *Conversely, if $\mathcal{S}$ is invariant under $B^{-1}A$ and $\mathcal{U} = B\mathcal{S}$, then $\mathcal{U}$ is invariant under AB^{-1}, and $(\mathcal{S}, \mathcal{U})$ is a deflating pair for (A, B).*

The next proposition gives a useful matrix characterization of deflating pairs.

Proposition 6.6.2. *Let $S_1, U_1 \in \mathbb{C}^{n \times k}$ be matrices whose columns form bases for $\mathcal{S}$ and $\mathcal{U}$, respectively; $\mathcal{S} = \mathcal{R}(S_1)$ and $\mathcal{U} = \mathcal{R}(U_1)$.*

(a) *Then (S, U) is a deflating pair for (A, B) if and only if there exist matrices E_{11} and F_{11} such that*

$$AS_1 = U_1 E_{11} \quad and \quad BS_1 = U_1 F_{11}. \tag{6.6.1}$$

(b) *If (S, U) is a deflating pair, then every eigenvalue of the pair (E_{11}, F_{11}) is an eigenvalue of (A, B). If x is an eigenvector of (E_{11}, F_{11}) with eigenvalue λ, then $S_1 x$ is an eigenvector of (A, B) with eigenvalue λ.*

(c) *If (S, U) is a deflating pair, then if $v \in S$ is an eigenvector of (A, B) with eigenvalue λ, and $v = S_1 x$, then x is an eigenvector of (E_{11}, F_{11}) with eigenvalue λ.*

If (S, U) is a deflating pair for (A, B), then the eigenvalues of (E_{11}, F_{11}) in Proposition 6.6.2 are called the *eigenvalues of (A, B) associated with (S, U)*.

Proposition 6.6.3. *Let (S, U) be a deflating pair for (A, B).*

(a) *Then $BS = U$ if and only if all of the eigenvalues of (A, B) associated with (S, U) are finite.*

(b) *If $BS = U$ and we take $U_1 = BS_1$, then the equations (6.6.1) simplify to the single equation*

$$AS_1 = BS_1 E_{11}.$$

The eigenvalues of E_{11} are the eigenvalues of (A, B) associated with (S, U).

Knowledge of deflating subspaces allows us to transform a matrix pair to block triangular form.

Proposition 6.6.4. *Let (S, U) be a k-dimensional deflating pair for (A, B), and let S_1, $U_1 \in \mathbb{C}^{n \times k}$ be matrices such that $S = \mathcal{R}(S_1)$ and $U = \mathcal{R}(U_1)$. Let S_2, $U_2 \in \mathbb{C}^{n \times (n-k)}$ be such that the matrices $S = \begin{bmatrix} S_1 & S_2 \end{bmatrix}$ and $U = \begin{bmatrix} U_1 & U_2 \end{bmatrix}$ are nonsingular. Let $E = U^{-1}AS$ and $F = U^{-1}BS$. Then*

$$E - \lambda F = \begin{bmatrix} E_{11} & E_{12} \\ 0 & E_{22} \end{bmatrix} - \lambda \begin{bmatrix} F_{11} & F_{12} \\ 0 & F_{22} \end{bmatrix}.$$

E_{11} and F_{11} are $k \times k$ and satisfy (6.6.1).

A pair that is in the block triangular form of (E, F) in Proposition 6.6.4 has $(\mathcal{E}_k, \mathcal{E}_k)$ as a deflating pair, where $\mathcal{E}_k = \text{span}\{e_1, \ldots, e_k\}$.

Deflating pairs are also known as *right deflating pairs*. A *left deflating pair* of subspaces for (A, B) is a pair that is right deflating for (A^*, B^*). The next proposition gives the matrix characterization of left deflating pairs.

Proposition 6.6.5. *Let $\hat{S}_1$, $\hat{U}_1 \in \mathbb{C}^{n \times k}$ be matrices whose columns form bases for $\hat{S}$ and $\hat{U}$, respectively.*

(a) *Then $(\hat{S}, \hat{U})$ is a left deflating pair for (A, B) if and only if there exist matrices $\hat{E}_{11}$ and $\hat{F}_{11}$ such that*

$$\hat{S}_1^* A = \hat{E}_{11} \hat{U}_1^* \quad and \quad \hat{S}_1^* B = \hat{F}_{11} \hat{U}_1^*. \tag{6.6.2}$$

(b) *If* $(\hat{S}, \hat{U})$ *is a left deflating pair, then every eigenvalue of the pair* $(\hat{E}_{11}, \hat{F}_{11})$ *is an eigenvalue of* (A, B). *If* x^* *is a left eigenvector of* $(\hat{E}_{11}, \hat{F}_{11})$ *with eigenvalue* λ, *then* $x^* \hat{S}_1^*$ *in a left eigenvector of* (A, B) *with eigenvalue* λ.

(c) *If* $(\hat{S}, \hat{U})$ *is a left deflating pair, then if* $v \in \hat{S}$, v^* *is a left eigenvector of* (A, B) *with eigenvalue* λ, *and* $v = S_1 x$, *then* x^* *is a left eigenvector of* $(\hat{E}_{11}, \hat{F}_{11})$ *with eigenvalue* λ.

If $(\hat{S}, \hat{U})$ is a left deflating pair for (A, B), then the eigenvalues of $(\hat{E}_{11}, \hat{F}_{11})$ in Proposition 6.6.5 are called the *eigenvalues of* (A, B) *associated with* $(\hat{S}, \hat{U})$.

Proposition 6.6.6. *The pair* (S, U) *is a right deflating pair for* (A, B) *if and only if* $(U^\perp, S^\perp)$ *is a left deflating pair. If* (S, U) *is a deflating pair, then the sets of eigenvalues associated with* (S, U) *and* $(U^\perp, S^\perp)$ *are complementary subsets of the spectrum of* (A, B), *counting multiplicity.*

Proof: We outline a proof here. (Some details are worked out in Exercise 6.6.6.) $AS \subseteq U$ if and only if $A^* U^\perp \subseteq S^\perp$. Since this is true for any matrix, it is true for B as well. This establishes the first part of the proposition.

The second part is perhaps most easily seen using matrices. If (S, U) is a deflating pair, then there is an equivalence transformation to block triangular form

$$U^{-1}(A - \lambda B)S = E - \lambda F = \begin{bmatrix} E_{11} & E_{12} \\ 0 & E_{22} \end{bmatrix} - \lambda \begin{bmatrix} F_{11} & F_{12} \\ 0 & F_{22} \end{bmatrix}, \qquad (6.6.3)$$

where $S = \begin{bmatrix} S_1 & S_2 \end{bmatrix}$ and $U = \begin{bmatrix} U_1 & U_2 \end{bmatrix}$, as described in Proposition 6.6.4. The eigenvalues of (E_{11}, F_{11}) are the eigenvalues of (A, B) associated with (S, U). The transformation (6.6.3) can be rewritten as

$$U^{-1}(A - \lambda B) = (E - \lambda F)S^{-1}. \qquad (6.6.4)$$

Define matrices $V = \begin{bmatrix} V_1 & V_2 \end{bmatrix}$ and $W = \begin{bmatrix} W_1 & W_2 \end{bmatrix}$ by $V^* = S^{-1}$ and $W^* = U^{-1}$. Then $V^* S = I$, which implies that $\mathcal{R}(V_2) = S^\perp$. Similarly $\mathcal{R}(W_2) = U^\perp$. Equation (6.6.4) can be written in block form as

$$\begin{bmatrix} W_1^* \\ W_2^* \end{bmatrix} (A - \lambda B) = \begin{bmatrix} E_{11} - \lambda F_{11} & E_{12} - \lambda F_{12} \\ 0 & E_{22} - \lambda F_{22} \end{bmatrix} \begin{bmatrix} V_1^* \\ V_2^* \end{bmatrix},$$

the second equation of which implies

$$W_2^* A = E_{22} V_2^* \quad \text{and} \quad W_2^* B = F_{22} V_2^*.$$

Since $\mathcal{R}(W_2) = U^\perp$ and $\mathcal{R}(V_2) = S^\perp$, these equations show, by Proposition 6.6.5, that $(U^\perp, S^\perp)$ is a left deflating pair for (A, B) (reproving part one), and the associated eigenvalues are the eigenvalues of (E_{22}, F_{22}). Since the eigenvalues of (E_{11}, F_{11}) and (E_{22}, F_{22}) are complementary subsets of the spectrum of (A, B), the proof is complete. $\square$

Finally we have the following very compact characterization.

Proposition 6.6.7. *Let $S_1 \in \mathbb{C}^{n \times k}$ and $W_2 \in \mathbb{C}^{n \times (n-k)}$ be matrices such that $\mathcal{S} = \mathcal{R}(S_1)$ and $\mathcal{U}^\perp = \mathcal{R}(W_2)$. Then $(\mathcal{S}, \mathcal{U})$ is a deflating pair for (A, B) if and only if*

$$W_2^* A S_1 = 0 \quad and \quad W_2^* B S_1 = 0.$$

Swapping Blocks

At this point we find it convenient to use the pencil notation. Whether we speak of a matrix pair (E, F) or a matrix pencil $E - \lambda F$, we are speaking of the same thing. Consider the following swapping problem. A block triangular matrix pencil

$$E - \lambda F = \begin{bmatrix} E_{11} & E_{12} \\ 0 & E_{22} \end{bmatrix} - \lambda \begin{bmatrix} F_{11} & F_{12} \\ 0 & F_{22} \end{bmatrix} \tag{6.6.5}$$

is given. Suppose that F is nonsingular, so that all of the eigenvalues are finite. Suppose further that the eigenvalues of $E_{11} - \lambda F_{11}$ are $\rho_1, \ldots, \rho_m$, the eigenvalues of $E_{22} - \lambda F_{22}$ are $\tau_1, \ldots, \tau_k$, and $\rho_i \neq \tau_j$ for all i and j. The problem is to make a unitary equivalence that transforms (6.6.5) to

$$\hat{E} - \lambda \hat{F} = \begin{bmatrix} \hat{E}_{11} & \hat{E}_{12} \\ 0 & \hat{E}_{22} \end{bmatrix} - \lambda \begin{bmatrix} \hat{F}_{11} & \hat{F}_{12} \\ 0 & \hat{F}_{22} \end{bmatrix}, \tag{6.6.6}$$

where $\hat{E}_{11} - \lambda \hat{F}_{11}$ has eigenvalues $\tau_1, \ldots, \tau_k$, and $\hat{E}_{22} - \lambda \hat{F}_{22}$ has eigenvalues $\rho_1, \ldots, \rho_m$.

We considered a problem like this in connection with standard eigenvalue problems in Section 4.8. There we discussed the interchange of blocks in a block triangular matrix in order to compute an invariant subspace associated with specified eigenvalues. We found that we could do this if we could solve a Sylvester equation (2.4.4). The block swapping problem that we are studying here can be useful in a similar context. Once we have transformed a pair to triangular or quasi-triangular form by the QZ algorithm, if we wish to compute a deflating pair of subspaces associated with a specified set of eigenvalues, we must move those eigenvalues or the blocks containing those eigenvalues to the upper left-hand corner of the pair. This requires swapping small blocks. The need to swap blocks will also arise in a different context in Section 7.3. As we shall see, this sort of swap requires solving a *generalized Sylvester equation*.[6]

Proceeding as in Section 4.8, we begin by transforming by

$$\begin{bmatrix} 0 & I \\ I & 0 \end{bmatrix}$$

on left and right to transform (6.6.5) to

$$\begin{bmatrix} E_{22} & 0 \\ E_{12} & E_{11} \end{bmatrix} - \lambda \begin{bmatrix} F_{22} & 0 \\ F_{12} & F_{11} \end{bmatrix}.$$

Now the main-diagonal blocks are in the right place at least, but the off-diagonal blocks are not. In Section 4.8 (standard eigenvalue problem) we were able to make a similarity

---

[6]Another use for generalized Sylvester equations is the study of the sensitivity of deflating subspaces [195, 196].

transformation (2.4.3) to eliminate the off-diagonal block. This required solving a Sylvester equation (2.4.4); see Theorem 2.4.2 and Exercise 2.4.5. In our present situation we have two blocks to annihilate, but we have greater flexibility, as we are not restricted to similarity transformations. Using (2.4.3) as a guide, it seems reasonable to seek $X, Y \in \mathbb{C}^{m \times k}$ such that

$$
\begin{bmatrix} I & 0 \\ Y & I \end{bmatrix} \begin{bmatrix} E_{22} - \lambda F_{22} & 0 \\ E_{12} - \lambda F_{12} & E_{11} - \lambda F_{11} \end{bmatrix} \begin{bmatrix} I & 0 \\ X & I \end{bmatrix}
$$

$$
= \begin{bmatrix} E_{22} - \lambda F_{22} & 0 \\ 0 & E_{11} - \lambda F_{11} \end{bmatrix}. \tag{6.6.7}
$$

One easily checks that X and Y achieve this if and only if

$$
E_{11} X + Y E_{22} = -E_{12} \quad \text{and} \quad F_{11} X + Y F_{22} = -F_{12}. \tag{6.6.8}
$$

These are the *generalized Sylvester equations*. This is a system of $2mk$ linear equations in $2mk$ unknowns.

Theorem 6.6.8. *The generalized Sylvester equations (6.6.8) have a unique solution if and only if the pairs (E_{11}, F_{11}) and (E_{22}, F_{22}) have disjoint spectra.*

The proof (Exercise 6.6.9) is similar to that for the standard Sylvester equation (Exercises 2.4.4 and 2.4.5); the ideas are exactly the same, but the details are a bit more complicated. Exercise 6.6.9 also sketches a method for solving generalized Sylvester equations efficiently in practice. A number of variants have appeared in the literature [58, 82, 122, 124, 125]. All use generalizations of the Bartels–Stewart algorithm to triangularize the system and render the subsequent solution process trivial.

Once we have X and Y, we can perform the transformation (6.6.7). However, we prefer not to do this literally because the transforming matrices could be ill conditioned. Instead we make two decompositions,

$$
\begin{bmatrix} I & 0 \\ X & I \end{bmatrix} = QR = \begin{bmatrix} Q_{11} & Q_{12} \\ Q_{21} & Q_{22} \end{bmatrix} \begin{bmatrix} R_{11} & R_{12} \\ 0 & R_{22} \end{bmatrix} \tag{6.6.9}
$$

and

$$
\begin{bmatrix} I & 0 \\ Y & I \end{bmatrix} = SZ = \begin{bmatrix} S_{11} & S_{12} \\ 0 & S_{22} \end{bmatrix} \begin{bmatrix} Z_{11} & Z_{12} \\ Z_{21} & Z_{22} \end{bmatrix}, \tag{6.6.10}
$$

where Q and Z are unitary and R and S are upper triangular. Then, letting

$$
\begin{bmatrix} \hat{E}_{11} - \lambda \hat{F}_{11} & \hat{E}_{12} - \lambda \hat{F}_{12} \\ \hat{E}_{21} - \lambda \hat{F}_{21} & \hat{E}_{22} - \lambda \hat{F}_{22} \end{bmatrix}
$$

$$
= \begin{bmatrix} Z_{11} & Z_{12} \\ Z_{21} & Z_{22} \end{bmatrix} \begin{bmatrix} E_{22} - \lambda F_{22} & 0 \\ E_{12} - \lambda F_{12} & E_{11} - \lambda F_{11} \end{bmatrix} \begin{bmatrix} Q_{11} & Q_{12} \\ Q_{21} & Q_{22} \end{bmatrix}, \tag{6.6.11}
$$

we have $\hat{E}_{21} = \hat{F}_{21} = 0$; the eigenvalues of $(\hat{E}_{11}, \hat{F}_{11})$ are those of (E_{22}, F_{22}), i.e., $\tau_1, \ldots, \tau_k$, and the eigenvalues of $(\hat{E}_{22}, \hat{F}_{22})$ are those of (E_{11}, F_{11}), i.e., $\rho_1, \ldots, \rho_m$. These facts are verified in Exercise 6.6.10. The desired exchange, from (6.6.5) to (6.6.6), has been accomplished.

This is a theoretical result. In practice roundoff errors cause $\hat{E}_{21}$ and $\hat{F}_{21}$ to be nonzero, and if they are too far from zero we must reject the swap. To improve performance an additional QR or RQ decomposition can be applied to $\hat{F}$ to enforce block triangularity. Then the same transformation can be applied to $\hat{E}$ [122]. Finally, $\| \hat{E}_{21} \|$ is checked as a stability check. If this is small enough, we set it to zero. If it is not small enough, we must reject the exchange.

Relationship to Deflating Subspaces

The important part of (6.6.7) is the (2,1) block, which can be written as

$$\begin{bmatrix} Y & I \end{bmatrix} \begin{bmatrix} E_{22} & 0 \\ E_{12} & E_{11} \end{bmatrix} \begin{bmatrix} I \\ X \end{bmatrix} = 0, \qquad \begin{bmatrix} Y & I \end{bmatrix} \begin{bmatrix} F_{22} & 0 \\ F_{12} & F_{11} \end{bmatrix} \begin{bmatrix} I \\ X \end{bmatrix} = 0,$$

$$(6.6.12)$$

and is equivalent to the generalized Sylvester equations (6.6.8). If we let $S_1 = \begin{bmatrix} I \\ X \end{bmatrix}$, $S = \mathcal{R}(S_1)$, $W_2 = \begin{bmatrix} Y^* \\ I \end{bmatrix}$, and $\mathcal{U} = \mathcal{R}(W_2)^\perp$ and compare with Proposition 6.6.7, we see that $(S, \mathcal{U})$ is a deflating pair for the pencil

$$\begin{bmatrix} E_{22} & 0 \\ E_{12} & E_{11} \end{bmatrix} - \lambda \begin{bmatrix} F_{22} & 0 \\ F_{12} & F_{11} \end{bmatrix}. \tag{6.6.13}$$

Thus solving the generalized Sylvester equation amounts to computing a deflating pair of subspaces. Exercise 6.6.11 shows that the eigenvalues associated with $(S, \mathcal{U})$ are the eigenvalues of (E_{22}, F_{22}), while the eigenvalues associated with $(\mathcal{U}^\perp, S^\perp)$ are those of (E_{11}, F_{11}). The decompositions (6.6.9) and (6.6.10) guarantee that the columns of $\begin{bmatrix} Q_{11} \\ Q_{21} \end{bmatrix}$ span the same space as the columns of $\begin{bmatrix} I \\ X \end{bmatrix}$, and the rows of $\begin{bmatrix} Z_{21} & Z_{22} \end{bmatrix}$ span the same space as the rows of $\begin{bmatrix} Y & I \end{bmatrix}$. These equalities guarantee that the transformation (6.6.11) has the desired effect (cf. Proposition 6.6.4).

Exercises

6.6.1. Prove Proposition 6.6.1.

6.6.2. Prove Proposition 6.6.2.

6.6.3. Prove Proposition 6.6.3.

6.6.4. Prove Proposition 6.6.4. (Rewrite the equation $E = U^{-1}AS$ as $AS = UE$. Partition S, U, and E appropriately, and show that $E_{21} = 0$. Do the same for the equation $F = U^{-1}BS$.)

6.6.5. Prove Proposition 6.6.5.

6.6.6. This exercise works out some details of the proof of Proposition 6.6.6.

(a) Use Proposition 1.2.2 to show that $AS \subseteq \mathcal{U}$ if and only if $A^*\mathcal{U}^\perp \subseteq S^\perp$.

(b) Show that $V^*S = I$, $V_2^*S_1 = 0$, and prove that $\mathcal{R}(V_2) = S^\perp$. Similarly prove $\mathcal{R}(W_2) = \mathcal{U}^\perp$.

6.6.7. Prove Proposition 6.6.7.

6.6.8. Verify that (6.6.7) holds if and only if the generalized Sylvester equations (6.6.8) hold.

6.6.9. This exercise proves Theorem 6.6.8.

(a) Apply the generalized Schur theorem (Theorem 6.1.2) to the pairs (E_{11}^T, F_{11}^T) and (E_{22}, F_{22}) to show that the generalized Sylvester equations (6.6.8) can be replaced by an equivalent system in which E_{11} and F_{11} are lower triangular and E_{22} and F_{22} are upper triangular.

(b) Using properties of Kronecker products (Exercises 2.4.3 and 2.4.4), show that the generalized Sylvester equations (6.6.8) can be rewritten as

$$\left[\begin{array}{cc} I \otimes E_{11} & E_{22}^T \otimes I \\ I \otimes F_{11} & F_{22}^T \otimes I \end{array} \right] \left[\begin{array}{c} \text{vec}(X) \\ \text{vec}(Y) \end{array} \right] = - \left[\begin{array}{c} \text{vec}(E_{12}) \\ \text{vec}(F_{12}) \end{array} \right]. \tag{6.6.14}$$

(c) From part (a) we can assume that E_{11}, F_{11}, E_{22}^T, and F_{22}^T are all lower triangular in (6.6.14). Assume this for the remainder of the exercise. Show that each of the four blocks in the coefficient matrix of (6.6.14) is lower triangular.

(d) Transform (6.6.14) to an equivalent system by doing a perfect shuffle of the rows and the columns. Thus rows (or columns) that were ordered $1, 2, \ldots, 2n$ get reordered as $1, n+1, 2, n+2, \ldots, n, 2n$. Show that the resulting system is block lower triangular with 2×2 blocks. Show that each main-diagonal block has the form

$$\left[\begin{array}{cc} e_1 & e_2 \\ f_1 & f_2 \end{array} \right],$$

where e_i/f_i is an eigenvalue of the pair (E_{ii}, F_{ii}) for $i = 1, 2$. Deduce that the coefficient matrix is nonsingular if and only if (E_{11}, F_{11}) and (E_{22}, F_{22}) have no eigenvalues in common. This completes the proof of Theorem 6.6.8.

(e) Describe an algorithm to solve the block triangular generalized Sylvester equations. (You will probably find it easier to work with the unshuffled version.) This is a generalization of the (complex version of the) Bartels–Stewart algorithm.

6.6.10. Suppose $\hat{E} - \lambda\hat{F}$ is given by (6.6.11).

(a) Show that

$$\left[\begin{array}{cc} \hat{E}_{11} - \lambda\hat{F}_{11} & \hat{E}_{12} - \lambda\hat{F}_{12} \\ \hat{E}_{21} - \lambda\hat{F}_{21} & \hat{E}_{22} - \lambda\hat{F}_{22} \end{array} \right]$$

$$= \left[\begin{array}{cc} S_{11} & S_{12} \\ 0 & S_{22} \end{array} \right]^{-1} \left[\begin{array}{cc} E_{22} - \lambda F_{22} & 0 \\ 0 & E_{11} - \lambda F_{11} \end{array} \right] \left[\begin{array}{cc} R_{11} & R_{12} \\ 0 & R_{22} \end{array} \right]^{-1}.$$

(b) Deduce that $\hat{E}_{21} - \lambda\hat{F}_{21} = 0$.

(c) Show that $\hat{E}_{11} - \lambda\hat{F}_{11} = S_{11}^{-1}(E_{22} - \lambda F_{22})R_{11}^{-1}$. Deduce that the eigenvalues of $\hat{E}_{11} - \lambda\hat{F}_{11}$ are $\tau_1, \ldots, \tau_k$.

(d) Show that $\hat{E}_{22} - \lambda \hat{F}_{22} = S_{22}^{-1}(E_{11} - \lambda F_{11})R_{22}^{-1}$. Deduce that the eigenvalues of $\hat{E}_{22} - \lambda \hat{F}_{22}$ are $\rho_1, \ldots, \rho_m$.

6.6.11.

(a) Show that if the columns of $\begin{bmatrix} I \\ X \end{bmatrix}$ span a deflating subspace $\mathcal{S}$ of the pencil (6.6.13), then the eigenvalues associated with $\mathcal{S}$ are the eigenvalues of (E_{22}, F_{22}).

(b) Show that if the columns of $\begin{bmatrix} Y^* \\ I \end{bmatrix}$ span a left deflating subspace $\mathcal{U}^\perp$ of the pencil (6.6.13), then the eigenvalues associated with $\mathcal{U}^\perp$ are the eigenvalues of (E_{11}, F_{11}).

Chapter 7
Inside the Bulge

The purpose of shifts is to accelerate convergence. In an implicit GR iteration the shifts are used only at the very beginning, to help determine the transformation that creates the initial bulge. After that the shifts are forgotten, as the bulge is chased to the bottom of the matrix. In the end, if the shifts were good, significant progress toward convergence will have been made. Progress is felt mostly in the rapid diminution of certain subdiagonal entries near the bottom of the matrix. Thus the shifts are loaded in at the top, and their effect is felt at the bottom. It is therefore reasonable to ask how the information about the shifts is transported in the bulge from the top to the bottom of the matrix.

In this chapter we identify the mechanism by which the shifts are transmitted through the matrix during a bulge-chase. We then demonstrate by numerical experiments that the mechanism works well for iterations of low degree. However, for iterations of high degree we find that roundoff errors cause the shift information to be propagated inaccurately. We call this phenomenon *shift blurring*. It provides an explanation for why the multishift QR algorithm proposed in [13] did not work as well as expected.

Another interesting application of the shift transmission mechanism is a procedure for passing bulges through each other. So far in this book we have focused on chasing bulges from the top to the bottom of a matrix. However, it is also possible to chase a bulge from bottom to top. We might sometimes find it useful to chase bulges in both directions at once. When the bulges collide, is there a way to pass them through each other so that the information that they carry does not get mixed up? Yes, there is, and we will develop it in Section 7.3.

Throughout this chapter we will work in the context of the generalized eigenvalue problem, deriving results about the standard eigenvalue problem by specialization. Thus we will start out talking about GZ iterations and later specialize to GR iterations. We draw substantially from [216, 217, 219, 220, 223].

7.1 Transmission of Shifts

Whether we speak of a matrix pair (A, B) or a matrix pencil $A - \lambda B$, we are speaking of the same object. In this chapter we find it convenient to use the pencil terminology. Suppose that $A - \lambda B$ is *properly* Hessenberg-triangular. By this we mean that A is properly upper Hessenberg and B is nonsingular and upper triangular. In an implicit GZ iteration of degree

m applied to $A - \lambda B$, the shifts $\rho_1, \ldots, \rho_m$ are used to compute a vector

$$x = \alpha(AB^{-1} - \rho_1 I) \cdots (AB^{-1} - \rho_m I)e_1 \qquad (7.1.1)$$

that determines the transformation that creates the initial bulge; see (6.3.5). This is the only place where the shifts are used, and yet somehow the information that good shifts convey is transmitted from top to bottom of the pencil. In this section we present the mechanism by which this happens.

Equation (7.1.1) can be rewritten as

$$x = \alpha p(AB^{-1})e_1, \qquad (7.1.2)$$

where $p(z) = (z - \rho_1) \cdots (z - \rho_m)$. It is clear that x is determined up to a scalar multiple by p. It is worth noting that the converse is true as well.

Proposition 7.1.1. *Let C be a properly upper Hessenberg matrix, and let $x \in \mathbb{C}^n$ be a nonzero vector. Then there is a unique monic polynomial p of degree $n - 1$ or less such that $x = \alpha p(C)e_1$ for some nonzero scalar α. If $x_{m+1} \neq 0$ and $x_i = 0$ for all $i > m + 1$, then the degree of p is exactly m.*

The easy proof is worked out in Exercise 7.1.3.

For the coming analysis it is convenient to embed the pencil $A - \lambda B$ in an *augmented pencil* $\tilde{A} - \lambda \tilde{B}$, obtained by adjoining a column on the left and a row on the bottom:

$$\tilde{A} - \lambda \tilde{B} = \begin{bmatrix} x & A - \lambda B \\ 0 & y^T \end{bmatrix}, \qquad (7.1.3)$$

where we will consider various choices of x and y^T. Notice that the main diagonal of $\tilde{A} - \lambda \tilde{B}$ corresponds to the subdiagonal of $A - \lambda B$.

First suppose $x = e_1$ and $y^T = e_n^T$. Then $\tilde{A}$ is upper triangular with all main-diagonal entries nonzero, and $\tilde{B}$ is strictly upper triangular. This implies that the characteristic polynomial of $\tilde{A} - \lambda \tilde{B}$ is a nonzero constant. Thus the pencil has only the eigenvalue ∞ with algebraic multiplicity $n + 1$. Each main-diagonal entry $a_{j+1,j} - \lambda 0$ signals an infinite eigenvalue.

In order to analyze a GZ iteration of degree m with shifts $\rho_1, \ldots, \rho_m$, consider (7.1.3) with x as in (7.1.1) or (7.1.2). That is, take

$$x = \alpha p(AB^{-1})e_1 \quad \text{and} \quad y^T = e_n^T, \qquad (7.1.4)$$

where $p(z) = (z - \rho_1) \cdots (z - \rho_m)$. Now the eigenvalues of (7.1.3) are not all infinite. With this choice of x, the augmented pencil is no longer upper triangular. However, recalling that all of the entries of x below x_{m+1} are zero, we see that (7.1.3) is block triangular, having the form

$$\begin{bmatrix} \tilde{A}_{22} - \lambda \tilde{B}_{22} & * \\ 0 & \tilde{A}_{33} - \lambda \tilde{B}_{33} \end{bmatrix}, \qquad (7.1.5)$$

where $\tilde{A}_{22} \in \mathbb{C}^{(m+1) \times (m+1)}$ would be upper triangular, except that its first column is $\tilde{x}$, the nonzero part of x, $\tilde{A}_{33}$ is upper triangular, and $\tilde{B}_{22}$ and $\tilde{B}_{33}$ are strictly upper triangular. The spectrum of $\tilde{A} - \lambda \tilde{B}$ is the union of the spectra of $\tilde{A}_{22} - \lambda \tilde{B}_{22}$ and $\tilde{A}_{33} - \lambda \tilde{B}_{33}$. The spectrum of $\tilde{A}_{33} - \lambda\tilde{B}_{33}$ consists of ∞ with multiplicity $n - m$, but that of $\tilde{A}_{22} - \lambda\tilde{B}_{22}$ is much more interesting.

We will also use the notation $E_0 - \lambda F_0 = \tilde{A}_{22} - \lambda\tilde{B}_{22}$ and refer to this pencil as the *initial bulge pencil*. Note that

$$
E_0 = \begin{bmatrix}
x_1 & a_{1,1} & \cdots & a_{1,m-1} & a_{1,m} \\
x_2 & a_{2,1} & \cdots & a_{2,m-1} & a_{2,m} \\
\vdots & \vdots & & \vdots & \vdots \\
x_m & 0 & \cdots & a_{m,m-1} & a_{m,m} \\
x_{m+1} & 0 & \cdots & 0 & a_{m+1,m}
\end{bmatrix}
$$

and

$$
F_0 = \begin{bmatrix}
0 & b_{1,1} & \cdots & b_{1,m-1} & b_{1,m} \\
0 & 0 & \cdots & b_{2,m-1} & b_{2,m} \\
\vdots & \vdots & & \vdots & \vdots \\
0 & 0 & \cdots & 0 & b_{m,m} \\
0 & 0 & \cdots & 0 & 0
\end{bmatrix}.
$$

By inspecting (7.1.1) or (6.3.5) we easily see (Exercise 7.1.4) that x is completely determined by $\rho_1, \ldots, \rho_m$, the first m columns of A, and the first m columns of B. It is exactly these columns that appear in the initial bulge pencil. Exercise 7.1.4 shows that the initial bulge pencil is regular with m finite eigenvalues.

Theorem 7.1.2. *The eigenvalues of the initial bulge pencil $E_0 - \lambda F_0$ are the shifts ρ_1, $\rho_2, \ldots, \rho_m$, together with ∞.*

Proof: We sketch the proof here, leaving the details to Exercise 7.1.5. Because F_0 is singular, ∞ is an eigenvalue. A corresponding eigenvector is e_1. Now we show that each shift ρ_i is an eigenvalue. Equation (7.1.1) can be rewritten as $x = (A - \rho_i B)B^{-1}\tilde{p}(AB^{-1})e_1$, where $\tilde{p}$ is a polynomial of degree $m - 1$. Thus $x = (A - \rho_i B)z$, where z is a vector such that $z_i = 0$ for $i > m$. Let $\tilde{z} \in \mathbb{C}^m$ denote the "nonzero part" of z, that consisting of the first m components. Bearing in mind that the vectors x and z both end with a lot of zeros, we find easily that the equation $x = (A - \rho B)z$ is equivalent to $(E_0 - \rho_i F_0)\begin{bmatrix} -1 \\ \tilde{z} \end{bmatrix} = 0$. Thus ρ_i is an eigenvalue of the initial bulge pencil.

If $\rho_1, \ldots, \rho_m$ are distinct, the proof is now complete, since ∞ is also an eigenvalue, and the $(m + 1) \times (m + 1)$ pencil $E_0 - \lambda F_0$ cannot have more than $m + 1$ eigenvalues. If $\rho_1, \ldots, \rho_m$ are not distinct, the proof can be completed by a continuity argument: We can perturb each ρ_i slightly to $\rho_{i,\epsilon} = \rho_i + \epsilon\sigma_i$ in such a way that $\rho_{1,\epsilon}, \ldots, \rho_{m,\epsilon}$ are distinct if ϵ is sufficiently small. This perturbation causes a small perturbation in x, which in turn causes a small perturbation in the bulge pencil. The perturbed bulge pencil has the $m + 1$ distinct eigenvalues $\rho_{1,\epsilon}, \ldots, \rho_{m,\epsilon}, \infty$. As $\epsilon \to 0$, the perturbed bulge pencil tends to the original bulge pencil, and its eigenvalues tend to $\rho_1, \ldots, \rho_m, \infty$. Therefore, by continuity of eigenvalues, the eigenvalues of the unperturbed bulge pencil must be $\rho_1, \ldots, \rho_m$ and ∞. $\square$

Theorem 7.1.2 implies that the eigenvalues of the augmented pencil $\tilde{A} - \lambda\tilde{B}$ given by (7.1.5) are ∞ repeated $n - m + 1$ times and $\rho_1, \ldots, \rho_m$. Now let us consider what happens to this pencil during an implicit GZ iteration that transforms $A - \lambda B$ to $\hat{A} - \lambda \hat{B}$. Each transformation that is applied to $A - \lambda B$ has a corresponding transformation on the augmented pencil $\tilde{A} - \lambda \tilde{B}$. Under each such transformation the new augmented pencil is equivalent to the old one, so the eigenvalues are preserved.

Consider the very first transformation, that of $A - \lambda B$ to $A_{1/2} - \lambda B_{1/2} = G_1^{-1}(A - \lambda B)$. This corresponds to multiplying $\tilde{A} - \lambda \tilde{B}$ on the left by $\text{diag}\{G_1, 1\}$. Recall that this transformation was designed so that $G_1 e_1 = \eta x$ for some nonzero η or, equivalently, $G_1^{-1} x = \eta^{-1} e_1$. Thus the transformed augmented pencil is

$$\left[\begin{array}{cc} \eta^{-1} e_1 & A_{1/2} - \lambda B_{1/2} \\ 0 & e_n^T \end{array} \right].$$

The "bulge" that was caused by x in the first column has now been eliminated, and a bulge has been created in the upper left-hand corner of $A_{1/2} - \lambda B_{1/2}$. By working with the augmented pencil, we are able to see that the initial transformation, which we had previously viewed as a bulge-creating transformation, can also be viewed as a bulge-chasing transformation.

The transformation from $A_{1/2} - \lambda B_{1/2}$ to $A_1 - \lambda B_1 = (A_{1/2} - \lambda B_{1/2})Z_1$ corresponds to multiplication of the augmented pencil on the right by $\text{diag}\{1, Z_1\}$ to produce

$$\left[\begin{array}{cc} \eta^{-1} e_1 & A_1 - \lambda B_1 \\ 0 & e_n^T \end{array} \right].$$

The transformation removes a column from the "B" bulge and adds a row to the "A" bulge, as explained in Section 6.3. It does not touch the first column or the last row of the augmented pencil. Subsequent transformations push the bulge downward. None of these touches the first column of the augmented pencil. None of them touch the bottom row either, until the bulge reaches the bottom of the pencil.

At intermediate stages of the bulge-chase, the augmented pencil has the form

$$\left[\begin{array}{cc} \eta^{-1} e_1 & A_i - \lambda B_i \\ 0 & e_n^T \end{array} \right]. \tag{7.1.6}$$

It is not upper triangular, because there is a bulge in $A_i - \lambda B_i$, but it is block upper triangular. Recycling the notation, we can write the augmented pencil as

$$\left[\begin{array}{ccc} \tilde{A}_{11} - \lambda \tilde{B}_{11} & * & * \\ & \tilde{A}_{22} - \lambda \tilde{B}_{22} & * \\ & & \tilde{A}_{33} - \lambda \tilde{B}_{33} \end{array} \right], \tag{7.1.7}$$

where $\tilde{A}_{22} - \lambda \tilde{B}_{22}$ is the part that contains the bulge. $\tilde{A}_{11} - \lambda \tilde{B}_{11}$ is $i \times i$ and upper triangular with constant, nonzero main-diagonal entries. Thus its characteristic polynomial is a constant, and its eigenvalues are ∞, repeated i times. Similarly, $\tilde{A}_{33} - \lambda \tilde{B}_{33}$ has only the eigenvalue ∞, repeated $n - m - i$ times. Since the eigenvalues of the augmented pencil are invariant under equivalence transformations, we know that the eigenvalues of (7.1.7) are ∞, with multiplicity $n - m + 1$ and $\rho_1, \ldots, \rho_k$. Thus we can deduce that the eigenvalues of $\tilde{A}_{22} - \lambda \tilde{B}_{22}$ must be $\rho_1, \ldots, \rho_m$ and ∞. Since this is the part of the pencil that carries the bulge, we call it the ith *bulge pencil*, and we introduce the notation

$E_i - \lambda F_i = \tilde{A}_{22} - \lambda \tilde{B}_{22}$. We have proved the following theorem, which elucidates the mechanism by which information about the shifts is transmitted in the bulge.

Theorem 7.1.3. *In an implicit GZ iteration driven of degree m with shifts $\rho_1, \ldots, \rho_m$, the eigenvalues of the bulge pencil $E_i - \lambda F_i$ at every step are $\rho_1, \ldots, \rho_m$ and ∞. In other words, the characteristic polynomial $\det(E_i - \lambda F_i)$ is a multiple of $p(z) = (z - \rho_1) \cdots (z - \rho_m)$.*

The Final Configuration

Once the GZ iteration is complete, the augmented pencil has the form

$$\left[\begin{array}{cc} \eta^{-1} e_1 & \hat{A} - \lambda \hat{B} \\ 0 & z^T \end{array} \right],$$

where $\hat{A} - \lambda \hat{B} = G^{-1}(A - \lambda B)Z$, and $z^T = e_n^T Z$. This is the *final configuration*. The bulge has been chased from $\hat{A} - \lambda \hat{B}$, but the augmented pencil still has a bulge, which has been compressed into z^T. Because of the Hessenberg-triangular form, the first $n - m - 1$ entries of z^T are zero. Thus the final augmented pencil has the form

$$\left[\begin{array}{cc} \tilde{A}_{11} - \lambda \tilde{B}_{11} & * \\ & \tilde{A}_{22} - \lambda \tilde{B}_{22} \end{array} \right],$$

where $\tilde{A}_{22} - \lambda \tilde{B}_{22}$ has the nonzero part of z^T in its last row. It is the final bulge pencil for the GZ iteration; its eigenvalues are, of course, $\rho_1, \ldots, \rho_m$ and ∞.

Application to the Standard Eigenvalue Problem

In the standard eigenvalue problem ($B = I$), if we take $Z_i = G_i$ at every step, then the GZ algorithm reduces to the GR algorithm. Therefore Theorems 7.1.2 and 7.1.3 hold for the standard eigenvalue problem as well. In this case the matrix F_i in the bulge pencil is a nonprincipal submatrix of the identity matrix:

$$F_i = \left[\begin{array}{ccccc} 0 & 1 & 0 & & \\ & 0 & 1 & \ddots & \\ & & 0 & \ddots & 0 \\ & & & \ddots & 1 \\ & & & & 0 \end{array} \right].$$

The Step-by-Step Viewpoint

We proved Theorem 7.1.3 by looking at the eigenvalues of the augmented pencil. A second approach, which provides additional insight, is to prove the result by induction. After i steps of the GZ iteration, the pencil has been transformed to $A_i - \lambda B_i$, and the augmented pencil has the form (7.1.6), which we have also written as (7.1.7), where $\tilde{A}_{22} - \lambda \tilde{B}_{22} = E_i - \lambda F_i$ is the bulge pencil. E_i and F_i are the nonprincipal submatrices of A_i and B_i, respectively, occupying rows $i + 1$ through $i + m + 1$ and columns i through $i + m$. In the case $m = 2$ it has the form

$$E_i - \lambda F_i = \begin{bmatrix} a_{i+1,i} & a_{i+1,i+1} & a_{i+1,i+2} \\ a_{i+2,i} & a_{i+2,i+1} & a_{i+2,i+2} \\ a_{i+3,i} & a_{i+3,i+1} & a_{i+3,i+2} \end{bmatrix} - \lambda \begin{bmatrix} 0 & b_{i+1,i+1} & b_{i+1,i+2} \\ 0 & b_{i+2,i+1} & b_{i+2,i+2} \\ 0 & 0 & 0 \end{bmatrix}.$$

We could have written $a_{i+1,i}^{(i)}$ instead of $a_{i+1,i}$, for example, to indicate the fact that this entry belongs to A_i and not A. We have left off the superscripts to avoid clutter.

The initial bulge pencil is $E_0 - \lambda F_0$, which comes from the upper left-hand corner of the initial augmented pencil. From Theorem 7.1.2 we know that its eigenvalues are $\rho_1, \ldots, \rho_m$ and ∞. If we can show that for every i the eigenvalues of $E_{i+1} - \lambda F_{i+1}$ are the same as those of $E_i - \lambda F_i$, then we will have proved Theorem 7.1.3 by induction: Every $E_i - \lambda F_i$ has eigenvalues $\rho_1, \ldots, \rho_m$ and ∞.

Suppose, therefore, that the eigenvalues of $E_i - \lambda F_i$ are $\rho_1, \ldots, \rho_m$ and ∞. The next step in the bulge-chase is the transformation $A_i - \lambda B_i \to G_{i+1}^{-1}(A_i - \lambda B_i) = A_{i+1/2} - \lambda B_{i+1/2}$, which removes a column from the "A" bulge and adds a row to the bottom of the "B" bulge. The corresponding operation on the bulge pencil is $E_i - \lambda F_i \to \tilde{G}_{i+1}^{-1}(E_i - \lambda F_i)$, creating zeros in the first column of E_i and destroying zeros in the bottom row of F_i. Since this is an equivalence transform on the pencil, it does not alter the eigenvalues. In the case $m = 2$ we have

$$\begin{bmatrix} a_{i+1,i} & a_{i+1,i+1} & a_{i+1,i+2} \\ 0 & a_{i+2,i+1} & a_{i+2,i+2} \\ 0 & a_{i+3,i+1} & a_{i+3,i+2} \end{bmatrix} - \lambda \begin{bmatrix} 0 & b_{i+1,i+1} & b_{i+1,i+2} \\ 0 & b_{i+2,i+1} & b_{i+2,i+2} \\ 0 & b_{i+3,i+1} & b_{i+3,i+2} \end{bmatrix}.$$

Notice that at this intermediate stage, the bulge pencil is in block triangular form; the infinite eigenvalue has been exposed in the upper left-hand corner. We can deflate out the infinite eigenvalue by deleting the first row and column. Let us call the resulting subpencil the *intermediate bulge pencil* and denote it

$$E_{i+1/2} - \lambda F_{i+1/2}. \tag{7.1.8}$$

It is $m \times m$, it is the nonprincipal subpencil of $A_{i+1/2} - \lambda B_{i+1/2}$ taken from rows $i + 2, \ldots, i + m + 1$ and columns $i + 1, \ldots, i + m$, and its eigenvalues are $\rho_1, \ldots, \rho_m$. In the case $m = 2$ we have

$$E_{i+1/2} - \lambda F_{i+1/2} = \begin{bmatrix} a_{i+2,i+1} & a_{i+2,i+2} \\ a_{i+3,i+1} & a_{i+3,i+2} \end{bmatrix} - \lambda \begin{bmatrix} b_{i+2,i+1} & b_{i+2,i+2} \\ b_{i+3,i+1} & b_{i+3,i+2} \end{bmatrix}.$$

The next step in the bulge-chase is the transformation $A_{i+1/2} - \lambda B_{i+1/2} \to (A_{i+1/2} - \lambda B_{i+1/2})Z_{i+1} = A_{i+1} - \lambda B_{i+1}$, removing a column from the "B" bulge and adding a row to the bottom of the "A" bulge. Before we can discuss the action of this step on the bulge pencil, we need to adjoin a row and a column to the bottom of the intermediate bulge pencil. The resulting pencil has the form

$$\begin{bmatrix} & & & * \\ & E_{i+1/2} & & \vdots \\ & & & * \\ 0 & \cdots & 0 & a_{i+m+2,i+m+1} \end{bmatrix} - \lambda \begin{bmatrix} & & & * \\ & F_{i+1/2} & & \vdots \\ & & & * \\ 0 & \cdots & 0 & 0 \end{bmatrix},$$

or, in the case $m = 2$,

$$
\begin{bmatrix} a_{i+2,i+1} & a_{i+2,i+2} & a_{i+2,i+3} \\ a_{i+3,i+1} & a_{i+3,i+2} & a_{i+3,i+3} \\ 0 & 0 & a_{i+4,i+3} \end{bmatrix} - \lambda \begin{bmatrix} b_{i+2,i+1} & b_{i+2,i+2} & b_{i+2,i+3} \\ b_{i+3,i+1} & b_{i+3,i+2} & b_{i+3,i+3} \\ 0 & 0 & 0 \end{bmatrix}.
$$

This extension adjoins an infinite eigenvalue, so the the resulting pencil has eigenvalues $\rho_1, \ldots, \rho_m$ and ∞. Right multiplication by $\tilde{Z}_{i+1}$ transforms this pencil to $E_{i+1} - \lambda F_{i+1}$. In the case $m = 2$ we have

$$
E_{i+1} - \lambda F_{i+1} = \begin{bmatrix} a_{i+2,i+1} & a_{i+2,i+2} & a_{i+2,i+3} \\ a_{i+3,i+1} & a_{i+3,i+2} & a_{i+3,i+3} \\ a_{i+4,i+1} & a_{i+4,i+2} & a_{i+4,i+3} \end{bmatrix} - \lambda \begin{bmatrix} 0 & b_{i+2,i+2} & b_{i+2,i+3} \\ 0 & b_{i+3,i+2} & b_{i+3,i+3} \\ 0 & 0 & 0 \end{bmatrix}.
$$

Since this is an equivalence transform on the pencil, it leaves the eigenvalues unchanged. Therefore $E_{i+1} - \lambda F_{i+1}$ has eigenvalues $\rho_1, \ldots, \rho_m$ and ∞. We have proved Theorem 7.1.3 by induction.

Transmission of Shifts in Practice

Theorems 7.1.2 and 7.1.3 are true in exact arithmetic. In [217] we showed that they hold up well in the presence of roundoff errors if m is small but not if m is large. In numerous experiments we computed the finite eigenvalues of bulge pencils and compared them with the shifts, which they should equal. These studies were concerned with the QR algorithm for the standard eigenvalue problem for nonsymmetric real matrices. We found that when $m = 2$ the bulge eigenvalues generally agreed with the shifts to about 15 decimal places. When $m = 12$ the agreement was only to about 5 decimal places, and when $m = 24$, there was hardly any agreement at all. We found further that as m was increased, the bulge pencil eigenvalues became increasingly ill conditioned. A tiny relative perturbation in the bulge pencil causes a huge change in its eigenvalues. Thus the shifts are not well specified by the bulge pencil; they are blurred, so to speak. Thus we call this phenomenon *shift blurring*. Because of shift blurring, QR iterations with large m do not converge quadratically as they should. See [217] for more details. Kressner [135] has related shift blurring to the notorious ill conditioning of the pole-placement problem of control theory.

    Here we present one set of experiments on a random complex properly Hessenberg-triangular pencil of order 300. We performed a single iteration of degree m of the QZ algorithm with various choices of m. We computed the eigenvalues of the first intermediate bulge pencil $E_{1/2} - \lambda F_{1/2}$ and compared them with the shifts. We also checked the last intermediate bulge pencil before the bulge is chased off of the bottom of the matrix. The results are shown in Table 7.1. We see that for small values of m the bulge pencil eigenvalues approximate the intended shifts very well. The error is a modest multiple of the unit roundoff $u \approx 10^{-16}$. However, as m is increased, the errors soon become quite large until at $m = 20$ the error is more than 100%. Notice that the damage is done at the beginning of the iteration; the errors do not grow significantly during the iteration. The shift blurring for large m seriously impedes convergence. (However, as Kressner [135] has pointed out, the use of aggressive early deflation [45] can help.)

    Because of these and many other experiments, we do not recommend the use of the QR or QZ algorithm or any GZ algorithm of any kind with m larger than 8 or so.

Table 7.1. *Shift errors for a* 300×300 *pencil*

	Maximum relative shift error	
	At start	At end
m	of iteration	of iteration
1	4.6×10^{-16}	1.8×10^{-15}
2	1.3×10^{-15}	3.0×10^{-15}
3	3.2×10^{-14}	3.1×10^{-14}
4	2.1×10^{-14}	1.5×10^{-14}
5	2.1×10^{-13}	2.4×10^{-13}
6	2.1×10^{-11}	2.1×10^{-11}
8	1.1×10^{-9}	1.1×10^{-9}
10	3.0×10^{-8}	2.8×10^{-8}
12	3.9×10^{-7}	4.5×10^{-7}
16	6.5×10^{-3}	6.6×10^{-3}
20	2.2×10^{-0}	2.4×10^{-0}

Table 7.2. *Apparent shift errors for a* 300×300 *pencil* $(m = 2)$

Iteration	Apparent relative shift error		$|a_{n-1,n-2}|$	$|a_{n,n-1}|$
	At start of iteration	At end of iteration		
1	2.1×10^{-15}	2.8×10^{-15}	8.7×10^{-1}	9.9×10^{-1}
2	3.3×10^{-15}	4.7×10^{-15}	5.1×10^{-1}	5.6×10^{-1}
3	2.9×10^{-15}	4.4×10^{-15}	1.3×10^{-1}	1.3×10^{-1}
4	1.2×10^{-14}	6.3×10^{-15}	1.1×10^{-2}	1.0×10^{-2}
5	1.2×10^{-14}	1.5×10^{-14}	4.5×10^{-5}	7.7×10^{-5}
6	1.7×10^{-15}	1.3×10^{-12}	6.7×10^{-10}	6.9×10^{-9}
7	1.0×10^{-14}	1.3×10^{-00}	1.6×10^{-19}	2.7×10^{-13}

In Table 7.1 we showed only the result of a single iteration. On subsequent iterations the results are more or less the same. Table 7.2 shows the history of the first seven iterations on another random 300×300 pencil. These are iterations of degree $m = 2$. The values of $|a_{n-1,n-2}|$ and $|a_{n,n-1}|$ after each iteration are given, so that we can monitor convergence. The column for $|a_{n-1,n-2}|$ shows a strong signal of quadratic convergence; after seven iterations it is time to deflate out two eigenvalues. The errors in the bulge pencil eigenvalues are on the order of 10^{-14} at every step, except that on the seventh iteration (and to a much lesser extent on the sixth) the error in the bulge pencil eigenvalues at the end of the iteration is suddenly very large. It appears that the shifts have been lost right at the moment of convergence. However, closer inspection shows that this effect is only apparent, not real.

Table 7.3. *Pencil with* $a_{150,149} = a_{151,150} = a_{152,151} = 10^{-60}$ *(m = 2)*

| Iteration | Apparent relative shift error | | $|a_{n-1,n-2}|$ | $|a_{n,n-1}|$ |
	At start of iteration	At end of iteration		
1	4.6×10^{-16}	6.6×10^{-15}	6.2×10^{-1}	1.7×10^{-0}
2	1.1×10^{-15}	1.3×10^{-14}	7.1×10^{-1}	1.2×10^{-1}
3	8.4×10^{-16}	8.7×10^{-15}	2.7×10^{-1}	9.4×10^{-4}
4	2.0×10^{-15}	1.6×10^{-14}	2.2×10^{-2}	3.2×10^{-7}
5	1.4×10^{-15}	3.1×10^{-13}	4.4×10^{-5}	8.2×10^{-14}
6	1.1×10^{-15}	1.6×10^{-00}	8.2×10^{-11}	7.0×10^{-23}

Looking at the intermediate bulge pencil at the bottom of the matrix, we find that

$$|E| - \lambda|F| \approx \begin{bmatrix} 10^{-10} & 10^{-00} \\ 10^{-18} & 10^{-18} \end{bmatrix} - \lambda \begin{bmatrix} 10^{-10} & 10^{-00} \\ 10^{-18} & 10^{-09} \end{bmatrix}.$$

We have indicated only the magnitudes of the entries, because the important point here is that this pencil is badly out of scale. This is the result of the extremely small values of $a_{n-1,n-2}$ and $a_{n,n-1}$, and it has as a consequence that MATLAB's `eig` function is unable to compute its eigenvalues accurately. If one rescales the pencil by multiplying on the left by $D_1 = \text{diag}\{1, 10^9\}$ and on the right by $D_2 = \text{diag}\{10^9, 1\}$, one obtains

$$D_1(|E| - \lambda|F|)D_2 \approx \begin{bmatrix} 10^{-01} & 10^{-00} \\ 10^{-00} & 10^{-00} \end{bmatrix} - \lambda \begin{bmatrix} 10^{-01} & 10^{-00} \\ 10^{-00} & 10^{-09} \end{bmatrix},$$

which is much better scaled. When we compute the eigenvalues of this rescaled pencil and compare them with the intended shifts, we find that the error is 7.0×10^{-15}. Thus the shifts have not been lost after all.

Another interesting experiment involves pencils with some extremely tiny subdiagonal entries. We took another random 300×300 pencil and modified it so that $a_{150,149} = a_{151,150} = a_{152,151} = 10^{-60}$. Normally one would declare such tiny subdiagonal entries to be effectively zero, set them to zero, and decouple the eigenvalue problem. This is certainly the right move in the interest of efficiency. For years it was widely thought that it was also necessary for the functioning of the algorithm. Surely such tiny entries would "wash out" the QZ iteration and render it ineffective. We were able to show in [216] that this bit of conventional wisdom is wrong. Table 7.3 shows what happens in the the first six iterations on a pencil with $a_{150,149} = a_{151,150} = a_{152,151} = 10^{-60}$.

The columns for $|a_{n-1,n-2}|$ and $|a_{n,n-1}|$ show that once again we have rapid quadratic convergence. The tiny subdiagonal entries have not at all hindered the convergence of the algorithm, nor have they interfered with the transmission of the shifts. In all cases the shift error at the end of the iteration is tiny, except on the sixth iteration. But here again we have only an apparent loss of shifts caused by the bulge pencil being seriously out of scale. If we should rescale the bulge pencil appropriately and then compute its eigenvalues, we would find that the shifts have not been lost.

Table 7.4. *Pencil with $b_{100,100} = b_{150,150} = b_{200,200} = 0$ (m = 2)*

| Iteration | Apparent relative shift error | | $|a_{n-1,n-2}|$ | $|a_{n,n-1}|$ |
	At start of iteration	At end of iteration		
1	3.6×10^{-16}	7.8×10^{-15}	6.4×10^{-1}	9.1×10^{-1}
2	1.7×10^{-15}	7.6×10^{-15}	1.1×10^{-0}	1.1×10^{-1}
3	6.6×10^{-16}	5.6×10^{-15}	3.6×10^{-1}	2.9×10^{-3}
4	1.6×10^{-15}	1.3×10^{-14}	9.9×10^{-2}	3.4×10^{-6}
5	8.7×10^{-16}	4.3×10^{-15}	5.5×10^{-3}	4.9×10^{-12}
6	1.9×10^{-15}	5.0×10^{-15}	3.8×10^{-5}	2.1×10^{-22}
7	5.6×10^{-16}	9.1×10^{-03}	2.0×10^{-9}	3.1×10^{-30}

If we had checked the bulge pencil at the moment when it was passing through the tiny subdiagonal entries, we would have found that at that point is was grotesquely out of scale. An attempt by `eig` to compute its eigenvalues might cause one to think that the shifts have been lost. But again, an appropriate rescaling would show that they have not. This is so, as long as the numbers are not so tiny that they cause an underflow. If we repeat the experiment with, say, $a_{150,149} = a_{151,150} = a_{152,151} = 10^{-120}$, so that their product, which is 10^{-360}, is below the underflow level for IEEE standard arithmetic, we find that the ensuing underflow *does* kill the QZ iterations. The shifts are not transmitted effectively, nor do the iterations converge.

Our last experiment shows that the shift transmission mechanism is not impaired by the presence of infinite eigenvalues. We modified yet another 300×300 Hessenberg-triangular pencil so that $b_{100,100} = b_{150,150} = b_{200,200} = 0$. This pencil has at least one and probably three infinite eigenvalues. As we noted in Section 6.5, if we apply the QZ algorithm to this pencil, the zeros on the main diagonal will be pushed steadily upward until they (that is, the infinite eigenvalues) can be deflated from the top. We claimed in Section 6.5 that the presence of the infinite eigenvalues does not in any way impair the convergence of the QZ algorithm. Now we are in a position to demonstrate this by numerical experiment. First of all, if you examine carefully the proofs of Theorems 7.1.2 and 7.1.3, you will find that there is nothing about them that requires that the main-diagonal entries of B be nonzero, except that the top entries $b_{11}, \ldots, b_{mm}$ have to be nonzero in order to compute the initial vector x. Thus we expect the shift transmission mechanism to work well in this case too. Table 7.4 shows what actually happens. Again we see strong quadratic convergence. We should have deflated out an eigenvalue after iteration six, but for this experiment we did not. The shifts are transmitted effectively.

Exercises

7.1.1.

(a) Download the file `blurtest.m` from the website[1] and figure out what it does. It performs an implicit QR iteration and is identical to `downchase.m`, except that

[1] www.siam.org/books/ot101

some code has been inserted that computes the intermediate bulge pencil (7.1.8) at the beginning and end of the bulge-chase. Download the file `blurgo.m`, which is a simple driver program for `blurtest.m`, and figure out what it does.

(b) Run the code with a fairly large n, say $n = 500$, and various values of m between 1 and 30 or 40. Notice that the shifts are transmitted effectively when m is small, and shift blurring sets in as m is increased.

7.1.2.

(a) Modify the code `blurgo.m` so that the Hessenberg matrix has a few tiny entries (on the order of 10^{-60} or so) along the subdiagonal. (You will have to modify or eliminate the stopping criterion as well.) Run your modified code with $n = 500$ and small values of m, and demonstrate that the tiny entries do not interfere with the shift-transmission mechanism or convergence of the algorithm.

(b) Show that if A has two consecutive very tiny subdiagonal entries, and these entries are small enough to cause underflow in the bulge, the shift-transmission mechanism breaks down and the algorithm fails to converge.

7.1.3. This exercise works out the proof of Proposition 7.1.1.

(a) Show that the vectors $e_1, Ce_1, C^2e_1, \ldots, C^{n-1}e_1$ are linearly independent.

(b) Deduce that there are unique coefficients $a_0, \ldots, a_{n-1}$, not all zero, such that $x = (a_0 I + a_1 C + a_2 C^2 + \cdots + a_{n-1} C^{n-1})e_1$. Show that there is a unique polynomial $\tilde{p}$ of degree $< n$ such that $x = \tilde{p}(C)e_1$. Suppose that the degree of $\tilde{p}$ is k. Show that there is a unique monic polynomial p of degree $< n$ such that $x = \alpha p(C)e_1$ for some nonzero α (in fact $\alpha = a_k$ and $p = \alpha_k^{-1} \tilde{p}$).

(c) Show that if $x_{m+1} \neq 0$ and $x_i = 0$ for all $i > m + 1$, then the degree of p is m.

7.1.4. Let $A - \lambda B$ be a properly Hessenberg-triangular pencil, and let $E_0 - \lambda F_0$ be the initial bulge pencil, as defined in the text.

(a) Show that the outcome of the computation of x in (7.1.1) (and (6.3.5)) is completely determined by $\rho_1, \ldots, \rho_m$, the first m columns of A, and the first m columns of B.

(b) Compute x_{m+1} and deduce that $x_{m+1} \neq 0$.

(c) Use the definition of the determinant and the fact that $x_{m+1} \neq 0$ to show that the degree of the characteristic polynomial of $E_0 - \lambda F_0$ is exactly m.

Therefore the initial bulge pencil is regular and has m finite eigenvalues and one infinite eigenvalue.

7.1.5. This exercise fills in the details of the proof of Theorem 7.1.2.

(a) Show that ∞ is an eigenvalue of the bulge pencil $E_0 - \lambda F_0$ with associated eigenvector e_1. (Show that 0 is an eigenvalue of the inverted pencil $F_0 - \mu E_0$ with eigenvector e_1.)

(b) Bearing in mind that the factors in (7.1.1) commute, show that (7.1.1) can be rewritten as $x = (A - \rho_i B) B^{-1} \tilde{p}(AB^{-1}) e_1$, where $\tilde{p}$ is a polynomial of degree $m - 1$.

(c) Let $z = B^{-1} \tilde{p}(AB^{-1}) e_1$. Show $z_i = 0$ for $i > m$.

(d) Let $\tilde{z} \in \mathbb{C}^m$ consist of the first m components of z. Show that the equation $x = (A - \rho_i B) z$ implies that $\begin{bmatrix} -1 \\ \tilde{z} \end{bmatrix}$ is an eigenvector of $E_0 - \lambda F_0$ with associated eigenvalue ρ_i.

(e) Show that if $\rho_1, \ldots, \rho_m$ are not distinct, it is always possible to find $\sigma_1, \ldots, \sigma_m$ such that $\rho_1 + \epsilon \sigma_1, \ldots, \rho_m + \epsilon \sigma_m$ are distinct for all ϵ sufficiently close to 0.

(f) We proved continuity of the eigenvalues only for the standard eigenvalue problem (Theorem 2.7.1). Show that the argument in the proof of Theorem 2.7.1 can also be applied to the generalized eigenvalue problem considered in the proof of Theorem 7.1.2.

7.2   Bulge-Chasing in Reverse

This section breaks the continuity of the development somewhat, but it is necessary for the section that follows. Chapters 3 through 6, as well as Section 7.1, have been dominated by GR decompositions. One might reasonably ask why we chose GR rather than RG. Indeed, there is an entirely analogous theory of RG algorithms based on RG decompositions (Section 3.4), which we now consider briefly.[2] Given a matrix A whose eigenvalues we would like to know and a polynomial q, a generic RG iteration driven by q is summarized by the two equations

$$q(A) = RG, \qquad \hat{A} = GAG^{-1}, \qquad (7.2.1)$$

where R is upper triangular and G is nonsingular. Of course the RG decomposition here is far from unique. The new iterate $\hat{A}$ is similar to A, so it has the same eigenvalues. Repeated RG iterations with good choices of q will drive the matrix toward block triangular form, revealing the eigenvalues. As we shall see, the RG algorithm is the GR algorithm with time reversal. How can it be, then, that both the GR and the RG algorithms reveal the eigenvalues? Shouldn't RG rather bury them? It turns out that (with fixed q, for example) the GR algorithm yields the eigenvalues in a certain order on the main diagonal, and the RG algorithm gives them in the reverse order.

Suppose that $q(A)$ is nonsingular. Then so is R and, letting $\tilde{R} = R^{-1}$ and $\tilde{G} = G^{-1}$, we can rewrite (7.2.1) as

$$q(A)^{-1} = \tilde{G} \tilde{R}, \qquad \hat{A} = \tilde{G}^{-1} A \tilde{G}.$$

This is a GR iteration driven by the rational function $1/q$. Thus we conclude that an RG iteration driven by q is equivalent to a GR iteration driven by $1/q$. Recalling the convergence theory, we know that the convergence rate of GR iterations driven by q is determined (in part) by ratios $|q(\lambda_{k+1})/q(\lambda_k)|$ determined by the eigenvalues $q(\lambda_1), \ldots, q(\lambda_n)$. On the other

---

[2]There are also analogous GL and LG theories, where L is lower triangular. It was a matter of chance that the numerical linear algebra community settled on GR rather than one of these others.

hand, for GR iterations driven by $1/q$, the relevant eigenvalues are $q(\lambda_n)^{-1}, \ldots, q(\lambda_1)^{-1}$. The magnitudes of the eigenvalues are reversed, the ratios are reversed, and the eigenvalues converge at the same rates but in the opposite order.

If we can implement GR iterations implicitly by bulge-chasing, it must also be possible to implement RG iterations in an analogous way. In view of our observation that RG iterations are GR iterations in reverse, it should not be surprising that RG iterations can be effected by bulge-chasing from bottom to top.

We will now switch to the generalized eigenvalue problem. The theory for the standard eigenvalue problem can be deduced from the more general theory by setting $B = I$.

Reverse GZ Iterations

Consider a pencil $A - \lambda B$, where B is nonsingular. Recalling that an explicit GZ iteration driven by p is given by (6.3.3) and (6.3.4), we define an explicit *reverse GZ iteration* driven by q by

$$q(AB^{-1}) = RG, \qquad q(B^{-1}A) = SZ \tag{7.2.2}$$

and

$$\hat{A} - \lambda \hat{B} = G(A - \lambda B)Z^{-1}, \tag{7.2.3}$$

where R and S are upper triangular and G and Z are nonsingular. We resist the urge to speak of "ZG" iterations. If $q(AB^{-1})$ and $q(B^{-1}A)$ are nonsingular, then we can invert the equations in (7.2.2) and (letting $\tilde{G} = G^{-1}$, etc.) deduce that a reverse GZ iteration driven by q is exactly a (forward) GZ iteration driven by $1/q$.

In order to do an *implicit* reverse GZ iteration, we need a pencil that is in proper Hessenberg-triangular form, so let us assume that $A - \lambda B$ has been transformed to this form by the algorithm described in Section 6.2, for example. Recall that this means that A is properly upper Hessenberg and B is upper triangular and nonsingular.

The implicit (forward) GZ iteration succeeds by getting the first column of G right (Proposition 6.3.1). Similarly the implicit reverse GZ iteration succeeds by getting the bottom row of Z right.

To get the reverse GZ iteration started, we need a vector y^T proportional to the bottom row of $q(B^{-1}A)$. Suppose that q has degree k and $q(z) = (z - \tau_1) \cdots (z - \tau_k)$. Then

$$y^T = \beta e_n^T (B^{-1}A - \tau_1 I) \cdots (B^{-1}A - \tau_k I), \tag{7.2.4}$$

where β is any nonzero constant. This vector can be computed in $O(k^3)$ work by an algorithm analogous to (6.3.5). Because of the proper Hessenberg-triangular form of (A, B), the first $n - k - 1$ entries of y are zero, and $y_{n-k} \neq 0$.

Let Z_1 be an elimination matrix such that $\gamma_1 y^T = e_n^T Z_1$ for some nonzero γ_1. The last row of Z_1 is proportional to the last row of $q(B^{-1}A)$. Because of the form of y^T, Z_1 can take the form

$$Z_1 = \begin{bmatrix} I_{n-k-1} & \\ & \tilde{Z}_1 \end{bmatrix}.$$

The first transformation of the reverse GZ iteration is

$$A_{1/2} = AZ_1^{-1}, \qquad B_{1/2} = BZ_1^{-1}.$$

The action of Z_1^{-1} affects the last $k + 1$ columns of A and B, disturbing the Hessenberg-triangular form. In the case $k = 2$, $A_{1/2}$ and $B_{1/2}$ have the forms

$$
\begin{bmatrix}
a & a & a & a & a \\
a & a & a & a & a \\
  & a & a & a & a \\
  &   & a & a & a \\
  &   & + & a & a
\end{bmatrix}
\quad \text{and} \quad
\begin{bmatrix}
b & b & b & b & b \\
  & b & b & b & b \\
  &   & b & b & b \\
  &   & + & b & b \\
  &   & + & + & b
\end{bmatrix},
$$

respectively. The plus symbols indicate bulges that have crept into the matrices.

The rest of the iteration consists of returning the pair to Hessenberg-triangular form by chasing the bulges from bottom to top by an algorithm analogous to the second algorithm discussed in Section 6.2. The first task is to return the bottom row of $B_{1/2}$ to upper triangular form. This requires a QR decomposition of $B_{1/2}$, which is inexpensive because $B_{1/2}$ is already close to triangular form. Write $B_{1/2}$ in the partitioned form

$$
B_{1/2} = \begin{bmatrix} B_{11} & B_{12} \\ & B_{22} \end{bmatrix},
$$

where B_{22} is $(k + 1) \times (k + 1)$ and B_{11} is upper triangular. If $B_{22} = Q_{22} R_{22}$ is a QR decomposition of B_{22}, then a QR decomposition of $B_{1/2}$ is given by

$$
B_{1/2} = \begin{bmatrix} B_{11} & B_{12} \\ & B_{22} \end{bmatrix} = \begin{bmatrix} I & \\ & Q_{22} \end{bmatrix} \begin{bmatrix} B_{11} & B_{12} \\ & R_{22} \end{bmatrix}.
$$

This costs only $O(k^3)$ flops. Let $\tilde{G}_1 \in \mathbb{C}^{(k+1) \times (k+1)}$ be an elimination matrix whose last row is proportional to the last row of Q_{22}^*. This guarantees that the last row of $\tilde{G}_1$ is orthogonal to all columns of B_{22} but the last. Thus the transformation $B_{22} \to \tilde{G}_1 B_{22}$ returns the bottom row of B_{22} to triangular form. Let $G_1 = \text{diag}\{I_{n-k-1}, \tilde{G}_1\}$, and let

$$
A_1 = G_1 A_{1/2}, \qquad B_1 = G_1 B_{1/2}.
$$

The bottom row of B_1 is now in upper triangular form. The transforming matrix acts on the last $k + 1$ rows of $A_{1/2}$, with the result that one column is added to the left of the bulge. In the case $k = 2$, A_1 and B_1 have the forms

$$
\begin{bmatrix}
a & a & a & a & a \\
a & a & a & a & a \\
  & a & a & a & a \\
  & + & a & a & a \\
  & + & + & a & a
\end{bmatrix}
\quad \text{and} \quad
\begin{bmatrix}
b & b & b & b & b \\
  & b & b & b & b \\
  &   & b & b & b \\
  &   & + & b & b \\
  &   &   &   & b
\end{bmatrix},
$$

respectively. The bulge in B has been shrunk at the expense of expanding the bulge in A.

The next step transforms the bottom row of A_1 to Hessenberg form by a transformation

$$
A_{3/2} = A_1 Z_2^{-1}, \qquad B_{3/2} = B_1 Z_2^{-1},
$$

where $Z_2 = \text{diag}\{I_{n-k-2}, \tilde{Z}_2, 1\}$. The transformation $B_{3/2} = B_1 Z_2^{-1}$ acts on columns $n - k - 1, \ldots, n - 1$ of B, adding a column to the bulge. In the case $k = 2$, $A_{3/2}$ and $B_{3/2}$

have the forms

$$
\begin{bmatrix}
a & a & a & a & a \\
a & a & a & a & a \\
  & a & a & a & a \\
  & + & a & a & a \\
  &   &   & a & a
\end{bmatrix}
\quad \text{and} \quad
\begin{bmatrix}
b & b & b & b & b \\
  & b & b & b & b \\
  & + & b & b & b \\
  & + & + & b & b \\
  &   &   &   & b
\end{bmatrix},
$$

respectively. The bulge in A has been shrunk and the bulge in B expanded. Both bulges have now been moved up one row and over one column from where they were originally. The pattern should now be clear. The next transformation, G_2, clears out row $(n-1)$ of $B_{3/2}$. This requires another QR decomposition, which is again inexpensive. G_2 acts on rows $n-k-1, \ldots, n-1$ of $A_{3/2}$, adding a column to the left of the bulge. Next a transformation Z_3^{-1} clears out row $n-1$ of A_2, adding a column to the left of the bulge in B_2. Continuing in this manner, we chase the bulges upward and eventually off the top of the matrices. Once this happens, the implicit reverse GZ iteration is complete. This takes a total of $n-1$ transformations on left and right. The result is

$$
\hat{A} = A_{n-1} = G_{n-1} \cdots G_1 A Z_1^{-1} \cdots Z_{n-1}^{-1},
$$

$$
\hat{B} = B_{n-1} = G_{n-1} \cdots G_1 B Z_1^{-1} \cdots Z_{n-1}^{-1}.
$$

Letting

$$
G = G_{n-1} \cdots G_2 G_1 \quad \text{and} \quad Z = Z_{n-1} \cdots Z_2 Z_1, \tag{7.2.5}
$$

we have

$$
\hat{A} - \lambda \hat{B} = G(A - \lambda B)Z^{-1},
$$

as in (7.2.3). To show that a reverse GZ iteration has been effected, we must show that the matrices G and Z of (7.2.5) satisfy equations of the form (7.2.2). To this end we begin by noting that the form of the matrices $Z_2, \ldots, Z_{n-1}$ implies that the last row of Z is the same as the last row of Z_1. Thus the last row of Z is proportional to the last row of $q(B^{-1}A)$; that is,

$$
e_n^T Z = \beta e_n^T q(B^{-1}A)
$$

for some nonzero β.

To draw the desired conclusion we need an analogue of Proposition 6.3.1, and for this we need to introduce *left* Krylov matrices and prove analogues of Lemmas 4.5.2, 4.5.3, and 4.5.4. Given an $n \times n$ matrix C and a row vector v^T, the *left Krylov matrix* $\tilde{\kappa}(v^T, C)$ is the $n \times n$ matrix with rows $v^T C^{n-1}, v^T C^{n-2}, \ldots, v^T C, v^T$, from top to bottom. The following lemmas are now easily proved.

Lemma 7.2.1. *Suppose* $\hat{C}Z = ZC$. *Then for any* $w \in \mathbb{C}^n$,

$$
\tilde{\kappa}(w^T, \hat{C})Z = \tilde{\kappa}(w^T Z, C).
$$

Lemma 7.2.2. *For any* $C \in \mathbb{C}^{n \times n}$ *and any* $w \in \mathbb{C}^n$,

$$
\tilde{\kappa}(w^T, C)q(C) = \tilde{\kappa}(w^T q(C), C).
$$

Lemma 7.2.3. *Let* $C \in \mathbb{C}^{n \times n}$ *be an upper Hessenberg matrix. Then*

(a) $\tilde{\kappa}(e_n^T, C)$ is upper triangular.

(b) $\tilde{\kappa}(e_n^T, C)$ is nonsingular if and only if C is properly upper Hessenberg.

With the help of these lemmas we can prove the following analogue of Proposition 6.3.1. (See Exercise 7.2.4.)

Proposition 7.2.4. *Suppose that the matrix pair (A, B) is properly Hessenberg-triangular, and suppose further that*

$$\hat{A} = GAZ^{-1} \quad and \quad \hat{B} = GBZ^{-1},$$

where G and Z are nonsingular and the pair $(\hat{A}, \hat{B})$ is also Hessenberg-triangular. Finally, suppose $e_n^T Z = \beta e_n^T q(B^{-1}A)$ for some $\beta \neq 0$. Then there exist upper triangular matrices R and S such that

$$q(AB^{-1}) = RG \quad and \quad q(B^{-1}A) = SZ.$$

Since the reverse bulge-chase described above satisfies all of the hypotheses of Proposition 7.2.4, we deduce that it effects a reverse GZ iteration (7.2.2), (7.2.3) implicitly.

Transmission of Shifts in the Reverse *GZ* Iteration

As one would expect, the mechanism by which information about the shifts is transmitted in a reverse GZ bulge-chase is the same as it is for a forward GZ iteration. Let us consider once again the augmented pencil

$$\tilde{A} - \lambda \tilde{B} = \begin{bmatrix} x & A - \lambda B \\ 0 & y^T \end{bmatrix}, \tag{7.2.6}$$

but this time we take

$$x = e_1 \quad and \quad y^T = \beta e_n^T q(B^{-1}A). \tag{7.2.7}$$

With this choice of x and y, the augmented pencil is not upper triangular, but it is block triangular, having the form

$$\begin{bmatrix} \tilde{A}_{11} - \lambda \tilde{B}_{11} & * \\ 0 & \tilde{A}_{22} - \lambda \tilde{B}_{22} \end{bmatrix}, \tag{7.2.8}$$

where $\tilde{A}_{22} \in \mathbb{C}^{(k+1)\times(k+1)}$ would be upper triangular, except that its bottom is $\tilde{y}^T$, the nonzero part of y^T; $\tilde{A}_{11}$ is upper triangular; and $\tilde{B}_{11}$ and $\tilde{B}_{22}$ are strictly upper triangular. The spectrum of $\tilde{A} - \lambda \tilde{B}$ is the union of the spectra of $\tilde{A}_{11} - \lambda \tilde{B}_{11}$ and $\tilde{A}_{22} - \lambda \tilde{B}_{22}$. The spectrum of $\tilde{A}_{11} - \lambda \tilde{B}_{11}$ consists of ∞ with multiplicity $n - k$. Based on what we know about the forward GZ iteration, we would expect that the spectrum of $\tilde{A}_{22} - \lambda \tilde{B}_{22}$ must be ∞ and the shifts $\tau_1, \ldots, \tau_k$, the zeros of q.

We will use the notation $\tilde{E}_0 - \lambda \tilde{F}_0 = \tilde{A}_{22} - \lambda \tilde{B}_{22}$ and call this pencil the *initial bulge pencil* for the implicit reverse GZ iteration. In analogy with Theorem 7.1.2 we have the following result, which is proved in Exercise 7.2.5.

Theorem 7.2.5. *The eigenvalues of the initial bulge pencil $\tilde{E}_0 - \lambda \tilde{F}_0$ are the shifts $\tau_1, \ldots, \tau_k$, together with ∞.*

Thanks to this theorem we know that the augmented pencil has eigenvalues ∞ with algebraic multiplicity $n - k + 1$ and $\tau_1, \ldots, \tau_k$ initially. During the implicit reverse GZ iteration, each transformation that is applied to the pencil has a corresponding transformation on the augmented pencil. Each of these is a strict equivalence, so the eigenvalues of the augmented pencil are preserved. The first transformation maps $A - \lambda B$ to $A_{1/2} - \lambda B_{1/2} = (A - \lambda B)Z_1^{-1}$. The corresponding transformation on the augmented pencil multiplies $\tilde{A} - \lambda \tilde{B}$ on the right by $\mathrm{diag}\{1, Z_1^{-1}\}$ Since this transformation was designed so that $e_n^T Z_1 = \zeta y^T$ for some nonzero ζ or, equivalently, $y^T Z_1^{-1} = \zeta^{-1} e_n^T$, the transformed augmented pencil is

$$\left[\begin{array}{cc} e_1 & A_{1/2} - \lambda B_{1/2} \\ 0 & \zeta^{-1} e_n^T \end{array} \right].$$

The transformation from $A_{1/2} - \lambda B_{1/2}$ to $A_1 - \lambda B_1 = G_1(A_{1/2} - \lambda B_{1/2})$ corresponds to multiplication of the augmented pencil on the left by $\mathrm{diag}\{G_1, 1\}$ to produce

$$\left[\begin{array}{cc} e_1 & A_1 - \lambda B_1 \\ 0 & \zeta^{-1} e_n^T \end{array} \right].$$

The transformation removes a row from the "B" bulge and adds a column to the "A" bulge. It does not touch the first column or the last row of the augmented pencil. Subsequent transformations push the bulge upward. None of these touches the bottom row of the augmented pencil. None of them touches the first column either, until the bulge reaches the top of the pencil.

At intermediate stages of the bulge chase, the augmented pencil has the form

$$\left[\begin{array}{cc} e_1 & A_i - \lambda B_i \\ 0 & \zeta^{-1} e_n^T \end{array} \right]. \tag{7.2.9}$$

It is not upper triangular, because there is a bulge in $A_i - \lambda B_i$; we can rewrite it as

$$\left[\begin{array}{ccc} \tilde{A}_{11} - \lambda \tilde{B}_{11} & * & * \\ & \tilde{A}_{22} - \lambda \tilde{B}_{22} & * \\ & & \tilde{A}_{33} - \lambda \tilde{B}_{33} \end{array} \right], \tag{7.2.10}$$

where the $(k + 1) \times (k + 1)$ subpencil $\tilde{A}_{22} - \lambda \tilde{B}_{22}$ is the part that contains the bulge. We will call this the ith bulge pencil for the upward bulge-chase and denote it by $\tilde{E}_i - \lambda \tilde{F}_i$. In analogy with Theorem 7.1.3, we have the following result (Exercise 7.2.6).

Theorem 7.2.6. *In an implicit reverse GZ iteration of degree k with shifts $\tau_1, \ldots, \tau_k$, the eigenvalues of the bulge pencil $\tilde{E}_i - \lambda \tilde{F}_i$ at every step are $\tau_1, \ldots, \tau_k$ and ∞. In other words, the characteristic polynomial $\det(\tilde{E}_i - \lambda \tilde{F}_i)$ is a multiple of $q(z) = (z - \tau_1) \cdots (z - \tau_k)$.*

One way to prove this result is by induction on i, moving from $\tilde{E}_i - \lambda \tilde{F}_i$ to $\tilde{E}_{i+1} - \lambda \tilde{F}_{i+1}$ via an intermediate bulge pencil $\tilde{E}_{i+1/2} - \lambda \tilde{F}_{i+1/2}$. In the process we find that the intermediate pencil, which is $k \times k$, has eigenvalues $\tau_1, \ldots, \tau_k$.

Once the reverse GZ iteration is complete, the augmented pencil is in its *final configuration*, which looks like

$$\begin{bmatrix} w & \hat{A} - \lambda\hat{B} \\ 0 & \zeta^{-1}e_n^T \end{bmatrix} = \begin{bmatrix} \tilde{A}_{22} - \lambda\tilde{B}_{22} & * \\ & \tilde{A}_{33} - \lambda\tilde{B}_{33} \end{bmatrix}, \qquad (7.2.11)$$

where $\hat{A} - \lambda\hat{B} = G(A - \lambda B)Z^{-1}$ and $w = Ge_1$. The bulge has been chased from $\hat{A} - \lambda\hat{B}$, but the augmented pencil still has a bulge, which has been compressed into w. $\tilde{A}_{22} - \lambda\tilde{B}_{22}$ is the part of the augmented pencil that is not upper triangular. It is $(k+1) \times (k+1)$, and its first column consists of the nonzero part of w. It is the final bulge pencil for the reverse GZ iteration, and its eigenvalues are, of course, $\tau_1, \ldots, \tau_k$ and ∞.

The pencil $\hat{A} - \lambda\hat{B}$ will always be Hessenberg-triangular, and generically it will be *properly* Hessenberg-triangular. Let us assume that it is. Then by Proposition 7.1.1, the vector w can be expressed in the form $w = \alpha\hat{q}(\hat{A}\hat{B}^{-1})e_1$ for some uniquely determined monic polynomial $\hat{q}$ of degree k. Suppose $\hat{q}(z) = (z - \hat{\tau}_1)\cdots(z - \hat{\tau}_k)$. Then, arguing as in the proof of Theorem 7.1.2, we find that the eigenvalues of the final bulge pencil $\tilde{A}_{22} - \lambda\tilde{B}_{22}$ in (7.2.11) must be $\hat{\tau}_1, \ldots, \hat{\tau}_k$ and ∞. But we already know that they are $\tau_1, \ldots, \tau_k$ and ∞. Thus $\hat{q} = q$, and $w = \alpha q(AB^{-1})e_1$.

We conclude that the final configuration (7.2.11) of a reverse GZ bulge-chase of degree k with shifts $\tau_1, \ldots \tau_k$ is exactly an initial configuration for a forward GZ bulge-chase of the same degree with the same shifts. If we now proceed with this downward bulge-chase, it will undo the upward bulge-chase that was just completed.

Similarly the final configuration of a forward GZ iteration is exactly an initial configuration for a backward GZ iteration with the same shifts. If we execute the backward iteration, it will undo the forward iteration that was just done.

These reversibility results depend upon the assumption that the final pencil $\hat{A} - \lambda\hat{B}$ is properly upper Hessenberg. Of course, if it is not properly Hessenberg-triangular, that is good news, because we get to decouple the problem. In Section 4.6 we studied the question of when an exact zero appears on the subdiagonal in the case of GR iterations for the standard eigenvalue problem. We found that this happens if and only if one or more of the shifts is an eigenvalue of the matrix. Since GZ iterations effect GR iterations on AB^{-1} and $B^{-1}A$, these findings for the standard eigenvalue problem also apply to the generalized problem. Thus a GZ iteration will result in a deflation if and only if at least one of the shifts is an eigenvalue of $A - \lambda B$. In this and only this case, the GZ iterations cannot be undone by a reverse GZ iteration. The same applies to the undoing of reverse GZ iterations as well, of course.

Exercises

7.2.1. Show that the explicit reverse GZ iteration (7.2.2), (7.2.3) automatically effects RG iterations on AB^{-1} and $B^{-1}A$.

7.2.2. This exercise is concerned with the details of the computation of y^T in (7.2.4), assuming that the pair (A, B) is properly Hessenberg-triangular.

    (a) Write an algorithm like (6.3.5) that computes the vector y^T given by (7.2.4).

(b) Show that $y_i = 0$ for $i = 1, \ldots, n - k - 1$, and that the nonzero part of y can be computed in $O(k^3)$ work.

(c) Show that $y_{n-k} \neq 0$.

7.2.3. Prove the following:

    (a) Lemma 7.2.1 (the analogue of Lemma 4.5.2).

    (b) Lemma 7.2.2 (the analogue of Lemma 4.5.3).

    (c) Lemma 7.2.3 (the analogue of Lemma 4.5.4).

7.2.4. This exercise proves Proposition 7.2.4.

    (a) Show that $\hat{B}^{-1}\hat{A} = Z(B^{-1}A)Z^{-1}$, and hence $(\hat{B}^{-1}\hat{A})Z = Z(B^{-1}A)$. Then apply Lemma 7.2.1 to deduce that

$$\tilde{\kappa}(e_n^T, \hat{B}^{-1}\hat{A})Z = \tilde{\kappa}(e_n^T Z, B^{-1}A).$$

    (b) Use the hypothesis $e_n^T Z = \beta e_n^T q(B^{-1}A)$ along with Lemma 7.2.2 to show that

$$\tilde{\kappa}(e_n^T, \hat{B}^{-1}\hat{A})Z = \beta\tilde{\kappa}(e_n^T, B^{-1}A)q(B^{-1}A).$$

    (c) Deduce that

$$q(B^{-1}A) = SZ, \quad \text{where} \quad S = \beta^{-1}\tilde{\kappa}(e_n^T, B^{-1}A)^{-1}\tilde{\kappa}(e_n^T, \hat{B}^{-1}\hat{A}).$$

    Use Lemma 7.2.3 to show that S exists and is upper triangular. Of course, Proposition 1.3.3 is also used here.

    (d) Show that $q(B^{-1}A)B^{-1} = B^{-1}q(AB^{-1})$.

    (e) Show that $G = \hat{B}ZB^{-1}$. Use this fact, the fact from part (d), and the hypothesis $e_n^T Z = \beta e_n^T q(B^{-1}A)$ to show that

$$e_n^T G = \delta e_n^T q(AB^{-1}),$$

    where $\delta = \hat{b}_{nn}\beta b_{nn}^{-1} \neq 0$. Bear in mind that $\hat{B}$ and B^{-1} are upper triangular.

    (f) Show that $\hat{A}\hat{B}^{-1} = G(AB^{-1})G^{-1}$, and hence $(\hat{A}\hat{B}^{-1})G = G(AB^{-1})$. Then apply Lemma 7.2.1 to deduce that

$$\tilde{\kappa}(e_n^T, \hat{A}\hat{B}^{-1})G = \tilde{\kappa}(e_n^T G, AB^{-1}).$$

    (g) Use the conclusions of parts (e) and (f) along with Lemma 7.2.2 to show that

$$\tilde{\kappa}(e_n^T, \hat{A}\hat{B}^{-1})G = \delta\tilde{\kappa}(e_n^T, AB^{-1})q(AB^{-1}).$$

(h) Deduce that

$$q(AB^{-1}) = RG, \quad \text{where} \quad R = \delta^{-1}\tilde{\kappa}(e_n^T, AB^{-1})^{-1}\tilde{\kappa}(e_n^T, \hat{A}\hat{B}^{-1}).$$

Use Lemma 7.2.3 to show that R exists and is upper triangular.

7.2.5. This exercise proves Theorem 7.2.5.

(a) Show that ∞ is an eigenvalue of the bulge pencil $\tilde{E}_0 - \lambda\tilde{F}_0$ with associated left eigenvector e_{k+1}^T.

(b) Let τ_i be any one of the shifts. Bearing in mind that the factors in (7.2.4) commute, show that this equation can be rewritten as $y^T = e_n^T \tilde{q}(B^{-1}A)(A - \tau_i B)$, where $\tilde{q}$ is a polynomial of degree $k - 1$.

(c) Let $z^T = e_n^T \tilde{q}(B^{-1}A)B^{-1}$. Show $z_i = 0$ for $i < n - k + 1$.

(d) Let $\tilde{z} \in \mathbb{C}^k$ consist of the last k components of z. Show that the equation $y^T = z^T(A - \tau_i B)$ implies that $\begin{bmatrix} \tilde{z}^T & -1 \end{bmatrix}$ is a left eigenvector of $\tilde{E}_0 - \lambda\tilde{F}_0$ with associated eigenvalue τ_i.

(e) Explain why what you have done to this point proves that the spectrum of $\tilde{E}_0 - \lambda\tilde{F}_0$ consists precisely of $\tau_1, \ldots, \tau_k$ and ∞ in the case when $\tau_1, \ldots, \tau_k$ are distinct.

(f) Devise a continuity argument that finishes the proof in the case when $\tau_1, \ldots, \tau_k$ are not distinct.

7.2.6. Prove Theorem 7.2.6 by two different methods.

(a) Consider the eigenvalues of the whole augmented pencil.

(b) Prove the result by induction on i by moving from $\tilde{E}_i - \lambda\tilde{F}_i$ to $\tilde{E}_{i+1} - \lambda\tilde{F}_{i+1}$ through an intermediate bulge pencil $\tilde{E}_{i+1/2} - \lambda\tilde{F}_{i+1/2}$.

7.3 Passing Bulges through Each Other

As we have seen, a reverse GZ iteration driven by q is the same as a (forward) GZ iteration driven by $1/q$. Therefore, if we do a GZ iteration driven by p followed by a reverse GZ iteration by q, this is the same as a GZ iteration driven by the rational function $r = p/q$. If we do the reverse iteration driven by q first, then do the forward iteration by p, we get the same result. It is natural to ask whether we can do both at once. Can we start a bulge at each end of the matrix, chase the bulges toward each other, pass them through each other, and complete the iteration? In this section we will see that the answer is yes.

We will continue to assume that p and q have degrees m and k, respectively, and that their zeros are $\rho_1, \ldots, \rho_m$ and $\tau_1, \ldots, \tau_k$, respectively. Before we get into the analysis, we should note at once that there is no point in doing such an iteration unless the zeros of p are well separated from those of q. In the extreme case of nonseparation we would have $p = q$, $r = p/p = 1$, and the iteration would accomplish nothing. More generally, if some of the ρ_i are close to some of the τ_i, the iteration will be partially self-cancelling.

On the other hand, if $\rho_1, \ldots, \rho_m$ are well separated from $\tau_1, \ldots, \tau_k$, then eigenvalues near $\rho_1, \ldots, \rho_m$ will be extracted at the bottom of the matrix while eigenvalues near $\tau_1, \ldots, \tau_k$ will be extracted at the top. The two processes will not interfere with each other at all.

To analyze a bidirectional bulge-chase, consider the augmented pencil

$$\begin{bmatrix} x & A - \lambda B \\ 0 & y^T \end{bmatrix}$$

with

$$x = \alpha p(AB^{-1})e_1 \quad \text{and} \quad y^T = \beta e_n^T q(B^{-1}A).$$

Simultaneously we start bulges at top and bottom and chase them toward each other. At some point the bulges will meet, and we have to pass them through each other. We do not insist that they meet in the middle of the matrix. We can imagine that one bulge might have traveled much further than the other, and they could meet anywhere. Once we have pushed the two bulges to the point where they are about to collide, we can do an additional half step on the upper bulge and deflate an infinite eigenvalue from the top to leave an intermediate bulge pencil like (7.1.8). Similarly we can deflate an infinite eigenvalue from the bottom of the lower bulge. Then we have two adjacent intermediate bulge pencils

$$\begin{bmatrix} E_{11} - \lambda F_{11} & E_{12} - \lambda F_{12} \\ 0 & E_{22} - \lambda F_{22} \end{bmatrix},$$

where $E_{11} - \lambda F_{11}$ is $m \times m$ and has eigenvalues $\rho_1, \ldots, \rho_m$, and $E_{22} - \lambda F_{22}$ is $k \times k$ and has eigenvalues $\tau_1, \ldots, \tau_k$. Our task is to transform this to a form

$$\begin{bmatrix} \hat{E}_{11} - \lambda \hat{F}_{11} & \hat{E}_{12} - \lambda \hat{F}_{12} \\ 0 & \hat{E}_{22} - \lambda \hat{F}_{22} \end{bmatrix},$$

where $\hat{E}_{11} - \lambda \hat{F}_{11}$ is $k \times k$ and has eigenvalues $\tau_1, \ldots, \tau_k$.

We showed in Section 6.6 that a swap of this type requires the solution of the generalized Sylvester equations (6.6.8). This is a system of $2mk$ linear equations, and it always has a unique solution if $\rho_1, \ldots, \rho_m$ are disjoint from $\tau_1, \ldots, \tau_k$, as is the case here. Since m and k are generally small, we can solve the system by a standard method. Alternatively we can use a specialized method (see [122] and Exercise 6.6.9).

The exchange always works in exact arithmetic. In practice, as we mentioned in Section 6.6, roundoff errors cause $\hat{E}_{21}$ to be nonzero. If $\| \hat{E}_{21} \|$ is small enough, we set it to zero. We then complete the bidirectional QZ iteration by pushing the two bulges away from each other until they are pushed off the ends of the pencil.

If, on the other hand, $\| \hat{E}_{21} \|$ is not small enough, we must reject the swap, and both bulges must be chased off in the same direction. This nullifies one of the bulge-chases, and it is a waste of effort, so we hope that it happens very rarely.

Theoretical Verification of *GZ* Iteration

Suppose that we have accomplished the interchange and chased the bulges off the end of the pencil. Call the resulting pencil $\hat{A} - \lambda \hat{B}$. The final augmented matrix is

$$\begin{bmatrix} z & \hat{A} - \lambda \hat{B} \\ 0 & w^T \end{bmatrix},$$

where z has nonzeros in at most its first $k + 1$ positions, and w^T has nonzeros in at most its last $m + 1$ positions. We can also write this pencil as

$$\begin{bmatrix} \tilde{A}_{11} - \lambda \tilde{B}_{11} & * & * \\ & \tilde{A}_{22} - \lambda \tilde{B}_{22} & * \\ & & \tilde{A}_{33} - \lambda \tilde{B}_{33} \end{bmatrix},$$

where $\tilde{A}_{11} - \lambda \tilde{B}_{11}$ is $(k + 1) \times (k + 1)$ and has $\tilde{z}$, the nonzero part of z, as its first column. Its characteristic polynomial is proportional to q. Similarly $\tilde{A}_{33} - \lambda \tilde{B}_{33}$ is $(m + 1) \times (m + 1)$ and has $\tilde{w}^T$, the nonzero part of w^T, as its bottom row. Its characteristic polynomial is proportional to p.

From what we know about the transport of information in bulges, the procedure we have just described ought to effect a GZ iteration driven by the rational function $r = p/q$, but we have not yet proved that it does. We will prove it in the generic case, when none of the shifts $\rho_1, \ldots, \rho_m$ or $\tau_1, \ldots, \tau_k$ is an eigenvalue of the pencil $A - \lambda B$, which implies that $\hat{A}$ is properly upper Hessenberg.

Since $\hat{A}\hat{B}^{-1}$ is properly upper Hessenberg, we can apply Proposition 7.1.1 with C replaced by $\hat{A}\hat{B}^{-1}$ to deduce that there is a unique polynomial $\tilde{q}$ of degree k such that $z = \tilde{q}(\hat{A}\hat{B}^{-1})e_1$. Reasoning as in the proof of Theorem 7.1.2, we conclude that the characteristic polynomial of $\tilde{A}_{11} - \lambda \tilde{B}_{11}$ is proportional to $\tilde{q}$. But we already know that that characteristic polynomial is proportional to q. Thus $\tilde{q}$ is proportional to q, and we can write $z = \gamma q(\hat{A}\hat{B}^{-1})e_1$ for some nonzero γ. Applying a similar argument involving $\tilde{A}_{33} - \lambda \tilde{B}_{33}$, we can show that $w^T = \delta p(\hat{B}^{-1}\hat{A})$ for some nonzero constant δ (Exercise 7.3.3). Summarizing the initial and final conditions, we have

$$x = \alpha p(AB^{-1})e_1, \qquad y^T = \beta e_n^T q(B^{-1}A), \tag{7.3.1}$$

$$z = \gamma q(\hat{A}\hat{B}^{-1})e_1, \qquad w^T = \delta e_n^T p(\hat{B}^{-1}\hat{A}). \tag{7.3.2}$$

The transformation from $A - \lambda B$ to $\hat{A} - \lambda \hat{B}$ is an equivalence, so

$$\hat{A} - \lambda \hat{B} = G^{-1}(A - \lambda B)Z$$

for some nonsingular G and Z. The corresponding transformation of the augmented pencil is

$$\begin{bmatrix} z & \hat{A} - \lambda \hat{B} \\ 0 & w^T \end{bmatrix} = \begin{bmatrix} G^{-1} & \\ & 1 \end{bmatrix} \begin{bmatrix} x & A - \lambda B \\ 0 & y^T \end{bmatrix} \begin{bmatrix} 1 & \\ & Z \end{bmatrix},$$

which implies

$$x = Gz \quad \text{and} \quad w^T = y^T Z. \tag{7.3.3}$$

Now we are ready to prove that the bidirectional bulge-chase effects a GZ iteration driven by $r = p/q$.

Theorem 7.3.1. *Suppose that $A - \lambda B$ and $\hat{A} - \lambda \hat{B}$ are equivalent properly Hessenberg-triangular pencils satisfying $\hat{A} - \lambda \hat{B} = G^{-1}(A - \lambda B)Z$. Let p and q be nonzero polynomials, and let $r = p/q$. Suppose that x, y, z, and w are vectors satisfying the relationships* (7.3.1), (7.3.2), (7.3.3). *Then there exist upper triangular matrices R and S such that*

$$r(AB^{-1}) = GR \quad \text{and} \quad r(B^{-1}A) = ZS.$$

Thus the equivalence transformation $\hat{A} - \lambda\hat{B} = G^{-1}(A - \lambda B)Z$ is an iteration of the GZ algorithm driven by the rational function $r = p/q$.

This theorem generalizes both Proposition 6.3.1 and Proposition 7.2.4; taking $q = 1$, we get Proposition 6.3.1, and taking $p = 1$, we get Proposition 7.2.4.

Proof: Since $\hat{A}\hat{B}^{-1} = G^{-1}(AB^{-1})G$, we have $G(\hat{A}\hat{B}^{-1}) = (AB^{-1})G$ and

$$Gq(\hat{A}\hat{B}^{-1}) = q(AB^{-1})G.$$

Using this relationship and the first equation from each of (7.3.1), (7.3.2), and (7.3.3), we have

$$\alpha p(AB^{-1})e_1 = x = Gz = \gamma Gq(\hat{A}\hat{B}^{-1})e_1 = \gamma q(AB^{-1})Ge_1.$$

Consequently

$$r(AB^{-1})e_1 = \tilde{\gamma}Ge_1,$$

where $\tilde{\gamma} = \gamma/\alpha \neq 0$.

Using this last result along with Lemmas 4.5.3 and 4.5.2, we have

$$r(AB^{-1})\kappa(AB^{-1}, e_1) = \kappa(AB^{-1}, r(AB^{-1})e_1)$$
$$= \tilde{\gamma}\kappa(AB^{-1}, Ge_1) = \tilde{\gamma}G\kappa(\hat{A}\hat{B}^{-1}, e_1).$$

Lemma 4.5.3 was stated for a polynomial p, but here we have applied it with a rational function r, which is clearly okay.

We conclude that

$$r(AB^{-1}) = GR,$$

where $R = \tilde{\gamma}\kappa(AB^{-1}, e_1)\kappa(\hat{A}\hat{B}^{-1}, e_1)^{-1}$. The existence and triangularity of R follow from Lemma 4.5.4 and the fact that $\hat{A}\hat{B}^{-1}$ is properly upper Hessenberg, which is a consequence of the genericness assumption. The existence of $r(AB^{-1}) = q(AB^{-1})^{-1}p(AB^{-1})$ also relies on genericness, since this guarantees that $q(AB^{-1})$ has an inverse.

The proof that there is an upper triangular S such that $r(B^{-1}A) = ZS$ follows similar lines. Since $\hat{B}^{-1}\hat{A} = Z^{-1}(B^{-1}A)Z$, we have $Z(\hat{B}^{-1}\hat{A}) = (B^{-1}A)Z$,

$$Zq(\hat{B}^{-1}\hat{A}) = q(B^{-1}A)Z \quad \text{and} \quad Zr(\hat{B}^{-1}\hat{A}) = r(B^{-1}A)Z. \tag{7.3.4}$$

Using this and the second equation from each of (7.3.1), (7.3.2), and (7.3.3), we have

$$\delta e_n^T p(\hat{B}^{-1}\hat{A}) = w^T = y^T Z = \beta e_n^T q(B^{-1}A)Z = \beta e_n^T Zq(\hat{B}^{-1}\hat{A}),$$

so

$$e_n^T r(\hat{B}^{-1}\hat{A}) = \tilde{\beta}e_n^T Z,$$

where $\tilde{\beta} = \beta/\delta \neq 0$. Using this along with Lemmas 7.2.1 and 7.2.2, we obtain

$$\tilde{\kappa}(e_n^T, \hat{B}^{-1}\hat{A})r(\hat{B}^{-1}\hat{A}) = \tilde{\kappa}(e_n^T r(\hat{B}^{-1}\hat{A}), \hat{B}^{-1}\hat{A})$$
$$= \tilde{\beta}\tilde{\kappa}(e_n^T Z, \hat{B}^{-1}\hat{A}) = \tilde{\beta}\tilde{\kappa}(e_n^T, B^{-1}A)Z.$$

Thus

$$r(\hat{B}^{-1}\hat{A}) = SZ, \tag{7.3.5}$$

where $Z = \tilde{\beta}\tilde{\kappa}(e_n^T, \hat{B}^{-1}\hat{A})^{-1}\tilde{\kappa}(e_n^T, B^{-1}A)$. The existence and triangularity of S follow from Lemma 7.2.3 and the fact that $\hat{B}^{-1}\hat{A}$ is properly upper Hessenberg. Now multiplying (7.3.5) by Z on the left and Z^{-1} on the right and using the second equation in (7.3.4), we have

$$r(B^{-1}A) = ZS.$$

This completes the proof. $\square$

Application to the Standard Eigenvalue Problem

The standard eigenvalue problem is the special case of the generalized eigenvalue problem obtained by taking $B = I$, so every algorithm for the generalized eigenvalue problem can also be applied to the standard problem. If the implicit forward and reverse GZ iterations are run with $B_0 = I$ and $G_i = Z_i$ at every step, these algorithms reduce to the implicit GR and RG iterations, respectively. In this case the choice $G_i = Z_i$ serves to chase the bulge in the "B" matrix at every step, because it ensures that $B_i = I$ for every integer value of i.

The shift transmission mechanism is the same for the standard problem as for the generalized problem. The procedure for swapping bulges outlined here for the generalized problem can also be applied to the standard problem with one proviso. In the standard case we want the "B" matrix to equal I after each complete step. After the bulge swap, this is definitely not the case. In fact, the bulge pencils $\hat{E}_{11} - \lambda\hat{F}_{11}$ and $\hat{E}_{22} - \lambda\hat{F}_{22}$ in

$$\left[\begin{array}{cc} \hat{E}_{11} - \lambda\hat{F}_{11} & \hat{E}_{12} - \lambda\hat{F}_{12} \\ 0 & \hat{E}_{22} - \lambda\hat{F}_{22} \end{array} \right]$$

are intermediate (midstep) bulge pencils. To return "B" to the identity matrix, we must first push the two pencils each a half step away from the other. We adjoin a row and a column to the bottom of the lower pencil (adjoining an eigenvalue ∞) and apply a transformation on the right (a "Z" transformation) that transforms the first column of $\hat{F}_{22}$ to zero. Similarly we adjoin a row and column to the top of the upper pencil and apply a transformation on the left (a "G" transformation) that transforms the bottom row of $\hat{F}_{11}$ to zero. After this we are no longer at midstep, but the "B" matrix is not yet equal to I. At this point the original pencil $A - \lambda B$ has been transformed to a form $\tilde{A} - \lambda\tilde{B}$ in which $\tilde{B} = \mathrm{diag}\{I, \check{B}, I\}$. $\check{B}$ is the part in the bulge region that has been disturbed. It is of dimension $m + k + 1$, and its fine structure, as a result of the half steps just described, is

$$\check{B} = \left[\begin{array}{ccc} C_{11} & C_{12} & C_{13} \\ 0 & C_{22} & C_{23} \\ 0 & 0 & C_{33} \end{array} \right],$$

where C_{11} is $k \times k$, C_{22} is 1×1, and C_{33} is $m \times m$.

We restore the $\tilde{B}$ to the identity by simply multiplying the pencil on the right by $\tilde{B}^{-1}$. This is inexpensive because $\check{B}$ is small and block triangular. Because of the block triangular structure, this transformation effects an equivalence transformation on each of the bulge pencils. The upper pencil is multiplied by the inverse of $\left[\begin{array}{cc} C_{11} & C_{12} \\ 0 & C_{22} \end{array} \right]$, and the lower by the inverse of $\begin{bmatrix} C_{22} & C_{23} \\ 0 & C_{33} \end{bmatrix}$. Thus the shifts are preserved. In the case where all of the transformations are taken to be unitary (the "QR" case), the transformation is stable, since $\check{B}$ is unitary.

Once the identity matrix has been restored, the iteration is completed by chasing the upper bulge to the top by an RG bulge-chase and the lower bulge to the bottom by a GR bulge-chase.

Although the description has not been easy, the process is actually quite simple. To summarize, the bulges are pushed together, then they are exchanged by the method that we developed for the generalized problem. Once the bulges have been exchanged, the "B" matrix is restored to the identity, and the bulge-chases are completed.

When one compares the procedure presented here with the earlier one developed in [219], one sees that in [219] we felt compelled to keep "$B = I$," which forced us to introduce and analyze a new class of Sylvester-like equations. Here the presentation is simpler because we are able to connect our problem with previously developed theory and take advantage of that theory.

Exercises

7.3.1.

(a) Download the file `bulgecrash.m` from the website[3] and take a look at it. It does implicit QR and RQ iterations simultaneously by passing bulges through each other. Also download the driver program `swapgo.m` and the auxiliary routine `swap.m`.

(b) Run the `swapgo.m` code with n around 40 and $md = mu = 3$. md and mu are the degrees of the downward and upward bulges, respectively. You can try some other values of md and mu, and they do not have to be equal. If you take either of them bigger than 3, you will observe that the bulge exchanges are rejected frequently.

7.3.2. Modify `bulgecrash.m` so that the same shifts are passed in both directions. What happens then? Also try the case where just one (of two or more) shifts is the same in both directions.

7.3.3. Prove the following analogue of Proposition 7.1.1: Let C be a properly upper Hessenberg matrix, and let $w \in \mathbb{C}^n$ be a nonzero vector. Then there is a unique monic polynomial p of degree $< n$ such that $w^T = \alpha e_n^T p(C)$ for some nonzero scalar α. If $w_{n-m} \neq 0$ and $w_i = 0$ for all $i < n - m$, then the degree of p is exactly m.

---

[3]www.siam.org/books/ot101

Chapter 8

Product Eigenvalue Problems

8.1   Introduction

This chapter draws heavily from [224]. Many eigenvalue problems are most naturally viewed as product eigenvalue problems. The eigenvalues of a matrix A are wanted, but A is not given explicitly. Instead it is presented as a product of several factors: $A = A_k \cdots A_1$. Usually more accurate results are obtained by working with the factors rather than forming A explicitly. Perhaps the best known example is the singular value decomposition (SVD), Theorem 1.7.1. Every matrix $B \in \mathbb{C}^{n \times m}$ can be decomposed as

$$B = U \Sigma V^*,$$

where $U \in \mathbb{C}^{n \times n}$ and $V \in \mathbb{C}^{m \times m}$ are unitary and $\Sigma \in \mathbb{R}^{n \times m}$ is diagonal:

$$\Sigma = \begin{bmatrix} \sigma_1 & & & & \\ & \sigma_2 & & & \\ & & \ddots & & \\ & & & \sigma_r & \\ \hline & & & & 0 \\ & & & & & \ddots \end{bmatrix}, \qquad \sigma_1 \geq \sigma_2 \geq \cdots \geq \sigma_r > 0,$$

where r is the rank of B. Once we have computed the SVD, we immediately have the spectral decompositions

$$BB^* = U(\Sigma \Sigma^T)U^* \quad \text{and} \quad B^*B = V(\Sigma^T \Sigma)V^*$$

implicitly. Thus the SVD of B gives the complete eigensystems of BB^* and B^*B without forming these products explicitly. From the standpoint of accuracy, this is the right way to compute these eigensystems, since information about the smaller eigenvalues would be lost when BB^* and B^*B were computed in floating point arithmetic.

Example 8.1.1. *It is easy to generate small problems in MATLAB that illustrate the advantage of working with B instead of B^*B. As an extreme case consider the MATLAB code*

```
format long e
randn('state',123);
[P,R] = qr(randn(3));
[Q,R] = qr(randn(2));
S = diag([1, 1.23456789e-10]);
B = P(:,1:2)*S*Q';
Eigenvalues = eig(B'*B)
Singular_Values = svd(B)
```

This generates a 3×2 matrix with random singular vectors and specified singular values $\sigma_1 = 1$ and $\sigma_2 = 1.23456789 \times 10^{-10}$. It then computes the eigenvalues of B^*B by the command eig. *These should be the squares of the singular values. Finally, it computes the singular values of B using the* svd *command. Here's the output:*

```
Eigenvalues =

    -2.775557561562891e-17
     9.999999999999998e-01


Singular_Values =

     9.999999999999999e-01
     1.234567850536871e-10
```

The large eigenvalue is correct to full (IEEE standard [160] double) precision, but the tiny one is completely off; it's not even positive. This is not the fault of the eig *command. By the time B^*B had been computed, the damage was done. The* svd *computation came out much better. The large singular value is correct to full precision and the tiny one to eight decimal places.*

Let us now return to the general problem $A = A_k A_{k-1} \cdots A_1$. It often happens that some of the A_j are given in inverse form. That is, we might have, for example, $A = A_2 B_2^{-1} A_1 B_1^{-1}$, where A_1, A_2, B_1, and B_2 are given. We wish to compute eigenvalues of A without forming the inverses.

Many important eigenvalue problems present themselves most naturally as generalized eigenvalue problems $(A - \lambda B)v = 0$. This is the simplest instance of an eigenvalue problem involving an inverse, as it is equivalent to the eigenvalue problem for AB^{-1} or $B^{-1}A$. However, as we know from Chapter 6, it is best to solve the problem without forming either of these products or even inverting B. Again accuracy can be lost through the inversion of B or formation of the product. Sometimes B does not have an inverse.

In addition to the SVD problem and the generalized eigenvalue problem, there are many other examples of product eigenvalue problems. A rich variety will be discussed in this chapter. Some of these problems are normally considered standard eigenvalue problems for which the matrix has some special structure. We shall see that in many cases we can profit by reformulating the structured problem as a product eigenvalue problem.

8.2 The *GR* Algorithm for a Product of Matrices

Consider a product

$$A = A_k A_{k-1} \cdots A_1 \in \mathbb{C}^{n \times n}. \tag{8.2.1}$$

In most of our applications, all of the factors A_i are square, but they do not have to be. A is closely related to the cyclic matrix

$$
C = \begin{bmatrix}
& & & & A_k \\
A_1 & & & & \\
& A_2 & & & \\
& & \ddots & & \\
& & & A_{k-1} &
\end{bmatrix}. \tag{8.2.2}
$$

The following theorem is well known and easy to prove.

Theorem 8.2.1. *The nonzero complex number λ is an eigenvalue of A if and only if its kth roots $\lambda^{1/k}$, $\lambda^{1/k}\omega$, $\lambda^{1/k}\omega^2$, ..., $\lambda^{1/k}\omega^{k-1}$ are all eigenvalues of C. Here $\omega = e^{2\pi i/k}$, and $\lambda^{1/k}$ denotes any one of the kth roots of λ.*

The proof is worked out, along with the relationships between the eigenvectors, in Exercise 8.2.1.

The study of product GR iterations is most easily undertaken by considering a generic GR iteration on the cyclic matrix C. What sort of polynomial f should we use to drive the iteration? The cyclic structure of C implies that if τ is an eigenvalue, then so are $\tau\omega$, $\tau\omega^2, \ldots, \tau\omega^{k-1}$. If we wish to preserve the cyclic structure, we must seek to extract all of these eigenvalues simultaneously. This means that if we use a shift μ (approximating τ, say), we must also use $\mu\omega, \ldots, \mu\omega^{k-1}$ as shifts. Thus our driving polynomial f must have a factor

$$(z - \mu)(z - \mu\omega)(z - \mu\omega^2) \cdots (z - \mu\omega^{k-1}) = z^k - \mu^k.$$

If we wish to apply shifts $\mu_1, \ldots, \mu_m$, along with the associated $\mu_i\omega^j$, we should take

$$f(z) = (z^k - \mu_1^k)(z^k - \mu_2^k) \cdots (z^k - \mu_m^k).$$

Thus the principle of structure preservation dictates that $f(C)$ should be a polynomial in C^k: $f(C) = p(C^k)$.

Clearly C^k has the block diagonal form

$$
C^k = \begin{bmatrix}
A_k A_{k-1} \cdots A_1 & & & \\
& A_1 A_k \cdots A_2 & & \\
& & \ddots & \\
& & & A_{k-1} \cdots A_1 A_k
\end{bmatrix},
$$

and so $f(C)$ is also block diagonal.

Let us consider the case $k = 3$ for illustration. It is convenient to modify the notation slightly: In place of A_1, A_2, and A_3 we write A_{21}, A_{32}, and A_{13}, respectively. Then

$$
C = \begin{bmatrix}
& & A_{13} \\
A_{21} & & \\
& A_{32} &
\end{bmatrix}
$$

and

$$
C^3 = \begin{bmatrix} A_{13}A_{32}A_{21} & & \\ & A_{21}A_{13}A_{32} & \\ & & A_{32}A_{21}A_{13} \end{bmatrix},
$$

and thus

$$
f(C) = p(C^3) = \begin{bmatrix} p(A_{13}A_{32}A_{21}) & & \\ & p(A_{21}A_{13}A_{32}) & \\ & & p(A_{32}A_{21}A_{13}) \end{bmatrix}.
$$

To effect an iteration of the generic GR algorithm we must now obtain a decomposition $f(C) = GR$. The obvious way to do this is to decompose the blocks separately and assemble the result. Say $p(A_{13}A_{32}A_{21}) = G_1 R_1$, $p(A_{21}A_{13}A_{32}) = G_2 R_2$, and $p(A_{32}A_{21}A_{13}) = G_3 R_3$. Then $f(C) = p(C^3) = GR$, where

$$
G = \begin{bmatrix} G_1 & & \\ & G_2 & \\ & & G_3 \end{bmatrix} \quad \text{and} \quad R = \begin{bmatrix} R_1 & & \\ & R_2 & \\ & & R_3 \end{bmatrix}.
$$

The GR iteration is completed by a similarity transformation $\hat{C} = G^{-1}CG$. Clearly

$$
\hat{C} = \begin{bmatrix} & & \hat{A}_{13} \\ \hat{A}_{21} & & \\ & \hat{A}_{32} & \end{bmatrix} = \begin{bmatrix} & & G_1^{-1}A_{13}G_3 \\ G_2^{-1}A_{21}G_1 & & \\ & G_3^{-1}A_{32}G_2 & \end{bmatrix}. \quad (8.2.3)
$$

The cyclic structure has been preserved.

Now consider what has happened to the product $A = A_{13}A_{32}A_{21}$. We easily check that $\hat{A}_{13}\hat{A}_{32}\hat{A}_{21} = G_1^{-1}(A_{13}A_{32}A_{21})G_1$. The equations

$$
p(A_{13}A_{32}A_{21}) = G_1 R_1, \qquad \hat{A}_{13}\hat{A}_{32}\hat{A}_{21} = G_1^{-1}(A_{13}A_{32}A_{21})G_1
$$

together imply that a generic GR iteration driven by p has been effected on the product $A_{13}A_{32}A_{21}$. Similarly we have

$$
p(A_{21}A_{13}A_{32}) = G_2 R_2, \qquad \hat{A}_{21}\hat{A}_{13}\hat{A}_{32} = G_2^{-1}(A_{21}A_{13}A_{32})G_2
$$

and

$$
p(A_{32}A_{21}A_{13}) = G_3 R_3, \qquad \hat{A}_{32}\hat{A}_{21}\hat{A}_{13} = G_3^{-1}(A_{32}A_{21}A_{13})G_3,
$$

and so the GR iteration on C implicitly effects GR iterations on the products $A_{13}A_{32}A_{21}$, $A_{21}A_{13}A_{32}$, and $A_{32}A_{21}A_{13}$ simultaneously. These are iterations of degree m with shifts $\mu_1^k, \ldots, \mu_m^k$, whereas the iteration on C is of degree mk with shifts $\mu_i \omega^j$, $i = 1, \ldots, m$, $j = 0, \ldots, k - 1$.

This looks like a lot of work; it seems to require that we compute

$$
p(A_{1k} \cdots A_{32}A_{21}), \qquad p(A_{21}A_{1k} \cdots A_{32}),
$$

and so on. Fortunately there is a way around all this work; we can do the iteration implicitly by bulge-chasing. To start the bulge-chase, all that is needed is the first column of

$p(A_{1k} \cdots A_{32}A_{21})$, and this can be computed with relative ease. We never form any of the products. We work on the factors separately, carrying out transforms of the form

$$\hat{A}_{j+1,j} = G_{j+1}^{-1} A_{j+1,j} G_j, \qquad (8.2.4)$$

as indicated in (8.2.3). If any of the factors is presented in inverse form, we perform the equivalent transformation on the inverse. That is, if we actually have in hand $B_{j,j+1} = A_{j+1,j}^{-1}$, instead of $A_{j+1,j}$, we do the transformation

$$\hat{B}_{j,j+1} = G_j^{-1} B_{j,j+1} G_{j+1}$$

instead of (8.2.4).

Exercises

8.2.1. This exercise leads to the proof of Theorem 8.2.1 and more. Let $A = A_k \cdots A_1$, and let C be the cyclic matrix given by (8.2.2).

(a) Let x be an eigenvector of C with associated nonzero eigenvalue τ. Partition x conformably with C as

$$x = \begin{bmatrix} x_1 \\ x_2 \\ \vdots \\ x_k \end{bmatrix}.$$

Use the equation $Cx = x\tau$ to determine relationships between the components $x_1, \ldots, x_k$. Deduce that $x_1, \ldots, x_k$ are all nonzero. Show that x_1 is an eigenvector of A with eigenvalue $\lambda = \tau^k$. *Conclusion:* Every eigenvalue of C is a kth root of an eigenvalue of A.

(b) Show that x_2 is also an eigenvector of a certain product matrix, and so are $x_3, \ldots, x_k$, all with eigenvalue $\lambda = \tau^k$.

(c) Let x be as in part (a), let α be any kth root of unity, and let

$$y = \begin{bmatrix} y_1 \\ y_2 \\ \vdots \\ y_k \end{bmatrix} = \begin{bmatrix} x_1\alpha^{k-1} \\ x_2\alpha^{k-2} \\ \vdots \\ x_k\alpha^0 \end{bmatrix}.$$

Show that y is an eigenvector of C with associated eigenvalue $\alpha\tau$. *Conclusion:* If one kth root of λ is an eigenvalue of C, then so is every kth root of λ.

(d) Conversely, let z_1 be an eigenvector of $A = A_k \cdots A_1$ associated with a nonzero eigenvalue λ, and let τ be any kth root of λ. Use z_1 to construct an eigenvector of C with associated eigenvalue τ. *Conclusion:* Every kth root of an eigenvalue of A is an eigenvalue of C.

(e) Show that 0 is an eigenvalue of both A and C if and only if at least one of the factors A_j has a nonzero null space. What do eigenvectors look like? Show that if not all of the blocks A_j are square, then 0 must be an eigenvalue.

8.3   Implicit Product *GR* Iterations

Reduction to Condensed Form

Now consider a product $A = A_k \cdots A_1$. If we wish to effect product GR iterations efficiently, we must transform the factors A_i to a condensed form of some sort. Notice that if
we make A_k upper Hessenberg and all other A_i upper triangular, then the product A will be
upper Hessenberg. We will indicate how such a transformation can be effected [111] in the
case $k = 3$. Again it is convenient to work with the cyclic matrix

$$C = \begin{bmatrix} & & A_{13} \\ A_{21} & & \\ & A_{32} & \end{bmatrix},$$

which we can write schematically as

$$\begin{bmatrix}
 & & & & & & & & x & x & x & x \\
 & & & & & & & & x & x & x & x \\
 & & & & & & & & x & x & x & x \\
 & & & & & & & & x & x & x & x \\
z & z & z & z & & & & & & & & \\
z & z & z & z & & & & & & & & \\
z & z & z & z & & & & & & & & \\
z & z & z & z & & & & & & & & \\
 & & & & y & y & y & y & & & & \\
 & & & & y & y & y & y & & & & \\
 & & & & y & y & y & y & & & & \\
 & & & & y & y & y & y & & & &
\end{bmatrix} \tag{8.3.1}$$

in the case where the factors are 4×4. The first step of the reduction applies an elimination
matrix on the left to zero out the first column of A_{21} (the z-matrix) below the main diagonal.
To complete a similarity transformation on C, we apply the inverse of the elimination matrix
on the right. Since the left transformation acted only on the second block row of C, the
inverse acts only on the second block column. That is, it affects the y-matrix only. The
second step of the reduction applies an elimination matrix on the left to zero out the first
column of the y-matrix below the main diagonal. Completing the similarity transformation
on C, we apply the inverse transformation on the right. This affects only the x-matrix. The
next step creates zeros in the first column of the x-matrix. But now we have to be cautious
and zero out only the entries below the subdiagonal. That is, we apply a transformation
on the left that operates on rows 2, 3, and 4 and creates zeros in positions $(3, 1)$ and $(4, 1)$
of the x-matrix. Now when we complete the similarity transformation we operate only on
columns 2, 3, and 4. This affects only the z-matrix, and it does not disturb the zeros that
were previously created there, because it does not touch the first column. At this point the

matrix has the form

$$
\begin{bmatrix}
 & & & & & & & & x & x & x & x \\
 & & & & & & & & x & x & x & x \\
 & & & & & & & & 0 & x & x & x \\
 & & & & & & & & 0 & x & x & x \\
z & z & z & z & & & & & & & & \\
0 & z & z & z & & & & & & & & \\
0 & z & z & z & & & & & & & & \\
0 & z & z & z & & & & & & & & \\
 & & & & y & y & y & y & & & & \\
 & & & & 0 & y & y & y & & & & \\
 & & & & 0 & y & y & y & & & & \\
 & & & & 0 & y & y & y & & & &
\end{bmatrix}.
$$

We now continue the reduction by making another round of the matrix, creating zeros in the second column of the z-matrix, the y-matrix, and the x-matrix in turn. At each step our choice of how many zeros to create is just modest enough to guarantee that we do not destroy any of the zeros that we had created previously. After $n - 1$ times around the matrix, we have reduced all of the blocks to triangular form, except for A_k, which is upper Hessenberg. The condensed C matrix looks like

$$
\begin{bmatrix}
 & & & & & & & & x & x & x & x \\
 & & & & & & & & x & x & x & x \\
 & & & & & & & & & x & x & x \\
 & & & & & & & & & & x & x \\
z & z & z & z & & & & & & & & \\
 & z & z & z & & & & & & & & \\
 & & z & z & & & & & & & & \\
 & & & z & & & & & & & & \\
 & & & & y & y & y & y & & & & \\
 & & & & & y & y & y & & & & \\
 & & & & & & y & y & & & & \\
 & & & & & & & y & & & &
\end{bmatrix} \tag{8.3.2}
$$

in our special case.

This procedure looks like a nice generalization of the standard reduction to Hessenberg form outlined in Section 3.3, but Kressner [136] has shown that it is actually just an instance of the standard reduction. To see this, consider what happens when we do a perfect shuffle of the rows and columns of (8.3.1). That is, we reorder the rows and columns in the order

$1, 5, 9, 2, 6, 10, 3, \ldots$. The result is

$$
\left[\begin{array}{cccc|cccc|cccc|cccc}
 & x & & & & x & & & & x & & & & x & & \\
z & & & & z & & & & z & & & & z & & & \\
 & & y & & & & y & & & & y & & & & y & \\
\hline
 & x & & & & x & & & & x & & & & x & & \\
z & & & & z & & & & z & & & & z & & & \\
 & & y & & & & y & & & & y & & & & y & \\
\hline
 & x & & & & x & & & & x & & & & x & & \\
z & & & & z & & & & z & & & & z & & & \\
 & & y & & & & y & & & & y & & & & y & \\
\hline
 & x & & & & x & & & & x & & & & x & & \\
z & & & & z & & & & z & & & & z & & & \\
 & & y & & & & y & & & & y & & & & y & \\
\end{array}\right].
\qquad (8.3.3)
$$

Now think about applying the standard reduction to upper Hessenberg form to this matrix. First we zero out the first column below the subdiagonal. Since there are only a few nonzeros in that column, the transformation needs to act only on rows 2, 5, 8, and 11, making zeros in positions $(5, 1)$, $(8, 1)$, and $(11, 1)$. This transformation affects only the elements labelled z. The similarity transformation is completed by an operation on columns 2, 5, 8, and 11. This touches only elements labelled y. The second step is to zero out the second column below the subdiagonal. Since there are only a few nonzeros in this column, the transformation acts only on rows 3, 6, 9, and 12, and it touches only entries labelled y. The similarity transformation is completed by an operation on columns 3, 6, 9, and 12, which touches only entries labelled x. Continuing in this manner, we reduce the matrix to upper Hessenberg form, ending with

$$
\left[\begin{array}{cccc|cccc|cccc|cccc}
 & x & & & & x & & & & x & & & & x & & \\
z & & & & z & & & & z & & & & z & & & \\
 & & y & & & & y & & & & y & & & & y & \\
\hline
 & x & & & & x & & & & x & & & & x & & \\
 & & & & z & & & & z & & & & z & & & \\
 & & y & & & & y & & & & y & & & & y & \\
\hline
 & & & & & x & & & & x & & & & x & & \\
 & & & & z & & & & z & & & & z & & & \\
 & & y & & & & y & & & & y & & & & y & \\
\hline
 & & & & & & & & & x & & & & x & & \\
 & & & & & & & & z & & & & z & & & \\
 & & & & & & y & & & & y & & & & y & \\
\end{array}\right].
\qquad (8.3.4)
$$

A moment's thought reveals that this is just the shuffled version of (8.3.2). Thus (8.3.2) is just a disguised upper Hessenberg matrix. Moreover, one easily checks that the operations in the reduction of (8.3.1) to (8.3.2) are identical to the operations in the reduction of the shuffled matrix (8.3.3) to the Hessenberg form (8.3.4). Thus the reduction of (8.3.1) to (8.3.2) is really just a very special instance of the standard reduction to Hessenberg form.

The Product Implicit *GR* Step

Once we have the upper Hessenberg matrix (8.3.4), we can perform implicit GR steps by bulge-chasing. It turns out to be easier to think about the unshuffled version (8.3.2). We will continue to use the case $k = 3$ for illustration, although what we have to say is valid for any positive integer k. Now let

$$C = \begin{bmatrix} & & H_{13} \\ T_{21} & & \\ & T_{32} & \end{bmatrix}$$

denote the cyclic "Hessenberg" matrix (8.3.2). From Section 4.5 we know that implicit GR steps can be effected only on *properly* Hessenberg matrices. The pattern (8.3.4) shows clearly what that means in this case. The Hessenberg matrix H_{13} must be properly upper Hessenberg, and the triangular matrices must be nonsingular. Notice that these are necessary and sufficient conditions for the product matrices $H_{13}T_{32}T_{21}$, $T_{21}H_{13}T_{32}$, and $T_{32}T_{21}H_{13}$ to be properly upper Hessenberg. The case when H_{13} is not properly upper Hessenberg is easy to deal with; there is an obvious reduction to two smaller product eigenvalue problems. One of the triangular matrices has a zero on the main diagonal if and only if C is singular. In this case it is possible to deflate k zero eigenvalues *and* break the problem into two smaller subproblems. This deflation procedure is more easily understood after a discussion of the implicit product GR step, so read this section before working out the details of the deflation in Exercise 8.3.4. For the rest of this section we will assume that the Hessenberg factor H_{13} is properly upper Hessenberg and that the triangular factors are nonsingular.

We know from Section 8.2 that if we want to execute a GR iteration on C with a shift μ, then we should also apply shifts $\mu\omega$ and $\mu\omega^2$ ($\omega = e^{2\pi i/3}$) at the same time to preserve the cyclic structure. In other words, the polynomial that drives the GR iteration should have the form $p(C^3)$. The degree of p is the number of triples of shifts, and the degree of the GR iteration on C is three times the degree of p. To keep the discussion simple, let us suppose that we are going to apply just one triple of shifts: μ, $\mu\omega$, $\mu\omega^2$. Thus we will perform a GR iteration of degree 3 on C. As we observed in Section 8.2, this is equivalent to a GR iteration of degree one with shift μ^3 applied to the product matrices $H_{13}T_{32}T_{21}$, $T_{21}H_{13}T_{32}$, and $T_{32}T_{21}H_{13}$ simultaneously. We have

$$p(C^3) = C^3 - \mu^3 I.$$

From Section 4.5 we know that the key to performing an implicit GR iteration is to get the first column of the transformation matrix right, and this is proportional to the first column of $p(C^3)$. However, the observations of Section 4.5 apply to Hessenberg matrices, so we should really be working with the shuffled version (8.3.4) instead of (8.3.2). But the shuffled and unshuffled versions (of C, C^3, $p(C^3)$, etc.) have the same first column, up to a permutation of the entries, so it is okay to work with the unshuffled version. Recall that

$$C^3 = \begin{bmatrix} H_{13}T_{32}T_{21} & & \\ & T_{21}H_{13}T_{32} & \\ & & T_{32}T_{21}H_{13} \end{bmatrix},$$

so the first column of $p(C^3) = C^3 - \mu^3 I$ is essentially the same as the first column of the shifted product $H_{13}T_{32}T_{21} - \mu^3 I$. This column has only two nonzero entries, and it is trivial to compute, since the two T matrices are upper triangular. Proceeding as in Section 4.5, we build a transforming matrix G_0 whose first column is proportional to the first column of $p(C^3)$. We apply G_0^{-1} to C on the left, affecting only the first two rows. When we apply G_0 on the right, it affects the first two columns, creating a bulge, denoted by a plus sign, in the "z" part of the matrix:

$$
\left[
\begin{array}{cccccccccccc}
 & & & & & & & & x & x & x & x \\
 & & & & & & & & x & x & x & x \\
 & & & & & & & & & x & x & x \\
 & & & & & & & & & & x & x \\
z & z & z & z & & & & & & & & \\
+ & z & z & z & & & & & & & & \\
 & & z & z & & & & & & & & \\
 & & & z & & & & & & & & \\
 & & & & y & y & y & y & & & & \\
 & & & & & y & y & y & & & & \\
 & & & & & & y & y & & & & \\
 & & & & & & & y & & & & \\
\end{array}
\right] .
$$

The rest of the GR iteration consists of returning the matrix to the Hessenberg-triangular form (8.3.2) by the algorithm outlined earlier in this section. The first transformation acts on rows 5 and 6, returning the bulge entry to zero. When the similarity transformation is completed by applying a transformation to columns 5 and 6, a new bulge is created in the y-matrix. The next transformation acts on rows 9 and 10, annihilating this bulge. Completing the similarity transformation, we create a new bulge in the x-matrix. Next we apply a transformation to rows 2 and 3, annihilating the new bulge. When we complete the similarity transformation, the bulge shows up in the $(3, 2)$ position of the x-matrix. We have now chased the bulge once around the cycle, and it has ended up down and over one position from where it began. Continuing this process, we chase the bulge around and around the matrix until it falls off the bottom.

We have described a cyclic GR step of degree 3. Iterations of degree 6, 9, or higher generally follow the same pattern, except that the bulges are bigger, and bulges exist simultaneously in all of the submatrices (Exercise 8.3.3). This is called the periodic GR algorithm [111]. It appears to be a generalization of the standard GR bulge-chase, but the observation of Kressner [136] shows that it is really just a special case. The reader is invited to check how these operations look in the shuffled coordinates (Exercises 8.3.2 and 8.3.3). One easily sees that the periodic GR algorithm is just an ordinary (though very special) bulge-chase of degree 3 (or, in general, $3m$) on the upper Hessenberg matrix (8.3.4).

This is an important observation. It implies that everything that is true for the bulge-chasing algorithm of Section 4.5 is true for the periodic GR algorithm described here. There is no need to prove a bunch of new results for cyclic matrices. For example, we deduce immediately from Theorem 4.5.5 that the bulge-chasing algorithm described here does indeed effect an iteration of the generic GR algorithm on the cyclic matrix C. There is no need to prove a new result just for the cyclic case.

The Case $k = 2$

We have used the case $k = 3$ for illustration. In some applications k is 3 or larger, but in most of our applications we will have $k = 2$. Therefore we pause to consider what the matrices look like in that case. We have $A = A_2 A_1$. The corresponding cyclic matrix is

$$C = \begin{bmatrix} & A_2 \\ A_1 & \end{bmatrix},$$

which looks like

$$\begin{bmatrix} & & & x & x & x \\ & & & x & x & x \\ & & & x & x & x \\ \hline y & y & y & & & \\ y & y & y & & & \\ y & y & y & & & \end{bmatrix}.$$

The Hessenberg form is

$$\begin{bmatrix} & & & x & x & x \\ & & & x & x & x \\ & & & & x & x \\ \hline y & y & y & & & \\ & y & y & & & \\ & & y & & & \end{bmatrix}.$$

The shuffled versions are

$$\begin{bmatrix} & x & & x & & x \\ y & & & y & & y \\ & x & & x & & x \\ y & & & y & & y \\ & x & & x & & x \\ y & & y & & y & \end{bmatrix} \quad \text{and} \quad \begin{bmatrix} & x & & x & & x \\ y & & & y & & y \\ & x & & x & & x \\ & & & y & & y \\ & & & x & & x \\ & & & & y & \end{bmatrix}.$$

In this case the GR iterations are always of even degree, with the shifts occurring in $\pm\mu$ pairs. A typical driving polynomial looks like

$$f(C) = p(C^2) = (C^2 - \mu_1^2 I)(C^2 - \mu_2^2 I)\cdots(C^2 - \mu_m^2 I).$$

Exercises

8.3.1. Check that the operations in the reduction of (8.3.1) to (8.3.2) are identical to the operations in the reduction of the shuffled matrix (8.3.3) to the Hessenberg form (8.3.4).

8.3.2. Consider the periodic GR step of degree 3 as described above. Show that in the shuffled coordinate system this is exactly a standard implicit GR iteration driven by $p(C^3) = C^3 - \mu^3 I$.

8.3.3. Describe a bulge-chasing periodic GR step of degree $3m$ on the cyclic matrix C, driven by $p(C^3) = (C^3 - \mu_1^3)\cdots(C^3 - \mu_m^3)$. How does this look in the shuffled coordinate system?

Convince yourself that this just a special case of the bulge-chasing GR iteration described in Section 4.5.

8.3.4. In this exercise we work out the details of the preliminary deflations before the implicit product GR iteration. We continue to assume that

$$C = \begin{bmatrix} & & H_{13} \\ T_{21} & & \\ & T_{32} & \end{bmatrix};$$

that is, we consider the case $k = 3$. However, the following observations are valid for any k.

(a) Show that if any of the subdiagonal entries of the Hessenberg matrix H_{13} is zero, then the product eigenvalue problem can be reduced to two smaller product eigenvalue problems.

(b) Show that if a main-diagonal entry of T_{21} or T_{32} is zero, then zero is an eigenvalue of C.

(c) If a main-diagonal entry of T_{21} or T_{32} is zero, then $k = 3$ zero eigenvalues can be deflated from C. The deflation is effected by two bulge-chases, one starting from the top and one starting from the bottom. For definiteness let us assume that the mth main-diagonal entry of T_{32} is zero and that m is neither 1 nor n. Let h_{ij} denote the (i, j) entry of H_{13} (or a matrix obtained from H_{13} by subsequent transformations). Begin an implicit triple GR step on C with shifts 0, 0, and 0. Show that the first transformation changes h_{21} to zero. Continue the bulge-chase. Show that when the bulge returns to H_{13}, the submatrix

$$\begin{bmatrix} h_{21} & h_{22} \\ h_{31} & h_{32} \end{bmatrix}$$

(in the vicinity of the bulge) has rank one. Deduce that the subsequent transformation changes h_{32} to zero. More generally, show that each time the bulge returns to H_{13}, the submatrix

$$\begin{bmatrix} h_{j+1,j} & h_{j+1,j+1} \\ h_{j+2,j} & h_{j+2,j+1} \end{bmatrix}$$

in the vicinity of the bulge has rank one, and consequently the subsequent transformation changes $h_{j+2,j+1}$ to zero.

(d) Show that the bulge-chase dies when it reaches the zero entry in T_{32} and that at this point we have $h_{m,m-1} = 0$.

(e) Initiate a second bulge-chase from the bottom of H_{13} by applying a transformation on the right (to the last two columns) that transforms $h_{n,n-1}$ to zero. Complete the similarity transformation to get a bulge at the bottom of T_{32}. Chase the bulge upward. This is an implicit RG iteration (Section 7.2) with shifts 0, 0, and 0. Show that each time the bulge returns to H_{13}, the submatrix

$$\begin{bmatrix} h_{j+1,j} & h_{j+1,j+1} \\ h_{j+2,j} & h_{j+2,j+1} \end{bmatrix}$$

in the vicinity of the bulge has rank one, and consequently the subsequent transformation changes $h_{j+1,j}$ to zero.

(f) Show that the bulge-chase dies when it reaches the zero entry in T_{32} and that at this point we have $h_{m+1,m} = 0$. If we consider the shuffled form (8.3.4), there are now three zeros on the subdiagonal of the Hessenberg matrix.

(g) Work out how these bulge-chases look in the shuffled coordinate system. Show that in the end, the configuration in the "middle" of the shuffled matrix is

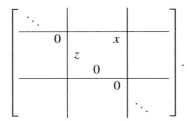

How is the picture changed if the zero is in T_{21} instead of T_{32}? Show that the eigenvalues of

$$\begin{bmatrix} & & x \\ z & & \\ & 0 & \end{bmatrix}$$

are 0, 0, and 0. Show that the product eigenvalue problem can now be reduced to two smaller product eigenvalue problems without the zero eigenvalues.

(h) How is the procedure simplified if the zero in T_{32} (or T_{21}) is in the $(1, 1)$ or (n, n) position?

(i) How is the deflation procedure changed if the total number of zeros on the main diagonals of T_{21} and T_{32} is greater than one?

(j) Think of a better way to organize this exercise.

The deflation procedure discussed here assumes that all of the factors are given directly; i.e., none appears as an inverse. If inverse factors are present, and one of them (assumed triangular) has a zero on the main diagonal, then there are infinite eigenvalues. These can be removed by a procedure like the one outlined in Section 6.5 for the generalized eigenvalue problem. All of the deflation procedures are outlined in [111].

8.4 The Singular Value Decomposition

We begin our discussion of special cases with the SVD problem. Suppose we wish to compute the SVD of a matrix $B \in \mathbb{C}^{n \times m}$. This is equivalent to computing the eigensystems of B^*B and BB^* or the cyclic matrix

$$C = \begin{bmatrix} & B^* \\ B & \end{bmatrix}.$$

We can assume without loss of generality that $n \geq m$, since otherwise we can reverse the roles of B and B^*.

Reduction to Bidiagonal Form

We begin by reducing B to bidiagonal form. All eliminations are done with unitary transformations, since the matrices U and V in the decomposition $B = U \Sigma V^*$ have to be unitary. The reduction of Golub and Kahan [97] and Golub and Reinsch [98] proceeds as follows. First a unitary elimination matrix $U_1 \in \mathbb{C}^{n \times n}$ is chosen so that zeros are introduced into the first column:

$$U_1^* B = \begin{bmatrix} \alpha_1 & * & \cdots & * \\ \hline 0 & * & \cdots & * \\ \vdots & \vdots & & \vdots \\ 0 & * & \cdots & * \end{bmatrix}.$$

The computation can always be normalized so that α_1 is real and nonnegative. Let us assume that that has been done. Next a unitary

$$V_1 = \begin{bmatrix} 1 & 0^T \\ 0 & \tilde{V}_1 \end{bmatrix} \in \mathbb{C}^{m \times m}$$

is chosen so that zeros are introduced in the first row,

$$U_1^* B V_1 = \begin{bmatrix} \alpha_1 & \beta_1 & 0 & \cdots & 0 \\ 0 & * & * & \cdots & * \\ 0 & * & * & \cdots & * \\ \vdots & \vdots & \vdots & & \vdots \\ 0 & * & * & \cdots & * \end{bmatrix},$$

and $\beta_1 \geq 0$. The form of V_1 guarantees that it does not destroy the zeros that were introduced by U_1^*. The next transformation is a

$$U_2 = \begin{bmatrix} 1 & 0^T \\ 0 & \tilde{U}_2 \end{bmatrix} \in \mathbb{C}^{n \times n}$$

such that $U_2^* U_1^* B V_1$ has zeros in the second column below the $(2, 2)$ position. This is followed by a V_2 that introduces zeros in the second row, and so on. After m transformations on the left and $m - 2$ transformations on the right, we have reduced B to a bidiagonal matrix

$$\tilde{B} = U_m^* \cdots U_1^* B V_1 \cdots V_{m-2}.$$

The form of $\hat{B}$ is

$$\tilde{B} = \begin{bmatrix} \alpha_1 & \beta_1 & & & \\ & \alpha_2 & \ddots & & \\ & & \ddots & \beta_{m-1} & \\ & & & \alpha_m & \\ 0 & \cdots & & & 0 \end{bmatrix},$$

where all α_j and β_j are real and nonnegative. The zeros at the bottom represent a block of $n - m$ zero rows.

Let $\tilde{U} = U_1 U_2 \cdots U_m$ and $\tilde{V} = V_1 V_2 \cdots V_{m-2}$. Then

$$B = \tilde{U}\tilde{B}\tilde{V}^*,$$

and we are well on our way to an SVD. If we can compute the SVD of $\tilde{B}$, we will immediately have the SVD of B.

The foregoing reduction is easily related to the reduction of the cyclic matrix

$$C = \begin{bmatrix} & B^* \\ B & \end{bmatrix} \qquad (8.4.1)$$

to Hessenberg-triangular form, as described in Section 8.3. If we perform that reduction on C entirely with unitary transformations, the relationship between the two nonzero blocks is maintained: Each is always the conjugate transpose of the other. This is so because

$$\begin{bmatrix} U^* & \\ & V^* \end{bmatrix} \begin{bmatrix} & B^* \\ B & \end{bmatrix} \begin{bmatrix} U & \\ & V \end{bmatrix} = \begin{bmatrix} & U^*B^*V \\ V^*BU & \end{bmatrix}.$$

Thus, once B has been reduced to upper triangular form and B^* simultaneously to upper Hessenberg form, the resulting $\tilde{B}$ must be bidiagonal. Clearly the operations that are performed on the B^* block are essentially the same as those are performed on the B block, and there is no practical reason to perform both. If one performs only the operations on B, then one is doing exactly the Golub–Kahan reduction to bidiagonal form, which we have just described above.

The result of Kressner [136] implies that we can relate the reduction to bidiagonal form to the reduction of the shuffled cyclic matrix to tridiagonal form. Let C be the cyclic matrix (8.4.1), and consider its shuffled form. We have to be a bit careful now about what we mean by a perfect shuffle, since it may be that $n \neq m$. However, this causes no difficulties. We just shuffle in the order $1, m+1, 2, m+2$, and so on, until we run out of rows (columns) from one block or the other. Then we simply append the remaining rows (columns). In the 3×2 case, the shuffled version of (8.4.1) is

$$\left[\begin{array}{cc|cc|c} 0 & \overline{b_{11}} & 0 & \overline{b_{21}} & \overline{b_{31}} \\ b_{11} & 0 & b_{12} & 0 & 0 \\ \hline 0 & \overline{b_{12}} & 0 & \overline{b_{22}} & \overline{b_{32}} \\ b_{21} & 0 & b_{22} & 0 & 0 \\ \hline b_{31} & 0 & b_{32} & 0 & 0 \end{array} \right].$$

If we reduce the shuffled matrix to upper Hessenberg form by the method of Section 3.3 using only unitary transformations, then the Hermitian structure is preserved, so the resulting matrix is actually tridiagonal. The result has the form

$$\left[\begin{array}{cc|cc|c} 0 & \alpha_1 & 0 & 0 & 0 \\ \alpha_1 & 0 & \beta_1 & 0 & 0 \\ \hline 0 & \beta_1 & 0 & \alpha_2 & 0 \\ 0 & 0 & \alpha_2 & 0 & 0 \\ \hline 0 & 0 & 0 & 0 & 0 \end{array} \right].$$

This is the shuffled version of

$$
\left[\begin{array}{c|ccc}
 & \alpha_1 & 0 & 0 \\
 & \beta_1 & \alpha_2 & 0 \\
\hline
\alpha_1 & \beta_1 & & \\
0 & \alpha_2 & & \\
0 & 0 & &
\end{array}\right]
=
\left[\begin{array}{cc}
 & \tilde{B}^T \\
\tilde{B} &
\end{array}\right],
\tag{8.4.2}
$$

in which the blocks are bidiagonal.

A variant on the Golub–Kahan reduction was suggested by Chan [56]. This yields substantial savings if $n \gg m$. First perform a QR decomposition: $B = QR$. In other words, attack B from the left first. Then $R = \begin{bmatrix} \hat{R} \\ 0 \end{bmatrix}$ has a big $((n-m) \times m)$ block of zeros that stays zero as the small matrix $\hat{R}$ is reduced to bidiagonal form. (See Exercise 8.4.1.)

The QR Algorithm for the SVD

Once the reduction to bidiagonal form has been been achieved, we might as well assume that $\tilde{B}$ is square, since its nontrivial part is square. There is a block of zeros that can be disregarded; for example, if $\tilde{B}$ is 3×2, we can get rid of the bottom row of zeros. This is the same as dropping the last row and column of (8.4.2).

Let us now assume that B is square and bidiagonal. In the 3×3 case we have

$$
B = \left[\begin{array}{ccc}
\alpha_1 & \beta_1 & \\
 & \alpha_2 & \beta_2 \\
 & & \alpha_3
\end{array}\right].
$$

The α_i and β_i are all nonnegative. If any of the β_i are zero, we can immediately reduce the problem to two smaller subproblems. If any of the α_i is zero, then zero must be a singular value, and it can be deflated immediately. The procedure is described in Exercise 8.4.2. We will assume therefore that all of the α_i and β_i are positive. If we embed B in the cyclic matrix C of (8.4.1) and apply the periodic QR algorithm to C, the transpose relationship of the two blocks is preserved. Each operation on the B block is matched by an equivalent operation on the B^T block. Any bulge that appears on the bottom side of the B block is mirrored by a bulge on the top side of the B^T block, and vice versa. Since the operations on both blocks are the same, half the flops can be saved by working on just one of the blocks. Since we are in the case $k = 2$, the QR iterations should be of even degree $2s$, and the shifts should be in $\pm \mu$ pairs.

Let us briefly describe the case $s = 1$, which is the Golub–Kahan [97] QR algorithm for the SVD. Pick a pair of shifts $\pm \mu$.[1] We start the iteration by computing $x = (C^2 - \mu^2)e_1$, the first column of $C^2 - \mu^2$. Only the top two entries of x are nonzero, and they are easily seen to be $\alpha_1^2 - \mu^2$, $\beta_1 \alpha_1$. Let Q be an orthogonal matrix (e.g., a rotator) whose first column is proportional to x. The bulge-chase is set in motion by transforming C to $Q^T C Q$.

---

[1] Some shift choices: $\mu = 0$, $\mu = \alpha_m$, $\mu = $ the square root of the eigenvalue of the lower right 2×2 submatrix of $B^T B$ or $B B^T$ that is closer to α_m^2.

This alters only the first two rows and columns of C, i.e., the first two rows of B^T and (equivalently) the first two columns of B. From this point on, we will cease to mention B^T. The transformation $B \to BQ$ alters the first two columns of B, creating a bulge in position $(2, 1)$. The next step is to create a rotator (say) that acts on rows 1 and 2 of B to annihilate the bulge. This creates a new bulge in position $(1, 3)$. Now create a rotator that acts on columns 2 and 3 to annihilate the $(1, 3)$ bulge. This creates a new bulge in position $(3, 2)$. Next a rotator acting on columns 2 and 3 annihilates this bulge and creates a new bulge in position $(2, 4)$. Continuing in this manner, alternating left and right transformations, we chase the bulge from top to bottom and off the lower right-hand corner of the matrix.

It is worthwhile to think about this algorithm as it applies to B only, and as it relates to the cyclic matrix C. It is also worthwhile to think about the shuffled version of the algorithm. If we do a perfect shuffle of the rows and columns of

$$
C =
\left[
\begin{array}{ccc|ccc}
 & & & \alpha_1 & & \\
 & & & \beta_1 & \alpha_2 & \\
 & & & & \beta_2 & \alpha_3 \\
\hline
\alpha_1 & \beta_1 & & & & \\
 & \alpha_2 & \beta_2 & & & \\
 & & \alpha_3 & & & \\
\end{array}
\right],
$$

we obtain the irreducible, symmetric, tridiagonal matrix

$$
\left[
\begin{array}{cc|cc|cc}
 & \alpha_1 & & & & \\
\alpha_1 & & \beta_1 & & & \\
\hline
 & \beta_1 & & \alpha_2 & & \\
 & & \alpha_2 & & \beta_2 & \\
\hline
 & & & \beta_2 & & \alpha_3 \\
 & & & & \alpha_3 & \\
\end{array}
\right].
\tag{8.4.3}
$$

If we apply the double-shift QR algorithm to this matrix with paired shifts $\pm\mu$ and throw out the redundant operations, we are doing exactly the Golub–Kahan QR algorithm. In the 1968 paper [96] Golub stopped just short of making this observation.

Repeated iterations of the Golub–Kahan QR step with well-chosen shifts will cause β_{m-1} to converge to zero cubically, and α_m will converge rapidly to a singular value. After a deflation, we can go after the next singular value, and so on. If we perform QR iterations of higher degree $2s$, we are going after s singular values at once, and we will often be able to deflate more than one singular value at a time.

Exercises

8.4.1. Consider the reduction of $B \in \mathbb{C}^{n \times m}$ to bidiagonal form by the Golub–Kahan reduction.

(a) Show that the Golub–Kahan reduction costs about $4nm^2 - \frac{4}{3}m^3$ flops.

(b) Consider the Chan [56] variant in which a decomposition $B = QR$ is done first. Show that the QR decomposition costs about $2nm^2 - \frac{2}{3}m^3$ flops, and the subsequent reduction of $\hat{R}$ to bidiagonal form costs about $\frac{8}{3}m^3$ flops, assuming that the triangular structure of $\hat{R}$ is not exploited. Show that this variant saves flops if $n > \frac{5}{3}m$.

(c) Show that the Chan variant cuts the flop count roughly in half if $n \gg m$.

8.4.2. Let B be a bidiagonal matrix

$$
\begin{bmatrix}
\alpha_1 & \beta_1 & & & \\
 & \alpha_2 & \ddots & & \\
 & & \ddots & \beta_{m-1} & \\
 & & & \alpha_m &
\end{bmatrix}
$$

with $\alpha_i = 0$ for some i. In this exercise we work out how to deflate a zero singular value from B. For the sake of argument we will assume $1 < i < m$. (The special cases $i = 1$ and $i = m$ are easier.)

(a) We begin with a series of transformations that transforms β_i to zero. Show that a rotator (from the left) that acts on rows i and $i + 1$ can set β_i to zero while introducing only one new nonzero entry in the matrix, in position $(i, i + 2)$. In particular, why is it important that $\alpha_i = 0$?

(b) Show that a second rotator, acting on rows i and $i + 2$, can annihilate the new $(i, i + 2)$ entry while leaving α_i and β_i zero. A new nonzero entry is produced in position $(i, i + 3)$. Devise an algorithm that uses additional rotators to continue to push the unwanted nonzero entry toward the edge of the matrix, resulting in the end in a bidiagonal matrix in which both α_i and β_i are zero.

(c) Devise an algorithm similar to that in part (b), applying rotators on the right (columns instead of rows), which transforms β_{i-1} to zero.

(d) Once we have $\alpha_i = 0$, $\beta_i = 0$, and $\beta_{i-1} = 0$, what reductions can we make?

(e) What is the relationship of the procedure outlined here to that of Exercise 8.3.4?

8.4.3. Flesh out the details of the Golub–Kahan QR step for the SVD. Think about how this relates to the periodic QR step on the cyclic matrix C.

8.4.4. Work out the details of the double-shift QR algorithm on the shuffled version of C and note that it is just (two copies of) the Golub–Kahan QR step.

8.5 Accurate Computation of Singular Values

Demmel and Kahan [74] made a simple but important improvement to the SVD algorithm of Golub and Kahan. First they made the observation (which was already implicit in Kahan's 1966 technical report [126]) that the entries of the bidiagonal B determine the singular values to high relative accuracy. This means that if we make a small relative perturbation in each of the α_i and β_i in B, the relative perturbation in each of the singular values is comparably small. Thus if we change each entry of B in the fifteenth decimal place, this will cause a change in each singular value in about the fifteenth decimal place. This is true even for very tiny singular values. For example, if a singular value has magnitude about 10^{-60}, the magnitude of the perturbation will be about 10^{-75}. Thus there is the possibility of computing all singular values, even the very tiny ones, to high relative accuracy.

Let $T = B^T B$. Then the entries of T do not determine the eigenvalues of T (nor the singular values of B) to high relative accuracy, as Exercise 8.5.2 shows. Thus it is better to work with B than with T.

Demmel and Kahan showed how to compute all singular values with high relative accuracy: In the Golub-Kahan QR algorithm with a zero-shift, the arithmetic can be reorganized so that the entire step is performed without any subtractions. See Exercise 8.5.3. All additions in the step involve only positive numbers. Since there is no possibility of cancellation, all quantities are computed to high relative precision. Therefore the zero-shift QR algorithm can be used to extract the tiniest singular values to high relative accuracy. This will take only a few iterations. The convergence of the QR algorithm depends upon ratios of singular values σ_{j+1}/σ_j. If $\sigma_{j+1}, \ldots, \sigma_m$ are tiny, and σ_j is not, the ratio σ_{j+1}/σ_j will be favorable. Once the tiny singular values have been extracted, the others can be found by the standard shifted algorithm.

We must state an important caveat here. This algorithm can compute singular values of a bidiagonal matrix to high relative accuracy, but it cannot do the same for a general matrix. If the matrix must first be reduced to bidiagonal form, the roundoff errors in the preliminary reduction will normally destroy the accuracy of the tiny singular values. Exercise 8.5.1 suggests some simple MATLAB experiments that confirm this.

The dqd Algorithm

A second high-accuracy method for the SVD is the differential quotient-difference (dqd) algorithm, which was already introduced in Section 4.4. Here we derive it independently as an instance of the periodic GR algorithm. Consider the cyclic matrix

$$C = \begin{bmatrix} & B^T \\ B & \end{bmatrix}$$

and its shuffled variant

$$\begin{bmatrix} 0 & \alpha_1 & & & & & \\ \alpha_1 & 0 & \beta_1 & & & & \\ & \beta_1 & 0 & \alpha_2 & & & \\ & & \alpha_2 & 0 & \beta_2 & & \\ & & & \beta_2 & 0 & \alpha_3 & \\ & & & & \alpha_3 & 0 & \beta_3 \\ & & & & & \beta_3 & \ddots \\ & & & & & & \ddots \end{bmatrix}.$$

Now let us intentionally break the symmetry. It is easy to show (Exercise 4.4.3) that there is a diagonal similarity transformation that maps this matrix to the form

$$
\begin{bmatrix}
0 & 1 & & & & & & \\
\alpha_1^2 & 0 & 1 & & & & & \\
& \beta_1^2 & 0 & 1 & & & & \\
& & \alpha_2^2 & 0 & 1 & & & \\
& & & \beta_2^2 & 0 & 1 & & \\
& & & & \alpha_3^2 & 0 & 1 & \\
& & & & & \beta_3^2 & & \ddots \\
& & & & & & & \ddots
\end{bmatrix}.
\tag{8.5.1}
$$

If we now deshuffle this matrix, we obtain a cyclic matrix

$$
\tilde{C} = \begin{bmatrix} & L \\ R & \end{bmatrix} =
\begin{bmatrix}
& & & & 1 & & & \\
& & & & \beta_1^2 & 1 & & \\
& & & & & \beta_2^2 & 1 & \\
& & & & & & \ddots & \ddots \\
\alpha_1^2 & 1 & & & & & & \\
& \alpha_2^2 & 1 & & & & & \\
& & \alpha_3^2 & \ddots & & & & \\
& & & \ddots & & & &
\end{bmatrix}.
$$

Compare this with the result of Exercise 4.4.5. We have

$$
\begin{bmatrix} & L \\ R & \end{bmatrix} =
\begin{bmatrix} D^{-1} & \\ & E^{-1} \end{bmatrix}
\begin{bmatrix} & B^T \\ B & \end{bmatrix}
\begin{bmatrix} D & \\ & E \end{bmatrix},
$$

where D and E are as in Exercise 4.4.5. Since Since $\tilde{C}$ is similar to C, its eigenvalues are plus and minus the singular values of B. Thus we can find the singular values of B by solving the eigenvalue problem for $\tilde{C}$. The data in $\tilde{C}$ determines the singular values to high relative accuracy, since it is essentially the same as the data in B. Thus we have the possibility of creating a high-accuracy algorithm using $\tilde{C}$.

Exercise 4.4.6 showed that the 1's in the superdiagonal of (8.5.1) are preserved by the LR algorithm, so let us apply the periodic LR algorithm to $\tilde{C}$. Since we are in the case $k = 2$, we do iterations of degree 2. Since we would like to compute the tiny singular values to high relative accuracy, we use shifts ± 0. Let $r_j = \alpha_j^2$ and $l_j = \beta_j^2$. For simplicity we drop the tilde from C. Then, in the case $m = 4$,

$$C = \left[\begin{array}{cccc|cccc} & & & & 1 & & & \\ & & & & l_1 & 1 & & \\ & & & & & l_2 & 1 & \\ & & & & & & l_3 & 1 \\ \hline r_1 & 1 & & & & & & \\ & r_2 & 1 & & & & & \\ & & r_3 & 1 & & & & \\ & & & r_4 & & & & \end{array}\right].$$

The LR iteration is set in motion by a transformation whose first column is proportional to the first column of

$$C^2 - 0^2 I = \left[\begin{array}{cc} LR & \\ & RL \end{array}\right],$$

which is $r_1 \left[\begin{array}{ccccc} 1 & l_1 & 0 & \cdots & 0 \end{array}\right]^T$. The LR algorithm performs similarity transformations $\tilde{C} \to \tilde{L}^{-1}\tilde{C}\tilde{L}$, where $\tilde{L}$ is a unit lower triangular matrix built up from Gaussian elimination transforms (with no pivoting). Thus our first transform will have the form

$$\tilde{L} = \left[\begin{array}{cc|c} 1 & & \\ l_1 & 1 & \\ \hline & & I_{n-2} \end{array}\right], \qquad \tilde{L}^{-1} = \left[\begin{array}{cc|c} 1 & & \\ -l_1 & 1 & \\ \hline & & I_{n-2} \end{array}\right].$$

The transformation $C \to \tilde{L}^{-1}C$ subtracts l_1 times the first row from the second row. The sole effect of this is to transform the entry l_1 to 0 in C. This zero is important; it is the consequence of using a zero shift. The transformation $\tilde{L}^{-1}C \to \tilde{L}^{-1}C\tilde{L}$, completing the similarity transformation, adds l_1 times the second column to the first column, creating a bulge to the left of r_2. In the case $n = 4$ the result looks like this:

$$\left[\begin{array}{cccc|cccc} & & & & 1 & & & \\ & & & & 0 & 1 & & \\ & & & & & l_2 & 1 & \\ & & & & & & l_3 & 1 \\ \hline \hat{r}_1 & 1 & & & & & & \\ b & r_2 & 1 & & & & & \\ & & r_3 & 1 & & & & \\ & & & r_4 & & & & \end{array}\right],$$

where $\hat{r}_1 = r_1 + l_1$, and $b = r_2 l_1$.

We pause to recall the general rule of Gaussian elimination similarity transformations: If the left transformation subtracts t times row i from row j, then the right transformation (completing the similarity transformation) adds t times column j to column i.

The rest of the LR iteration consists of chasing the bulge from the matrix. Accordingly, the second transformation subtracts $b/\hat{r}_1$ times row $m + 1$ from row $m + 2$ to annihilate the bulge. It will soon become clear that a good name for this multiplier is $\hat{l}_1$. Thus we are subtracting $\hat{l}_1 = b/\hat{r}_1$ times row $m + 1$ from row $m + 2$. We then complete the similarity transformation by adding $\hat{l}_1$ times column $m + 2$ to column $m + 1$ to yield

$$
\begin{bmatrix}
1 & & & & & & & \\
\hat{l}_1 & 1 & & & & & & \\
l_2\hat{l}_1 & l_2 & 1 & & & & & \\
& & l_3 & 1 & & & & \\
\hline
\hat{r}_1 & 1 & & & & & & \\
0 & \tilde{r}_2 & 1 & & & & & \\
& & r_3 & 1 & & & & \\
& & & r_4 & & & &
\end{bmatrix},
$$

where $\tilde{r}_2 = r_2 - \hat{l}_1$. The entry $l_2\hat{l}_1$ is the new bulge. The subtraction in the formula for $\tilde{r}_2$ is unwelcome. Fortunately there is a second formula that requires only multiplication and division:

$$
\tilde{r}_2 = r_2 - \hat{l}_1 = r_2 - \frac{r_2 l_1}{r_1 + l_1} = \frac{r_2 r_1}{\hat{r}_1}.
$$

To chase the new bulge forward we must subtract the appropriate multiple of row 2 from row 3. Clearly the appropriate multiple is l_2. It is important that the submatrix

$$
\begin{bmatrix}
\hat{l}_1 & 1 \\
l_2\hat{l}_1 & l_2
\end{bmatrix}
$$

has rank one. Therefore, when we subtract l_2 times the second row from the third row, we zero out both the bulge and the entry l_2. We complete the similarity transformation by adding l_2 times column 3 to column 2. The result is

$$
\begin{bmatrix}
1 & & & & & & & \\
\hat{l}_1 & 1 & & & & & & \\
0 & 0 & 1 & & & & & \\
& & l_3 & 1 & & & & \\
\hline
\hat{r}_1 & 1 & & & & & & \\
0 & \hat{r}_2 & 1 & & & & & \\
r_3 l_2 & r_3 & 1 & & & & & \\
& & r_4 & & & & &
\end{bmatrix},
$$

where $\hat{r}_2 = \tilde{r}_2 + l_2$. The next step works out just the same as the previous step: We subtract $\hat{l}_2 = r_3 l_2 / \hat{r}_2$ times row $n + 2$ from row $n + 3$ to chase the bulge, and we add $\hat{l}_2$ times column $n + 3$ to column $n + 2$ to create a new bulge $l_3\hat{l}_2$ to the left of l_3. The step creates an intermediate quantity $\tilde{r}_3 = r_3 - \hat{l}_2$, which fortunately can be computed by the alternate formula $\tilde{r}_3 = r_3\hat{r}_2 / \hat{r}_2$. Also, the 2×2 submatrix in the vicinity of the new bulge has rank one, which guarantees that the transformation that chases the bulge further will also set the entry l_3 to zero. The algorithm continues in this fashion until the bulge has been chased off the bottom. The complete algorithm is quite succinct:

$$
\begin{aligned}
&\tilde{r}_1 \leftarrow r_1 \\
&\text{for } j = 1, \ldots, m-1 \\
&\quad \left[\begin{array}{l}
\hat{r}_j \leftarrow \tilde{r}_j + l_j \\
\hat{l}_j \leftarrow l_j (r_{j+1}/\hat{r}_j) \\
\tilde{r}_{j+1} \leftarrow \tilde{r}_j (r_{j+1}/\hat{r}_j)
\end{array} \right. \\
&\hat{r}_m \leftarrow \tilde{r}_m
\end{aligned}
\tag{8.5.2}
$$

Comparing this with (4.4.10), we see that this is just the dqd algorithm with zero-shift. As we remarked back in Chapter 4, cancellations cannot occur in this algorithm. The only sum in the loop adds two positive numbers. Consequently, the algorithm can compute the singular values of a bidiagonal matrix to high relative accuracy. Equivalently, it can compute the eigenvalues of a tridiagonal matrix to high relative accuracy.

This approach to deriving the dqd algorithm has the advantage that the auxiliary quantities $\tilde{r}_j$ appear in a natural way, while the development in Chapter 4 introduced them without motivation. The disadvantage of this approach is that it does not show how to introduce shifts.

Exercises

8.5.1. In this exercise, which was suggested by Cleve Moler, we have MATLAB compute the singular values of

$$
B = \begin{bmatrix}
1 & \beta & & \\
& 1 & \beta & \\
& & 1 & \beta \\
& & & 1
\end{bmatrix},
$$

where $\beta = 1234567$, two different ways.

(a) Show that for any matrix $A \in \mathbb{C}^{m \times m}$, $|\det(A)| = \sigma_1 \cdots \sigma_m$, the product of the singular values. Deduce that the product of the singular values of B is 1.

(b) Use MATLAB to compute the singular values of B and the product of the singular values. For example, you could use the following commands:

```
B = toeplitz([1 0 0 0],[1 1234567 0 0])
format long e
sv = svd(B)
svprod = prod(sv)
```

Observe that B has three large singular values and one tiny one. Since MATLAB uses the Demmel–Kahan variant of the Golub–Kahan QR algorithm, we expect all of the singular values to be computed to high relative accuracy. The three large ones are surely accurate, but is the tiny one accurate as well? Explain why the product calculation confirms that the tiny singular value has also been computed to high relative accuracy.

(c) Now have MATLAB compute the singular values of B^T by, e.g.,

```
svtrans = svd(B')
```

What happened this time? Notice that the three large singular values are still computed to full precision, but the tiny one is completely inaccurate. MATLAB doesn't notice that B^T is the transpose of an upper bidiagonal matrix, so it simply transforms it to upper bidiagonal form. This transformation wipes out the tiny singular value.

8.5.2. Let ϵ be a tiny positive number, and let

$$T = \begin{bmatrix} 1 + \epsilon & 1 \\ 1 & 1 + \epsilon \end{bmatrix}.$$

(a) Show that the eigenvalues of T are $2 + \epsilon$ and ϵ.

(b) Show that doubling ϵ causes a 100% change in one of the eigenvalues. Thus a tiny perturbation of the entries of T can cause a large relative perturbation of one of the eigenvalues. We conclude that the entries of T do not determine the eigenvalues to high relative accuracy.

8.5.3. This exercise indicates the changes that need to be made in the Golub–Kahan QR step with zero shift so that the entire iteration is done with no subtractions. Let B be bidiagonal with all α_i and β_i positive. Consider a Golub–Kahan QR step on B with shift $\mu = 0$.

(a) Show that the first transforming matrix of the step has a first column that is proportional to the first column of B^T. Deduce that the second column of the transforming matrix is orthogonal to the first column of B^T. Show that after the first transformation, B has been transformed to the form

$$\begin{bmatrix} \tilde{\alpha}_1 & 0 & & \\ b_{21} & \tilde{\alpha}_2 & \beta_2 & \\ & & \alpha_3 & \\ & & & \ddots \end{bmatrix}.$$

In particular, β_1 has been transformed to zero.

(b) Write explicit formulas for the entries of the transforming matrix and for $\tilde{\alpha}_1$, $\tilde{\alpha}_2$, and the bulge b_{21}, and show that all of these formulas involve only multiplications, divisions, extraction of square roots (which can be done with high relative accuracy), and additions of positive numbers. Thus no cancellations occur, and all of these quantities are computed to high relative precision. Show that $\tilde{\alpha}_1$, $\tilde{\alpha}_2$, and b_{21} are all positive.

(c) The second transformation acts on rows 1 and 2 to annihilate bulge b_{21} and transform the matrix to the form

$$\begin{bmatrix} \hat{\alpha}_1 & \tilde{\beta}_1 & b_{13} & \\ 0 & \check{\alpha}_2 & \tilde{\beta}_2 & \\ & & \alpha_3 & \\ & & & \ddots \end{bmatrix}.$$

Write explicit formulas for the entries of the transformation matrix and for $\hat{\alpha}_1$, $\tilde{\beta}_1$, b_{13}, $\check{\alpha}_2$, and $\tilde{\beta}_2$, and show that the operations involve no subtractions. Thus all quantities are computed to high relative precision. Show that all of the computed quantities are positive.

(d) Show that the submatrix

$$\begin{bmatrix} \tilde{\beta}_1 & b_{13} \\ \check{\alpha}_2 & \tilde{\beta}_2 \end{bmatrix}$$

has rank one, and deduce that the third transformation results in the form

$$\begin{bmatrix} \hat{\alpha}_1 & \hat{\beta}_1 & 0 & \\ & \hat{\alpha}_2 & 0 & \\ & b_{32} & \tilde{\alpha}_3 & \\ & & & \ddots \end{bmatrix}.$$

Show that again the computations involve no subtractions, all quantities are computed to high relative precision, and all of the computed quantities are positive. The difference between the computation here and that in the general Golub–Kahan iteration is that here we know that the entry in the position of β_2 is going to become zero, so we simply set it to zero; we do not need to compute it (and commit a roundoff error).

(e) Check that all subsequent steps are like the previous ones. Thus the zero-shift Golub–Kahan iteration can be performed to high relative precision.

8.6 Eigenvalue Problems Related to the SVD

The Hermitian Eigenvalue Problem

Any Hermitian or real symmetric eigenvalue problem can be transformed into an SVD problem. If $A \in \mathbb{C}^{n \times n}$ is a Hermitian matrix, it can be converted to a positive definite one by adding a sufficiently large shift. Specifically, if all of the eigenvalues of A are greater than $-\tau$, then all of the eigenvalues of $A + \tau I$ are positive. This implies that $A + \tau I$ is positive definite, so it must have an upper triangular Cholesky factor R: $A + \tau I = R^* R$ (Theorem 1.4.1). If we now compute the SVD of R, this reveals the complete eigensystem of $A + \tau I$ and hence of A. Of course, if A happens to have some tiny (in magnitude) eigenvalues, we may not be able to compute them to high relative accuracy. Each singular value σ of R yields an eigenvalue $\lambda = \sigma^2 - \tau$ of A. If λ is very close to zero, some of the good digits of σ^2 will be cancelled off when τ is subtracted from it. Moreover, if an unnecessarily large τ is used, the accuracy of all of the eigenvalues can suffer. If τ is several orders of magnitude larger than $\|A\|$, the accuracy of all of the computed eigenvalues will be diminished accordingly. One way to get an adequate shift τ is to use the Gerschgorin disk theorem (Theorem 2.7.3).

Zero-Diagonal Symmetric Problem

The tridiagonal matrix (8.4.3) is special; all of its main-diagonal entries are zero. Matrices of this type arise in certain applications, for example, the computation of sample points for

Gaussian quadrature formulas [94, 101]. Deshuffling the matrix, we find that its eigenvalue problem is equivalent to the singular value problem for a bidiagonal matrix B. For each singular value σ, the special tridiagonal matrix has eigenvalues $\pm\sigma$.

Obviously this observation can be extended to symmetric, tridiagonal matrices with constant main-diagonal τ. A shift by τ puts us in the zero-diagonal case [213].

The Skew Symmetric Eigenvalue Problem

The skew-symmetric eigenvalue problem was analyzed by Ward and Gray [213]. If we wish to compute the eigenvalues of a skew-symmetric matrix $A \in \mathbb{R}^{n \times n}$, we can begin by reducing it to upper Hessenberg form. The principle of preservation of structure dictates that we should use orthogonal transformation matrices to preserve skew-symmetry. Thus the upper Hessenberg form is actually tridiagonal. Moreover, the main-diagonal entries are zero. In the 6×6 case, the matrix looks like this:

$$
\left[\begin{array}{cc|cc|cc}
 & \alpha_1 & & & & \\
-\alpha_1 & & -\beta_1 & & & \\
\hline
 & \beta_1 & & \alpha_2 & & \\
 & & -\alpha_2 & & -\beta_2 & \\
\hline
 & & & \beta_2 & & \alpha_3 \\
 & & & & -\alpha_3 &
\end{array}\right]. \tag{8.6.1}
$$

We have labelled the entries with a bit of foresight and a peek at (8.4.3). Deshuffling this matrix, we obtain

$$
\left[\begin{array}{ccc|ccc}
 & & & \alpha_1 & & \\
 & & & \beta_1 & \alpha_2 & \\
 & & & & \beta_2 & \alpha_3 \\
\hline
-\alpha_1 & -\beta_1 & & & & \\
 & -\alpha_2 & -\beta_2 & & & \\
 & & -\alpha_3 & & &
\end{array}\right],
$$

and it becomes clear that the skew-symmetric tridiagonal eigenvalue problem is equivalent to the SVD problem for a bidiagonal matrix B. If we apply the double-shift QR algorithm to (8.6.1) with paired shifts $\pm i\mu$, we are doing exactly the Golub–Kahan QR algorithm with shift μ on B. For each singular value σ of B, the tridiagonal matrix has a complex conjugate pair of eigenvalues $\pm i\sigma$.

The SVD of a Product

Various generalizations of the least-squares problem require the computation of the SVD of a product [69, 70]. The product SVD problem also arises in the computation of Lyapunov exponents of dynamical systems [1]. Thus we consider the problem of computing the SVD of $B = B_k B_{k-1} \cdots B_1$, given the factors $B_1, \ldots, B_k$.[2] Jacobi-like methods have been preferred for solving problems of this type; however explicit QR-like algorithms have been

---

[2]For some of the factors, we might be given B_i^{-1} instead of B_i.

proposed in [1, 158, 193]. Golub, Sølna, and Van Dooren [99] proposed an implicit QR-like scheme. This is not so straightforward as it might at first seem. To understand the difficulties, consider the case of a product of two matrices $B = B_2 B_1$. The SVD problem for $B_2 B_1$ is equivalent to the eigenvalue problems for the products $B_1^T B_2^T B_2 B_1$ and $B_2 B_1 B_1^T B_2^T$, so it would seem that we should build an algorithm based on the cyclic matrix

$$\begin{bmatrix} & & B_1^T \\ B_1 & & \\ & B_2 & \\ & & B_2^T & \end{bmatrix}.$$

If we reduce this matrix to the Hessenberg-triangular form by the algorithm described in Section 8.3 (using orthogonal matrices, as this is an SVD problem), we lose some of the important structure. We obtain

$$\begin{bmatrix} & & H_4 \\ T_1 & & \\ & T_2 & \\ & & T_3 & \end{bmatrix},$$

where $H_4 \neq T_1^T$ and $T_3 \neq T_2^T$. The algorithm does not realize that certain of the blocks are related and that these relationships should be maintained. What is needed is a reduction to a form

$$\begin{bmatrix} & & \hat{B}_1^T \\ \hat{B}_1 & & \\ & \hat{B}_2 & \\ & & \hat{B}_2^T & \end{bmatrix},$$

in which, say, $\hat{B}_1$ is bidiagonal and $\hat{B}_2$ is diagonal, making $\hat{B} = \hat{B}_2 \hat{B}_1$ also bidiagonal. Such a reduction may be hard to come by.

Golub, Sølna, and Van Dooren took a different approach. They transformed $B = B_k \cdots B_1$ by an orthogonal equivalence transformations to a form $\hat{B} = \hat{B}_k \ldots \hat{B}_1$ in which each of the factors $\hat{B}_i$ is upper triangular, none of the $\hat{B}_i$ is bidiagonal, but the product $\hat{B}$ is bidiagonal. In floating point arithmetic the computed $\hat{B}$ will not be bidiagonal. However, one can compute its main diagonal and superdiagonal and ignore the rest. The method works better than expected; see [99] for details.

Exercises

8.6.1. Let $A \in \mathbb{C}^{n \times n}$ be a Hermitian indefinite matrix. Explain how to use the Gerschgorin disk theorem to determine a real number τ such that $A + \tau I$ is positive-definite.

8.7 The Generalized Eigenvalue Problem

We discussed the generalized eigenvalue problem $(A - \lambda B)v = 0$ in detail in Chapter 6. We now take a second look, viewing it as a product eigenvalue problem. If B is nonsingular, the generalized eigenvalue problem is equivalent to the product eigenvalue problem for $B^{-1}A$ and AB^{-1}. The corresponding cyclic matrix is

$$C = \begin{bmatrix} & A \\ B^{-1} & \end{bmatrix}. \tag{8.7.1}$$

Operations on C entail operations on B^{-1}, but we need not work with B^{-1} explicitly. If we need to effect, say, the transformation $G_2^{-1}B^{-1}G_1$, we will instead perform the equivalent transformation $G_1^{-1}BG_2$.

As we saw in Chapter 6, allowable transformations on matrix pencils or pairs are equivalences $(A, B) \to (G^{-1}AZ, G^{-1}BZ)$, where G and Z are nonsingular. These correspond to similarity transformations on AB^{-1}, $B^{-1}A$, and the cyclic matrix C of (8.7.1).

To begin with, imagine that we have B^{-1} explicitly. If we shuffle the rows and columns of C in (8.7.1) as in Section 8.3, then reduce C to upper Hessenberg form by the standard algorithm of Section 3.3, then unshuffle, we arrive at a form where A is upper Hessenberg and B^{-1} (hence also B) is upper triangular. Thus we have reduced the pair (A, B) to Hessenberg-triangular form.

In practice we never shuffle. If we think about how the algorithm would be executed on the unshuffled matrix, we see that the algorithm first transforms the first column of B^{-1} to upper triangular form, then the first column of A to Hessenberg form, then the second column of B^{-1} to triangular form, and so on. In practice we don't have B^{-1}, so we have to work with B instead. This changes the details of the reduction. If we want to do a transformation $B^{-1} \to Z^{-1}B^{-1}$, transforming the first column of B^{-1} to triangular form, we have to do instead the equivalent transformation $B \to BZ$, transforming the first column of B to upper triangular form. The difficulty here is to figure out what Z should be without knowing B^{-1}. In Section 6.2 we showed how to do this, and if we follow the procedure described there, then the algorithm that we are describing here turns out to be exactly the second of the reduction algorithms described in Section 6.2, whose first step is given by (6.2.1).

The second reduction of Section 6.2 requires an extra RQ decomposition at each step and is therefore very inefficient, so in practice one should use the first reduction. Whichever reduction one uses, one arrives at a Hessenberg triangular form (H, T), corresponding to a cyclic matrix

$$\hat{C} = \begin{bmatrix} & H \\ T^{-1} & \end{bmatrix}. \tag{8.7.2}$$

One can now once again imagine shuffling the rows and columns of $\hat{C}$ to make the matrix upper Hessenberg and running the implicit GR algorithm of degree $2m$ with shifts in $\pm\mu$ pairs on the shuffled matrix. But once again, we do not actually do the shuffle. If we carry out the equivalent operations in the unshuffled matrix, then we are just doing the periodic GR algorithm of Section 8.3 in the case $k = 2$. This involves chasing bulges through H and T^{-1}, but once again we do not really have T^{-1}. Instead we chase the bulge through T. With this slight adjustment, we are now doing exactly the GZ algorithm of Section 6.3.

Van Loan's Extension

In 1975 Van Loan [207] published the VZ algorithm, which computes the eigenvalues of a matrix of the form $AB(CD)^{-1}$. In the case when $B = C = D = I$, it reduces to the QR

algorithm on the single matrix A. When $B = D = I$, it reduces to the QZ algorithm on the pencil $A - \lambda C$. When $C = D = I$ and $A = B^T$, it reduces to the Golub–Kahan QR algorithm for the SVD. When $A = B^T$ and $C = D^T$, it yields an algorithm for computing eigenvalues of pencils of the form $B^T B - \lambda D^T D$.

If we apply QR iterations of degree four or eight to the cyclic matrix

$$\begin{bmatrix} & & & A \\ C^{-1} & & & \\ & D^{-1} & & \\ & & B & \end{bmatrix},$$

we obtain the VZ algorithm. It is understood that the inverse matrices are handled as outlined for the generalized eigenvalue problem above and in Sections 6.2 and 6.3.

Periodic Control Systems

Hench and Laub [111] and Bojanczyk, Golub, and Van Dooren [41] independently studied discrete-time periodic control systems that lead to eigenvalue problems of the form

$$E_k^{-1} F_k \cdots E_2^{-1} F_2 E_1^{-1} F_1 \tag{8.7.3}$$

and developed the periodic QZ algorithm for solving them. That algorithm is equivalent to the QR algorithm applied to a cyclic matrix of period $2k$, as described in Section 8.3, where the inverse matrices are handled just as for the generalized eigenvalue problem.

The product (8.7.3) is a formal product only, as many of the E_i are normally singular matrices. Fortunately the periodic QZ algorithm functions just fine even when some of the E_i are not invertible. We demonstrated this for the generalized eigenvalue problem in Section 6.5.

The periodic QZ algorithm also plays a role in the stable numerical solution of Hamiltonian eigenvalue problems arising from continuous-time optimal control problems [27]. Kressner [133] carefully studied the implementational details of the periodic QZ algorithm.

Exercises

8.7.1. Show that the equivalence transform $(A, B) \rightarrow (G^{-1}AZ, G^{-1}BZ)$, where G and Z are nonsingular, induces similarity transformations on AB^{-1}, $B^{-1}A$, and the cyclic matrix $C = \begin{bmatrix} & A \\ B^{-1} & \end{bmatrix}$.

8.8 The Hamiltonian Eigenvalue Problem

Recall from Section 1.4 that a matrix $\mathcal{H} \in \mathbb{R}^{2n \times 2n}$ is *Hamiltonian* if it has the form

$$\mathcal{H} = \begin{bmatrix} A & K \\ N & -A^T \end{bmatrix}, \quad \text{where } K = K^T, N = N^T. \tag{8.8.1}$$

A more compact equivalent statement is that $\mathcal{H}$ is Hamiltonian if and only if $(J\mathcal{H})^T = J\mathcal{H}$, where

$$J = \begin{bmatrix} 0 & I \\ -I & 0 \end{bmatrix}. \tag{8.8.2}$$

A matrix is skew-Hamiltonian if $(J\mathcal{H})^T = -J\mathcal{H}$. We already know how to handle the skew-Hamiltonian eigenvalue problem. The proof of Theorem 3.3.3 shows how to reduce the matrix to skew-Hamiltonian Hessenberg form

$$\begin{bmatrix} B & K \\ 0 & B^T \end{bmatrix},$$

where B is upper Hessenberg. Then the QR algorithm can be applied to B to reduce the matrix further to *skew-Hamiltonian Schur form*

$$\begin{bmatrix} T & \tilde{K} \\ 0 & T^T \end{bmatrix},$$

where T is quasi-triangular, revealing the eigenvalues (Theorem 3.3.5).

Since the skew-Hamiltonian problem is so easy, it seems natural to try to solve the Hamiltonian problem by transforming it to a skew-Hamiltonian problem. From Proposition 1.4.3 we know that if $\mathcal{H}$ is Hamiltonian, then $\mathcal{H}^2$ is skew-Hamiltonian. Thus one can simply compute $\mathcal{H}^2$, calculate its eigenvalues by the method already described, and take their square roots to get the eigenvalues of $\mathcal{H}$. Just such a method was proposed by Van Loan [208]. Its advantages are simplicity and efficiency. Unfortunately the squaring causes loss of accuracy, which is most noticeable in the eigenvalues that are smallest in magnitude. Moreover, the method does not yield eigenvectors or invariant subspaces.

In light of what we now know about product eigenvalue problems, we might consider working with the cyclic matrix

$$\begin{bmatrix} 0 & \mathcal{H} \\ \mathcal{H} & 0 \end{bmatrix},$$

which has the same eigenvalues as $\mathcal{H}$, instead of $\mathcal{H}^2$. It turns out that a slightly different matrix works even better. Following Kressner [134], let

$$M = \begin{bmatrix} 0 & \mathcal{H} \\ -\mathcal{H} & 0 \end{bmatrix}.$$

If $\pm\lambda$ are eigenvalues of $\mathcal{H}$, then $\pm i\lambda$ are double eigenvalues of M. If our Hamiltonian matrix $\mathcal{H}$ has the form

$$\mathcal{H} = \begin{bmatrix} A & K \\ N & -A^T \end{bmatrix}, \qquad K = K^T, \quad N = N^T,$$

then

$$M = \left[\begin{array}{cc|cc} & & A & K \\ & & N & -A^T \\ \hline -A & -K & & \\ -N & A^T & & \end{array} \right].$$

Doing a perfect shuffle of the blocks, we obtain the permutationally similar matrix

$$
\tilde{M} = \left[
\begin{array}{cc|cc}
0 & A & 0 & K \\
-A & 0 & -K & 0 \\
\hline
0 & N & 0 & -A^T \\
-N & 0 & A^T & 0
\end{array}
\right],
\tag{8.8.3}
$$

which is skew-Hamiltonian. Now consider a further permutation similarity in which we do a perfect shuffle of the rows and columns of $\left[\begin{smallmatrix} 0 & A \\ -A & 0 \end{smallmatrix}\right]$, and similarly for each of the other blocks in $\tilde{M}$. This yields a matrix

$$
\hat{M} = \left[
\begin{array}{cc}
W_A & W_K \\
W_N & W_A^T
\end{array}
\right],
$$

which is again skew-Hamiltonian. Now consider the transformation of this matrix to skew-Hamiltonian Hessenberg form as in Theorem 3.3.3. One easily checks that the many zeros in $\hat{M}$ (half the matrix) are preserved by the reduction. It is perhaps easier to think about the reduction in the partially shuffled coordinate system of (8.8.3). First we rewrite $\tilde{M}$ as

$$
\tilde{M} = \left[
\begin{array}{cc|cc}
0 & A_2 & 0 & K \\
-A_1 & 0 & -K^T & 0 \\
\hline
0 & N & 0 & -A_1^T \\
-N^T & 0 & A_2^T & 0
\end{array}
\right].
\tag{8.8.4}
$$

Of course $A_1 = A_2 = A$, but A_1 and A_2 get transformed to different matrices during the reduction. Moreover, $N = N^T$ and $K = K^T$ initially, but these relationships are not preserved by the reduction. However, all of the relationships indicated in (8.8.4) are preserved during the reduction because the reduction preserves skew-Hamiltonian form.

The first step of the reduction is an orthogonal symplectic transformation acting on block rows 2 and 4, creating zeros in the entire first column of $-N^T$ and in the first column of $-A_1$ except for the $(1, 1)$ entry. When the similarity transformation is completed by acting on block columns 2 and 4, zeros are created in the first row of matrices N and $-A_1^T$. (These operations are, of course, redundant. In an actual implementation, one copy each of A_1, A_2, N, and K would be kept, and redundant operations would be skipped.) The second step acts on block rows (and columns) 1 and 3 and creates zeros in the entire first column of N and in the first column of A_2, except for the first two entries. The third step acts on the second column of $-A_1$ and $-N^T$, and so on. After $2n$ steps, the matrix $\tilde{M}$ has been transformed to the form

$$
\check{M} = \left[
\begin{array}{cc|cc}
0 & \tilde{S} & 0 & \tilde{B} \\
\tilde{T} & 0 & -\tilde{B}^T & 0 \\
\hline
0 & 0 & 0 & \tilde{T}^T \\
0 & 0 & \tilde{S}^T & 0
\end{array}
\right],
\tag{8.8.5}
$$

where $\tilde{T}$ is upper triangular, $\tilde{S}$ is upper Hessenberg, and $\tilde{B}$ is full. The entire similarity transformation from $\tilde{M}$ to $\check{M}$ has the form $\tilde{M} = \tilde{Q} \check{M} \tilde{Q}^T$, where the orthogonal symplectic matrix $\check{Q}$ has the form

$$
\check{Q} = \left[\begin{array}{cc|cc}
\tilde{U}_1 & 0 & \tilde{U}_2 & 0 \\
0 & \tilde{V}_1 & 0 & \tilde{V}_2 \\
\hline
-\tilde{U}_2 & 0 & \tilde{U}_1 & 0 \\
0 & -\tilde{V}_2 & 0 & \tilde{V}_1
\end{array} \right].
$$

If we do a perfect shuffle of the block rows and columns (undoing the perfect shuffle that was performed earlier), $\check{Q}$ is transformed to

$$
\left[\begin{array}{cc|cc}
\tilde{U}_1 & \tilde{U}_2 & & \\
-\tilde{U}_2 & \tilde{U}_1 & & \\
\hline
& & \tilde{V}_1 & \tilde{V}_2 \\
& & -\tilde{V}_2 & \tilde{V}_1
\end{array} \right],
$$

$\check{M}$ is transformed to

$$
\left[\begin{array}{cc|cc}
& & \tilde{S} & \tilde{B} \\
& & 0 & \tilde{T}^T \\
\hline
\tilde{T} & -\tilde{B}^T & & \\
0 & \tilde{S}^T & &
\end{array} \right],
$$

and $\tilde{M}$ is transformed to $M = \begin{bmatrix} 0 & \mathcal{H} \\ -\mathcal{H} & 0 \end{bmatrix}$. This shows that the similarity transformation $\check{M} = \check{Q} \check{M} \check{Q}^T$ can be decomposed into two equivalence transformations: Let

$$
\tilde{U} = \left[\begin{array}{cc} \tilde{U}_1 & \tilde{U}_2 \\ -\tilde{U}_2 & \tilde{U}_1 \end{array} \right], \qquad \tilde{V} = \left[\begin{array}{cc} \tilde{V}_1 & \tilde{V}_2 \\ -\tilde{V}_2 & \tilde{V}_1 \end{array} \right],
$$

$$
\tilde{R}_1 = \left[\begin{array}{cc} \tilde{S} & \tilde{B} \\ 0 & \tilde{T}^T \end{array} \right], \quad \text{and} \quad \tilde{R}_2 = \left[\begin{array}{cc} -\tilde{T} & \tilde{B}^T \\ 0 & -\tilde{S}^T \end{array} \right].
$$

Then

$$
\mathcal{H} = \tilde{U} \tilde{R}_1 \tilde{V}^T \quad \text{and} \quad \mathcal{H} = \tilde{V} \tilde{R}_2 \tilde{U}^T.
$$

The matrices $\tilde{U}$, $\tilde{V} \in \mathbb{R}^{2n \times 2n}$ are orthogonal and symplectic. This is one version of the so-called *symplectic URV decomposition*, which dates from [35]. The construction that we have sketched here shows that the symplectic *URV* decomposition can be computed by a backward stable direct algorithm in $O(n^3)$ work.

Theorem 8.8.1 (symplectic *URV* decomposition I). *Let $\mathcal{H} \in \mathbb{R}^{2n \times 2n}$ be a Hamiltonian matrix. Then there exist orthogonal, symplectic $\tilde{U}$, $\tilde{V} \in \mathbb{R}^{2n \times 2n}$, an upper-triangular $\tilde{T} \in \mathbb{R}^{n \times n}$, upper Hessenberg $\tilde{S} \in \mathbb{R}^{n \times n}$, and $\tilde{B} \in \mathbb{R}^{n \times n}$ such that*

$$
\mathcal{H} = \tilde{U} \tilde{R}_1 \tilde{V}^T \quad \text{and} \quad \mathcal{H} = \tilde{V} \tilde{R}_2 \tilde{U}^T,
$$

where

$$
\tilde{R}_1 = \left[\begin{array}{cc} \tilde{S} & \tilde{B} \\ 0 & \tilde{T}^T \end{array} \right] \quad \text{and} \quad \tilde{R}_2 = \left[\begin{array}{cc} -\tilde{T} & \tilde{B}^T \\ 0 & -\tilde{S}^T \end{array} \right].
$$

This decomposition can be computed by a backward stable direct algorithm in $O(n^3)$ work.

From Theorem 8.8.1 we see that $\mathcal{H}^2 = \tilde{U} \tilde{R}_1 \tilde{R}_2 \tilde{U}^T$, so the eigenvalues of $\mathcal{H}$ are the square roots of the eigenvalues of $\tilde{R}_1 \tilde{R}_2$. Since

$$\tilde{R}_1 \tilde{R}_2 = \begin{bmatrix} -\tilde{S}\tilde{T} & * \\ 0 & -(\tilde{S}\tilde{T})^T \end{bmatrix},$$

the eigenvalues of $\mathcal{H}$ are the square roots of the eigenvalues of $-\tilde{S}\tilde{T}$. This can also be seen from looking at the form of $\check{M}$ in (8.8.5). The eigenvalues of $\check{M}$ are the same as those of M, and these are i times the eigenvalues of $\mathcal{H}$ (in duplicate). The eigenvalues of $\check{M}$ are the eigenvalues of $\begin{bmatrix} 0 & \tilde{S} \\ \tilde{T} & 0 \end{bmatrix}$, which are the square roots of the eigenvalues of $\tilde{S}\tilde{T}$. Thus we see (for the second time) that the eigenvalues of $\mathcal{H}$ are the square roots of the eigenvalues of $-\tilde{S}\tilde{T}$. The periodic QR algorithm, the QR version of the algorithm sketched in Section 8.3 (and see [111]) can be used to transform $\begin{bmatrix} 0 & \tilde{S} \\ \tilde{T} & 0 \end{bmatrix}$ to the form $\begin{bmatrix} 0 & S \\ T & 0 \end{bmatrix}$, where T is triangular and S is quasi-triangular, yielding the eigenvalues. (The product ST is then in Wintner–Murnaghan form. See Theorem 2.2.6.)

Suppose that the transformation is given by

$$\begin{bmatrix} 0 & \tilde{S} \\ \tilde{T} & 0 \end{bmatrix} = \begin{bmatrix} Y & 0 \\ 0 & Z \end{bmatrix} \begin{bmatrix} 0 & S \\ T & 0 \end{bmatrix} \begin{bmatrix} Y^T & 0 \\ 0 & Z^T \end{bmatrix},$$

where Y and Z are orthogonal. If we let $X = \text{diag}\{Y, Z, Y, Z\}$ and $\check{M} = X^T \check{M} X$, we have

$$\dot{M} = \begin{bmatrix} 0 & S & 0 & B \\ T & 0 & -B^T & 0 \\ \hline 0 & 0 & 0 & T^T \\ 0 & 0 & S^T & 0 \end{bmatrix},$$

where $B = Y^T \tilde{B} Z$. Letting $Q = \tilde{Q}X$, we have $\tilde{M} = Q\dot{M}Q^T$. Moreover,

$$Q = \begin{bmatrix} U_1 & 0 & U_2 & 0 \\ 0 & V_1 & 0 & V_2 \\ \hline -U_2 & 0 & U_1 & 0 \\ 0 & -V_2 & 0 & V_1 \end{bmatrix},$$

where $U_i = \tilde{U}_i Y$ and $V_i = \tilde{V}_i Z$ for $i = 1, 2$. If we now do a shuffle of the block rows and columns of $\dot{M}$ and Q, just as we did for $\check{M}$ and $\tilde{Q}$ above, we find that the similarity transformation $\tilde{M} = Q\dot{M}Q^T$ can be broken into two equivalence transformations. Let

$$U = \begin{bmatrix} U_1 & U_2 \\ -U_2 & U_1 \end{bmatrix}, \qquad V = \begin{bmatrix} V_1 & V_2 \\ -V_2 & V_1 \end{bmatrix},$$

$$R_1 = \begin{bmatrix} S & B \\ 0 & T^T \end{bmatrix}, \quad \text{and} \quad R_2 = \begin{bmatrix} -T & B^T \\ 0 & -S^T \end{bmatrix}.$$

Then

$$\mathcal{H} = U R_1 V^T \quad \text{and} \quad \mathcal{H} = V R_2 U^T.$$

This is another version of the symplectic URV decomposition.

Theorem 8.8.2 (symplectic URV decomposition II). *Let $\mathcal{H} \in \mathbb{R}^{2n \times 2n}$ be a Hamiltonian matrix. Then there exist orthogonal symplectic U, $V \in \mathbb{R}^{2n \times 2n}$, an upper triangular $T \in \mathbb{R}^{n \times n}$, quasi-triangular $S \in \mathbb{R}^{n \times n}$, and $B \in \mathbb{R}^{n \times n}$ such that*

$$\mathcal{H} = U R_1 V^T \quad and \quad \mathcal{H} = V R_2 U^T,$$

where

$$R_1 = \begin{bmatrix} S & B \\ 0 & T^T \end{bmatrix} \quad and \quad R_2 = \begin{bmatrix} -T & B^T \\ 0 & -S^T \end{bmatrix}.$$

This decomposition can be computed by a backward stable algorithm in $O(n^3)$ work. The eigenvalues of $\mathcal{H}$ are the square roots of the eigenvalues of the quasi-triangular matrix $-ST$.

The algorithm to compute this decomposition is not direct, as the periodic QR algorithm is needed for the final step. The algorithms sketched to this point in this section are implemented in HAPACK [33].

Hamiltonian Schur Form

The algorithm behind Theorem 8.8.2 yields the eigenvalues of the Hamiltonian matrix $\mathcal{H}$, but it does not give eigenvectors or invariant subspaces. From the columns of U or V we can extract invariant subspaces of $\mathcal{H}^2$, but not of $\mathcal{H}$.

A Hamiltonian matrix is in *Hamiltonian Schur form* if it has the form

$$\begin{bmatrix} T & K \\ 0 & -T^T \end{bmatrix},$$

where T is quasi-triangular (and $K^T = K$). If a Hamiltonian matrix $\mathcal{H} \in \mathbb{R}^{2n \times 2n}$ has no eigenvalues on the imaginary axis, then exactly n are in the left half plane; the other n are their mirror images in the right half plane. The following theorem states that any such $\mathcal{H}$ is orthogonally, symplectically similar to a matrix in Hamiltonian Schur form. The transformation can be constructed so that all of the eigenvalues of T are in the left half plane. One way to get at invariant subspaces of $\mathcal{H}$ is to compute the Hamiltonian Schur form.

Theorem 8.8.3 (Hamiltonian Schur form). *Let $\mathcal{H} \in \mathbb{R}^{2n \times 2n}$ be a Hamiltonian matrix that has no eigenvalues on the imaginary axis. Then there is an orthogonal, symplectic matrix W such that $\mathcal{H} = W \mathcal{M} W^{-1}$, where*

$$\mathcal{M} = \begin{bmatrix} T & K \\ 0 & -T^T \end{bmatrix}$$

is in Hamiltonian Schur form, and the eigenvalues of T all lie in the left half plane.

Theorem 8.8.3 was proved in [165] by a nonconstructive method. In the rest of this section we outline a constructive proof from [226], which was adapted from the construction given in [57].

A subspace $S \subseteq \mathbb{R}^{2n}$ is called *isotropic* if $x^T J y = 0$ for all x, $y \in S$. If $X \in \mathbb{R}^{2n \times j}$ is a matrix such that $S = \mathcal{R}(X)$, then S is isotropic if and only if $X^T J X = 0$. Given a symplectic matrix $S \in \mathbb{R}^{2n \times 2n}$, let S_1 and S_2 denote the first n and last n columns of S, respectively. Then the relationship $S^T J S = J$ implies that $\mathcal{R}(S_1)$ and $\mathcal{R}(S_2)$ are both isotropic. We will need the following lemma.

Lemma 8.8.4. *Let $S \subseteq \mathbb{R}^{2n}$ be a subspace that is invariant under the Hamiltonian matrix $\mathcal{H}$. Suppose that all of the eigenvalues of $\mathcal{H}$ associated with S satisfy $\Re(\lambda) < 0$. Then S is isotropic.*

Proof: Let X be a matrix with linearly independent columns such that $S = \mathcal{R}(X)$. We need to show that $X^T J X = 0$. Invariance of S implies that $\mathcal{H}X = XC$ for some C whose eigenvalues all satisfy $\Re(\lambda) < 0$. Multiplying on the left by $X^T J$, we get $X^T J X C = X^T (J\mathcal{H})X$. Since $J\mathcal{H}$ is symmetric, we deduce that $X^T J X C$ is symmetric, so that $X^T J X C + C^T X^T J X = 0$. Letting $Y = X^T J X$, we have $YC + C^T Y = 0$. The Sylvester operator $Y \to YC + C^T Y$ has eigenvalues $\lambda_j + \mu_k$, where λ_j and μ_k are eigenvalues of C and C^T, respectively. Since the eigenvalues $\lambda_j + \mu_k$ all lie in the open left half plane, they are all nonzero. Thus the Sylvester operator is nonsingular, and the homogeneous Sylvester equation $YC + C^T Y = 0$ has only the solution $Y = 0$. (See Theorem 2.4.2.) Therefore $X^T J X = 0$. $\square$

Let $\mathcal{H} \in \mathbb{R}^{2n \times 2n}$ be a Hamiltonian matrix that has no eigenvalues on the imaginary axis. As a first step toward computing the Hamiltonian Schur form of $\mathcal{H}$ we compute the symplectic *URV* decomposition given by Theorem 8.8.2. Taking the orthogonal symplectic matrix U from that decomposition, let $\hat{\mathcal{H}} = U^T \mathcal{H} U$. Then $\hat{\mathcal{H}}$ is a Hamiltonian matrix with the property that $\hat{\mathcal{H}}^2$ is in skew-Hamiltonian Schur form. For notational simplicity let us now drop the hat and assume that $\mathcal{H}$ itself is a Hamiltonian matrix such that $\mathcal{H}^2$ is in skew-Hamiltonian Schur form.

We will produce (at $O(n^2)$ cost) an orthogonal symplectic Q such that the Hamiltonian matrix $Q^T \mathcal{H} Q$ can be partitioned as

$$Q^T \mathcal{H} Q = \left[\begin{array}{cc|cc} \hat{A}_1 & * & * & * \\ 0 & \hat{A} & * & \hat{N} \\ \hline 0 & 0 & -\hat{A}_1^T & 0 \\ 0 & \hat{K} & * & -\hat{A}^T \end{array} \right], \qquad (8.8.6)$$

where $\hat{A}_1$ is either 1×1 with a negative real eigenvalue or 2×2 with a complex-conjugate pair of eigenvalues in the left half plane, and the remaining Hamiltonian matrix

$$\hat{\mathcal{H}} = \left[\begin{array}{cc} \hat{A} & \hat{N} \\ \hat{K} & -\hat{A}^T \end{array} \right] \qquad (8.8.7)$$

has the property that its square is in skew-Hamiltonian Schur form. Now we can apply the same procedure to $\hat{\mathcal{H}}$ to deflate out another eigenvalue or two, and repeating the procedure as many times as necessary, we get the Hamiltonian Schur form of $\mathcal{H}$ in $O(n^3)$ work.

The Method in the Simplest Case

To keep the discussion simple let us assume at first that all of the eigenvalues of $\mathcal{H}$ are real. Then

$$\mathcal{H}^2 = \begin{bmatrix} B & M \\ 0 & B^T \end{bmatrix},$$

where B is upper triangular and has positive entries on the main diagonal. It follows that e_1 is an eigenvector of $\mathcal{H}^2$ with associated eigenvalue b_{11}. Generically e_1 will *not* be an eigenvector of $\mathcal{H}$, and we will assume this. Nongeneric cases are discussed in detail in [57] and [226]. We will discuss only the generic case here. Thus the vectors e_1 and $\mathcal{H}e_1$ are linearly independent. Since e_1 *is* an eigenvector of $\mathcal{H}^2$, the two-dimensional space span$\{e_1, \mathcal{H}e_1\}$ is invariant under $\mathcal{H}$. In fact

$$\mathcal{H}\begin{bmatrix} e_1 & \mathcal{H}e_1 \end{bmatrix} = \begin{bmatrix} e_1 & \mathcal{H}e_1 \end{bmatrix} \begin{bmatrix} 0 & b_{11} \\ 1 & 0 \end{bmatrix}.$$

The associated eigenvalues are $\lambda = -\sqrt{b_{11}}$ and $-\lambda = \sqrt{b_{11}}$, and the eigenvectors are $(\mathcal{H} + \lambda I)e_1$ and $(\mathcal{H} - \lambda I)e_1$, respectively.

Our objective is to build an orthogonal symplectic Q that causes a deflation as explained above. Let

$$x = (\mathcal{H} + \lambda I)e_1, \tag{8.8.8}$$

the eigenvector of $\mathcal{H}$ associated with eigenvalue λ. We will build an orthogonal symplectic Q that has its first column proportional to x. This will guarantee that

$$Q^T \mathcal{H} Q = \left[\begin{array}{cc|cc} \lambda & * & * & * \\ 0 & \hat{A} & * & \hat{N} \\ \hline 0 & 0 & -\lambda & 0 \\ 0 & \hat{K} & * & -\hat{A}^T \end{array} \right], \tag{8.8.9}$$

thereby deflating out eigenvalues $\pm\lambda$. We have to be careful how we construct Q, as we must also guarantee that the deflated matrix $\hat{\mathcal{H}}$ of (8.8.7) has the property that its square is in skew-Hamiltonian Schur form. For this it is necessary and sufficient that $Q^T \mathcal{H}^2 Q$ be in skew-Hamiltonian Schur form. We will outline the algorithm first; then we will demonstrate that it has the desired properties.

Outline of the Algorithm, Real-Eigenvalue Case

Partition the eigenvector x from (8.8.8) as

$$x = \begin{bmatrix} v \\ w \end{bmatrix},$$

where $v, w \in \mathbb{R}^n$. We begin by introducing zeros into the vector w.

Let $U_{1,2}$ be a rotator in the $(1, 2)$ plane such that $U_{1,2}^T w$ has a zero in position 1. Then let $U_{2,3}$ be a rotator in the $(2, 3)$ plane such that $U_{2,3}^T U_{1,2}^T w$ has zeros in positions 1 and 2. Continuing in this manner, produce $n - 1$ rotators such that

$$U_{n-1,n}^T \cdots U_{2,3}^T U_{1,2}^T w = \gamma e_n,$$

where $\gamma = \pm\|w\|_2$. Let

$$U = U_{1,2}U_{2,3}\cdots U_{n-1,n}.$$

Then $U^T w = \gamma e_n$. Let

$$Q_1 = \begin{bmatrix} U & \\ & U \end{bmatrix},$$

and let

$$x^{(1)} = \begin{bmatrix} v^{(1)} \\ w^{(1)} \end{bmatrix} = Q_1^T \begin{bmatrix} v \\ w \end{bmatrix} = \begin{bmatrix} U^T v \\ \gamma e_n \end{bmatrix}.$$

Let Q_2 be an orthogonal symplectic rotator acting in the $(n, 2n)$ plane that annihilates the γ. In other words,

$$Q_2 x^{(1)} = \begin{bmatrix} v^{(2)} \\ 0 \end{bmatrix}.$$

Next let $Y_{n-1,n}$ be a rotator in the $(n-1, n)$ plane such that $Y_{n-1,n}^T v^{(2)}$ has a zero in the nth position. Then let $Y_{n-2,n-1}$ be a rotator in the $(n-2, n-1)$ plane such that $Y_{n-2,n-1}^T Y_{n-1,n}^T v^{(2)}$ has zeros in positions $n-1$ and n. Continuing in this manner, produce rotators $Y_{n-3,n-2}, \ldots, Y_{1,2}$ such that $Y_{1,2}^T \cdots Y_{n-1,n}^T v^{(2)} = \beta e_1$, where $\beta = \pm\|v^{(2)}\|$. Letting $Y = Y_{n-1,n}\cdots Y_{1,2}$, we have $Y^T v^{(1)} = \beta e_1$. Let

$$Q_3 = \begin{bmatrix} Y & \\ & Y \end{bmatrix},$$

and

$$Q = Q_1 Q_2 Q_3.$$

This is the desired orthogonal symplectic transformation matrix.

Each of U and Y is a product of $n-1$ rotators. Thus Q is a product of $2(n-1)$ symplectic double rotators and one symplectic single rotator. The transformation $\mathcal{H} \to Q^T \mathcal{H} Q$ is effected by applying the rotators in the correct order.

Why the Algorithm Works (Real-Eigenvalue Case)

The transforming matrix Q was designed so that $Q^T x = Q_3^T Q_2^T Q_1^T x = \beta e_1$. Thus $Q e_1 = \beta^{-1} x$; the first column of Q is proportional to x. Therefore we get the deflation shown in (8.8.9).

The more challenging task is to show that $Q^T \mathcal{H}^2 Q$ is in skew-Hamiltonian Schur form. To this end, let

$$\mathcal{H}_1 = Q_1^T \mathcal{H} Q_1, \qquad \mathcal{H}_2 = Q_2^T \mathcal{H}_1 Q_2, \qquad \text{and} \qquad \mathcal{H}_3 = Q_3^T \mathcal{H}_2 Q_3.$$

Then

$$\mathcal{H}_3 = Q^T \mathcal{H} Q,$$

so our objective is to show that $\mathcal{H}_3^2$ is in skew-Hamiltonian Schur form. We will do this by showing that the skew-Hamiltonian Schur form is preserved at every step of the algorithm.

We start with the transformation from $\mathcal{H}$ to $\mathcal{H}_1$. Since

$$\mathcal{H}_1^2 = Q_1^T \mathcal{H}^2 Q_1 = \left[\begin{array}{cc} U^T B U & U^T M U \\ 0 & U^T B^T U \end{array} \right],$$

it suffices to study the transformation $B \rightarrow U^T B U$. Since U is a product of rotators, the transformation from B to $U^T B U$ can be effected by applying the rotators one after another. We will show that each of these rotators preserves the upper triangular structure of B, effecting a swap of two eigenvalues.

We will find it convenient to use computer programming notation here. We will treat B as an array whose entries can be changed. As we apply the rotators, B will gradually be transformed to $U^T B U$, but we will continue to refer to the matrix as B rather than giving it a new name each time we change it.

Our argument uses the fact that w is an eigenvector of B^T: The equation $\mathcal{H}x = x\lambda$ implies $\mathcal{H}^2 x = x\lambda^2$, which implies $B^T w = w\lambda^2$. The transformations in B correspond to transformations in w. For example, when we change B to $U_{1,2}^T B U_{1,2}$ to get a new "B," we also change w to $U_{1,2}^T w$, and this is our new "w." Thus we are also treating w as an array. It will start out as the true w and be transformed gradually to the final "w," which is γe_n.

Let's start with the original B and w. Generically w_1 will be nonzero, and we will assume this. (For the nongeneric cases, see [57] and [226]). Thus the rotator $U_{1,2}$ is nontrivial, and the operation $w \leftarrow U_{1,2}^T w$ gives a new w that satisfies $w_1 = 0$ and $w_2 \neq 0$. The transformation $B \leftarrow U_{1,2}^T B U_{1,2}$ recombines the first two rows of B and the first two columns, resulting in a matrix that is upper triangular, except that its $(2,1)$ entry could be nonzero. The new B and w still satisfy the relationship $B^T w = w\lambda^2$. Looking at the first component of this equation, remembering that $w_1 = 0$, we have $b_{2,1} w_2 = 0$. Thus $b_{2,1} = 0$, and B is upper triangular. Moreover, the second component of the equation implies $b_{2,2} w_2 = w_2 \lambda^2$, so $b_{2,2} = \lambda^2$. Since we originally had $b_{1,1} = \lambda^2$, the similarity transformation has reversed the positions of the first two eigenvalues in B.

Now consider the second step: $B \leftarrow U_{2,3}^T B U_{2,3}$. Since $w_2 \neq 0$, the rotator $U_{2,3}$ is nontrivial. and the transformation $w \leftarrow U_{2,3}^T w$ gives a new w that satisfies $w_1 = w_2 = 0$ and $w_3 \neq 0$. The corresponding transformation $B \leftarrow U_{2,3}^T B U_{2,3}$ operates on rows and columns 2 and 3, and gives a new B that is upper triangular, except that the $(3,2)$ entry could be nonzero. The new B and w continue to satisfy the relationship $B^T w = w\lambda^2$. Looking at the second component of this equation, we see that $b_{3,2} w_3 = 0$. Therefore $b_{3,2} = 0$, and B is upper triangular. Looking at the third component of the equation, we find that $b_{3,3} = \lambda^2$. Thus another eigenvalue swap has occurred.

Clearly this same argument works for all stages of the transformation. We conclude that the final B, which is really $U^T B U$, is upper triangular. We have $b_{nn} = \lambda^2$. Each of the transforming rotators is nontrivial and moves the eigenvalue λ^2 downward. In the end, λ^2 is at the bottom of the matrix, and each of the other eigenvalues has been moved up one position. We have

$$\mathcal{H}_1^2 = \left[\begin{array}{cc} B^{(1)} & M^{(1)} \\ 0 & B^{(1)T} \end{array} \right],$$

where $B^{(1)}$ is upper triangular.

Now consider the transformation from $\mathcal{H}_1$ to $\mathcal{H}_2 = Q_2^T \mathcal{H}_1 Q_2$, which is effected by a single symplectic rotator in the $(n, 2n)$ plane. The 2×2 submatrix of $\mathcal{H}_1^2$ extracted from rows and columns n and $2n$ is

$$\begin{bmatrix} b_{nn} & 0 \\ 0 & b_{nn} \end{bmatrix}.$$

(Recall that the "M" matrix is skew symmetric.) This submatrix is a multiple of the identity matrix, so it remains unchanged under the transformation from $\mathcal{H}_1^2$ to $\mathcal{H}_2^2$. An inspection of the other entries in rows and columns n and $2n$ of $\mathcal{H}_1^2$ shows that no unwanted new nonzero entries are produced by the transformation. We have

$$\mathcal{H}_2 = \begin{bmatrix} B^{(2)} & M^{(2)} \\ 0 & B^{(2)T} \end{bmatrix},$$

where $B^{(2)}$ is upper triangular.

Now consider the transformation from $\mathcal{H}_2$ to $\mathcal{H}_3$. Since

$$\mathcal{H}_3^2 = Q_3^T \mathcal{H}_2^2 Q_3 = \begin{bmatrix} Y^T B^{(2)} Y & Y^T M^{(2)} Y \\ 0 & Y^T B^{(2)T} Y \end{bmatrix},$$

it suffices to study the transformation of $B^{(2)}$ to $Y^T B^{(2)} Y$.

At this point the eigenvector equation $\mathcal{H}^2 x = x \lambda^2$ has been transformed to $\mathcal{H}_2^2 x^{(2)} = x^{(2)} \lambda^2$ or

$$\begin{bmatrix} B^{(2)} & M^{(2)} \\ 0 & B^{(2)T} \end{bmatrix} \begin{bmatrix} v^{(2)} \\ 0 \end{bmatrix} = \begin{bmatrix} v^{(2)} \\ 0 \end{bmatrix} \lambda^2,$$

which implies $B^{(2)} v^{(2)} = v^{(2)} \lambda^2$. The nth component of this equation is $b_{n,n}^{(2)} v_n^{(2)} = v_n^{(2)} \lambda^2$, implying that $b_{n,n}^{(2)} = \lambda^2$. In fact, we already knew this. The entry $v_n^{(2)}$ cannot be zero, unless the original w was zero. We showed above that if $w \neq 0$, then we must have $b_{n,n}^{(2)} = \lambda^2$.

Returning to our computer programming style of notation, we now write B and v in place of $B^{(2)}$ and $v^{(2)}$ and consider the transformation from B to $Y^T B Y$, where $B v = v \lambda^2$. Since Y is a product of rotators, the transformation from B to $Y^T B Y$ can be effected by applying the rotators one after another. We will show that each of these rotators preserves the upper triangular structure of B, effecting a swap of two eigenvalues.

The first transforming rotator is $Y_{n-1,n}$, which acts in the $(n-1, n)$ plane and is designed to set v_n to zero. Since $v_n \neq 0$, the rotator $Y_{n-1,n}$ is nontrivial. The transformation $B \leftarrow Y_{n-1,n}^T B Y_{n-1,n}$ leaves B in upper triangular form, except that the $(n, n-1)$ entry could be nonzero. The transformation $v \leftarrow Y_{n-1,n}^T v$ gives a new v with $v_n = 0$ and $v_{n-1} \neq 0$. The new B and v still satisfy $B v = v \lambda^2$. The nth component of this equation is $b_{n,n-1} v_{n-1} = 0$, which implies $b_{n,n-1} = 0$. Thus B is upper triangular. Moreover, the $(n-1)$th component of the equation is $b_{n-1,n-1} v_{n-1} = v_{n-1} \lambda^2$, which implies $b_{n-1,n-1} = \lambda^2$. Thus the eigenvalue λ^2 has been swapped upward.

The next rotator, $Y_{n-2,n-1}$, acts in the $(n-2, n-1)$ plane and sets v_{n-1} to zero. The transformation $B \leftarrow Y_{n-2,n-1}^T B Y_{n-2,n-1}$ leaves B in upper triangular form, except that the $(n-1, n-2)$ entry could be nonzero. The transformation $v \leftarrow Y_{n-2,n-1}^T v$ gives a new v with $v_n = v_{n-1} = 0$ and $v_{n-2} \neq 0$. The equation $B v = v \lambda^2$ still holds. Its $(n-1)$st component is $b_{n-1,n-2} v_{n-2} = 0$, which implies $b_{n-1,n-2} = 0$. Thus B is still upper triangular. Furthermore the $(n-2)$nd component of the equation is $b_{n-2,n-2} v_{n-2} = v_{n-2} \lambda^2$, which implies $b_{n-2,n-2} = \lambda^2$. Thus the eigenvalue λ^2 has been swapped one more position upward.

Clearly this argument works at every step, and our final B, which is actually $B^{(3)}$, is upper triangular. Thus $\mathcal{H}_3^2$ has skew-Hamiltonian Hessenberg form.

The first part of the algorithm marches λ^2 from top to bottom of B, moving each other eigenvalue upward. Then the last part of the algorithm exactly reverses the process, returning all eigenvalues of B to their original positions.

Accommodating Complex Eigenvalues

Now we drop the assumption that the eigenvalues are all real. Then in the skew-Hamiltonian Schur form

$$\mathcal{H}^2 = \begin{bmatrix} B & M \\ 0 & B^T \end{bmatrix},$$

B is quasi-triangular; that is, it is block triangular with 1×1 and 2×2 blocks on the main diagonal. Each 2×2 block houses a complex-conjugate pair of eigenvalues of $\mathcal{H}^2$. Each 1×1 block is a positive real eigenvalue of $\mathcal{H}^2$. There are no nonpositive real eigenvalues because, by assumption, $\mathcal{H}$ has no purely imaginary eigenvalues. Suppose that there are l diagonal blocks of dimensions $n_1, \ldots, n_l$. Thus n_i is either 1 or 2 for each i.

As before, our objective is to build an orthogonal symplectic Q that causes a deflation of the form (8.8.6), as explained previously. How we proceed depends upon whether n_1 is 1 or 2. If $n_1 = 1$, we do exactly as described for the real-eigenvalue case above.

Now suppose $n_1 = 2$. In this case span$\{e_1\}$ is not invariant under $\mathcal{H}^2$ but span$\{e_1, e_2\}$ is. It follows that span$\{e_1, e_2, \mathcal{H}e_1, \mathcal{H}e_2\}$ is invariant under $\mathcal{H}$. We will make the generically valid assumption that this space has dimension 4. Letting $E = \begin{bmatrix} e_1 & e_2 \end{bmatrix}$, we have

$$\mathcal{H}\begin{bmatrix} E & \mathcal{H}E \end{bmatrix} = \begin{bmatrix} E & \mathcal{H}E \end{bmatrix}\begin{bmatrix} 0 & C \\ I & 0 \end{bmatrix},$$

where $C = \begin{bmatrix} b_{11} & b_{12} \\ b_{21} & b_{22} \end{bmatrix}$ has complex-conjugate eigenvalues μ and $\overline{\mu}$. Thus the eigenvalues of $\mathcal{H}$ associated with the invariant subspace span$\{e_1, e_2, \mathcal{H}e_1, \mathcal{H}e_2\}$ are λ, $-\lambda$, $\overline{\lambda}$, and $-\overline{\lambda}$, where $\lambda^2 = \mu$. Without loss of generality assume $\Re(\lambda) < 0$.

From span$\{e_1, e_2, \mathcal{H}e_1, \mathcal{H}e_2\}$ extract a two-dimensional subspace $\mathcal{S}_2$, invariant under $\mathcal{H}$, associated with the eigenvalues λ and $\overline{\lambda}$. This subspace is necessarily isotropic by Lemma 8.8.4. We will build a symplectic orthogonal Q whose first two columns span $\mathcal{S}_2$. This will guarantee that we get a deflation as in (8.8.6), where $\hat{A}_1$ is 2×2 and has λ and $\overline{\lambda}$ as its eigenvalues. Of course Q must have other properties that guarantee that the skew-Hamiltonian form of the squared matrix is preserved.

Outline of the Algorithm, Complex-Eigenvalue Case

The procedure is very similar to that in the case of a real eigenvalue. Let

$$X = \begin{bmatrix} V \\ W \end{bmatrix}$$

denote a $2n \times 2$ matrix whose columns are an orthonormal basis of $\mathcal{S}_2$. We can choose the basis so that $w_{11} = 0$. Isotropy implies $X^T J X = 0$. The first part of the algorithm applies a sequence of rotators that introduce zeros into W. We will use the notation $U_{i,i+1}^{(j)}$, $j = 1, 2$, to denote a rotator acting in the $(i, i+1)$ plane such that $U_{i,i+1}^{(j)T}$ introduces a zero in position $w_{i,j}$. (Here we are using the computer programming notation again, referring to the (i, j) position of an array that started out as W but has subsequently been transformed by some rotators.) The rotators are applied in the order $U_{2,3}^{(1)}, U_{1,2}^{(2)}, U_{3,4}^{(1)}, U_{2,3}^{(2)}, \ldots, U_{n-1,n}^{(1)}, U_{n-2,n-1}^{(2)}$. The rotators are carefully ordered so that the zeros that are created are not destroyed by subsequent rotators. We can think of the rotators as coming in pairs. The job of the pair $U_{i+1,i+2}^{(1)}, U_{i,i+1}^{(2)}$ together is to zero out the ith row of W.[3] Let

$$U = U_{2,3}^{(1)} U_{1,2}^{(2)} \cdots U_{n-1,n}^{(1)} U_{n-2,n-1}^{(2)}.$$

Then $U^T W = W^{(1)}$, where

$$W^{(1)} = \begin{bmatrix} 0 & 0 \\ \vdots & \vdots \\ 0 & 0 \\ 0 & w_{n-1,2}^{(1)} \\ w_{n,1}^{(1)} & w_{n,2}^{(1)} \end{bmatrix}.$$

Let

$$Q_1 = \begin{bmatrix} U & \\ & U \end{bmatrix},$$

and let

$$X^{(1)} = Q_1^T X = \begin{bmatrix} V^{(1)} \\ W^{(1)} \end{bmatrix}.$$

Next let Q_2 denote an orthogonal symplectic matrix acting on rows $n-1, n, 2n-1$, $2n$ such that Q_2^T zeros out the remaining nonzero entries in $W^{(1)}$, that is,

$$X^{(2)} = Q_2^T X^{(1)} = \begin{bmatrix} V^{(2)} \\ 0 \end{bmatrix}.$$

This can be done stably by a product of four rotators. Starting from the configuration

$$\begin{bmatrix} v_{n-1,1} & v_{n-1,2} \\ v_{n,1} & v_{n,2} \\ 0 & w_{n-1,2} \\ w_{n,1} & w_{n,2} \end{bmatrix}$$

(leaving off the superscripts to avoid notational clutter), a symplectic rotator in the $(n, 2n)$ plane sets the $w_{n,1}$ entry to zero. Then a symplectic double rotator acting in the $(n-1, n)$ and $(2n-1, 2n)$ planes transforms the $w_{n-1,2}$ and $v_{n,1}$ entries to zero simultaneously, as we explain below. Finally a symplectic rotator acting in the $(n, 2n)$ plane zeros out the $w_{n,2}$ entry. Q_2^T is the product of these rotators.

---

[3]Actually $U_{i+1,i+2}^{(1)}$ zeros out the entry $w_{i+1,1}$. This makes it possible for $U_{i,i+1}^{(2)}$ to zero out the entry $w_{i,1}$, thereby completing the task of forming zeros in row i.

The simultaneous appearance of two zeros is a consequence of isotropy. The condition $X^T J X = 0$, which was valid initially, continues to hold as X is transformed, since each of the transforming matrices is symplectic. The isotropy forces $w_{n-1,2}$ to zero when $v_{n,1}$ is set to zero, and vice versa.

The final part of the algorithm applies a sequence of rotators that introduce zeros into $V^{(2)}$. Notice that the $(n, 1)$ entry of $V^{(2)}$ is already zero. We will use the notation $Y_{i-1,i}^{(j)}$, $j = 1, 2$, to denote a rotator acting in the $(i - 1, i)$ plane such that $Y_{i-1,i}^{(j)T}$ introduces a zero in position $v_{i,j}$ (actually the transformed $v_{i,j}$). The rotators are applied in the order $Y_{n-2,n-1}^{(1)}$, $Y_{n-1,n}^{(2)}$, $Y_{n-3,n-2}^{(1)}$, $Y_{n-2,n-1}^{(2)}$, $\ldots$, $Y_{1,2}^{(1)}$, $Y_{2,3}^{(2)}$. Again the rotators are ordered so that the zeros that are created are not destroyed by subsequent rotators, and again the rotators come in pairs. The job of the pair $Y_{i-2,i-1}^{(1)}$, $Y_{i-1,i}^{(2)}$ together is to zero out the ith row of V. Let

$$Y = Y_{n-2,n-1}^{(1)} Y_{n-1,n}^{(2)} \cdots Y_{1,2}^{(1)} Y_{2,3}^{(2)}.$$

Then $Y^T V^{(2)} = V^{(3)}$, where

$$V^{(3)} = \begin{bmatrix} v_{11}^{(3)} & v_{12}^{(3)} \\ 0 & v_{22}^{(3)} \\ 0 & 0 \\ \vdots & \vdots \\ 0 & 0 \end{bmatrix}.$$

Let

$$Q_3 = \begin{bmatrix} Y & \\ & Y \end{bmatrix},$$

and let

$$X^{(3)} = Q_3^T X^{(2)} = \begin{bmatrix} V^{(3)} \\ 0 \end{bmatrix}.$$

Let

$$Q = Q_1 Q_2 Q_3.$$

This is the desired orthogonal symplectic transformation matrix. Q is a product of approximately $8n$ rotators; the transformation $\mathcal{H} \to Q^T \mathcal{H} Q$ is effected by applying the rotators in order.

Why the Algorithm Works

The transforming matrix Q was designed so that $Q^T X = Q_3^T Q_2^T Q_1^T X = X^{(3)}$. Thus $Q X^{(3)} = X$. Since $\mathcal{R}(X^{(3)}) = \text{span}\{e_1, e_2\}$, we have $Q \, \text{span}\{e_1, e_2\} = \mathcal{S}_2$; the first two columns of Q span the invariant subspace of $\mathcal{H}$ associated with λ and $\bar{\lambda}$. Therefore we get the deflation shown in (8.8.6).

Again the more challenging task is to show that $Q^T \mathcal{H}^2 Q$ is in skew-Hamiltonian Schur form. Let

$$\mathcal{H}_1 = Q_1^T \mathcal{H} Q_1, \qquad \mathcal{H}_2 = Q_2^T \mathcal{H}_1 Q_2, \qquad \text{and} \qquad \mathcal{H}_3 = Q_3^T \mathcal{H}_2 Q_3,$$

so that

$$\mathcal{H}_3 = Q^T \mathcal{H} Q,$$

as before. We must show that $\mathcal{H}_3^2$ is in skew-Hamiltonian Schur form.

We will cover the cases $n_1 = 2$ and $n_1 = 1$. In the case $n_1 = 2$, X denotes an $n \times 2$ matrix such that the two-dimensional space $\mathcal{R}(X)$ is invariant under $\mathcal{H}$. In the case $n_1 = 1$, X will denote an eigenvector x. For this X, the one-dimensional space $\mathcal{R}(X)$ is invariant under $\mathcal{H}$.

Since $\mathcal{R}(X)$ is invariant under $\mathcal{H}$, we have $\mathcal{H}X = X\Lambda$, for some $n_i \times n_i$ matrix Λ. If $n_i = 1$, then $\Lambda = \lambda$, a negative real eigenvalue. If $n_i = 2$, Λ has complex eigenvalues λ and $\overline{\lambda}$ lying in the left half plane. Consequently $\mathcal{H}^2 X = X\Lambda^2$ or

$$\begin{bmatrix} B & M \\ 0 & B^T \end{bmatrix} \begin{bmatrix} V \\ W \end{bmatrix} = \begin{bmatrix} V \\ W \end{bmatrix} \Lambda^2,$$

which implies

$$B^T W = W\Lambda^2.$$

Thus $\mathcal{R}(W)$ is invariant under B^T.

Recalling that B is quasi-triangular with main-diagonal blocks of dimension n_i, $i = 1, \ldots, l$, we introduce the partition

$$B = \begin{bmatrix} B_{11} & B_{12} & \cdots & B_{1l} \\ & B_{22} & & B_{2l} \\ & & \ddots & \vdots \\ & & & B_{ll} \end{bmatrix},$$

where B_{ii} is $n_i \times n_i$. Partitioning W conformably with B, we have

$$W = \begin{bmatrix} W_1 \\ \vdots \\ W_l \end{bmatrix},$$

where W_i is $n_i \times n_1$.

We start with the transformation from $\mathcal{H}$ to $\mathcal{H}_1$, for which it suffices to study the transformation from B to $U^T B U$. Again we will use computer programming notation, treating B and W as arrays whose entries get changed as the rotators are applied.

The equation $B^T W = W\Lambda^2$ implies that

$$\begin{bmatrix} B_{11}^T & 0 \\ B_{12}^T & B_{22}^T \end{bmatrix} \begin{bmatrix} W_1 \\ W_2 \end{bmatrix} = \begin{bmatrix} W_1 \\ W_2 \end{bmatrix} \Lambda^2. \tag{8.8.10}$$

In particular, $B_{11}^T W_1 = W_1 \Lambda^2$, which reminds us that the eigenvalues of B_{11} are the same as those of Λ^2. Assume that W_1 is nonsingular, as is generically true. This implies that the equation $B_{11}^T W_1 = W_1 \Lambda^2$ is a similarity relationship between B_{11}^T and Λ^2.

The algorithm begins by creating zeros in the top of W. Proceed until the first n_2 (1 or 2) rows of W have been transformed to zero. In the case $n_1 = 1$ (resp., $n_1 = 2$), this will require n_2 rotators (resp., *pairs* of rotators). We apply these rotators to B as well, effecting

an orthogonal symplectic similarity transformation. The action of these rotators is confined
to the first $n_1 + n_2$ rows and columns of B. The resulting new "B" remains quasi-triangular,
except that the block B_{21} could now be nonzero. At this point the relationship (8.8.10) will
have been transformed to

$$
\begin{bmatrix} B_{11}^T & B_{21}^T \\ B_{12}^T & B_{22}^T \end{bmatrix} \begin{bmatrix} 0 \\ W_2 \end{bmatrix} = \begin{bmatrix} 0 \\ W_2 \end{bmatrix} \Lambda^2. \tag{8.8.11}
$$

Here we must readjust the partition in order to retain conformability. The block of zeros is
$n_2 \times n_1$, so the new W_2 must be $n_1 \times n_1$. Since the original W_1 was nonsingular, the new W_2
must also be nonsingular. The new B_{11} and B_{22} are $n_2 \times n_2$ and $n_1 \times n_1$, respectively. The
top equation of (8.8.11) is $B_{21}^T W_2 = 0$, which implies $B_{21} = 0$. Thus the quasi-triangularity
is preserved. Moreover, the second equation of (8.8.11) is $B_{22}^T W_2 = W_2 \Lambda^2$, which implies
that B_{22} has the eigenvalues of Λ^2 (and the new B_{11} must have the eigenvalues of the old
B_{22}). Thus the eigenvalues have been swapped.

Now, proceeding inductively, assume that the first $n_2 + n_3 \cdots + n_k$ rows of W have
been transformed to zero. At this point we have a transformed equation $B^T W = W \Lambda^2$.
Assume inductively that B has quasi-triangular form, where B_{kk} is $n_1 \times n_1$ and has the
eigenvalues of Λ^2. For $i = 1, \ldots, k - 1$, B_{ii} is $n_{i+1} \times n_{i+1}$ and has the same eigenvalues
as the original $B_{i+1,i+1}$ had. For $i = k + 1, \ldots, l$, B_{ii} has not yet been touched by the
algorithm. Our current W has the form

$$
W = \begin{bmatrix} 0 \\ W_k \\ \vdots \\ W_l \end{bmatrix},
$$

where the zero block has $n_2 + \cdots + n_k$ rows, W_k is $n_1 \times n_1$ and nonsingular, and $W_{k+1}, \ldots, W_l$
have not yet been touched by the algorithm. From the current version of the equation
$B^T W = W \Lambda^2$ we see that

$$
\begin{bmatrix} B_{kk}^T & 0 \\ B_{k,k+1}^T & B_{k+1,k+1}^T \end{bmatrix} \begin{bmatrix} W_k \\ W_{k+1} \end{bmatrix} = \begin{bmatrix} W_k \\ W_{k+1} \end{bmatrix} \Lambda^2.
$$

The next step of the algorithm is to zero out the next n_{k+1} rows of W. The situation is
exactly the same as it was in the case $k = 1$. Arguing exactly as in that case, we see that
the quasi-triangularity is preserved, and the eigenvalues are swapped.

At the end of the first stage of the algorithm we have

$$
W = \begin{bmatrix} 0 \\ \vdots \\ 0 \\ W_l \end{bmatrix},
$$

where W_l is $n_1 \times n_1$ and nonsingular. This W is what we previously called $W^{(1)}$. The
corresponding B (which is actually $B^{(1)}$) is quasi-triangular. B_{ll} is $n_1 \times n_1$ and has the

eigenvalues of Λ^2. Each other B_{ii} is $n_{i+1} \times n_{i+1}$ and has the eigenvalues of the original $B_{i+1,i+1}$. At this point we have transformed $\mathcal{H}$ to $\mathcal{H}_1 = Q_1^T \mathcal{H} Q_1$. We have

$$\mathcal{H}_1^2 = \left[\begin{array}{cc} B^{(1)} & M^{(1)} \\ 0 & B^{(1)T} \end{array} \right],$$

where $B^{(1)}$ is quasi-triangular.

Now consider the transformation from $\mathcal{H}_1$ to $\mathcal{H}_2 = Q_2^T \mathcal{H}_1 Q_2$. We have to show that the skew-Hamiltonian Schur form of $\mathcal{H}_1^2$ is preserved by the transformation. If we write $\mathcal{H}_1^2$ in block form conformably with the blocking of $B^{(1)}$, we have $2l$ block rows and block columns. The action of the transformation affects only block rows and columns l and $2l$. If we consider the block triangular form of $B^{(1)}$, we find that zero blocks are combined only with zero blocks, and no unwanted nonzero entries are introduced except possibly in block $(2l, l)$.

To see that the $(2l, l)$ block also remains zero, examine our main equation

$$\left[\begin{array}{cc} B^{(1)} & M^{(1)} \\ 0 & B^{(1)T} \end{array} \right] \left[\begin{array}{c} V^{(1)} \\ W^{(1)} \end{array} \right] = \left[\begin{array}{c} V^{(1)} \\ W^{(1)} \end{array} \right] \Lambda^2,$$

bearing in mind that almost all of the entries of $W^{(1)}$ are zeros. Writing the equation in partitioned form and examining the lth and $2l$th block rows, we see that they amount to just

$$\left[\begin{array}{cc} B_{ll}^{(1)} & M_{ll}^{(1)} \\ 0 & B_{ll}^{(1)T} \end{array} \right] \left[\begin{array}{c} V_l^{(1)} \\ W_l^{(1)} \end{array} \right] = \left[\begin{array}{c} V_l^{(1)} \\ W_l^{(1)} \end{array} \right] \Lambda^2. \qquad (8.8.12)$$

Every matrix in this picture has dimension $n_1 \times n_1$. The similarity transformation by Q_2 changes (8.8.12) to

$$\left[\begin{array}{cc} B_{ll}^{(2)} & M_{ll}^{(2)} \\ F & B_{ll}^{(2)T} \end{array} \right] \left[\begin{array}{c} V_l^{(2)} \\ 0 \end{array} \right] = \left[\begin{array}{c} V_l^{(2)} \\ 0 \end{array} \right] \Lambda^2, \qquad (8.8.13)$$

where F is the block in question. Notice that $V_l^{(2)}$ cannot avoid being nonsingular. The second block equation in (8.8.13) is $F V_l^{(2)} = 0$, which implies $F = 0$, as desired. The first block equation is $B_{ll}^{(2)} V_l^{(2)} = V_l^{(2)} \Lambda^2$, which says that $B_{ll}^{(2)}$ is similar to Λ^2 and therefore has the same eigenvalues. We conclude that the skew-Hamiltonian Schur form is preserved in stage 2 of the algorithm.

Now consider the transformation from $\mathcal{H}_2$ to $\mathcal{H}_3$, for which it suffices to study the transformation from $B^{(2)}$ to $B^{(3)} = Y^T B^{(2)} Y$. At this point the main equation $\mathcal{H}^2 X = X \Lambda^2$ has been transformed to $\mathcal{H}_2^2 X^{(2)} = X^{(2)} \Lambda^2$ or

$$\left[\begin{array}{cc} B^{(2)} & M^{(2)} \\ 0 & B^{(2)T} \end{array} \right] \left[\begin{array}{c} V^{(2)} \\ 0 \end{array} \right] = \left[\begin{array}{c} V^{(2)} \\ 0 \end{array} \right] \Lambda^2,$$

which implies $B^{(2)} V^{(2)} = V^{(2)} \Lambda^2$, so $\mathcal{R}(V^{(2)})$ is invariant under $B^{(2)}$.

Returning to our computer programming style of notation, we now write B and V in place of $B^{(2)}$ and $V^{(2)}$ and consider the transformation from B to $Y^T B Y$, where $B V = V \Lambda^2$.

B is quasi-triangular. The block B_{ii} is $n_{i+1} \times n_{i+1}$ for $i = 1, \ldots, l-1$. The block B_{ll} is $n_1 \times n_1$ and has the eigenvalues of Λ^2. If we partition V conformably with B, we have

$$V = \begin{bmatrix} V_1 \\ \vdots \\ V_l \end{bmatrix},$$

where each V_i has its appropriate dimension. In particular V_l is $n_1 \times n_1$ and is nonsingular. The equation $BV = V\Lambda^2$ and the quasi triangularity of B imply that

$$\begin{bmatrix} B_{l-1,l-1} & B_{l-1,l} \\ 0 & B_{ll} \end{bmatrix} \begin{bmatrix} V_{l-1} \\ V_l \end{bmatrix} = \begin{bmatrix} V_{l-1} \\ V_l \end{bmatrix} \Lambda^2. \tag{8.8.14}$$

The third stage of the algorithm is wholly analogous with the first stage, except that we work upward instead of downward. We begin by transforming the bottom n_l rows of V to zero. In the case $n_1 = 1$ (resp., $n_1 = 2$) this requires n_l rotators (resp., *pairs* of rotators). When we apply these rotators to B, their action is confined to the bottom $n_l + n_1$ rows and columns, so the new B remains quasi-triangular, except that the block $B_{l,l-1}$ could now be nonzero. At this point the relationship (8.8.14) will have been transformed to

$$\begin{bmatrix} B_{l-1,l-1} & B_{l-1,l} \\ B_{l,l-1} & B_{ll} \end{bmatrix} \begin{bmatrix} V_{l-1} \\ 0 \end{bmatrix} = \begin{bmatrix} V_{l-1} \\ 0 \end{bmatrix} \Lambda^2. \tag{8.8.15}$$

Just as in stage 1, we must readjust the partition in order to retain conformability. The block of zeros is $n_l \times n_1$, so the new V_{l-1} must be $n_1 \times n_1$, and it is nonsingular. The new $B_{l-1,l-1}$ and B_{ll} are $n_1 \times n_1$ and $n_l \times n_l$, respectively. The bottom equation of (8.8.15) is $B_{l,l-1}V_{l-1} = 0$, which implies $B_{l,l-1} = 0$. Thus the quasi-triangularity is preserved. Moreover, the top equation of (8.8.15) is $B_{l-1,l-1}V_{l-1} = V_{l-1}\Lambda^2$, which implies that $B_{l-1,l-1}$ has the eigenvalues of Λ^2 (and the new B_{ll} must have the eigenvalues of the old $B_{l-1,l-1}$). Thus the eigenvalues have been swapped.

This first step sets the pattern for all of stage 3. We will skip the induction step, which is just like the first step. Stage 3 preserves the quasi-triangular form of B and swaps the eigenvalues of Λ^2 back to the top. In the end we have

$$\mathcal{H}_3^2 = \begin{bmatrix} B^{(3)} & M^{(3)} \\ 0 & B^{(3)T} \end{bmatrix},$$

where

$$B^{(3)} = \begin{bmatrix} B_{11}^{(3)} & B_{12}^{(3)} & \cdots & B_{1l}^{(3)} \\ 0 & B_{22}^{(3)} & & B_{2l}^{(3)} \\ \vdots & & \ddots & \vdots \\ 0 & & & B_{ll}^{(3)} \end{bmatrix},$$

where each $B_{ii}^{(3)}$ is $n_i \times n_i$ and has the same eigenvalues as the original B_{ii}.

It seems as if we are back where we started. Indeed we are, as far as $\mathcal{H}^2$ is concerned. But we must not forget that our primary interest is in $\mathcal{H}$, not $\mathcal{H}^2$. Thanks to this transformation, span$\{e_1\}$ (in the case $n_1 = 1$) or span$\{e_1, e_2\}$ (in the case $n_1 = 2$) is now invariant under $\mathcal{H}_3$ as well as $\mathcal{H}_3^2$, so we get the deflation (8.8.6).

8.9 Pseudosymmetric, Hamiltonian, and Symplectic Matrices

The Pseudosymmetric Eigenvalue Problem

Recall that a real tridiagonal matrix A is *pseudosymmetric* if

$$
A = TD = \begin{bmatrix} a_1 & b_1 & & & \\ b_1 & a_2 & b_2 & & \\ & b_2 & a_3 & \ddots & \\ & & \ddots & \ddots & b_{n-1} \\ & & & b_{n-1} & a_n \end{bmatrix} \begin{bmatrix} d_1 & & & & \\ & d_2 & & & \\ & & d_3 & & \\ & & & \ddots & \\ & & & & d_n \end{bmatrix},
$$

where T is symmetric and D is a signature matrix; that is, it is diagonal with each d_j equal to 1 or -1. We can even choose D so that all of the b_j are nonnegative. In Section 4.9 (Exercise 4.9.7) we showed that among GR algorithms, the HR algorithm is one that preserves pseudosymmetric structure. In Section 6.4 we showed that the HR algorithm can also be formulated as an HZ algorithm for the pencil $T - \lambda D$.

Since A can be represented by the product TD, we can also treat the pseudosymmetric eigenvalue problem as a product eigenvalue problem. We will pursue that angle here. In addition to its intrinsic interest, this viewpoint will show us that Hamiltonian and symplectic eigenvalue problems can be reduced to pseudosymmetric eigenvalue problems in a natural way.

Consider a pseudosymmetric matrix $A = TD$. Placing the factors T and D in a cyclic matrix, we have

$$
C = \begin{bmatrix} & T \\ D & \end{bmatrix}. \tag{8.9.1}
$$

Illustrating with the 3×3 case, we have

$$
C = \left[\begin{array}{ccc|ccc} & & & a_1 & b_1 & \\ & & & b_1 & a_2 & b_2 \\ & & & & b_2 & a_3 \\ \hline d_1 & & & & & \\ & d_2 & & & & \\ & & d_3 & & & \end{array} \right].
$$

The shuffled version is the Hessenberg matrix

$$
\left[\begin{array}{cc|cc|cc} a_1 & & b_1 & & & \\ d_1 & & & & & \\ \hline b_1 & & a_2 & & b_2 & \\ & d_2 & & & & \\ \hline & & b_2 & & a_3 & \\ & & & d_3 & & \end{array} \right].
$$

Recall from Section 4.9 that an HR iteration on $A = TD$ has the form

$$p(A) = HR, \qquad \hat{A} = H^{-1}AH,$$

where R is upper triangular, and H has the special property

$$H^T DH = \hat{D},$$

$\hat{D}$ being another signature matrix. Then $\hat{A} = \hat{T}\hat{D}$, where $\hat{T}$ is tridiagonal and symmetric. Thus pseudosymmetric structure is preserved.

Equivalently we can do an HR iteration on the cyclic matrix C defined in (8.9.1). Here we are in the case $k = 2$, so the driving function $f(C)$ for the iteration must be a function of C^2, say $f(C) = p(C^2)$. Thus

$$f(C) = p(C^2) = \begin{bmatrix} p(TD) & \\ & p(DT) \end{bmatrix}. \tag{8.9.2}$$

The HR decomposition of $p(C^2)$ is built up from HR decompositions of $p(TD)$ and $p(DT)$. As we observed during our discussion of the HZ algorithm in Section 6.4, the two decompositions are related. If $p(TD) = HR$ with $H^T DH = \hat{D}$, then $p(DT) = \tilde{H}\tilde{R}$, where $\tilde{H} = DH\hat{D} = H^{-T}$ satisfies $\tilde{H}^T D\tilde{H} = \hat{D}$, and $\tilde{R} = \hat{D}RD$ is upper triangular (Exercise 6.4.1). Thus the HR decomposition of $f(C)$ looks like

$$\begin{bmatrix} p(TD) & \\ & p(DT) \end{bmatrix} = \begin{bmatrix} H & \\ & H^{-T} \end{bmatrix} \begin{bmatrix} R & \\ & \tilde{R} \end{bmatrix}. \tag{8.9.3}$$

Performing the similarity transformation with the H factor from this decomposition, we complete the HR iteration on C:

$$\hat{C} = \begin{bmatrix} & \hat{T} \\ \hat{D} & \end{bmatrix} = \begin{bmatrix} H^{-1} & \\ & H^T \end{bmatrix} \begin{bmatrix} & T \\ D & \end{bmatrix} \begin{bmatrix} H & \\ & H^{-T} \end{bmatrix} \tag{8.9.4}$$

$$= \begin{bmatrix} & H^{-1}TH^{-T} \\ H^T DH & \end{bmatrix}.$$

This implicitly effects HR iterations on TD and DT simultaneously:

$$p(TD) = HR, \qquad \hat{T}\hat{D} = H^{-1}(TD)H$$

and

$$p(DT) = \tilde{H}\tilde{R}, \qquad \hat{D}\hat{T} = \tilde{H}^{-1}(DT)\tilde{H}.$$

The transformations in (8.9.4) are identical with those of the HZ algorithm in (6.4.2).

In practice the HR iteration is implemented as in implicit HZ algorithm, as described in detail in Section 6.4.

The Hamiltonian Eigenvalue Problem

If $\mathcal{H}$ is Hamiltonian and $\mathcal{S}$ is symplectic, then $\mathcal{S}^{-1}\mathcal{H}\mathcal{S}$ is Hamiltonian (Exercise 2.2.17). Thus the Hamiltonian structure will be respected by GR algorithms for which the transforming matrix is symplectic. These are SR algorithms, which we discussed in Section 4.9.

In [53] Bunse-Gerstner and Mehrmann showed that every Hamiltonian matrix is similar by a symplectic similarity transformation to a matrix of the highly condensed form

$$\begin{bmatrix} E & T \\ D & -E \end{bmatrix},$$ (8.9.5)

where E and D are diagonal and T is tridiagonal (and symmetric). There is some slack in the reduction algorithm: The main-diagonal entries of E can be chosen arbitrarily, while those of D must be nonzero but can be rescaled up or down. Bunse-Gerstner and Mehrmann advocated choosing the entries with an eye to making the transformation as stable as possible. This is a laudable goal. A more aggressive approach is to choose the entries to make the condensed form (8.9.5) as simple as possible. If one decides to do this, then the logical choice for E is the zero matrix. We can also scale D so that each of its main-diagonal entries is either $+1$ or -1, that is, it is a signature matrix. With these choices, the condensed form becomes

$$C = \begin{bmatrix} & T \\ D & \end{bmatrix},$$ (8.9.6)

with T tridiagonal and symmetric and D a signature matrix. This is *exactly* the form of the cyclic matrix C (8.9.1) that arose in the discussion of pseudosymmetric matrices. Thus the Hamiltonian eigenvalue problem is equivalent to the pseudosymmetric product eigenvalue problem.

If we wish to apply the SR algorithm to the condensed C of (8.9.6) or to any Hamiltonian matrix, we must choose shifts in a way that respects Hamiltonian structure. The eigenvalues of a real Hamiltonian matrix occur in $\pm\lambda$ pairs (Exercise 2.1.11). Thus the shifts should also be chosen in plus-minus pairs. In other words, the driving function for a Hamiltonian SR iteration on C should be a function of C^2: $f(C) = p(C^2)$, just as in (8.9.2). Now consider the HR decomposition (8.9.3). From Proposition 3.1.4 we know that for any real nonsingular matrix B, the block-diagonal matrix

$$\begin{bmatrix} B & \\ & B^{-T} \end{bmatrix}$$

is symplectic. Therefore the HR decomposition (8.9.3) is also an SR decomposition, and the HR iteration (8.9.3), (8.9.4) is also an SR iteration on C. In conclusion, if the condensed form (8.9.6) for Hamiltonian matrices is used, then the Hamiltonian SR algorithm is identical to the HR algorithm for pseudosymmetric matrices [31]. We will put this fact to use when we discuss the Hamiltonian Lanczos process in Section 9.8.

The Symplectic Eigenvalue Problem

For symplectic matrices the structure-preserving GR algorithm is also the SR algorithm, as we noted in Section 4.9. We also observed there that every symplectic matrix can be reduced by a symplectic similarity transformation to the extremely condensed form

$$B = \begin{bmatrix} 0 & -D \\ D & DT \end{bmatrix},$$ (8.9.7)

where D is a signature matrix and T is symmetric and tridiagonal. The characteristics of D and T are the same as those possessed by the same symbols in the previous two scenarios of this section.

If we want to apply the SR algorithm to the condensed symplectic matrix B of (8.9.7), how should we choose the shifts so that the symplectic structure is respected? From Exercise 2.1.11 we know that the eigenvalues of a real symplectic matrix occur in λ, λ^{-1} pairs. Therefore we should choose shifts in ρ, ρ^{-1} pairs. An SR iteration on B has the form

$$f(B) = SR, \qquad \hat{B} = S^{-1}BS.$$

If the driving function f is a polynomial, then its factors should appear in pairs of the form $B - \rho I$, $B - \rho^{-1}I$. Multiplying these factors together, we obtain

$$B^2 - (\rho + \rho^{-1})B + I.$$

Notice that if we multiply this product by B^{-1}, we obtain an interesting result:

$$B^{-1}(B^2 - (\rho + \rho^{-1})B + I) = (B + B^{-1}) - (\rho + \rho^{-1})I,$$

which suggests that it might be advantageous to take f to be a rational function of B instead of a polynomial. If we want to apply shifts $\rho_1, \ldots, \rho_k$ and their inverses, we can take $f(B) = p(B + B^{-1})$, where p is the kth degree polynomial whose zeros are $\rho_1 + \rho_1^{-1}, \ldots, \rho_k + \rho_k^{-1}$.

This idea relies on the accessibility of B^{-1}. Fortunately it is a simple matter to invert any symplectic matrix; the equation $S^T J S = J$ implies $S^{-1} = -J S^T J$. You can easily check that

$$B^{-1} = \begin{bmatrix} TD & D \\ -D & 0 \end{bmatrix}. \tag{8.9.8}$$

This has the happy consequence that

$$B + B^{-1} = \begin{bmatrix} TD & \\ & DT \end{bmatrix} \tag{8.9.9}$$

and

$$f(B) = p(B + B^{-1}) = \begin{bmatrix} p(TD) & \\ & p(DT) \end{bmatrix}. \tag{8.9.10}$$

Compare this with (8.9.2). In order to do an iteration of the SR algorithm, we need an SR decomposition of $f(B)$. However, as we have already observed, the HR decomposition (8.9.3) is also an SR decomposition, because the matrix

$$S = \begin{bmatrix} H & \\ & H^{-T} \end{bmatrix}$$

is symplectic. We complete an SR iteration by performing a similarity transformation with this matrix:

$$\begin{aligned}
\hat{B} = S^{-1}BS &= \begin{bmatrix} H^{-1} & \\ & H^T \end{bmatrix} \begin{bmatrix} 0 & -D \\ D & DT \end{bmatrix} \begin{bmatrix} H & \\ & H^{-T} \end{bmatrix} \\
&= \begin{bmatrix} 0 & -H^{-1}DH^{-T} \\ H^T D H & (H^T DH)(H^{-1}TH^{-T}) \end{bmatrix} \\
&= \begin{bmatrix} 0 & -\hat{D} \\ \hat{D} & \hat{D}\hat{T} \end{bmatrix},
\end{aligned}$$

where

$$\hat{T} = H^{-1}TH^{-T} \quad \text{and} \quad \hat{D} = H^T DH.$$

In conclusion, if the condensed form (8.9.7) is used, the symplectic SR algorithm is equivalent to the pseudosymmetric HR algorithm [31]. We will exploit this connection when we discuss the symplectic Lanczos process in Section 9.9.

Exercises

8.9.1. Check the details of the discussion of the symplectic case. In particular, show that if B is a symplectic matrix of the form (8.9.7), then its inverse has the form (8.9.8). Deduce (8.9.9) and (8.9.10).

8.10 The Unitary Eigenvalue Problem

Let $U \in \mathbb{C}^{n \times n}$ be a unitary matrix that is properly upper Hessenberg. By a unitary diagonal similarity transformation we can make the subdiagonal entries real and positive. Thus we will assume that $u_{j+1,j} > 0$ for $j = 1, \ldots, n - 1$. Then U can be expressed as a product of matrices of a very simple form:

$$U = G_1 G_2 \cdots G_{n-1} G_n, \tag{8.10.1}$$

where

$$G_k = \text{diag}\left\{ I_{k-1}, \tilde{G}_k, I_{n-k-1} \right\}, \tag{8.10.2}$$

$$\tilde{G}_k = \begin{bmatrix} \gamma_k & \sigma_k \\ \sigma_k & -\overline{\gamma}_k \end{bmatrix}, \quad \sigma_k > 0, \quad |\gamma_k|^2 + \sigma_k^2 = 1, \tag{8.10.3}$$

for $k = 1, \ldots, n - 1$, and $G_n = \text{diag}\{I_{n-1}, \gamma_n\}$ with $|\gamma_n| = 1$. (See Exercise 8.10.2.) The matrices G_k resemble complex analogues of rotators, but in fact they are more like reflectors (Exercise 8.10.1). If U is real, then $G_1, \ldots, G_n$ are real.

The numbers $\gamma_1, \ldots, \gamma_n$ are called the *Schur parameters* of U, and $\sigma_1, \ldots, \sigma_{n-1}$ are the *complementary Schur parameters*. U is completely determined by these $2n - 1$ numbers. In fact, U is completely determined by $\gamma_1, \ldots, \gamma_n$, since $\sigma_k = \sqrt{1 - |\gamma_k|^2}$ for $k = 1, \ldots, n - 1$. However, in floating point arithmetic these formulas cannot compute σ_k accurately in cases when $|\gamma_k| \approx 1$, so we cannot dispense with $\sigma_1, \ldots, \sigma_{n-1}$ in practice. From this point on we will refer to the $2n - 1$ numbers $\gamma_1, \ldots, \gamma_n, \sigma_1, \ldots, \sigma_{n-1}$ collectively as *Schur parameters*.

If we work with U in the form (8.10.1), we can store U in the form of Schur parameters in $O(n)$ space instead of the usual $O(n^2)$. This being the case, we might reasonably hope to compute the eigenvalues of U in $O(n^2)$ work instead of the usual $O(n^3)$. This is a product eigenvalue problem, but it does not fit the pattern established in Section 8.3, in which one of the factors is upper Hessenberg and the others are upper triangular. Therefore we will have to come up with some other methodology. This has been done by a number of researchers in a variety of ways [5, 6, 7, 51, 52, 103, 105, 107, 181, 200]. Some of these methods are stable and some not.

Here we sketch a scheme [64, 63] for performing unitary QR iterations in $O(n)$ work per iteration. The scheme has several virtues: It can do multishift QR iterations of arbitrary degree, it is straightforward and easy to understand, and it is backward-stable [64].

Unitary Multishift QR Iteration

Recall that a multishift QR iteration of degree m begins with the selection of m shifts $\rho_1, \ldots, \rho_m$. Then a vector x proportional to $(U - \rho_1 I) \cdots (U - \rho_m I)e_1$ is formed. A unitary transformation Q_1 with first column proportional to x is built, and U is transformed to $U_1 = Q_1^* A Q_1$. The resulting matrix has a bulge (4.5.4) in the Hessenberg form extending out to position $(m + 2, 1)$. The rest of the iteration consists of returning the matrix to upper Hessenberg form by a sequence of unitary similarity transformations that chase the bulge to the bottom of the matrix.

In our current scenario we have U stored in product form: $U = G_1 G_2 \cdots G_n$, represented by $2n - 1$ parameters. At the end of the iteration we wish to have $\hat{U} = \hat{G}_1 \hat{G}_2 \cdots \hat{G}_n$, represented by a new set of $2n - 1$ parameters. How can we introduce and work with a bulge under these conditions? Our solution is to multiply together the first few of the G_i factors to build a leading submatrix of U that is big enough to accommodate the bulge. We then build the bulge and begin to chase it downward. As we do so, we must multiply in additional G_i factors to accommodate the progressing bulge. However, we also get to factor out matrices $\hat{G}_1, \hat{G}_2, \ldots$, from the top since, as soon as the bulge begins to move downward, we can begin to refactor the top part of the matrix, for which the iteration is complete. At any given point in the algorithm, the part of the matrix that contains the bulge can be stored in a work area of dimension $(m + 2) \times (m + 2)$. On each forward step we must factor in one new G_i at the bottom of the work area, and we get to factor out a $\hat{G}_j$ at the top. The total storage space needed by our algorithm is thus $O(n + m^2)$. Bear in mind that m is never very big in practice.

We begin by forming the product $W_0 = G_1 G_2 \cdots G_{m+1}$. The nontrivial part of W_0 can fit into an $(m + 2) \times (m + 2)$ array, and this is what goes into our work area initially. We have

$$U = W_0 G_{m+2} \cdots G_n.$$

The first m columns of W_0 are exactly the first m columns of U, and this is what we need to form the initial vector x and transformation Q_1. The nontrivial part of Q_1 is confined to the leading $(m + 1) \times (m + 1)$ submatrix, so Q_1 commutes with $G_{m+2}, \ldots, G_n$. Therefore

$$U_1 = Q_1^* U Q_1 = Q_1^* W_0 Q_1 G_{m+2} \cdots G_n.$$

Letting $W_1 = Q_1^* W_0 Q_1$, we have

$$U_1 = W_1 G_{m+2} \cdots G_n.$$

In this transformation, all the work takes place in the work area, which contains the nontrivial part of W_0 initially and the nontrivial part of W_1 in the end. At this point the work area is completely filled with the bulge.

The first bulge-chasing transformation Q_2 is constructed so that the first column of $Q_2^* W_1$ has the form $\begin{bmatrix} \hat{\gamma}_1 & \hat{\sigma}_1 & 0 & \cdots & 0 \end{bmatrix}^T$. Care is taken that $\hat{\sigma}_1 > 0$. The entries $\hat{\gamma}_1$

and $\hat{\sigma}_1$ will remain fixed for the remainder of the QR iteration. These are the entries that define $\hat{G}_1$, and we can now factor $\hat{G}_1$ out of the work area. Let $\tilde{W}_1 = \hat{G}_1^* Q_2^* W_1$. Then

$$Q_2^* U_1 = \hat{G}_1 \tilde{W}_1 G_{m+2} \cdots G_n.$$

The matrix $\tilde{W}_1$ now has the form

$$\tilde{W}_1 = \begin{bmatrix} 1 & & \\ & \hat{W}_1 & \\ & & I_{n-m-2} \end{bmatrix},$$

where $\hat{W}_1$ is $(m+1) \times (m+1)$. The part of $\tilde{W}_1$ that is stored in the work area is now

$$\begin{bmatrix} 1 & \\ & \hat{W}_1 \end{bmatrix}.$$

Clearly the first row and column are no longer needed, so we shift the data in the work area so that it now contains

$$\begin{bmatrix} \hat{W}_1 & \\ & 1 \end{bmatrix},$$

corresponding to rows and columns 2 through $m+3$ of $\tilde{W}$.

We have multiplied on the left by Q_2^*. We must now complete the similarity transformation by multiplying on the right by Q_2. The nontrivial part of Q_2 lies in rows and columns 2 through $m+2$, so Q_2 commutes with $G_{m+3}, \ldots, G_n$ but not with G_{m+2}. Since we have readjusted the work area, there is now room to multiply G_{m+2} into the work area. We have

$$
\begin{aligned}
U_2 &= Q_2^* U_1 Q_2 \\
&= \hat{G}_1 \tilde{W}_1 G_{m+2} Q_2 G_{m+3} \cdots G_n \\
&= \hat{G}_1 W_2 G_{m+3} \cdots G_n,
\end{aligned}
$$

where $W_2 = \tilde{W}_1 G_{m+2} Q_2$. The nontrivial part of W_2 is in rows and columns 2 through $m+3$ and just fits into the work area.

The next bulge-chasing transformation Q_3 is constructed so that the second column of $Q_3^* W_2$ has the form $\begin{bmatrix} * & \hat{\gamma}_2 & \hat{\sigma}_2 & 0 & \cdots & 0 \end{bmatrix}^T$, with care taken so that $\hat{\sigma}_2 > 0$. The interesting part of this column lies in the first column of the work area. Since the nontrivial part of Q_3 lies in rows and columns 3 through $m+3$, Q_3^* commutes with $\hat{G}_1$. Thus

$$Q_3^* U_2 = \hat{G}_1 Q_3^* W_2 G_{m+3} \cdots G_n.$$

Once we multiply the active part of Q_3^* into the work area, we can factor out $\hat{G}_2$, which is determined by $\hat{\gamma}_2$ and $\hat{\sigma}_2$. Letting $\tilde{W}_2 = \hat{G}_2^* Q_3^* W_2$, we have

$$Q_3^* W_2 = \hat{G}_1 \hat{G}_2 \tilde{W}_2 G_{m+3} \cdots G_n.$$

The portion of $\tilde{W}_2$ in the work area has the form

$$\begin{bmatrix} 1 & \\ & \hat{W}_2 \end{bmatrix}.$$

We reposition the data so that the work area contains

$$\left[\begin{array}{cc} \hat{W}_2 & \\ & 1 \end{array} \right],$$

corresponding to rows and columns 3 through $m+4$ of $\tilde{W}_2$. Now we have room to multiply G_{m+3} into the work area.

When we multiply on the right by Q_3 to complete the similarity transformation, we find that Q_3 commutes with $G_{m+4}, \ldots, G_n$, so

$$U_3 = Q_3^* U_2 Q_3 = \hat{G}_1 \hat{G}_2 W_3 G_{m+4} \cdots G_n,$$

where $W_3 = \tilde{W}_2 G_{m+3} Q_3$. The nontrivial part of W_3 is in rows and columns 3 through $m+4$ and just fits into the work area.

The pattern is now clear. Each step is like the one before, except that at the bottom the work area shrinks away, as there are no more G_j to feed in. In the end we have

$$\hat{U} = U_{n-1} = \hat{G}_1 \hat{G}_2 \cdots \hat{G}_n$$

stored in the form of $2n - 1$ Schur parameters.

Enforcement of Unitarity

One other important detail needs to be mentioned. Each new pair of Schur parameters $\hat{\gamma}_k$, $\hat{\sigma}_k$ satisfies $|\hat{\gamma}_k|^2 + \hat{\sigma}_k^2 = 1$ in principle, but in practice roundoff errors will cause this equation to be violated by a tiny amount. Therefore the following normalization step is required:

$$\begin{aligned} \nu &\leftarrow \left(|\hat{\gamma}_k|^2 + \hat{\sigma}_k^2 \right)^{1/2} \\ \hat{\gamma}_k &\leftarrow \hat{\gamma}_k/\nu \\ \hat{\sigma}_k &\leftarrow \hat{\sigma}_k/\nu. \end{aligned} \qquad (8.10.4)$$

This should be done even when $k = n$, taking $\hat{\sigma}_n = 0$. This enforcement of unitarity is essential to the stability of the algorithm. If it is not done, the matrix will (over the course of many iterations) drift away from being unitary, and the algorithm will fail. For a proof of backward stability see [64].

Operation Count

The bulk of the arithmetic in the algorithm occurs when the transformations Q_k^* and Q_k are multiplied into the work area on the left and right, respectively. Each Q_k is the product of a reflector followed by a diagonal phase-correcting transformation to enforce the condition $\hat{u}_{k,k-1} > 0$. The latter costs $O(m)$ arithmetic; the real work is in applying the reflector. Each of these is at most $(m + 1) \times (m + 1)$ (smaller at the very end of the iteration), and the cost of applying it efficiently to the work area on left or right is about $4m^2$ flops. Since the reflector is applied only to the small work area and not to the full Hessenberg matrix, the amount of arithmetic is $O(m^2)$ instead of $O(nm)$; this is where we realize our savings.

Since $n-1$ reflectors are applied (on left and right) in the whole iteration, the arithmetic cost is about $8nm^2$ flops.

If m is fixed and small, then we can say that the cost of an iteration is $O(n)$, in the sense that the arithmetic is bounded by $C_m n$, where C_m is independent of n. However, the fact that C_m grows like m^2 as m is increased shows that it will be inefficient to take m too large. Another important reason for keeping m fairly small is to avoid shift blurring (Section 7.1).

Shift Strategies

The most common way to obtain m shifts is to take the eigenvalues of the trailing $m \times m$ submatrix of U. This strategy is cubically convergent when it converges (see Theorem 5.2.9 and Example 5.2.10). However, it is not globally convergent, as the following well-known example shows. Let U be the unitary circulant shift matrix, which looks like

$$\begin{bmatrix} & & & 1 \\ 1 & & & \\ & 1 & & \\ & & 1 & \end{bmatrix}$$

in the 4×4 case. For any $m < n$, if we take the eigenvalues of the trailing submatrix as shifts, we get shifts $0, \ldots, 0$, which are equidistant from all of the eigenvalues. A QR iteration on U with these shifts goes nowhere.

Since the eigenvalues of a unitary matrix lie on the unit circle, it make sense to choose shifts that are on the unit circle. Two strategies come to mind. The first computes the eigenvalues of the trailing $m \times m$ submatrix and normalizes each of them by dividing it by its absolute value. If any of the tentative shifts happens to be zero, it is replaced by a random number on the unit circle. If we use this strategy on the circulant shift matrix, we get m random shifts.

A second strategy stems from the following observation. The last m rows of the Hessenberg matrix U are orthonormal. Since $u_{n-m+1,n-m} > 0$, the trailing $m \times m$ submatrix $U(n-m+1:n, n-m+1:n)$ is not unitary, but it is nearly unitary. Its rows are orthogonal, and they all have norm 1, except that the top row $U(n-m+1, n-m+1:n)$ has norm less than one. Unitarity can be restored by dividing this row by its norm. In the rare case when the whole top row is zero, a suitable first row can be generated by orthonormalizing a random row against rows 2 through m. The m eigenvalues of the modified matrix then give m shifts on the unit circle.

When this strategy is used on the circulant shift matrix, the orthonormalization process will generate a first row of the form $(0, \ldots, 0, \gamma)$ with $|\gamma| = 1$. The shifts are then the roots of the equation $z^m - \gamma = 0$, which are equally spaced points on the unit circle.

These two strategies work about equally well. Both are locally cubically convergent: As $u_{n-m+1,n-m} \to 0$, the trailing $m \times m$ submatrix becomes closer and closer to unitary. Its eigenvalues become ever closer to the unit circle, and normalizing them as in the first strategy moves them only slightly. On the other hand, if we modify the matrix as in the second strategy by normalizing its first row, that also moves the eigenvalues only slightly, because the rescaling factor is very close to 1. Thus both strategies behave asymptotically the same as the strategy that simply takes the eigenvalues of the trailing submatrix as shifts; that is, they converge cubically when they converge. We conjecture that both strategies converge globally.

Exercises

8.10.1. Let $G = \begin{bmatrix} \gamma & \sigma \\ \sigma & -\bar{\gamma} \end{bmatrix}$, where $|\gamma|^2 + \sigma^2 = 1$.

 (a) Compute $\det G$.

 (b) Compute G^{-1}. Deduce that G is unitary.

 (c) Compute the eigenvalues of G.

 (d) Show that if γ is real, the eigenvalues of G are ± 1, and G is a reflector.

8.10.2. Let $U \in \mathbb{C}^{n \times n}$ be a unitary upper Hessenberg matrix with $u_{j+1,j} > 0$, $j = 1, \ldots, n-1$.

 (a) Show that there is a unique matrix $G_1 = \operatorname{diag}\left\{ \tilde{G}_1, 1, \ldots, 1 \right\}$ with $\tilde{G}_1 = \begin{bmatrix} \gamma_1 & \sigma_1 \\ \sigma_1 & -\bar{\gamma}_1 \end{bmatrix}$,
 $|\gamma_1|^2 + \sigma_1^2 = 1$, and $\sigma_1 > 0$ such that the matrix $U_1 = G_1^{-1} U$ has e_1 as its first column. (This is the first step of a QR decomposition of U.)

 (b) Use the fact that U_1 is unitary to deduce that
$$U_1 = \begin{bmatrix} 1 & 0^T \\ 0 & \hat{U}_1 \end{bmatrix},$$
 where $\hat{U}_1 \in \mathbb{C}^{(n-1) \times (n-1)}$ is unitary.

 (c) Prove by induction that there are unique $G_1, \ldots, G_{n-1}$ of the form (8.10.2), (8.10.3) such that
$$G_{n-1}^{-1} \cdots G_1^{-1} U = G_n,$$
 where $G_n = \operatorname{diag}\{1, \ldots, 1, \gamma_n\}$ with $|\gamma_n| = 1$.

 (d) Deduce that U has a unique decomposition $U = G_1 G_2 \cdots G_n$ of the specified form.

 (e) Show that $\sigma_k = u_{k+1,k}$ for $k = 1, \ldots, n-1$.

8.11 The Totally Nonnegative Eigenvalue Problem

A matrix $A \in \mathbb{R}^{n \times n}$ is called *totally positive (TP)* if all of its minors of all sizes are positive. In other words, for any k, given any k rows and k columns of A, the corresponding $k \times k$ submatrix has a positive determinant. A matrix is *totally nonnegative (TN)* if all of its minors are nonnegative. TN and TP matrices arise in a variety of situations [48, 93, 128]. For example, the Vandermonde matrix of a positive increasing sequence is TP.

Every TN matrix has a decomposition into a product of bidiagonal matrices, the entries of which parametrize the family of TN matrices. We will not describe this decomposition in detail. In the 3×3 case it has the form

$$
\begin{bmatrix} 1 & & \\ & 1 & \\ & a_2 & 1 \end{bmatrix}
\begin{bmatrix} 1 & & \\ b_1 & 1 & \\ & b_2 & 1 \end{bmatrix}
\begin{bmatrix} c_1 & & \\ & c_2 & \\ & & c_3 \end{bmatrix}
\begin{bmatrix} 1 & d_1 & \\ & 1 & d_2 \\ & & 1 \end{bmatrix}
\begin{bmatrix} 1 & & \\ & 1 & e_2 \\ & & 1 \end{bmatrix}.
$$

In the parametrization of a TP matrix all of the parameters $a_2, b_1, \ldots, e_2$ are positive, but in a TN matrix certain of them can be zero. An interesting example is the TP *Pascal matrix* [46, 79]

$$
\begin{bmatrix} 1 & 1 & 1 \\ 1 & 2 & 3 \\ 1 & 3 & 6 \end{bmatrix},
$$

which has the bidiagonal decomposition

$$
\begin{bmatrix} 1 & & \\ & 1 & \\ & 1 & 1 \end{bmatrix}
\begin{bmatrix} 1 & & \\ 1 & 1 & \\ & 1 & 1 \end{bmatrix}
\begin{bmatrix} 1 & & \\ & 1 & \\ & & 1 \end{bmatrix}
\begin{bmatrix} 1 & 1 & \\ & 1 & 1 \\ & & 1 \end{bmatrix}
\begin{bmatrix} 1 & & \\ & 1 & 1 \\ & & 1 \end{bmatrix}.
$$

Koev [131] has shown recently that the parameters of the bidiagonal factorization determine the eigenvalues and singular values of a nonsingular TN matrix to high relative accuracy. Koev has also developed algorithms that compute the eigenvalues and singular values to high relative accuracy, given that the matrix is presented in factored form. This is another example of a product eigenvalue problem. We want the eigenvalues (or singular values) of the product of the bidiagonal factors, but we work with the factors themselves, never forming the product. Koev's algorithm for the eigenvalues performs a similarity transformation to tridiagonal form by a sequence of Neville eliminations. These are Gaussian eliminations in which a multiple of one row is subtracted from the row immediately below it to create a zero in that row. The key to high accuracy is this: The Neville elimination in the TN matrix is equivalent to setting one parameter to zero in the factored form. Thus the elimination incurs no error (and it is practically free). The similarity transformation is completed by adding a multiple of one column to an adjacent column. This is not free, but it can be done with high relative accuracy while working with the factored form. See [131] for details. Once the matrix is in (factored) tridiagonal form, it can be symmetrized, and its eigenvalues can be computed to high relative accuracy by either of the two methods discussed in Section 8.5, the zero-shift QR algorithm [74] or the dqd algorithm [87, 170, 182].

This is the first example of a class of (mostly) nonsymmetric matrices whose eigenvalues can be determined to high relative accuracy.

Chapter 9

Krylov Subspace Methods

9.1 Introduction

We now turn to methods for large matrices. If a matrix is really large, the computation of its complete spectrum is out of the question. Fortunately it often suffices to compute just a few of the eigenvalues. For example, if stability is at issue, one might just want to know the few eigenvalues with maximum real part or with maximum absolute value. If a certain frequency range is of interest, one might like to calculate the few eigenvalues that are closest to some specified target value τ. Eigenvalue methods for large problems are designed to perform tasks like these, and they do it by computing the invariant subspace associated with the desired eigenvalues.

Very large matrices that arise in applications are almost always sparse. That is, the vast majority of their entries are zero. Were this not so, it might not even be possible to store such a matrix, much less find its eigenvalues. For example, consider a matrix of order $n = 10^6$. If we store all of its entries in the conventional way, we have to store 10^{12} numbers. If we are storing, say, double-precision real numbers, that takes 8×10^{12} bytes, i.e., 8000 gigabytes. However, it might well happen that the matrix has only about ten nonzero entries per row, and then we can save the situation by using a special data structure that stores only the nonzero entries of the matrix. Then we have to store only about 10^7 numbers or 8×10^7 bytes. Of course there is also a bit of overhead involved: For each matrix entry that we store, we have to store its row and column number as well, so that we know where it belongs in the matrix. Multiplying by two to take the overhead into account we see that we will need about 16×10^7 bytes or 160 megabytes. This is a manageable file.

The methods that we have discussed so far all use similarity transformations. That approach is not possible for large sparse matrices, as each similarity transformation causes fill-in, the introduction of nonzeros in positions that previously contained zeros. After just a few similarity transformations, the matrix becomes completely full, and hence unstorable.

If we can't do similarity transformations, then what can we do? The one thing we can do very well with a large, sparse matrix is multiply it by a vector. Indeed, this operation can be done relatively quickly; the amount of work is proportional to the number of nonzeros in the matrix. If we multiply A by x to get Ax, we can then multiply A by that vector to get A^2x, and so on, so it is easy to build up a *Krylov sequence*

$$x, \ Ax, \ A^2x, \ A^3x, \ \ldots.$$

In this chapter we investigate *Krylov subspace methods* which build up Krylov subspaces

$$\mathcal{K}_j(A, x) = \mathrm{span}\{x, Ax, A^2x, \ldots, A^{j-1}x\}$$

and look for good approximations to eigenvectors and invariant subspaces within the Krylov spaces.

In many applications the matrix has structure that allows it to be stored even more compactly than we have indicated above. For example, the matrix may have submatrices with repeating patterns that do not need to be stored multiple times. In fact, it may not be necessary to store the matrix at all. If a subroutine that can compute Ax, given any input x, is provided, Krylov subspace methods can be used.

If we want to build a Krylov subspace, we need a starting vector x. What would be a good starting vector? If we could find one that was in the invariant subspace that we are looking for, then the Krylov spaces that it generates would lie entirely within that subspace. That's a bit much to ask for, but if we can get a vector that is close to the desired invariant subspace, perhaps we can get some good approximations. That also is a lot to ask for, at least initially. At the outset we might have no idea where the desired invariant subspace lies. In that case we might as well pick an x at random and get started. As we shall see, we will be able to replace it by a better x later.

After $j - 1$ steps of a Krylov subspace process, we will have built up a j-dimensional Krylov subspace $\mathcal{K}_j(A, x)$. The larger j is, the better is our chance that the space contains good approximations to desired eigenvectors. But now we have to make an important observation. Storing a j-dimensional space means storing j vectors, and the vectors take up a lot of space. Consider again the case $n = 10^6$. Then each vector consists of 10^6 numbers and requires 8 megabytes of storage. If we store a lot of vectors, we will quickly fill up our memory. Thus there is a limit to how big we can take j. In typical applications j is kept under 100, or at most a few hundreds.

Given the limitation on subspace size, we ordinarily resort to restarts. Say we are looking for an invariant subspace of some modest dimension $\hat{m}$. We pick m at least as big as $\hat{m}$ and preferably a bit bigger, e.g., $m = \hat{m} + 2$. We begin by generating a Krylov subspace $\mathcal{K}_k(A, x)$ of dimension k, where k is somewhat bigger than m, e.g., $k = 2m$ or $k = 5m$. Then we do an analysis that allows us to pick out a promising m-dimensional subspace of $\mathcal{K}_k(A, x)$, we discard all but m vectors, and we restart the process at step m. The details will be given in the next few sections. The restart is tantamount to picking a new, improved x and building up a new, improved Krylov space from that x. Since the process is organized so that we can start from step m instead of step 0, it is called an *implicit restart*. This gives improved efficiency and accuracy, compared with an explicit restart from step 0.

When the procedure works as intended (as it usually does), each restart gives us a better approximation to the desired invariant subspace. Repeated restarts lead to convergence.

Approximating Eigenvectors from Krylov Subspaces

We now consider the question of whether a given Krylov subspace $\mathcal{K}_j(A, x)$ contains good approximants to eigenvectors of A. We suppose $j \ll n$, as is usually the case in practical situations. We begin by establishing a simple characterization of $\mathcal{K}_j(A, x)$ in terms of polynomials.

$\mathcal{K}_j(A, x)$ is the set of all linear combinations

$$a_0 x + a_1 A x + a_2 A^2 x + \cdots + a_{j-1} A^{j-1} x. \qquad (9.1.1)$$

Given any coefficients $a_0, \ldots, a_{j-1}$, we can build a polynomial $q(z) = a_0 + a_1 z + a_2 z^2 + \cdots + a_{j-1} z^{j-1}$ of degree $j - 1$ or less. Then the linear combination in (9.1.1) can be written more compactly as $q(A)x$. Thus we have the following simple characterization of the Krylov subspace.

Proposition 9.1.1. *Let* $\mathcal{P}_{j-1}$ *denote the set of all polynomials of degree less than* j. *Then*

$$\mathcal{K}_j(A, x) = \left\{ q(A)x \mid q \in \mathcal{P}_{j-1} \right\}.$$

The question of whether $\mathcal{K}_j(A, x)$ contains good approximants to a given eigenvector v is therefore that of whether there are polynomials $q \in \mathcal{P}_{j-1}$ such the $q(A)x \approx v$. To keep the discussion simple, let us suppose that A is a semisimple matrix and has linearly independent eigenvectors $v_1, v_2, \ldots, v_n$ with associated eigenvalues $\lambda_1, \lambda_2, \ldots, \lambda_n$. Then

$$x = c_1 v_1 + c_2 v_2 + \cdots + c_n v_n$$

for some unknown coefficients $c_1, \ldots, c_n$. For any polynomial q we have

$$q(A)x = c_1 q(\lambda_1)v_1 + c_2 q(\lambda_2)v_2 + \cdots + c_n q(\lambda_n)v_n.$$

If we want to find a good approximant to, say, v_1, we should look for a polynomial q such that $|q(\lambda_1)|$ is large relative to $|q(\lambda_2)|, \ldots, |q(\lambda_n)|$. This is most easily achieved if λ_1 is well separated from the rest of the spectrum. Picture a situation in which all of the eigenvalues are in a cluster in the complex plane, except for λ_1, which is off at a distance. Then there exist polynomials of the desired form: Just build a q of degree j that has all of its zeros in the cluster; for example, take $q(z) = (z - z_0)^j$, where z_0 is somewhere in the "middle" of the cluster. Then $q(\lambda_1)$ will be much larger than $q(\lambda_2), \ldots, q(\lambda_n)$, even if j is quite small. Thus $\mathcal{K}_j(A, x)$ contains vectors that approximate v_1 well, unless c_1 happens to be unusually small.

The closer λ_1 is to the other eigenvalues, the harder it is to get a polynomial q with the desired properties. Now suppose that λ_1 is surrounded by other eigenvalues. Then it is normally impossible to build a q of low degree j such that $q(\lambda_1)$ is large and $q(\lambda_2), \ldots, q(\lambda_n)$ are small. Thus $\mathcal{K}_j(A, x)$ will not contain good approximants to v_1. We conclude that Krylov subspaces are best at approximating eigenvectors associated with eigenvalues on the periphery of the spectrum.

Rigorous results can be obtained with the help of Chebyshev polynomials, which are discussed in Exercises 9.1.2 and 9.1.4.[1] For more details see [100, 185], and the references therein, especially [127, 161, 183].

---

[1] Analyses of this type are most meaningful in the case when A is normal. Then the eigenvectors $v_1, \ldots, v_n$ can be taken to be orthonormal (Theorem 2.3.5), and it is natural to work in the coordinate system defined by them. If A is not normal, then $v_1, \ldots, v_n$ cannot be taken to be orthonormal, and the coordinate system they define is distorted. The distortion is measured by a factor $\kappa_2(V)$, the condition number of the matrix of eigenvectors, which appears in the rigorous bounds. Bounds can be obtained for defective matrices by considering the Jordan canonical form (Theorem 2.4.11), but again the coordinate system is distorted.

The function of restarts is to improve the coefficients $c_1, \ldots, c_n$. Let us suppose that $v_1, \ldots, v_m$ are the eigenvectors that we wish to find. Then, by restarting, we hope to replace x by a new starting vector

$$\hat{x} = \hat{c}_1 v_1 + \hat{c}_2 v_2 + \cdots + \hat{c}_n v_n$$

such that $\hat{c}_1, \ldots, \hat{c}_m$ are relatively larger and $\hat{c}_{m+1}, \ldots, \hat{c}_n$ are relatively smaller.

The Shift-and-Invert Strategy

As we have seen, Krylov subspace methods are good at producing the eigenvalues on the periphery of the spectrum. If we want to find eigenvalues in the interior, special action must be taken. Suppose we want to find the eigenvalues that are nearest to some target value $\tau \in \mathbb{C}$. If we shift by τ and invert, we get the matrix $(A - \tau I)^{-1}$, which has eigenpairs $((\lambda - \tau)^{-1}, v)$ corresponding to eigenpairs (λ, v) of A. Thus the eigenvalues of A that are closest to τ correspond to the eigenvalues of $(A - \tau I)^{-1}$ of greatest modulus. These are peripheral eigenvalues. If we apply a Krylov subspace method to $B = (A - \tau I)^{-1}$, we will get the desired eigenvalues and eigenvectors.

Care must be exercised in implementing this *shift-and-invert* strategy. We cannot afford to form the matrix $(A - \tau I)^{-1}$ explicitly, as it is not sparse. Fortunately we do not need $(A - \tau I)^{-1}$; all that's needed is a way of making the transformation $x \mapsto (A - \tau I)^{-1}x$ for any x. This can be achieved by solving $(A - \tau I)y = x$ for y, as then $y = (A - \tau I)^{-1}x$. If we compute an LU factorization with pivoting—$A = PLU$, where P is a permutation matrix, L is unit lower triangular, and U is upper triangular—we can use this factorization to backsolve for y, given any x. The factorization needs to be done only once, and then it can be used repeatedly. For sparse A it is typically the case that the factors L and U are much less sparse than A is, but they are (typically) still fairly sparse and can be stored compactly in a sparse data structure. However, they will (typically) take up much more memory than A does. The shift-and-invert strategy is a viable option if and only if there is enough available memory to store the L and U factors. When it is a viable option, it generally works very well.

Exercises

9.1.1. Let $A \in \mathbb{C}^{n \times n}$, and let $\tau \in \mathbb{C}$ be a number that is not an eigenvalue of A. Let $\mathcal{S}$ be a subspace of $\mathbb{C}^n$.

   (a) Show that $\mathcal{S}$ is invariant under $(A - \tau I)^{-1}$ if and only if $\mathcal{S}$ is invariant under A.

   (b) Show that (λ, v) is an eigenpair of A if and only if $((\lambda - \tau)^{-1}, v)$ is an eigenpair of $(A - \tau I)^{-1}$.

9.1.2. In this exercise we introduce and study the famous and very useful Chebyshev polynomials. A substantial reference on Chebyshev polynomials is [177]. We begin by studying the analytic map

$$x = \phi(z) = \frac{z + z^{-1}}{2}, \tag{9.1.2}$$

which has poles at zero and infinity and maps the rest of the complex plane onto the complex plane.

(a) Show that if z satisfies (9.1.2) for a given value of x, then z satisfies a certain quadratic equation. Solve that equation and deduce that

$$z = x \pm \sqrt{x^2 - 1}. \tag{9.1.3}$$

Thus each x is the image of two z values, which are distinct if $x \neq \pm 1$. Show that if $z = x + \sqrt{x^2 - 1}$, then $z^{-1} = x - \sqrt{x^2 - 1}$. The points ± 1 are images only of themselves. They are fixed points of ϕ. (*Remark:* ϕ is the map for Newton's method to compute $\sqrt{1}$.)

(b) For $k = 0, 1, 2, 3, \ldots$ the kth *Chebyshev polynomial* $T_k(x)$ is defined by

$$T_k(x) = \frac{z^k + z^{-k}}{2}, \quad \text{where} \quad x = \frac{z + z^{-1}}{2}, \tag{9.1.4}$$

that is,

$$T_k(x) = \phi(z^k), \quad \text{where} \quad x = \phi(z).$$

Use the definition (9.1.4) and some simple algebra to show that for $k = 1, 2, 3, \ldots$,

$$2x T_k(x) = T_{k+1}(x) + T_{k-1}(x).$$

This establishes the recurrence

$$T_{k+1}(x) = 2x T_k(x) - T_{k-1}(x), \quad k = 1, 2, 3, \ldots. \tag{9.1.5}$$

(c) This gives us a means of generating T_{k+1} from T_k and T_{k-1}. Show that the definition (9.1.4) gives $T_0(x) = 1$ and $T_1(x) = x$. Then use (9.1.5) to calculate $T_2, T_3, T_4,$ and T_5.

(d) Prove by induction on k that for $k = 0, 1, 2, 3, \ldots,$ T_k is a polynomial of degree k. Moreover, its leading coefficient is 2^{k-1}, except when $k = 0$. Show further that T_k is an even (odd) polynomial when k is even (odd).

(e) Substitute $z = e^{i\beta}$ into (9.1.4) and deduce that the Chebyshev polynomials satisfy

$$T_k(x) = \cos k\beta, \quad \text{where} \quad x = \cos \beta, \tag{9.1.6}$$

for x satisfying $-1 \leq x \leq 1$. In the following discussion, restrict β to the interval $[0, \pi]$, which the cosine function maps one-to-one onto $[-1, 1]$.

(f) Graph the functions $\cos \beta$, $\cos 2\beta$, and $\cos 3\beta$ for $0 \leq \beta \leq \pi$.

(g) Show that $T_k(x)$ has k real zeros in $(-1, 1)$. Since a kth degree polynomial can have no more than k zeros, we have now shown that all of the zeros of T_k are real and that they all lie in $(-1, 1)$.

(h) Show that $\sup_{-1 \leq x \leq 1} |T_k(x)| = 1$. Show that there are $k + 1$ points in $[-1, 1]$ at which $|T_k(x)| = 1$.

(i) We now know that $|T_k|$ is bounded by 1 in $[-1, 1]$ and that all of its zeros lie in that interval. It follows that T_k must grow rapidly as x moves away from $[-1, 1]$. This is already evident from definition (9.1.4): Show that when x is real and large, $x \approx z/2$ and $T_k(x) \approx z^k/2 \approx 2^{k-1}x$.

(j) More importantly, for a fixed $x > 1$, $T_k(x)$ grows exponentially as a function of k. To prove this we introduce another parametrization. Substitute $z = e^\alpha$ ($\alpha \geq 0$) into the definition (9.1.4) to obtain

$$T_k(x) = \cosh k\alpha, \quad \text{where} \quad x = \cosh \alpha, \tag{9.1.7}$$

for all x satisfying $1 \leq x < \infty$. For a given $x > 1$, there is a unique $\alpha > 0$ such that $x = \cosh \alpha$. Let $\rho = e^\alpha > 1$. Use (9.1.7) or (9.1.4) to show that $T_k(x) \geq \frac{1}{2}\rho^k$. Thus $T_k(x)$ grows exponentially as a function of k.

9.1.3. In this exercise we investigate how well eigenvectors associated with peripheral eigenvalues are approximated by Krylov subspaces in the case when the matrix has real eigenvalues. Let A be a semisimple matrix with (real) eigenvalues $\lambda_1 > \lambda_2 \geq \cdots \geq \lambda_n$. For example, A could be Hermitian. Assume also that $\lambda_2 > \lambda_n$.

(a) Show that the transformation

$$w = 1 - 2\frac{x - \lambda_2}{\lambda_n - \lambda_2}$$

maps $[\lambda_n, \lambda_2]$ onto $[-1, 1]$

(b) For $k = 0, 1, 2, \ldots$, define polynomials $q_k \in \mathcal{P}_k$ by

$$q_k(x) = T_k(w) = T_k\left(1 - 2\frac{x - \lambda_2}{\lambda_n - \lambda_2}\right),$$

where T_k is the Chebyshev polynomial defined in Exercise 9.1.2. Show that $|q_k(\lambda_i)| \leq 1$ for $i = 2, \ldots, n$ and

$$|q_k(\lambda_1)| \geq \frac{1}{2}\rho^m,$$

where

$$\rho \geq 1 + 2\frac{\lambda_1 - \lambda_2}{\lambda_2 - \lambda_n} > 1. \tag{9.1.8}$$

Use part (i) of Exercise 9.1.2.

(c) Let $v_1, \ldots, v_n$ be linearly independent eigenvectors associated with eigenvalues λ_1, $\ldots, \lambda_n$, respectively. Let x be a starting vector for a Krylov process satisfying

$$x = c_1 v_1 + \cdots + c_n v_n$$

with $c_1 \neq 0$. Define $\tilde{q}_k \in \mathcal{P}_k$ by $\tilde{q}_k(x) = q_k(x)/(c_1 q_k(\lambda_1))$. Let $w_k = \tilde{q}_{k-1}(A)x \in \mathcal{K}_k(A, x)$. Show that there is a constant C such that

$$\| w_k - v_1 \|_2 \leq C\rho^{-k}, \qquad k = 0, 1, 2, \ldots,$$

where ρ is as in (9.1.8). Thus, for large enough k, the Krylov subspace $\mathcal{K}_k(A, x)$ contains a vector that is close to the eigenvector v_1.

(d) Modify the arguments given above to show that $\mathcal{K}_k(A, x)$ contains vectors that are close to v_n under suitable assumptions.

(e) Suppose $\lambda_1 > \lambda_2 > \lambda_3 > \lambda_n$. For $k = 1, 2, 3, \ldots$ define a polynomial $\hat{q}_k \in \mathcal{P}_k$ by

$$\hat{q}_k(x) = (x - \lambda_1)T_{k-1}\left(1 - 2\frac{x - \lambda_3}{\lambda_n - \lambda_3}\right).$$

Use $\hat{q}_k$ to show that, for sufficiently large k, the Krylov subspace $\mathcal{K}_k(A, x)$ contains vectors that are close to v_2, assuming $c_2 \neq 0$.

(f) Suppose $\lambda_2 > \lambda_3 > \lambda_4 > \lambda_n$. Show that, for sufficiently large k, $\mathcal{K}_k(A, x)$ contains vectors that approximate v_3 well, assuming $c_3 \neq 0$.

9.1.4. In Exercise 9.1.2 we introduced the Chebyshev polynomials and studied their properties on the real line. We used these properties in Exercise 9.1.3 to investigate approximation properties of Krylov subspaces for semisimple matrices with real eigenvalues. If we want to obtain results for matrices with complex eigenvalues, we must investigate the properties of Chebyshev polynomials in the complex plane.

(a) Substitute $z = e^{\alpha + i\beta}$ into (9.1.4), where $\alpha > 0$ and $\beta \in (-\pi, \pi]$. Show that

$$x = \cosh \alpha \cos \beta + i \sinh \alpha \sin \beta$$

and

$$T_k(x) = \cosh k\alpha \cos k\beta + i \sinh k\alpha \sin k\beta.$$

This generalizes the results of parts (d) and (i) of Exercise 9.1.2.

(b) Let $u = \Re(x)$ and $v = \Im(x)$, so that $x = u + iv$. Suppose that α is held fixed and β is allowed to run from $-\pi$ to π. Show that x sweeps out an ellipse E_α with major semiaxis $u = \cosh \alpha$ and minor semiaxis $v = \sinh \alpha$, and meanwhile $T_k(x)$ sweeps out (k times) an ellipse $E_{k\alpha}$ with semiaxes $\cosh k\alpha$ and $\sinh k\alpha$. Notice that as $\alpha \to 0^+$, E_α becomes increasingly eccentric. In the limiting case $\alpha = 0$, E_α becomes the interval $[-1, 1]$ on the real axis.

(c) Show that T_k maps E_α (and its interior) to $E_{k\alpha}$ (and its interior). Thus

$$\max_{x \in E_\alpha} | T_k(x) | \leq \cosh k\alpha.$$

(d) These results can be used to get error bounds for Krylov subspace approximation, in case the spectrum is not real, as follows. Suppose that all of the eigenvalues but one lie in the ellipse E_{α_2}, except that there is a positive eigenvalue $\lambda_1 = \cosh \alpha_1$ to the right of E_{α_2}. That is, $\alpha_1 > \alpha_2 > 0$. Show that for $\lambda \in E_{\alpha_1}$,

$$\frac{|T_k(\lambda)|}{|T_k(\lambda_1)|} \leq \frac{\cosh k\alpha_2}{\cosh k\alpha_1} \leq 2 \left(e^{\alpha_2 - \alpha_1}\right)^k \to 0.$$

Deduce that the Krylov subspace $\mathcal{K}_k(A, x)$ contains good approximations to v_1 if k is large enough, provided that $c_1 \neq 0$. This result assumes that λ_1 is positive and real. However, similar bounds can be obtained for peripheral eigenvalues that are neither positive nor real since the polynomials can be shifted, scaled, and rotated. All we need is to find an ellipse that contains all eigenvalues but λ_1, such that λ_1 lies on the major semiaxis of the ellipse. Then we can get a convergence result.[2]

9.1.5. Using the definition (9.1.4) and inverse map (9.1.3), we immediately obtain the expression

$$T_k(x) = \frac{(x + \sqrt{x^2 - 1})^k + (x - \sqrt{x^2 - 1})^k}{2}. \tag{9.1.9}$$

(a) Use (9.1.9) to compute T_0, T_1, T_2, and T_3.

(b) Simplifying (9.1.9) by taking $w = \sqrt{x^2 - 1}$, we have

$$T_k(x) = \frac{(x + w)^k + (x - w)^k}{2},$$

where $x^2 - w^2 = 1$. Apply the binomial theorem to this expression, and observe that the odd powers of w cancel out. Conclude that

$$T_k(x) = \sum_{m=0}^{\lfloor k/2 \rfloor} \binom{k}{2m} x^{k-2m} w^{2m} = \sum_{m=0}^{\lfloor k/2 \rfloor} \binom{k}{2m} x^{k-2m} (x^2 - 1)^m.$$

This shows clearly that T_k is a polynomial of degree k.

(c) Show further that

$$T_k(x) = \sum_{j=0}^{\lfloor k/2 \rfloor} c_j x^{k-2j},$$

where

$$c_j = (-1)^j \sum_{m=j}^{\lfloor k/2 \rfloor} \binom{k}{2m} \binom{m}{j}.$$

9.1.6. Here we establish an orthogonality property for Chebyshev polynomials.

(a) Show that $\int_0^\pi \cos k\beta \cos j\beta \, d\beta = \delta_{kj} \frac{\pi}{2}$.

(b) Using (9.1.6), show that $\int_{-1}^1 T_k(x) T_j(x) (1 - x^2)^{-1/2} dx = \delta_{kj} \frac{\pi}{2}$.

---

[2] It must be acknowledged, however, that these results are much weaker than those one obtains in the real eigenvalue case, because the function $\cosh \alpha$ is flat at $\alpha = 0$ and not at $\alpha > 0$.

9.2 The Generic Krylov Process

If the vectors

$$x, Ax, A^2x, \ldots, A^{k-1}x$$

are linearly independent, they form a basis for the Krylov subspace $\mathcal{K}_k(A, x)$. From a numerical standpoint this is not a very good basis. As we know from our study of the power method, the vectors $A^j x$ (usually) point more and more in the direction of the dominant eigenvector as j increases. This means that for large (or even moderate) k, most of the vectors in the Krylov basis will point in about the same direction. Thus the basis is (usually) ill conditioned. Let us consider how we might build a well-conditioned basis. Start with $u_1 = c_1 x$, where c_1 is a scale factor. For example, we might take $c_1 = 1$, or we might choose c_1 so that $\| u_1 \| = 1$. To get another basis vector we can calculate Au_1, and then we can improve the basis by setting

$$u_2 = Au_1 - u_1 h_{11},$$

where h_{11} is chosen with an eye to making $\{u_1, u_2\}$ a better basis than $\{u_1, Au_1\}$ in some sense. For example, we could make u_2 orthogonal to u_1. This popular choice leads to the Arnoldi process, which will be described in Section 9.4. It is the best choice from the standpoint of numerical stability. Since we like to be able to control the norm of u_2, we set, more generally,

$$u_2 h_{21} = Au_1 - u_1 h_{11},$$

where h_{21} is a scale factor. We cannot afford to keep a lot of extra vectors lying around, so we keep u_2 and discard Au_1. When we build our next basis vector, we will start by forming Au_2 instead of the unavailable $A^2 u_1 = A(Au_1)$.

Now, proceeding by induction, let us suppose that we have generated $u_1, \ldots, u_j$, a basis for $\mathcal{K}_j(A, x)$. Suppose further that

$$\mathcal{K}_i(A, x) = \mathrm{span}\{u_1, \ldots, u_i\}, \quad i = 1, \ldots, j.$$

We now wish to find a vector u_{j+1} such that $\mathcal{K}_{j+1}(A, x) = \mathrm{span}\{u_1, \ldots, u_{j+1}\}$. Since $A^j u_1$ is unavailable, we use instead the vector $u_j \in \mathcal{K}_j(A, x) \setminus \mathcal{K}_{j-1}(A, x)$. We first compute Au_j and then form u_{j+1} by some normalization process:

$$u_{j+1} h_{j+1,j} = Au_j - \sum_{i=1}^{j} u_i h_{ij}. \tag{9.2.1}$$

We will admit only processes with the following characteristic: If it happens that $Au_j \in \mathrm{span}\{u_1, \ldots, u_j\}$, the h_{ij} should be chosen so that

$$Au_j = \sum_{i=1}^{j} u_i h_{ij}.$$

Then $h_{j+1,j} = 0$, u_{j+1} is indeterminate, and the process halts. We shall call such a procedure a *Krylov process*.

There are two widely used Krylov processes: the Arnoldi process (Section 9.4) and the unsymmetric Lanczos process (Section 9.6). The Arnoldi process simply chooses the h_{ij} so that the basis is orthonormal. We defer the description of the unsymmetric Lanczos process, which is more complicated. There is also a symmetric Lanczos process, which is a special case of both the Arnoldi and unsymmetric Lanczos processes. In Sections 9.8 and 9.9 we will develop special Krylov processes for Hamiltonian and symplectic matrices.

The two following propositions, which are easily proved (Exercises 9.2.1 and 9.2.2), state some basic facts about all Krylov processes.

Proposition 9.2.1. *If $x, Ax, A^2x, \ldots, A^{k-1}x$ are linearly independent, then every Krylov process starting with $u_1 = c_1 x$ produces a basis of $\mathcal{K}_k(A, x)$ such that*

$$\operatorname{span}\{u_1, \ldots, u_i\} = \mathcal{K}_i(A, x) \quad for \quad i = 1, \ldots, k.$$

Proposition 9.2.2. *If $x, Ax, A^2x, \ldots, A^{k-1}x$ are linearly independent but $A^k x \in \mathcal{K}_k(A, x)$, then $\mathcal{K}_k(A, x)$ is invariant under A. Every Krylov process starting with $u_1 = c_1 x$ produces a u_k such that $Au_k \in \mathcal{K}_k(A, x)$, causing termination at this step.*

The process (9.2.1) can be summarized neatly as a matrix equation. If we rewrite (9.2.1) as

$$Au_j = \sum_{i=1}^{j+1} u_i h_{ij},$$

we easily see that the following proposition holds.

Proposition 9.2.3. *Suppose that $x, Ax, A^2x, \ldots, A^{k-1}x$ are linearly independent, and let $u_1, \ldots, u_k$ be generated by a Krylov process starting with $u_1 = c_1 x$. Let U_k denote the $n \times k$ matrix whose columns are $u_1, \ldots, u_k$. Let $H_{k+1,k}$ denote the $(k+1) \times k$ upper Hessenberg matrix whose (i, j) entry h_{ij} is as in (9.2.1) if $i \le j + 1$, and $h_{ij} = 0$ if $i > j + 1$. Then*

$$AU_k = U_{k+1} H_{k+1,k}. \tag{9.2.2}$$

Let H_k denote the square upper Hessenberg matrix obtained by deleting the bottom row of $H_{k+1,k}$. Then

$$AU_k = U_k H_k + u_{k+1} h_{k+1,k} e_k^T, \tag{9.2.3}$$

where $e_k^T = [0\ 0\ \ldots\ 0\ 1] \in \mathbb{R}^k$.

If $\mathcal{K}_k(A, x)$ is invariant, then $h_{k+1,k} = 0$, and (9.2.3) becomes

$$AU_k = U_k H_k. \tag{9.2.4}$$

Finding an invariant subspace is a desired event. If $k \ll n$, as it is in practical situations, we can easily compute the eigenvalues and eigenvectors of H_k by, say, the QR algorithm. Every eigenpair of H_k yields an eigenpair of A: Given $H_k y = y\lambda$, let $v = U_k y$. Then $Av = v\lambda$ (Proposition 2.1.11).

If we temporarily ignore the problem of storage space, we can think about carrying on a Krylov process indefinitely. But in our finite-dimensional setting, every Krylov process must halt eventually. If $x, Ax, \ldots, A^{n-1}x$ are independent, then they span the space $\mathbb{C}^n$, and $A^n x$ must be a linear combination of them. The Krylov process will produce a basis

$u_1, \ldots, u_n$ and halt. If an invariant subspace is discovered along the way, we halt even sooner, but the process can be continued. If $\mathrm{span}\{u_1, \ldots, u_j\}$ is invariant, we can take u_{j+1} to be an appropriate vector that is not in $\mathrm{span}\{u_1, \ldots, u_j\}$. If we proceed in this way, we can always, in principle, build a basis $u_1, \ldots, u_n$. Then (9.2.4) becomes

$$AU = UH, \tag{9.2.5}$$

where U is $n \times n$ and nonsingular. Since $A = UHU^{-1}$, we conclude that every Krylov process, if carried to completion, produces a similarity transformation to an H that is in upper Hessenberg form. If we have found any invariant subspaces along the way, some of the $h_{j+1,j}$ will be zero, so H will be block triangular.

In real applications it is never possible to carry a Krylov process to completion, but now that we have considered the possibility, we can see that a partially completed Krylov process performs a partial similarity transformation. Looking at (9.2.2) or (9.2.3), we see that U_k consists of the first k columns of the transforming matrix U in (9.2.5), and H_k is the upper-left-hand $k \times k$ submatrix of H. Thus (9.2.2) and (9.2.3) represent partial similarity transformations.

It is a useful fact that Proposition 9.2.3 has a converse: If the vectors $u_1, \ldots, u_{k+1}$ satisfy an equation of the form (9.2.3), they must span nested Krylov subspaces.

Proposition 9.2.4. *Suppose that $u_1, \ldots, u_{k+1}$ are linearly independent vectors, U_k is the matrix whose columns are $u_1, \ldots, u_k$, H_k is properly upper Hessenberg, $h_{k+1,k} \neq 0$, and*

$$AU_k = U_k H_k + u_{k+1} h_{k+1,k} e_k^T .$$

Then

$$\mathrm{span}\{u_1, \ldots, u_j\} = \mathcal{K}_j(A, u_1), \qquad j = 1, \ldots, k+1.$$

To prove this result, one simply notes that the matrix equation implies that (9.2.1) holds for $j = 1, \ldots, k$. One then uses (9.2.1) to prove Proposition 9.2.4 by induction on j. (See Exercise 9.2.4.)

Quality of Eigenpair Approximants

After enough steps of a Krylov process, we expect the Krylov subspace $\mathcal{R}(U_k) = \mathcal{K}_k(A, x)$ to contain good approximants to eigenvectors associated with some of the peripheral eigenvalues of A. How do we locate these good approximants? We know that if $h_{k+1,k} = 0$, each eigenpair (μ, y) of H_k yields an eigenpair $(\mu, U_k y)$ of A. It is therefore reasonable to expect that if $h_{k+1,k}$ is near zero, then the eigenpairs of H_k will yield good approximations to eigenpairs of A. If k is not too large, then computing these quantities poses no problem. The next simple result shows that even if $h_{k+1,k}$ is not close to zero, some of the $(\mu, U_k y)$ may be excellent approximations of eigenpairs.

Proposition 9.2.5. *Suppose that we perform k steps of a Krylov process on A to generate U_{k+1} and $H_{k+1,k}$ related by (9.2.2) and (9.2.3). Let (μ, y) be an eigenpair of H_k, and let $w = U_k y$, so that (μ, w) is an approximate eigenpair of A. Then*

$$\| Aw - w\mu \| = \| u_{k+1} \| \, | h_{k+1,k} | \, | y_k |. \tag{9.2.6}$$

Here y_k is the last component of the eigenvector y.

This is easily proved by multiplying (9.2.3) by y and doing some obvious manipulations. The value of Proposition 9.2.5 is this: (μ, w) is an exact eigenpair of A if and only if the residual $Aw - w\mu$ is exactly zero. If the residual is tiny, we expect (μ, w) to be a good approximant of an eigenpair. The right-hand side of (9.2.6) shows how to compute the norm of the residual at virtually no cost. Practical Krylov processes always compute vectors satisfying $\| u_{k+1} \| \approx 1$, so we can ignore that factor. The factor $|h_{k+1,k}|$ is expected. The factor $| y_k |$ shows that even if $h_{k+1,k}$ is not small, the residual can nevertheless be tiny. If the bottom component of the eigenvector y is tiny, then the residual must be tiny. Practical experience has shown that this happens frequently. It often happens that peripheral eigenpairs are approximated excellently even though $h_{k+1,k}$ is not small. Of course we must remember that a tiny residual does not necessarily imply an excellent approximation. All it means is that (μ, w) is the exact eigenpair of some nearby matrix $A + E$ (Proposition 2.7.20). If the eigensystem of A is well conditioned, we can then infer that (μ, w) approximates an eigenpair of A well.

One further related problem merits discussion. If a Krylov subspace $\mathcal{R}(U_k)$ contains a vector that approximates an eigenvector v well, is it necessarily true that the eigenpair (λ, v) is well approximated by one of the eigenpairs $(\mu, U_k y)$ generated from the eigenpairs of H_k? Exercise 9.2.7 shows that the answer is usually but not always yes. Again (as in the previous paragraph) conditioning is an issue.

Notes and References

Around 1930 the engineer and scientist Alexei Nikolaevich Krylov used Krylov subspaces to compute the coefficients of a minimal polynomial, the roots of which are eigenvalues of a matrix [138]. Krylov's method is described in [83, § 42]. The modern era of Krylov subspace methods began with the Lanczos [143] and Arnoldi [11] processes, as well as the conjugate-gradient method [113] for solving positive-definite linear systems, around 1950.

Exercises

9.2.1. Prove Proposition 9.2.1 by induction on k:

(a) Show that the proposition holds for $k = 1$.

(b) Suppose $j < k$ and

$$\text{span}\{u_1, \ldots, u_i\} = \mathcal{K}_i(A, x) \quad \text{for} \quad i = 1, \ldots, j.$$

Show that $u_j \in \mathcal{K}_j(A, x) \setminus \mathcal{K}_{j-1}(A, x)$. Deduce that $Au_j \in \mathcal{K}_{j+1}(A, x) \setminus \mathcal{K}_j(A, x)$.

(c) Use (9.2.1) to show that $\text{span}\{u_1, \ldots, u_{j+1}\} \subseteq \mathcal{K}_{j+1}(A, x)$.

(d) Show that $\text{span}\{u_1, \ldots, u_{j+1}\} = \mathcal{K}_{j+1}(A, x)$. This completes the proof by induction.

9.2.2. Prove Proposition 9.2.2.

9.2.3. Prove Proposition 9.2.3.

9.2.4. Show that (9.2.3) implies that (9.2.1) holds for $j = 1, \ldots, k$. Then use (9.2.1) to prove by induction on j that $\text{span}\{u_1, \ldots, u_j\} \subseteq \mathcal{K}_j(A, u_1)$ for $j = 1, \ldots, k+1$. Get equality of subspaces by a dimension argument. This proves Proposition 9.2.4.

9.2.5. Consider the trivial Krylov process that takes $u_{j+1} = A u_j$, as long as the vectors remain linearly independent. Show that if this process is run until an invariant subspace $\mathcal{K}_i(A, x)$ is found, the resulting upper Hessenberg matrix H_i is a companion matrix.

9.2.6. Suppose that A, U, and B satisfy a relationship of the form

$$AU = UB + re_k^T$$

(cf. (9.2.3)), where U is $n \times k$ and has rank k, and r is some residual vector. Let $u_1, \ldots, u_k$ denote the columns of U. Show that B is upper Hessenberg if and only if

$$\text{span}\{u_1, \ldots, u_j\} = \mathcal{K}_j(A, u_1), \qquad j = 1, \ldots, k.$$

9.2.7. Proposition 9.2.3 summarizes the situation after k steps of a Krylov process. The eigenvalues of H_k are estimates of eigenvalues of A. If (μ, y) is an eigenpair of H_k, then $(\mu, U_k y)$ is an estimate of an eigenpair of A. We know that if $h_{k+1,k} = 0$, then $\mathcal{R}(U_k)$ is invariant under A, and $(\mu, U_k y)$ is an exact eigenpair of A (Proposition 2.1.11). It is therefore natural to expect that if $\mathcal{R}(U_k)$ is close to invariant, then each $(\mu, U_k y)$ will be a good approximation to an eigenpair. In this exercise we investigate a more general question. If $\mathcal{R}(U_k)$ contains a good approximation to an eigenvector v of A, is there an eigenpair (μ, y) of H_k such that $(\mu, U_k y)$ approximates the eigenpair (λ, v) well? In particular, does one of $\mu_1, \ldots, \mu_k$, the eigenvalues of H_k, approximate λ well?

Suppose that (λ, v) is an eigenpair of A satisfying $\| v \| = 1$, and that there is a vector $w \in \mathcal{R}(U_k)$ such that $\| v - w \| = \epsilon \ll 1$. There is a unique $y \in \mathbb{R}^k$ such that $w = U_k y$.

(a) The case when the Krylov process produces orthonormal vectors $u_1, u_2, \ldots$ is easiest to analyze. Assume this to be the case for now. Show that, under this assumption, $\| w \| = \| y \|$. Use (9.2.3) to show that $H_k = U_k^* A U_k$. Then show that $H_k y - y\lambda = U_k^*(Aw - w\lambda)$ and

$$\frac{\| H_k y - y\lambda \|}{\| y \|} \le \frac{\| Aw - w\lambda \|}{\| w \|}.$$

(b) Show that $\| Aw - w\lambda \| \le 2\| A \|\epsilon$. This result and the result of part (a) together show that the pair (λ, y) is an eigenpair of a matrix near H_k (Proposition 2.7.20). Now the question of whether (λ, y) really is close to an eigenpair of H_k is a question of conditioning. If all the eigenpairs of H_k are well conditioned, then one of those pairs has to be close to (λ, y). Thus we cannot say for sure that (λ, y) will always be close to an eigenpair of H_k, but in the well-conditioned case it will. Therefore, in the well-conditioned case, one of $\mu_1, \ldots, \mu_k$, the eigenvalues of H_k, must be close to λ.

(c) Now consider the case in which the vectors produced by the Krylov process are not orthonormal. Then U_k does not have orthonormal columns, but it does have full rank. Thus U_k has a finite condition number $\kappa(U_k) = \sigma_1/\sigma_k$, the ratio of largest and smallest singular values. Show that

$$\frac{\| H_k y - y\lambda \|}{\| y \|} \le \kappa(U_k) \frac{\| Aw - w\lambda \|}{\| w \|}.$$

Thus we draw the same conclusion as in parts (a) and (b), except that an extra factor $\kappa(U_k)$ has appeared. This is not a problem if $\kappa(U_k)$ is not too big.

9.3 Implicit Restarts

Suppose that we have run a Krylov process k steps, so that we have $k+1$ vectors $u_1, \ldots, u_{k+1}$, and the equations (9.2.2) and (9.2.3) hold. Unless we were really lucky, we will not have found an invariant subspace, at least not on the first try. Now we would like to use the information we have obtained so far to try to find a better starting vector and restart the process, hoping to get closer to an invariant subspace on the next try.

As we have already remarked, Krylov subspace methods are good at approximating eigenvalues on the periphery of the spectrum. Thus they are good at computing the eigenvalues of largest modulus, of largest real part, or of smallest real part, for example. For this discussion let us suppose that we are looking for the m eigenvalues of largest modulus and the associated invariant subspace. Then we would like our new starting vector to be rich in its components in the direction of the associated eigenvectors.

We will describe two closely related implicit restart processes here. The second might be superior to the first, but even if it is, the first process is worth discussing because it yields important insights.

First Method

The first method is modelled on Sorensen's implicitly restarted Arnoldi process [191], which is the basis of the popular ARPACK software [147]. Consider the equation

$$AU_k = U_k H_k + u_{k+1} h_{k+1,k} e_k^T \tag{9.3.1}$$

from Proposition 9.2.3. We know that if $h_{k+1,k} = 0$, then $\mathcal{R}(U_k)$ is invariant under A, and the eigenvalues of H_k are eigenvalues of A. Even if $h_{k+1,k} \neq 0$ (even if it is not close to zero), some of the largest eigenvalues of H_k may be good approximations to peripheral eigenvalues of A. Moreover, $\mathcal{R}(U_k)$ may contain good approximations to eigenvectors associated with peripheral eigenvalues. Since k is not too big, it is a simple matter to compute the eigenvalues of H_k by a GR algorithm. Let us call them $\mu_1, \mu_2, \ldots, \mu_k$ and order them so that $|\mu_1| \geq |\mu_2| \geq \cdots \geq |\mu_k|$.[3] The largest m values, $\mu_1, \ldots, \mu_m$, are the best approximations to the largest eigenvalues of A that we have so far.

Roughly speaking, our plan is to keep the portion of $\mathcal{R}(U_k)$ associated with $\mu_1, \ldots, \mu_m$ and discard the rest. To this end, we perform an iteration of an implicit GR algorithm of degree j on H_k, where $j = k - m$, using shifts $\nu_1, \ldots, \nu_j$ located in regions of the complex plane containing portions of the spectrum that we want to suppress. The most popular choice is $\nu_1 = \mu_{m+1}, \nu_2 = \mu_{m+2}, \ldots, \nu_j = \mu_k$. Thus we take as shifts the approximate eigenvalues that we wish to discard. We call this the *exact-shift* version of the implicit restart. In practice the GR iteration can be executed as a sequence of j iterations of degree

---

[3]This assumes that we are looking for the eigenvalues of largest modulus. If we were looking for the eigenvalues with largest real part, for example, we would order them so that $\Re\mu_1 \geq \cdots \geq \Re\mu_k$.

1 or $j/2$ iterations of degree 2, for example (see Exercise 4.9.1 or Theorem 5.2.1). After the iterations, H_k will have been transformed to

$$\hat{H}_k = G^{-1} H_k G, \tag{9.3.2}$$

where

$$p(H_k) = (H_k - v_1 I)(H_k - v_2 I) \cdots (H_k - v_j I) = G R. \tag{9.3.3}$$

If we now multiply (9.3.1) by G on the right and use (9.3.2) in the form $H_k G = G \hat{H}_k$, we obtain

$$A \hat{U}_k = \hat{U}_k \hat{H}_k + u_{k+1} h_{k+1,k} e_k^T G, \tag{9.3.4}$$

where

$$\hat{U}_k = U_k G.$$

The restart is accomplished by discarding all but the first m columns of (9.3.4). Let $\hat{U}_m$ denote the submatrix of $\hat{U}_k$ consisting of the first m columns, and let $\hat{H}_m$ denote the $m \times m$ principal submatrix of $\hat{H}_k$ taken from the upper left-hand corner. Consider also the form of $e_k^T G$, which is the bottom row of G. It is a crucial fact that G is j-Hessenberg (Exercise 9.3.2). This implies that $e_k^T G$ has zeros in its first $k - j - 1 = m - 1$ positions. Keep in mind also that $\hat{H}_k$ is upper Hessenberg. Let $\tilde{u}$ denote column $m + 1$ of $\hat{U}_k$, and $\tilde{h}$ the $(m + 1, m)$ entry of $\hat{H}_k$. Then, if we extract the first m columns of (9.3.4), we obtain

$$A \hat{U}_m = \hat{U}_m \hat{H}_m + \tilde{u}\tilde{h}e_m^T + u_{k+1} h_{k+1,k} g_{km} e_m^T. \tag{9.3.5}$$

Now define a vector $\hat{u}_{m+1}$ and scalar $\hat{h}_{m+1,m}$ so that[4]

$$\hat{u}_{m+1} \hat{h}_{m+1,m} = \tilde{u}\tilde{h} + u_{k+1} h_{k+1,k} g_{km}.$$

Then (9.3.5) becomes

$$A \hat{U}_m = \hat{U}_m \hat{H}_m + \hat{u}_{m+1} \hat{h}_{m+1,m} e_m^T. \tag{9.3.6}$$

In the unlikely event that $\hat{h}_{m+1,m} = 0$, the columns of $\hat{U}_m$ span an invariant subspace. If this is not the case, we start over, but not from scratch. Equation (9.3.6) is analogous to (9.3.1), except that it has only m columns. It could have been arrived at by m steps of a Krylov process, starting from a new starting vector $\hat{x} = c\hat{u}_1$. The implicit restart procedure does not start anew with $\hat{x}$; it starts from (9.3.6), which can also be written as

$$A \hat{U}_m = \hat{U}_{m+1} \hat{H}_{m+1,m},$$

in analogy with (9.2.2). The Krylov process is used to generate $\hat{u}_{m+2}, \hat{u}_{m+3}, \ldots, \hat{u}_{k+1}$ satisfying

$$A \hat{U}_k = \hat{U}_{k+1} \hat{H}_{k+1,k}.$$

Then another implicit restart can be done and a new Krylov subspace generated from a new starting vector. The restarts can be repeated until the desired invariant subspace is obtained.

---

[4]The vector $\hat{u}_{m+1}$ and scalar $\hat{h}_{m+1,m}$ are not uniquely determined. They would be if we were to insist that, for example, $\| \hat{u}_{m+1} \|_2 = 1$ and $\hat{h}_{m+1,m} > 0$.

Why the Method Works

The shifts $\nu_1, \ldots, \nu_j$ were used in the GR iteration that determined the transforming matrix G. Starting from (9.3.1), subtract $\nu_1 U_k$ from both sides to obtain

$$(A - \nu_1 I)U_k = U_k(H_k - \nu_1 I) + E_1, \qquad (9.3.7)$$

where $E_1 = u_{k+1}h_{k+1,k}e_k^T$ is a matrix that is zero except in its kth and last column. A similar equation holds for shifts $\nu_2, \ldots, \nu_k$. Now multiply (9.3.7) on the left by $A - \nu_2 I$ and use the ν_2 analogue of (9.3.7) to further process the resulting term involving $(A - \nu_2 I)U_k$. The result is

$$(A - \nu_2 I)(A - \nu_1 I)U_k = U_k(H_k - \nu_2 I)(H_k - \nu_1 I) + E_2, \qquad (9.3.8)$$

where $E_2 = E_1(H_k - \nu_1 I) + (A - \nu_2 I)E_1$ is a matrix that is zero except in its last two columns. Applying factors $A - \nu_3 I$, $A - \nu_4 I$, and so on, we easily find that

$$p(A)U_k = U_k p(H_k) + E_j, \qquad (9.3.9)$$

where E_j is zero except in the last j columns. In other words, the first m columns of E_j are zero. Here $p(z) = (z - \nu_1) \cdots (z - \nu_j)$, as before. Since $p(H_k) = GR$ and $U_k G = \hat{U}_k$, we can rewrite (9.3.9) as

$$p(A)U_k = \hat{U}_k R + E_j.$$

If we then extract the first m columns from this equation, bearing in mind that R is upper triangular, we obtain

$$p(A)U_m = \hat{U}_m R_m, \qquad (9.3.10)$$

where R_m is the $m \times m$ submatrix of R from the upper left-hand corner. This equation tells the whole story. If we focus on the first column, we find that

$$\hat{x} = \alpha\, p(A)x$$

for some constant α. The new starting vector is essentially $p(A)$ times the old one. Thus the restart effects an iteration of the power method driven by $p(A)$. For ease of discussion, let us suppose that A is semisimple with linearly independent eigenvectors $v_1, \ldots, v_n$ and eigenvalues $\lambda_1, \ldots, \lambda_n$. Then

$$x = c_1 v_1 + c_2 v_2 + \cdots + c_n v_n$$

for some (unknown) constants $c_1, \ldots, c_n$. With probability 1, all c_i are nonzero. The corresponding expansion of $\hat{x}$ is

$$\alpha^{-1}\hat{x} = c_1 p(\lambda_1)v_1 + c_2 p(\lambda_2)v_2 + \cdots + c_n p(\lambda_n)v_n.$$

The polynomial $p(z)$ is small in magnitude for z near the shifts $\nu_1, \ldots, \nu_j$ and large for z away from the shifts. Therefore, in the transformation from x to $\hat{x}$, the components corresponding to eigenvalues far from the shifts become enriched relative to those that are near the shifts. The peripheral eigenvalues furthest from the shifts receive the most enrichment. Thus the restart process reinforces the natural tendency of the Krylov process to bring out the peripheral eigenvalues. After a number of restarts we will have arrived at a starting vector

$$x \approx b_1 v_1 + \cdots + b_m v_m,$$

and we will have found the invariant subspace associated with the m eigenvalues of largest magnitude.

The implicit restart process uses different shifts on each restart. This makes the method adaptive and is essential to its success, but it also makes the analysis of convergence more difficult than it would be for a stationary power method. However, there has been some progress toward a rigorous convergence theory [25, 26].

Equation (9.3.10) shows that the implicit restart doesn't just do simple power iterations. It does nested subspace iterations. Bearing in mind that R_m is upper triangular, we find that

$$p(A)\text{span}\{u_1, \ldots, u_i\} = \text{span}\{\hat{u}_1, \ldots, \hat{u}_i\}, \qquad i = 1, \ldots, m. \tag{9.3.11}$$

These relationships are actually an automatic consequence of the fact that the subspaces in question are Krylov subspaces, as Exercise 9.3.5 shows.

Exact Shifts

If exact shifts $\nu_1 = \mu_{m+1}, \ldots, \nu_j = \mu_n$ are used in the process, then in $\hat{H}_k$ we have $\hat{h}_{m+1,m} = 0$, by Theorem 4.6.1. Letting $\tilde{G}$ denote the submatrix of G consisting of the first m columns, we see that $\mathcal{R}(\tilde{G})$ is the invariant subspace of H_k corresponding to eigenvalues $\mu_1, \ldots, \mu_m$. Note that $\hat{U}_m = U_k \tilde{G}$, so $\mathcal{R}(\hat{U}_m)$ is the subspace of $\mathcal{R}(U_k)$ corresponding to the eigenvalue approximations $\mu_1, \ldots, \mu_m$.

Second Method

Our second implicit restart method is modeled on the thick-restart Lanczos method of Wu and Simon [233] and Stewart's Krylov–Schur algorithm [198]. After k steps of a Krylov process we have

$$AU_k = U_k H_k + u_{k+1} h_{k+1,k} e_k^T. \tag{9.3.12}$$

Compute a similarity transformation $H_k = STS^{-1}$ with T block triangular:

$$T = \begin{bmatrix} T_{11} & T_{12} \\ 0 & T_{22} \end{bmatrix},$$

where T_{11} has eigenvalues $\mu_1, \ldots, \mu_m$ (the approximate eigenvalues we want to keep) and T_{22} has eigenvalues $\mu_{m+1}, \ldots, \mu_k$ (the approximate eigenvalues we want to discard). This can be done by a GR algorithm in conjunction with a block-swapping procedure as described in Section 4.8 to get the eigenvalues sorted appropriately. Multiply (9.3.12) on the right by S to obtain

$$A\tilde{U}_k = \tilde{U}_k T + u_{k+1} \tilde{z}^T, \tag{9.3.13}$$

where $\tilde{U}_k = U_k S$ and $\tilde{z}^T = h_{k+1,k} e_k^T S$. The restart is effected by retaining only the first m columns of (9.3.13):

$$A\tilde{U}_m = \tilde{U}_m T_{11} + u_{k+1} z^T, \tag{9.3.14}$$

where $\tilde{U}_m$ consists of the first m columns of $\tilde{U}_k$, and z consists of the first m entries of $\tilde{z}$.

Let $\tilde{S}$ denote the submatrix of S consisting of the first m columns. The similarity transformation $H_k S = ST$ implies that $\mathcal{R}(\tilde{S})$ is the invariant subspace of H_k corresponding to eigenvalues $\mu_1, \ldots, \mu_m$. This is the same as the space $\mathcal{R}(\tilde{G})$ in the exact-shift variant of the first method. Since $\tilde{U}_m = U_k \tilde{S}$, we see that $\mathcal{R}(\tilde{U}_m)$ is the same as the space $\mathcal{R}(\hat{U}_m)$ in the exact-shift variant of the first method. Thus the two methods retain the same space. This implies equivalence of the methods, as we will show at the end of the section.

In (9.3.14) we cannot claim that the first $m-1$ entries of z are zero as we could with $e_k^T G$ in the first method. In order to resume the Krylov process, we must return our configuration to a form like

$$A \hat{U}_m = \hat{U}_m \hat{H}_m + \hat{u}_{m+1} \hat{h}_{m+1,m} e_m^T,$$

in which z^T has been transformed to a multiple of e_m^T.[5] Let V_1 be an elimination matrix such that $z^T V_1^{-1} = \alpha e_m^T$ for some α. If we multiply (9.3.14) on the right by V_1^{-1}, we transform its last term to $u_{k+1} \alpha e_m^T = \hat{u}_{m+1} \hat{h}_{m+1,m} e_m^T$, where $\hat{u}_{m+1} = u_{k+1}$ and $\hat{h}_{m+1,m} = \alpha$.

The right multiplication by V_1^{-1} also transforms T_{11} to $V_1 T_{11} V_1^{-1}$. This is a full matrix, and we must reduce it to upper Hessenberg form. We do so, not column by column from left to right, but row by row from bottom to top, as described in Section 3.4. Elimination matrices $V_2 = \mathrm{diag}\{\tilde{V}_2, 1\}$, $V_3 = \mathrm{diag}\{\tilde{V}_3, 1, 1\}, \ldots, V_{m-1}$ introduce zeros in rows $m-1$, $m-2, \ldots, 3$, so that the matrix $\hat{H}_m = V_{m-1} \cdots V_1 T_{11} V_1^{-1} \cdots V_{m-1}^{-1}$ is upper Hessenberg. Let $V = V_{m-1} V_{m-2} \cdots V_1$. Since $z^T V_1^{-1} = \alpha e_m$, and $e_m^T V_j^{-1} = e_m^T$ for $j = 2, \ldots, m-1$, we also have $z^T V^{-1} = \alpha e_m^T$. In other words, the last row of V is proportional to z^T. This is an instance of Theorem 3.4.1, in which the roles of A, x, and G are played by T_{11}, z, and V, respectively.

Let $\hat{U}_m = \tilde{U}_m V^{-1}$, $\hat{u}_{m+1} = u_{k+1}$, and $\hat{h}_{m+1,m} = \alpha$. Multiplying (9.3.14) by V^{-1} on the right and using the fact that $z^T V^{-1} = \alpha e_m^T$, we obtain

$$A \hat{U}_m = \hat{U}_m \hat{H}_m + \hat{u}_{m+1} \hat{h}_{m+1,m} e_m^T, \qquad (9.3.15)$$

as desired. The columns of $\hat{U}_m$ span the same subspace as do those of $\tilde{U}_m$. This is the space corresponding to the eigenvalue estimates $\mu_1, \ldots, \mu_m$. Now that we have restored the Hessenberg form, the restart is complete. Now, starting from (9.3.15), which can also be written as

$$A \hat{U}_m = \hat{U}_{m+1} \hat{H}_{m+1,m},$$

we can use the Krylov process to generate $\hat{u}_{m+2}, \hat{u}_{m+3}, \ldots, \hat{u}_{k+1}$ satisfying

$$A \hat{U}_k = \hat{U}_{k+1} \hat{H}_{k+1,k}.$$

From here another implicit restart can be done. The restarts can be repeated until the desired invariant subspace is obtained.

---

[5]Actually it is possible to defer this transformation, as is done in both [233] and [198]. However, there seems to be no advantage to putting it off. Nothing is gained in terms of efficiency, and for the purpose of understanding the algorithm it is surely better to do the transformation right away.

Convergence and Locking

We still have to describe the process by which convergence is detected. A natural opportunity to check for convergence occurs at (9.3.14), which we rewrite as

$$A\tilde{U}_m = \tilde{U}_m B + u_{k+1} z^T, \tag{9.3.16}$$

replacing the symbol T_{11} by B. The matrix B is a submatrix of T, which was computed using some kind of GR algorithm. Therefore B will normally be quasi-triangular. For the sake of ease of discussion, let us assume at first that it is upper triangular.

Convergence can be checked by examining the entries of the vector z. If the first j entries of z are zero (or within a prescribed tolerance of zero), we can partition (9.3.16) as

$$A\begin{bmatrix} \tilde{U}_{m1} & \tilde{U}_{m2} \end{bmatrix} = \begin{bmatrix} \tilde{U}_{m1} & \tilde{U}_{m2} \end{bmatrix} \begin{bmatrix} B_{11} & B_{12} \\ 0 & B_{22} \end{bmatrix} + u_{k+1}\begin{bmatrix} 0 & z_2^T \end{bmatrix}, \tag{9.3.17}$$

where $\tilde{U}_{m1}$ is $n \times j$ and B_{11} is $j \times j$. The partition of B is valid because B is upper triangular. Equation (9.3.17) implies that $A\tilde{U}_{m1} = \tilde{U}_{m1}B_{11}$, whence $\mathcal{R}(\tilde{U}_{m1})$ is invariant under A, and the eigenvalues of B_{11} are eigenvalues of A. These are typically (though not necessarily) $\mu_1, \dots, \mu_j$.

We have assumed that B is upper triangular, which guarantees that no matter what the value of j is, the resulting partition of B has the form

$$\begin{bmatrix} B_{11} & B_{12} \\ 0 & B_{22} \end{bmatrix}.$$

Clearly we can make the same argument for quasi-triangular or more general block triangular B, so long as we refrain from using values of j that split a block.

Suppose that we have found an invariant subspace of dimension j, and we have the configuration (9.3.17). The next step in the process, as described above, is to build an elimination matrix V_1 such that $z^T V_1 = \alpha e_m^T$. But now, since the first j entries of z are already zero, we can take V_1 to have the form $V_1 = \text{diag}\{I_j, \hat{V}_1\}$. Then the operation $T_{11} = B \to BV_1$ does not touch the first j columns of B, nor does the operation $BV_1 \to V_1^{-1}BV_1 = M$ touch the first j rows. Thus the resulting matrix M has the form

$$M = \begin{bmatrix} B_{11} & M_{12} \\ 0 & M_{22} \end{bmatrix};$$

only the submatrix M_{22} is full and needs to be returned to Hessenberg form. This implies that the subsequent transformations $V_2, V_3, \dots$, which return M to upper Hessenberg form, all have the form $\text{diag}\{I_j, \hat{V}_i\}$, and so does the product V. Thus the transformation $\tilde{U}_m \to \tilde{U}_m V = \hat{U}_m$ does not touch the first j columns, the submatrix $\tilde{U}_{m1}$. These columns are now locked in and will remain untouched for the remainder of the computation. In the equation

$$A\hat{U}_m = \hat{U}_m \hat{H}_m + \hat{u}_{m+1}\hat{h}_{m+1,m}e_m^T,$$

the first j columns of $\hat{U}_m$ are $\tilde{U}_{m1}$, which span an invariant subspace. The $(j+1, j)$ entry of $\hat{H}_m$ is zero, so $\hat{H}_m$ is block triangular. Its upper left-hand $j \times j$ block is B_{11}, which houses j eigenvalues. B_{11} is also locked in and will remain unchanged for the rest of the computation.

Subsequent restarts will operate only on the "unconverged" parts of $\hat{U}_m$ and $\hat{H}_m$. Each restart has a chance to set more entries of z to zero, thereby enlarging the converged invariant subspace and causing more vectors to be locked in. Once a sufficiently large invariant subspace has been found, the computation is terminated.

Equivalence of the Two Methods

Here we show that the second method is equivalent to the exact-shift variant of the first method.[6] The key is the following simple lemma, which is proved in Exercise 9.3.6.

Lemma 9.3.1. *Let* $\mathcal{S} = \mathcal{K}_m(A, \hat{u}) = \mathcal{K}_m(A, \check{u})$, *and suppose that* $\mathcal{S}$ *is not invariant under* A. *Then* $\check{u} = \alpha\hat{u}$ *for some nonzero scalar* α.

Assume that no convergence or locking has taken place so far. In order to compare the two methods we rewrite (9.3.15) as

$$A\check{U}_m = \check{U}_m\check{H}_m + \check{u}_{m+1}\check{h}_{m+1,m}e_m^T. \tag{9.3.18}$$

We have observed that $\mathcal{R}(\check{U}_m) = \mathcal{R}(\tilde{U}_m)$ and that $\mathcal{R}(\tilde{U}_m) = \mathcal{R}(\hat{U}_m)$ if exact shifts are used in the first method, where $\hat{U}_m$ is as in (9.3.6). Since no convergence or locking has taken place, the matrix $\hat{H}_m$ in (9.3.6) is properly upper Hessenberg, and $\hat{h}_{m+1,m} \neq 0$. This implies by Proposition 9.2.4 that $\text{span}\{\hat{u}_1, \ldots, \hat{u}_j\} = \mathcal{K}_j(A, \hat{u}_1)$ for $j = 1, \ldots, m$. Moreover, none of these spaces is invariant. Applying the same reasoning to (9.3.18), we find that $\text{span}\{\check{u}_1, \ldots, \check{u}_j\} = \mathcal{K}_j(A, \check{u}_1)$ for $j = 1, \ldots, m$, and none of these spaces are invariant. We thus have $\mathcal{K}_m(A, \hat{u}_1) = \mathcal{R}(\hat{U}_m) = \mathcal{R}(\check{U}_m) = \mathcal{K}_m(A, \check{u}_1)$, and this space is not invariant. Thus by Lemma 9.3.1, $\hat{u}_1$ and $\check{u}_1$ are proportional, and

$$\text{span}\{\hat{u}_1, \ldots, \hat{u}_i\} = \mathcal{K}_i(A, \hat{u}_1) = \mathcal{K}_i(A, \check{u}_1) = \text{span}\{\check{u}_1, \ldots, \check{u}_i\}, \qquad i = 1, \ldots, m.$$

Thus the second restart method produces exactly the same sequence of Krylov subspaces as the first method with exact shifts.

For this analysis we have assumed that no convergence or locking have taken place. A more careful analysis shows that the equivalence continues to hold even after some vectors have been locked in. (See Exercise 9.3.7.)

Notes and References

Explicit restarts of Krylov processes were advocated by Saad [184] in 1984. The first implicit restarting method was due to Sorensen [191] in 1992. It is the basis of our first restart method, and it is the algorithm that underlies ARPACK [147]. The thick restart Lanczos process was introduced by Wu and Simon [233] for the symmetric case. (Precursors were [157] and [192].) Subsequently Stewart [198] introduced the Krylov–Schur algorithm, which reduces to the thick restart method in the symmetric case. These algorithms are the basis of our second restart method.

---

[6]I thank my former student Roden David for helping me understand this equivalence better.

Exercises

9.3.1.

(a) Verify (9.3.4).

(b) Verify (9.3.5).

9.3.2. Recall that a matrix G is m-Hessenberg if $g_{ij} = 0$ whenever $i > j + m$. Suppose that H is upper Hessenberg, that $p(H) = GR$, where p is a polynomial of degree j, and that R is upper triangular.

(a) Under the assumption that R is nonsingular, prove that G is j-Hessenberg. (See Exercise 4.5.5.)

(b) Assuming that R is singular, use Theorem 4.6.4 to prove that G is j-Hessenberg.

9.3.3.

(a) Verify (9.3.8).

(b) Prove (9.3.9) by induction.

9.3.4. Suppose that $v_1, \ldots, v_m$ are linearly independent eigenvectors of A, and let $x = c_1 v_1 + \cdots + c_m v_m$, where $c_1, \ldots, c_m$ are all nonzero. Prove that $\mathcal{K}_m(A, x) = \text{span}\{v_1, \ldots, v_m\}$. Thus the Krylov subspace is invariant.

9.3.5. Show that if $\hat{x} = p(A)x$ for some polynomial p, then

$$p(A)\mathcal{K}_i(A, x) = \mathcal{K}_i(A, \hat{x})$$

for $i = 0, 1, \ldots, n$.

9.3.6. This exercise works out the proof of Lemma 9.3.1.

(a) Under the hypotheses of Lemma 9.3.1, show that $\hat{u}, A\hat{u}, \ldots, A^{m-1}\hat{u}$ are linearly independent. Deduce that there are unique $c_1, \ldots, c_m$, not all zero, such that

$$\check{u} = c_1 \hat{u} + c_2 A\hat{u} + \cdots + c_m A^{m-1}\hat{u}. \tag{9.3.19}$$

(b) Show that $A^m \hat{u} \notin \mathcal{S}$.

(c) Let r be the largest integer such that $c_r \neq 0$ in (9.3.19). Note that if $r > 1$, then $A^{m-r+1}\check{u} \in \mathcal{S}$. But on the other hand,

$$A^{m-r+1}\check{u} = c_1 A^{m-r+1}\hat{u} + c_2 A^{m-r+2}\hat{u} + \cdots + c_r A^m \hat{u}.$$

Deduce that $A^m \hat{u} \in \mathcal{S}$, in contradiction to the result of part (b). Thus the assumption $r > 1$ must be false. Deduce that $r = 1$ and the lemma is true.

9.3.7. Consider (9.3.18) in the case when exactly s converged vectors have been locked in. Then $\check{h}_{s+1,s} = 0$, and $\check{h}_{j+1,j} \neq 0$ for $j = s+1, \ldots, m$. The space $\text{span}\{\check{u}_1, \ldots, \check{u}_s\}$ is invariant.

(a) Show that under these conditions

$$\text{span}\{\breve{u}_1, \ldots, \breve{u}_j\} = \text{span}\{\breve{u}_1, \ldots, \breve{u}_s\} \oplus \mathcal{K}_{j-s}(A, \breve{u}_{s+1})$$

for $j = s + 1, \ldots, m$.

(b) Show that $\text{span}\{\breve{u}_1, \ldots, \breve{u}_m\}$ is not invariant under A.

(c) Prove the following extension of Lemma 9.3.1: Suppose that the space

$$\text{span}\{\breve{u}_1, \ldots, \breve{u}_s\} = \text{span}\{\hat{u}_1, \ldots, \hat{u}_s\}$$

is invariant under A, and that the space

$$\begin{aligned}
\text{span}\{\breve{u}_1, \ldots, \breve{u}_m\} &= \text{span}\{\breve{u}_1, \ldots, \breve{u}_s\} \oplus \mathcal{K}_{m-s}(A, \breve{u}_{s+1}) \\
&= \text{span}\{\hat{u}_1, \ldots, \hat{u}_s\} \oplus \mathcal{K}_{m-s}(A, \hat{u}_{s+1}) \\
&= \text{span}\{\hat{u}_1, \ldots, \hat{u}_m\}
\end{aligned}$$

is not invariant under A. Then there is a nonzero constant α and a $v \in \text{span}\{\breve{u}_1, \ldots, \breve{u}_s\}$ such that

$$\breve{u}_{s+1} = \alpha \hat{u}_{s+1} + v.$$

(d) Using (9.3.18) and (9.3.6) with $\breve{h}_{s+1,s} = \hat{h}_{s+1,s} = 0$ and $\breve{u}_i = \hat{u}_i$ for $i = 1, \ldots, s$ (locked-in vectors untouched on this iteration) and the fact that $\mathcal{R}(\breve{U}_m) = \mathcal{R}(\hat{U}_m)$, show that

$$\text{span}\{\breve{u}_1, \ldots, \breve{u}_j\} = \text{span}\{\hat{u}_1, \ldots, \hat{u}_j\}, \qquad j = s + 1, \ldots, m.$$

Thus the two restart methods produce the same sequence of subspaces.

9.4   The Arnoldi and Symmetric Lanczos Processes

The Arnoldi process [11] is by far the most widely used Krylov process. Given $A \in \mathbb{C}^{n \times n}$ and nonzero starting vector $x \in \mathbb{C}^n$, the Arnoldi process begins with $u_1 = cx$, where $c = 1/\|x\|$. Given orthonormal vectors $u_1, \ldots, u_j$, it produces u_{j+1} as follows:

$$h_{ij} = \langle Au_j, u_i \rangle = u_i^* Au_j, \qquad i = 1, \ldots, j, \tag{9.4.1}$$

$$\hat{u}_{j+1} = Au_j - \sum_{i=1}^{j} u_i h_{ij}, \tag{9.4.2}$$

$$h_{j+1,j} = \|\hat{u}_{j+1}\|, \tag{9.4.3}$$

and if $h_{j+1,j} \neq 0$,

$$u_{j+1} = \hat{u}_{j+1}/h_{j+1,j}. \tag{9.4.4}$$

It is an easy matter to show that $\hat{u}_{j+1}$ is orthogonal to $u_1, \ldots, u_j$ if and only if the coefficients $h_{1j}, \ldots, h_{jj}$ are defined as shown in (9.4.1) (See Exercise 9.4.7). If $Au_j \notin$ span$\{u_1, \ldots, u_j\}$, then certainly $\hat{u}_{j+1} \neq 0$. Conversely, if $Au_j \in$ span$\{u_1, \ldots, u_j\}$, the choice of coefficients (9.4.1) forces $\hat{u}_{j+1} = 0$. In this case span$\{u_1, \ldots, u_j\} = \mathcal{K}_j(A, x)$ is invariant under A, and the Arnoldi process terminates. Thus the Arnoldi process is a Krylov process as defined in Section 9.2.

Letting $U_j = \begin{bmatrix} u_1 & \cdots & u_j \end{bmatrix}$, we can write (9.4.1) and (9.4.2) in matrix-vector form as

$$h_{1:j,j} = U_j^* Au_j \tag{9.4.5}$$

and

$$\hat{u}_{j+1} = Au_j - U_j h_{1:j,j}, \tag{9.4.6}$$

respectively. The Arnoldi process is an instance of the Gram–Schmidt process (Exercise 1.5.3), and as such it is vulnerable to roundoff errors [221, Section 3.4]. Over the course of several steps executed in floating point arithmetic, the orthogonality of the vectors will steadily deteriorate. A simple and practical remedy is to do the orthogonalization twice. One might even consider doing three or more orthogonalizations, but the unpublished advice of Kahan is that "twice is enough" [169]. This has been confirmed in practice and by theoretical analysis [95]. See also [2, 62, 116]. After two orthogonalizations, the vector $\hat{u}_{j+1}$ will be orthogonal to $u_1, \ldots, u_j$ to working precision. We do not mind the added cost of an additional orthogonalization because we never carry the Arnoldi run very far; we always keep the total number of vectors small. The following algorithm takes the need for reorthogonalization into account.

Arnoldi process with reorthogonalization.

$$
\begin{aligned}
&u_1 \leftarrow x/\|x\| \\
&U_1 \leftarrow [u_1] \\
&\text{for } j = 1, 2, 3, \ldots \\
&\qquad \left[\begin{aligned}
&u_{j+1} \leftarrow Au_j \\
&h_{1:j,j} \leftarrow U_j^* u_{j+1} \\
&u_{j+1} \leftarrow u_{j+1} - U_j h_{1:j,j} \quad \text{(orthogonalize)} \\
&\delta_{1:j} \leftarrow U_j^* u_{j+1} \\
&u_{j+1} \leftarrow u_{j+1} - U_j \delta_{1:j} \quad \text{(reorthogonalize)} \\
&h_{1:j,j} \leftarrow h_{1:j,j} + \delta_{1:j} \\
&h_{j+1,j} \leftarrow \|u_{j+1}\| \\
&\text{if } h_{j+1,j} = 0 \\
&\qquad \left[\begin{aligned}
&\text{set flag (span}\{u_1, \ldots, u_j\} \text{ is invariant)} \\
&\text{exit}
\end{aligned} \right. \\
&u_{j+1} \leftarrow u_{j+1}/h_{j+1,j} \\
&U_{j+1} \leftarrow \begin{bmatrix} U_j & u_{j+1} \end{bmatrix}
\end{aligned} \right.
\end{aligned}
\tag{9.4.7}
$$

Equations (9.4.2) and (9.4.4) together imply

$$Au_j = \sum_{i=1}^{j+1} u_i h_{ij},$$

and after k steps of the Arnoldi process, we have

$$AU_k = U_k H_k + u_{k+1} h_{k+1,k} e_k^T, \tag{9.4.8}$$

as explained in Proposition 9.2.3. The only thing we need to add to Proposition 9.2.3 in the Arnoldi case is that now the columns of U_k are orthonormal. Multiplying (9.4.8) by U_k^* on the left, we find that

$$H_k = U_k^* A U_k.$$

The eigenvalues of H_k are estimates of eigenvalues of A. In this case we call them *Ritz values* (Exercise 9.4.8).

Implicit Restarts

Since we normally do not have enough memory to store a great number of vectors, we normally have to do restarts. A second motive for restarting is that the computational cost of (9.4.1) and (9.4.2) increases linearly with j; each step takes more time than the previous one.

The two implicit restart methods described in Section 9.2 can both be applied to the Arnoldi process. The only caveat is that all transformations that are performed must be unitary in order to preserve the orthonormality of the vectors. Thus, for example, the filtering of approximate eigenvalues (now called Ritz values) must be done by the QR algorithm, not just any GR algorithm. In this context, the first restart method becomes the implicitly restarted Arnoldi method of Sorensen [191] (see also [147]) and the second method becomes the Krylov–Schur algorithm of Stewart [198].

The Symmetric Lanczos Process

Now assume that A is Hermitian ($A = A^*$). In this case the Arnoldi process undergoes significant simplification in principle and is called the *symmetric Lanczos process* [143]. The upper Hessenberg matrix $H_k = U_k^* A U_k$ is Hermitian because A is, so it is tridiagonal. The main-diagonal entries of H_k are, of course, real. Moreover, the definition (9.4.3) implies that the subdiagonal entries are real as well. Therefore H_k is a real symmetric tridiagonal matrix.

The fact that $h_{ij} = 0$ for $i < j - 1$ implies that most of the terms in the sum in (9.4.2) are zero. If we introduce new notation $\alpha_j = h_{jj}$ and $\beta_j = h_{j+1,j} = h_{j,j+1}$, (9.4.2) becomes

$$u_{j+1} \beta_j = Au_j - u_j \alpha_j - u_{j-1} \beta_{j-1}, \tag{9.4.9}$$

with the understanding that $u_0 \beta_0 = 0$. Based on this three-term recurrence, we have the following algorithm.

Symmetric Lanczos process without reorthogonalization.

$$
\begin{aligned}
&u_1 \leftarrow x/\|x\| \\
&\text{for } j = 1, 2, 3, \ldots \\
&\quad\left[\begin{array}{l}
u_{j+1} \leftarrow A u_j \\
\alpha_j \leftarrow u_j^* u_{j+1} \\
u_{j+1} \leftarrow u_{j+1} - u_j \alpha_j \\
\text{if } j > 1 \\
\quad\left[\ u_{j+1} \leftarrow u_{j+1} - u_{j-1}\beta_{j-1} \right. \\
\beta_j \leftarrow \|u_{j+1}\| \\
\text{if } \beta_j = 0 \\
\quad\left[\begin{array}{l}
\text{set flag (span}\{u_1, \ldots, u_j\} \text{ is invariant)} \\
\text{exit}
\end{array}\right. \\
u_{j+1} \leftarrow u_{j+1}/\beta_j
\end{array}\right.
\end{aligned}
\tag{9.4.10}
$$

The appeal of this algorithm is that, if we are willing to forego reorthogonalization, we can run it to much higher values of j. If we are just after eigenvalues, we do not need to save all the vectors that we generate. To compute u_{j+1}, we just need u_j and u_{j-1}. Moreover, the computational cost of each step is small and does not grow with j. The most expensive aspect is the multiplication of A by u_j. At any given step, the eigenvalues of the tridiagonal matrix H_j are estimates of eigenvalues of A. These can be computed cheaply by the QR algorithm or a number of other techniques.

The drawback of the procedure is that in floating point arithmetic the orthogonality is gradually lost. The consequences are interesting. Once an eigenvalue has been found, the algorithm "forgets" that it has found it and produces another one or more "ghost" copies. The use of the symmetric Lanczos algorithm without reorthogonalization was studied by Paige [161, 162, 163] and Cullum and Willoughby [59]. See also [154, 169]. Several partial reorthogonalization schemes, in which v_{j+1} is reorthogonalized only against selected v_i, have been proposed [169, 175, 189]. We will not pursue this idea.

There are interesting relationships between the symmetric Lanczos process, algorithms for generating orthogonal polynomials, and numerical integration formulas. See [215] and the references therein.

Preserving Orthogonality

The surest way to preserve orthogonality in the symmetric Lanczos process is to save all the vectors and orthogonalize against them. The runs are kept short, and restarts are done frequently. In this mode the symmetric Lanczos process differs little from the Arnoldi process.

Symmetric Lanczos process with reorthogonalization.

$$
\begin{aligned}
&u_1 \leftarrow x/\| x \| \\
&U_1 \leftarrow [u_1] \\
&\text{for } j = 1, 2, 3, \ldots \\
&\quad \left[\begin{array}{l}
u_{j+1} \leftarrow A u_j \\
\alpha_j \leftarrow u_j^* u_{j+1} \\
u_{j+1} \leftarrow u_{j+1} - u_j \alpha_j \\
\text{if } j > 1 \\
\quad \left[\begin{array}{l} u_{j+1} \leftarrow u_{j+1} - u_{j-1}\beta_{j-1} \end{array} \right. \\
\delta_{1:j} \leftarrow U_j^* u_{j+1} \\
u_{j+1} \leftarrow u_{j+1} - U_j \delta_{1:j} \quad \text{(reorthogonalize)} \\
\alpha_j \leftarrow \alpha_j + \delta_j \\
\beta_j \leftarrow \| u_{j+1} \| \\
\text{if } \beta_j = 0 \\
\quad \left[\begin{array}{l} \text{set flag (span}\{u_1, \ldots, u_j\} \text{ is invariant)} \\ \text{exit} \end{array} \right. \\
u_{j+1} \leftarrow u_{j+1}/\beta_j \\
U_{j+1} \leftarrow \left[\begin{array}{cc} U_j & u_{j+1} \end{array} \right]
\end{array} \right.
\end{aligned}
\tag{9.4.11}
$$

One might well ask why we should bother to use the symmetric Lanczos process in this mode. Why not just use the Arnoldi process? The answer is that the Lanczos process preserves the structure. The matrix

$$
H_k = \begin{bmatrix}
\alpha_1 & \beta_1 & & & \\
\beta_1 & \alpha_2 & \ddots & & \\
& \ddots & \ddots & \beta_{k-1} \\
& & \beta_{k-1} & \alpha_k
\end{bmatrix}
$$

that is produced is truly symmetric. When a restart is done, it can be done by the symmetric QR algorithm, which keeps the structure intact. The eigenvalues that are computed are real; there is no danger of a repeated real eigenvalue being confused for a pair of complex eigenvalues because of roundoff errors. The eigenvectors that are computed are orthogonal to working precision. These are properties that the eigensystem of a Hermitian matrix ought to have, but they can sometimes be lost if the structure-ignoring Arnoldi process is used. By the way, we do not feel overburdened by having to keep all the vectors around, since we need them anyway if we want to compute eigenvectors.

In this chapter we will meet many examples of structures (e.g., skew symmetric (Exercise 9.4.11), unitary, Hamiltonian, symplectic) for which there exist Krylov processes that have short recurrences. In no case are we able to exploit the short recurrence if we insist on preserving orthogonality. The methods are worth using nevertheless, simply because they preserve the structure.

Unitary Matrices and Cayley Transforms

Let U be a unitary matrix. The eigenvalue problem for U can be transformed to a Hermitian eigenvalue problem via a Cayley transform. This is an excellent option if U is not too large for the shift-and-invert methodology to be used [63].

All of the eigenvalues of U lie on the unit circle. Let τ be a target value on the unit circle that is not an eigenvalue of U, and suppose that we wish to find the eigenpairs of U associated with the eigenvalues that are closest to τ. The matrix

$$A = i(U + \tau I)(U - \tau I)^{-1} \tag{9.4.12}$$

is a Hermitian matrix whose largest eigenvalues correspond to the eigenvalues of U that are closest to τ (Exercise 9.4.10). Therefore the desired eigenpairs of U can be found quickly by applying the symmetric Lanczos process to A. Of course, we do not form the matrix A explicitly. We just need an LU factorization of $U - \tau I$, in order to effect the transformations $x \to Ax$. Thus this method is feasible whenever the shift-and-invert strategy is.

We refer to the transformation in (9.4.12) as a *Cayley transform*, and we call A a *Cayley transform* of U. We also refer to the inverse operation

$$U = \tau(A + iI)(A - iI)^{-1}$$

as a Cayley transform.

Exercises

9.4.1.

(a) Using (9.4.7) as a guide, write MATLAB code that implements the Arnoldi process with reorthogonalization. (Alternatively, download the file `arnoldi.m` from the website.[7])

(b) MATLAB supplies a sparse demonstration matrix called `west0479`. Try the following commands in MATLAB:

```
>> load west0479
>> A = west0479;
>> issparse(A)
>> size(A)
>> nnz(A) % number of nonzeros
>> spy(A)
```

Since this matrix is really not all that large, you can compute its "exact" eigenvalues using MATLAB's eig function: `lam = eig(full(A));` and you can plot them using `plot(real(lam),imag(lam),'r+')`, for example. Notice that this matrix has some outlying eigenvalues.

---

[7]www.siam.org/books/ot101

(c) Run 30 steps of the Arnoldi process, starting with a random initial vector. Compute $\| AU_j - U_{j+1}H_{j+1,j} \|$ (with $j = 30$) to check the correctness of your code. Also check the orthonormality of the columns of U by computing $\| I_{j+1} - U_{j+1}^*U_{j+1} \|$. Both of these residuals should be tiny.

(d) The Ritz values are the eigenvalues of the $j \times j$ Hessenberg matrix H_j. Compute these using the `eig` command, and plot the Ritz values and the exact eigenvalues together:

```
>> rl = real(lam);   il = imag(lam);
>> rr = real(ritz); ir = imag(ritz);
>> plot(rl,il,'r+',rr,ir,'bo')
```

As you can see, the outlying eigenvalues are approximated well by Ritz values, as we claimed in Section 9.1.

(e) If we had taken fewer Arnoldi steps, the Ritz values would not have been as good. Compute the Ritz values from 10-step and 20-step Arnoldi runs, and plot them together with the exact eigenvalues. For this it is not necessary to run the Arnoldi process again, since the desired Ritz values are eigenvalues of submatrices of the H that you computed on the first run.

9.4.2. Modify your Arnoldi code so that reorthogonalization is not done. Do 30 Arnoldi steps and then check the residual $\| I_{j+1} - U_{j+1}^*U_{j+1} \|$. Notice that the orthogonality has been significantly degraded. Repeat this with a run of 60 Arnoldi steps to see a complete loss of orthogonality.

9.4.3. To find the eigenvalues of `west0479` that are closest to a target τ, modify your Arnoldi code so that it does shift-and-invert. Instead of multiplying a vector by A at each step, multiply the vector by $(A - \tau I)^{-1}$. Do not form the matrix $(A - \tau I)^{-1}$. Instead perform a sparse LU decomposition: `[L R P] = lu(A-tau*speye(n));` and use the factors L, R, and P. The factorization has to be done only once for a given run; then the factors can be reused repeatedly. Try $\tau = 10$, $\tau = -10$, and other values. Complex values are okay as well.

9.4.4.

(a) Using (9.4.11) as a guide, write MATLAB code that implements the symmetric Lanczos process with reorthogonalization. (Alternatively, download the file `symlan.m` from the website.)

(b) Generate a sparse discrete negative Laplacian matrix using the commands

```
>> A = delsq(numgrid('H',40))
>> perm = symamd(A); A = A(perm,perm);
```

How big is A? Is A symmetric? Some informative commands:

```
>> issparse(A)
>> size(A)
>> nnz(A)
```

```
>> spy(A)
>> norm(A-A',1)
>> help delsq, help numgrid, help symamd, ...
```

A is a symmetric, positive-definite matrix, so its eigenvalues are real and positive. The smallest ones are of greatest physical interest. Again this matrix is not really large, so you can compute its "true" eigenvalues by `lam = eig(full(A));`.

(c) Run 20 steps of the symmetric Lanczos process applied to A, starting with a random initial vector. Compute $\| AU_j - U_{j+1}H_{j+1,j} \|$ (with $j = 20$) to check the correctness of your code. Also check the orthonormality of the columns of U by computing $\| I_{j+1} - U_{j+1}^* U_{j+1} \|$.

(d) The Ritz values are the eigenvalues of the $j \times j$ tridiagonal matrix H_j. Compute these using the `eig` command. Notice that the smallest Ritz values are not very good approximations of the smallest eigenvalues of A. Try some longer Lanczos runs to get better approximations.

(e) Now run 20 steps of the symmetric Lanczos process applied to A^{-1}. This is the shift-and-invert strategy with zero shift. Do not form A^{-1} explicitly; do an LU or Cholesky decomposition and use it to perform the transformations $x \rightarrow A^{-1}x$ economically. The inverses of the largest Ritz values are estimates of the smallest eigenvalues of A. Notice that we now have really good approximations.

(f) Modify your Lanczos code so that it does not do the reorthogonalization step. Then repeat the previous run. Notice that orthogonality is completely lost and the approximations do not look nearly so good. On closer inspection you will see that this is mainly due to the fact that the smallest eigenvalue has been found twice. Do a longer run of, say, 60 steps to observe more extensive duplication of eigenvalues.

9.4.5.

(a) Write a MATLAB program that does the symmetric Lanczos process with implicit restarts by the second restart method described in Section 9.2. Alternatively, download the file `relan.m` from the website.

(b) Apply your implicitly restarted Lanczos code to the matrix A of Exercise 9.4.4 to compute the smallest eight eigenvalues and corresponding eigenfunctions of A to as much accuracy as you can. Compute the residuals $\| Av - v\lambda \|$ to check the quality of your computed eigenpairs. Compare your computed eigenvalues with the true eigenvalues (from `eig(full(A))`) to make sure that they are correct.

(c) Apply your code to the matrix A^{-1}. Do not compute A^{-1} explicitly; use an LU or Cholesky decomposition to effect the transformations $x \rightarrow A^{-1}x$. Compute the smallest eight eigenvalues of A. As before, compute residuals and compare with the true eigenvalues. Notice that this iteration converged in many fewer iterations than the iteration with A did.

(d) In producing your test matrix A, you used the command `numgrid('H',40)` (and others). You can build a bigger test matrix by replacing the number 40 by a larger number. Build a matrix that is big enough that your computer runs out of memory when it tries to compute the eigenvalues by `eig(full(A))`. Repeat parts (b) and (c) using this larger matrix. (Of course, you will not be able to compare with the "true" eigenvalues from `eig`, but you can compare the results of (b) and (c) with each other.) In order to get convergence in (b), you may need to increase the dimension of the Krylov spaces that are being built (the parameter k, which is known as `nmax` in `relan.m`).

(e) Now make an even bigger matrix that is so big that you cannot use the shift-and-invert strategy because you do not have enough memory to store the factors of the LU or Cholesky decomposition. Compute the eight smallest eigenvalues by applying the implicitly restarted Lanczos scheme to A. This will take a long time, but if you take enough iterations and large enough Krylov subspaces, you will eventually get the eigenpairs.

9.4.6.

(a) Familiarize yourself with the built-in MATLAB function `eigs`, which computes eigenpairs by the implicitly restarted Arnoldi process, as implemented in ARPACK [147].

(b) Repeat Exercise 9.4.5 using `eigs`.

9.4.7. This exercise checks some basic properties of the Arnoldi process.

(a) Show that the vector $\hat{u}_{j+1}$ defined by (9.4.2) is orthogonal to u_k if and only if h_{kj} is defined as in (9.4.1).

(b) If $Au_j \in \text{span}\{u_1, \ldots, u_j\}$, then $Au_j = \sum_{i=1}^{j} c_i u_i$ for some choice of coefficients $c_1, \ldots, c_j$. Show that $c_k = u_k^* Au_j$ for $k = 1, \ldots, j$. Deduce that $\hat{u}_{j+1} = 0$ if and only if $Au_j \in \text{span}\{u_1, \ldots, u_j\}$.

(c) Verify (9.4.5) and (9.4.6).

(d) Show that $\hat{u}_{j+1} = (I - P_j)Au_j$, where $P_j = U_j U_j^*$, the orthoprojector of $\mathbb{C}^n$ onto $\mathcal{R}(U_j)$. Thus $\hat{u}_j$ is the orthogonal projection of Au_j onto $\mathcal{R}(U_j)^\perp$.

9.4.8. Let $A \in \mathbb{C}^{n \times n}$, and let $\mathcal{U}$ be a subspace of $\mathbb{C}^n$. Then a vector $u \in \mathcal{U}$ is called a *Ritz vector* of A associated with $\mathcal{U}$, $\theta \in \mathbb{C}$ is called a *Ritz value*, and (θ, u) is called a *Ritz pair* if

$$Au - u\theta \perp \mathcal{U}.$$

Let U be a matrix with orthonormal columns such that $\mathcal{U} = \mathcal{R}(U)$, and let $B = U^*AU$. Then each $u \in \mathcal{U}$ can be expressed in the form $u = Ux$, where $x = U^*u$. Prove that (θ, u) is a Ritz pair of A if and only if (θ, x) is an eigenpair of B.

9.4.9. In the symmetric Lanczos process, the Hessenberg matrices H_k are tridiagonal. Although this is a very easy result, it is nevertheless instructive to prove it by a second method. In this exercise we use Krylov subspaces to show that if $A = A^*$ in the Arnoldi process, then $h_{ij} = 0$ if $i < j - 1$. Recall that $\text{span}\{u_1, \ldots, u_i\} = \mathcal{K}_i(A, x)$. Thus $u_k \perp \mathcal{K}_i(A, x)$ if $k > i$. Assume $A = A^*$.

(a) Show that $h_{ij} = \langle Au_j, u_i \rangle = \langle u_j, Au_i \rangle$.

(b) Show that $Au_i \in \mathcal{K}_{i+1}(A, x)$.

(c) Deduce that $h_{ij} = 0$ if $i < j - 1$.

9.4.10. This exercise investigates Cayley transforms.

(a) Consider the transform $w \mapsto z$ of the extended complex plane $\mathbb{C} \cup \{\infty\}$ given by

$$z = \phi(w) = \frac{w + i}{w - i}.$$

This is the original Cayley transform, but we will refer to all of the other linear fractional transformations appearing in this exercise as Cayley transforms as well. Show that ϕ maps the extended real line one-to-one onto the unit circle, with $\phi(\infty) = 1$.

(b) Given a point τ on the unit circle, define

$$z = \phi_\tau(w) = \tau \frac{w + i}{w - i}.$$

Show that ϕ_τ maps the extended real line one-to-one onto the unit circle, with $\phi_\tau(\infty) = \tau$.

(c) Show that the inverse of ϕ_τ is given by

$$w = \phi_\tau^{-1}(z) = i \frac{z + \tau}{z - \tau}.$$

This transformation maps the unit circle onto the extended real line, with $\tau \mapsto \infty$.

(d) Let U be a unitary matrix, and let τ be a point on the unit circle that is not an eigenvalue of U. Let
$$A = i(U + \tau I)(U - \tau I)^{-1}.$$
Show that for each eigenpair (λ, v) of U, A has an eigenpair $(i(\lambda + \tau)/(\lambda - \tau), v)$. In particular, all of the eigenvalues of A are real. Moreover, the eigenvalues of U that are closest to τ are mapped to eigenvalues of A that have the largest modulus.

(e) Show that A is normal. Deduce that A is Hermitian.

9.4.11. What form does the Arnoldi process take when it is applied to a skew symmetric matrix $A = -A^T \in \mathbb{R}^{n \times n}$?

(a) What form does the equation $AU_k = U_{k+1}H_{k+1,k}$ (cf. Proposition 9.2.3) take in this case? That is, what is the form of $H_{k+1,k}$?

(b) What is the form of the short recurrence, and how are the coefficients computed? This is the *skew symmetric Lanczos process*.

(c) Show that $A^2 U_k = U_{k+2} H_{k+2,k+1} H_{k+1,k}$. Compute $H_{k+2,k+1} H_{k+1,k}$ and deduce that the skew symmetric Lanczos process is equivalent in principle to two independent Lanczos processes on the symmetric matrix A^2, one starting with u_1 and the other starting with $u_2 = A u_1 / \| A u_1 \|$.

9.5   The Unitary Arnoldi Process

In the previous section we discussed the unitary eigenvalue problem briefly, showing that a Cayley transform can be used to transform the problem to a Hermitian eigenvalue problem. In this section we consider an approach that works directly with the unitary matrix.

We have seen that if A is Hermitian, then the Arnoldi process admits significant simplification in principle. The same is true if A is unitary, although this is not so obvious. In this section we will develop the isometric Arnoldi process of Gragg [102, 104]. Here we will call it the unitary Arnoldi process, since all isometric matrices are unitary in the finite-dimensional case. Different but equivalent algorithms are developed in [40, 51, 84, 215, 218].[8]

Let $U \in \mathbb{C}^{n \times n}$ be a unitary matrix ($U^* = U^{-1}$). Then U^{-1} is as accessible as U is, so one could consider applying the Arnoldi process to U or U^{-1} with equal ease. It turns out that the significant simplification is achieved by doing both at once. If we apply the Arnoldi process to U with starting vector $v_1 = cx$, we produce an orthonormal sequence $v_1, v_2, v_3, \ldots$ such that

$$\mathrm{span}\{v_1, \ldots, v_j\} = \mathcal{K}_j(U, x), \qquad j = 1, 2, 3, \ldots .$$

The jth step of the process has the form

$$v_{j+1} h_{j+1,j} = U v_j - \sum_{i=1}^{j} v_i h_{ij}, \tag{9.5.1}$$

where the h_{ij} are chosen so that $v_{j+1} \perp \mathrm{span}\{v_1, \ldots, v_j\}$.

Now suppose that we apply the Arnoldi process to U^{-1} with the same starting vector $w_1 = v_1 = cx$. Then we generate an orthonormal sequence $w_1, w_2, w_3, \ldots$ such that

$$\mathrm{span}\{w_1, \ldots, w_j\} = \mathcal{K}_j(U^{-1}, x), \qquad j = 1, 2, 3, \ldots .$$

The jth step of the process has the form

$$w_{j+1} k_{j+1,j} = U^{-1} w_j - \sum_{i=1}^{j} w_i k_{ij}, \tag{9.5.2}$$

where the k_{ij} are chosen so that $w_{j+1} \perp \mathrm{span}\{w_1, \ldots, w_j\}$.

---

[8]The isometric Arnoldi process is also equivalent to Szegő's recursion [201] for orthogonal polynomials on the unit circle, the lattice algorithm of speech processing, and the Levinson–Durbin algorithm [78, 148] for solving the Yule–Walker equations. The relationships are discussed in [218] and elsewhere in the literature.

Notice that

$$\text{span}\{U^{j-1}w_j, \ldots, U^{j-1}w_1\} = U^{j-1}\text{span}\{U^{-(j-1)}x, \ldots, U^{-1}x, x\}$$

$$= \text{span}\{x, Ux, \ldots, U^{j-1}x\} = \mathcal{K}_j(U, x).$$

Thus $v_1, \ldots, v_j$ and $U^{j-1}w_j, \ldots, U^{j-1}w_1$ are orthonormal bases of the same space. The latter basis is orthonormal because unitary matrices preserve orthonormality. Therefore we could make v_{j+1} orthogonal to $\text{span}\{v_1, \ldots, v_j\}$ by orthogonalizing against $U^{j-1}w_j, \ldots,$ $U^{j-1}w_1$ instead of $v_1, \ldots, v_j$ in (9.5.1). This is where we realize our economy. Since $\text{span}\{U^{j-2}w_{j-1}, \ldots, U^{j-2}w_1\} = \text{span}\{x, Ux, \ldots, U^{j-2}x\} = \text{span}\{v_1, \ldots, v_{j-1}\}$, which is orthogonal to v_j, we see that $\text{span}\{U^{j-1}w_{j-1}, \ldots, U^{j-1}w_1\}$ is orthogonal to Uv_j, again using the fact that unitary matrices preserve orthogonality. Thus $U^{j-1}w_j$ is the only member of the basis $U^{j-1}w_j, \ldots, U^{j-1}w_1$ that is not already orthogonal to Uv_j, and we can replace (9.5.1) by

$$v_{j+1}h_{j+1,j} = Uv_j - U^{j-1}w_j\gamma_j, \tag{9.5.3}$$

where

$$\gamma_j = \langle Uv_j, U^{j-1}w_j \rangle. \tag{9.5.4}$$

Now it is natural to ask how one can economically produce the vector $U^{j-1}w_j$. Reversing the roles of U and U^{-1}, we find that the vectors $U^{-(j-1)}v_1, \ldots U^{-(j-1)}v_j$ span the same space as $w_j, \ldots, w_1$ and that all but $U^{-(j-1)}v_j$ are already orthogonal to $U^{-1}w_j$. Thus (9.5.2) can be replaced by

$$w_{j+1}k_{j+1,j} = U^{-1}w_j - U^{-(j-1)}v_j\delta_j, \tag{9.5.5}$$

where

$$\delta_j = \langle U^{-1}w_j, U^{-(j-1)}v_j \rangle. \tag{9.5.6}$$

One easily checks that $\delta_j = \overline{\gamma_j}$. Moreover, $k_{j+1,j} = h_{j+1,j} = \sqrt{1 - |\gamma_j|^2}$ by the orthogonality properties of the vectors in (9.5.3) and (9.5.5). Let $\tilde{v}_j = U^{j-1}w_j$ for $j = 1, 2, \ldots$. Then, multiplying (9.5.5) by U^j, we obtain

$$\tilde{v}_{j+1}\sigma_j = \tilde{v}_j - Uv_j\overline{\gamma_j},$$

where σ_j is a new name for $k_{j+1,j}$. Combining this with (9.5.3), we obtain the pair of two-term recurrences

$$\begin{aligned} v_{j+1}\sigma_j &= Uv_j - \tilde{v}_j\gamma_j, \\ \tilde{v}_{j+1}\sigma_j &= \tilde{v}_j - Uv_j\overline{\gamma_j}, \end{aligned} \tag{9.5.7}$$

where

$$\gamma_j = \langle Uv_j, \tilde{v}_j \rangle \quad \text{and} \quad \sigma_j = \sqrt{1 - |\gamma_j|^2}. \tag{9.5.8}$$

This is the unitary Arnoldi process [102, 104]. Equations (9.5.7) and (9.5.8) involve U only; U^{-1} has disappeared.[9] Of course, we could have defined $\tilde{w}_j = U^{-(j-1)}v_j$ and written down a pair of equivalent recurrences in which U^{-1} appears and U does not. For recurrences involving both U and U^{-1}, see [51, 215], for example.

In practical floating point computations, the unitary Arnoldi process must be modified in several ways to make it robust. First, the formula (9.5.8) is not a good way to compute σ_j because it is inaccurate when $|\gamma_j| \approx 1$. A better method is to compute

$$\hat{v}_{j+1} = Uv_j - \tilde{v}_j\gamma_j,$$

then let $\sigma_j = \|\hat{v}_{j+1}\|$ and $v_{j+1} = \sigma_j^{-1}\hat{v}_{j+1}$. This is just what we do in the standard Arnoldi process. In practice the σ_j and γ_j so produced will not exactly satisfy $\sigma_j^2 + |\gamma_j|^2 = 1$, so we must enforce unitarity by the steps

$$\nu \leftarrow \left(\sigma_j^2 + |\gamma_j^2|\right)^{1/2}$$
$$\gamma_j \leftarrow \gamma_j/\nu$$
$$\sigma_j \leftarrow \sigma_j/\nu$$

We took this same precaution in the unitary QR algorithm (cf. 8.10.4).

In principle the short recurrences (9.5.7) generate v_{j+1} from v_j and $\tilde{v}_j$; the earlier vectors are not needed. In practice roundoff errors will cause the same gradual loss of orthogonality as in any Krylov process unless v_{j+1} is specifically orthogonalized against all of the v_i. As before, this is not a problem if we do short runs with frequent implicit restarts. Moreover, we do not mind keeping all of the v_i around, as we need them anyway if we want to compute eigenvectors.

A worse problem is what to do about the auxiliary vectors $\tilde{v}_j$. The sequence $(w_j) = (U^{-j+1}\tilde{v}_j)$ is orthonormal in principle, and we should enforce orthonormality of that sequence as well. However, the sequence that we have in hand is $(\tilde{v}_j)$, not $(U^{-j+1}\tilde{v}_j)$, so we cannot do the orthonormalization. We have found [63, 65] that under these circumstances the second formula of (9.5.7) is not very robust; the accuracy of the $\tilde{v}_{j+1}$ is poor when σ_j is small. Fortunately there is another formula for $\tilde{v}_{j+1}$ that works much better in practice. Solving the first equation of (9.5.7) for Uv_j, substituting this into the second equation, using the formula $\sigma_j^2 + |\gamma_j^2| = 1$, and dividing by σ_j, we obtain

$$\tilde{v}_{j+1} = \tilde{v}_j\sigma_j - v_{j+1}\overline{\gamma_j}. \tag{9.5.9}$$

There is no possibility of cancellation in this formula. It takes the two orthogonal unit vectors $\tilde{v}_j$ and v_{j+1}, rescales them, and combines them to form a third unit vector $\tilde{v}_{j+1}$.

With these modifications we have the following algorithm.

---

[9]Since U^{-1} is not used in (9.5.7), (9.5.8), the process is applicable to isometric operators ($\|Ux\| = \|x\|$ on an infinite-dimensional Hilbert space) that are not unitary. Therefore Gragg called it the isometric Arnoldi process.

Unitary Arnoldi process with reorthogonalization.

$$v_1 \leftarrow x/\|x\|$$
$$V_1 \leftarrow [v_1]$$
$$\tilde{v} \leftarrow v_1$$
for $j = 1, 2, 3, \ldots$

$$
\begin{array}{l}
v_{j+1} \leftarrow U v_j \\
\gamma_j \leftarrow \tilde{v}^* v_{j+1} \\
v_{j+1} \leftarrow v_{j+1} - \tilde{v}\gamma_j \\
\delta_{1:j} \leftarrow V_j^* v_{j+1} \\
v_{j+1} \leftarrow v_{j+1} - V_j \delta_{1:j} \quad \text{(reorthogonalize)} \\
\gamma_j \leftarrow \gamma_j + \delta_j \\
\sigma_j \leftarrow \|v_{j+1}\| \\
\text{if } \sigma_j = 0 \\
\quad \left[\begin{array}{l} \text{set flag (span}\{v_1, \ldots, v_j\} \text{ is invariant)} \\ \text{exit} \end{array}\right. \\
v_{j+1} \leftarrow v_{j+1}/\sigma_j \\
\nu \leftarrow \sqrt{\sigma_j^2 + |\gamma_j^2|} \\
\sigma_j \leftarrow \sigma_j/\nu \\
\gamma_j \leftarrow \gamma_j/\nu \\
\tilde{v} \leftarrow \tilde{v}\sigma_j - v_{j+1}\overline{\gamma_j} \\
V_{j+1} \leftarrow \left[\begin{array}{cc} V_j & v_{j+1} \end{array}\right]
\end{array}
\tag{9.5.10}
$$

We have yet to discuss how the data produced by the unitary Arnoldi process can be used to yield information about the eigensystem of U. Just as in Section 9.4, we can rewrite (9.5.1) in the form

$$U V_j = V_{j+1} H_{j+1, j} = V_j H_j + v_{j+1} h_{j+1, j} e_j^T,$$

where the notation is analogous to that established in Section 9.4. If we want to use the unitary Arnoldi process to estimate eigenvalues, we need the entries h_{ij} of H_j or equivalent information. With (9.5.7) one gets instead the quantities γ_i and σ_i, $i = 1, 2, \ldots, j$. As we will show below, these are exactly the Schur parameters of H_j, as described in Section 8.10. Thus we can obtain H_j (or rather a closely related matrix) in product form and use an algorithm like the one described in Section 8.10 to compute its eigenvalues.

Comparing (9.5.1) with (9.5.7), it must be the case that

$$\tilde{v}_j \gamma_j = \sum_{i=1}^{j} v_i h_{ij}.$$

Since $v_1, \ldots, v_j$ are linearly independent, the h_{ij} are uniquely determined by this expression, and we can figure out what they are by establishing a recurrence for the $\tilde{v}_i$. To begin with, note that $\tilde{v}_1 = v_1$. We have already observed that

$$\tilde{v}_{j+1} = \tilde{v}_j \sigma_j - v_{j+1}\overline{\gamma_j}, \quad j = 1, 2, 3, \ldots. \tag{9.5.11}$$

Applying this result repeatedly, we can deduce that

$$\tilde{v}_j = v_1 \prod_{k=1}^{j-1} \sigma_k + \sum_{i=2}^{j} v_i \left(-\overline{\gamma}_{i-1} \prod_{k=i}^{j-1} \sigma_k \right) \tag{9.5.12}$$

for $j = 1, 2, 3, \ldots$, with the usual conventions that an empty product equals 1 and an empty sum equals zero. Introducing the simplifying notation

$$\pi_{ij} = \prod_{k=i}^{j-1} \sigma_k \tag{9.5.13}$$

(with $\pi_{jj} = 1$), equation (9.5.12) becomes

$$\tilde{v}_j = v_1 \pi_{1j} + \sum_{i=2}^{j} v_i \left(-\overline{\gamma}_{i-1} \pi_{ij} \right). \tag{9.5.14}$$

Rewriting the first equation in (9.5.7) as $U v_j = \tilde{v}_j \gamma_j + v_{j+1} \sigma_j$ and inserting the result from (9.5.14), we obtain

$$U v_j = v_1 \pi_{1j} \gamma_j + \sum_{i=2}^{j} v_i (-\overline{\gamma}_{i-1} \pi_{ij} \gamma_j) + v_{j+1} \sigma_j. \tag{9.5.15}$$

Since

$$U v_j = \sum_{i=1}^{j+1} v_i h_{ij},$$

equation (9.5.15) tells us immediately what the entries of $H_{j+1,j}$ are:

$$h_{ij} = \begin{cases} \pi_{ij} \gamma_j & \text{if } i = 1, \\ -\overline{\gamma}_{i-1} \pi_{ij} \gamma_j & \text{if } 2 \leq i \leq j, \\ \sigma_j & \text{if } i = j+1, \\ 0 & \text{if } i > j+1. \end{cases}$$

Thus

$$H_{j+1,j} = \begin{bmatrix} \gamma_1 & \sigma_1 \gamma_2 & \sigma_1 \sigma_2 \gamma_3 & \cdots & \sigma_1 \cdots \sigma_{j-1} \gamma_j \\ \sigma_1 & -\overline{\gamma}_1 \gamma_2 & -\overline{\gamma}_1 \sigma_2 \gamma_3 & \cdots & -\overline{\gamma}_1 \sigma_2 \cdots \sigma_{j-1} \gamma_j \\ 0 & \sigma_2 & -\overline{\gamma}_2 \gamma_3 & \cdots & -\overline{\gamma}_2 \sigma_3 \cdots \sigma_{j-1} \gamma_j \\ 0 & 0 & \sigma_3 & \cdots & -\overline{\gamma}_3 \sigma_4 \cdots \sigma_{j-1} \gamma_j \\ \vdots & \vdots & \ddots & \ddots & \vdots \\ 0 & 0 & \cdots & 0 & \sigma_j \end{bmatrix}. \tag{9.5.16}$$

For $k = 1, \ldots, j$, define matrices $G_k \in \mathbb{C}^{j+1 \times j+1}$ by (cf. Section 8.10)

$$G_k = \begin{bmatrix} I_{k-1} & & & \\ & \gamma_k & \sigma_k & \\ & \sigma_k & -\overline{\gamma}_k & \\ & & & I_{j-k} \end{bmatrix}. \tag{9.5.17}$$

Then one easily checks that

$$H_{j+1,j} = G_1 G_2 \cdots G_j I_{j+1,j}.$$

That is, $H_{j+1,j}$ consists of the first j columns of $G_1 \cdots G_j$.

Implicit Restarts

We describe a restart by the second (thick restart) method following [63, 65]. After k steps of the unitary Arnoldi process, we have

$$U V_k = V_{k+1} H_{k+1,k},$$

where $H_{k+1,k}$ is available in the form

$$H_{k+1,k} = G_1 G_2 \cdots G_k I_{k+1,k}.$$

To make an implicit restart we would normally compute the eigenvalues of H_k, the square matrix obtained by deleting the bottom row from $H_{k+1,k}$. Notice, however, that H_k is not unitary. The columns of $H_{k+1,k}$ are orthonormal because $H_{k+1,k}$ consists of the first k columns of the unitary matrix $G_1 \cdots G_k$. Thus the columns of H_k are orthogonal, but they are not orthonormal because the last column has norm less than 1 (unless $\sigma_k = 0$). We would like to have a unitary matrix to operate on so that we can manipulate it (almost) entirely in terms of its Schur parameters. To this end we make a modification of H_k. Let $\tilde{\gamma}_k = \gamma_k/|\gamma_k|$.[10] Let $\tilde{G}_1, \ldots, \tilde{G}_{k-1} \in \mathbb{C}^{k \times k}$ be the submatrices of $G_1, \ldots, G_{k-1}$ obtained by deleting the last row and column. Define $\tilde{G}_k = \mathrm{diag}\{1, \ldots, 1, \tilde{\gamma}_k\} \in \mathbb{C}^{k \times k}$ and

$$\tilde{H}_k = \tilde{G}_1 \ldots \tilde{G}_{k-1} \tilde{G}_k.$$

$\tilde{H}_k$ is the same as H_k, except that its last column has been rescaled to have norm 1. Now $\tilde{H}_k$ is unitary, and we can manipulate it in terms of its Schur parameters.

For our restart we will take the eigenvalues of $\tilde{H}_k$ as estimates of eigenvalues of U. Note that the eigenvalues of $\tilde{H}_k$ lie on the unit circle, while those of H_k lie inside but typically not on the unit circle. Once σ_k becomes small, as happens as we approach convergence after several restarts, the eigenvalues of H_k and $\tilde{H}_k$ are nearly the same; those of H_k lie just inside the unit circle.

Since H_k and $\tilde{H}_k$ differ only in their kth column, $H_k = \tilde{H}_k + p e_k^T$ for some $p \in \mathbb{C}^k$. Thus the relationship $U V_k = V_{k+1} H_{k+1,k} = V_k H_k + v_{k+1} \sigma_k e_k^T$ can be rewritten as

$$U V_k = V_k \tilde{H}_k + \tilde{p} e_k^T, \qquad (9.5.18)$$

where $\tilde{p} = p + v_{k+1}\sigma_k$.

Since $\tilde{H}_k$ is given in terms of its Schur parameters, the eigensystem of $\tilde{H}_k$ can be computed by the version of the QR algorithm presented in Section 8.10 or by any other stable method that works with the Schur parameters. Since U is normal, this yields a diagonalization

$$\tilde{H}_k = S D_k S^*,$$

---

[10]In the rare event $\gamma_k = 0$, take $\tilde{\gamma}_k = 1$ or $\tilde{\gamma}_k = e^{i\theta}$, where θ is random, for example.

where S is unitary and D_k is diagonal. We can suppose that the eigenvalues of D_k are sorted so that the m that we wish to retain are in the upper left-hand corner of D_k. Multiplying (9.5.18) on the right by S, we obtain

$$U\tilde{V}_k = \tilde{V}_k D_k + \tilde{p}\,\tilde{z}^T, \qquad (9.5.19)$$

where $\tilde{V}_k = V_k S$ and $\tilde{z}^T = e_k^T S$. To effect the restart we retain only the first m columns of (9.5.18):

$$U\tilde{V}_m = \tilde{V}_m D_m + \tilde{p}z^T, \qquad (9.5.20)$$

where D_m is the upper left-hand $m \times m$ submatrix of D_k, and z is the vector consisting of the first m components of $\tilde{z}$. Now we have to make a transformation that maps z^T to a multiple of e_m^T and returns D_m to upper Hessenberg form. $\tilde{H}_k$ was presented in terms of its Schur parameters, and so are D_m and D_k. We would also like the new Hessenberg matrix to be expressed in terms of Schur parameters. Thus we would like this final transformation to operate on the Schur parameters.

We begin by applying a unitary transformation on the right (rotator or reflector) that acts only on columns 1 and 2 and transforms the first entry of z^T to zero. When this same transformation is applied to D_m, it affects only the first two columns. When we complete the similarity transformation by a left multiplication, only the first two rows are affected. The resulting matrix has a nonzero entry in position (2, 1). It is upper Hessenberg, so it can be represented in terms of Schur parameters. Next a transformation is applied to columns 2 and 3 to create a zero in the second position of the (transformed) z^T vector. When this same transformation is applied to the (transformed) D_m matrix, it affects columns 2 and 3, creating a nonzero in the (3, 2) position. The matrix is still upper Hessenberg. When the similarity transformation is completed by transforming rows 2 and 3, a nonzero entry is created in position (3, 1). This is a bulge, and it needs to be eliminated. Fortunately it can be chased off the top of the matrix by the reverse (RQ variant) of the bulge-chasing procedure for unitary Hessenberg matrices described in Section 8.10. This just consists of a unitary operation on columns 1 and 2 to get rid of the bulge, followed by an operation on rows 1 and 2 that completes the similarity transformation. Since this transformation acts only on columns 1 and 2, it does not destroy the two zeros that have been created in z^T.

The next step is a transformation on columns 3 and 4 that creates a zero in the third position of the z vector. The corresponding operations on the D_m matrix create a bulge in position (4, 2). This too can be chased off of the top of the matrix. This requires operations in the (2, 3) plane, followed by operations in the (1, 2) plane. Because only the first three columns are touched, the zeros that have been created in z are left intact. Continuing in this manner, we transform z^T to a multiple of e_m^T and D_m to a unitary Hessenberg matrix $\hat{H}_m$ represented in terms of its Schur parameters. Let Q denote the product of all of the transformations involved in this process, so that $z^T Q = \alpha e_m^T$ and $Q^* D_m Q = \hat{H}_m$. Multiplying (9.5.20) on the right by Q and letting $\hat{V}_m = \tilde{V}_m Q$, we have

$$U\hat{V}_m = \hat{V}_m \hat{H}_m + \tilde{p}\,\alpha\, e_m^T. \qquad (9.5.21)$$

This is still not an Arnoldi configuration, as $\tilde{p}$ is not orthogonal to the columns of $\hat{V}_m$.[11] However, we can obtain an Arnoldi configuration by dropping the mth column from (9.5.21).

---

[11] Moreover, $\hat{H}_m$ is unitary, which is impossible in a unitary Arnoldi configuration unless $\mathcal{R}(\hat{V}_m)$ is invariant under U.

This gives

$$U \hat{V}_{m-1} = \hat{V}_m \hat{H}_{m,m-1} = \hat{V}_{m-1} \hat{H}_{m-1} + \hat{h}_m \hat{\sigma}_m e_m^T,$$

which is a configuration from which the Arnoldi process can be continued. This is exactly what would be obtained from $m - 1$ steps of the Arnoldi process with starting vector $\hat{v}_1$. We now recycle the notation, dropping the hats, and write

$$U V_{m-1} = V_{m-1} H_{m-1} + v_m \sigma_m e_m^T. \tag{9.5.22}$$

Recall that our unitary Arnoldi process uses the recurrences (9.5.7) and (9.5.9), so the first step after (9.5.22), to produce v_{m+1}, will be

$$\begin{aligned} v_{m+1} \sigma_m &= U v_m - \tilde{v}_m \gamma_m, \\ \tilde{v}_{m+1} &= \tilde{v}_m \sigma_m - v_{m+1} \overline{\gamma_m}, \end{aligned} \tag{9.5.23}$$

for which we need $\tilde{v}_m$. This we generate as follows. Replacing m by $m - 1$ in the first equation of (9.5.23) and solving for $\tilde{v}_{m-1}$, we have

$$\tilde{v}_{m-1} = (U v_{m-1} - v_m \sigma_{m-1})/\gamma_{m-1}.$$

Using this to compute $\tilde{v}_{m-1}$, we then can obtain $\tilde{v}_m$ by

$$\tilde{v}_m = \tilde{v}_{m-1} \sigma_{m-1} - v_m \overline{\gamma}_{m-1}.$$

The numbers γ_{m-1} and σ_{m-1} are available; they are Schur parameters that form part of the description of H_m. Once we have $\tilde{v}_m$, we can proceed with (9.5.23) and subsequent steps.

In Section 9.3 we justified implicit restarts by showing that they effect nested subspace iterations driven by $p(U)$, where p is a polynomial whose zeros are near parts of the spectrum that we want to suppress. The restart procedure outlined here is somewhat different because it involves a modification of the matrix H_k. However, the theory of Section 9.3 continues to be valid here since the modification to H_k affects only its last column. The key equation is (9.5.18), in which the error term $\tilde{p} e_k^T$ affects only the last column. More details are given in [63, 65].

Exercises

9.5.1. This exercise fills in some of the details in the development of the unitary Arnoldi process.

(a) Show that $\text{span}\{U^{-(j-1)}v_1, \ldots, U^{-(j-1)}v_j\} = \mathcal{K}_j(U^{-1}, x) = \text{span}\{w_1, \ldots, w_j\}$.

(b) Show that $U^{-1}w_j$ is orthogonal to $U^{-(j-1)}v_1, \ldots, U^{-(j-1)}v_{j-1}$.

(c) Using the fact that unitary matrices preserve the inner product, show that $\delta_j = \overline{\gamma_j}$, where γ_j and δ_j are as in (9.5.4) and (9.5.6), respectively.

(d) All of the vectors in (9.5.3) and (9.5.5) are unit vectors, and some of them are orthogonal to each other. Use the Pythagorean theorem to verify that $k_{j+1,j} = h_{j+1,j} = \sqrt{1 - |\gamma_j|^2}$. Thus, letting $\sigma_j = k_{j+1,j} = h_{j+1,j}$, we have $\sigma_j^2 + |\gamma_j|^2 = 1$.

(e) Let $\tilde{w}_j = U^{-(j-1)}v_j$. Write down a pair of recurrences for w_j and $\tilde{w}_j$ that are equivalent to (9.5.7) but involve U^{-1} and not U.

9.5.2. This exercise works out the details of the construction of the matrix $H_{j+1,j}$ from the parameters γ_i and σ_i.

(a) From (9.5.7) deduce that $Uv_j = \tilde{v}_j\gamma_j + v_{j+1}\sigma_j$. Substitute this into the second equation of (9.5.7) to deduce that

$$\tilde{v}_{j+1} = \tilde{v}_j\sigma_j - v_{j+1}\overline{\gamma_j}, \quad j = 1, 2, \ldots.$$

Bear in mind that $\sigma_j^2 + |\gamma_j|^2 = 1$.

(b) Deduce that

$$\tilde{v}_2 = v_1\sigma_1 - v_2\overline{\gamma}_1,$$
$$\tilde{v}_3 = v_1\sigma_1\sigma_2 - v_2\overline{\gamma}_1\sigma_2 - v_3\overline{\gamma}_2,$$
$$\tilde{v}_4 = v_1\sigma_1\sigma_2\sigma_3 - v_2\overline{\gamma}_1\sigma_2\sigma_3 - v_3\overline{\gamma}_2\sigma_3 - v_4\overline{\gamma}_3.$$

Prove by induction on j that

$$\tilde{v}_j = v_1 \prod_{k=1}^{j-1} \sigma_k + \sum_{i=2}^{j} v_i \left(-\overline{\gamma}_{i-1} \prod_{k=i}^{j-1} \sigma_k \right)$$

for $j = 1, 2, 3, \ldots$, which becomes

$$\tilde{v}_j = v_1\pi_{1j} + \sum_{i=2}^{j} v_j \left(-\overline{\gamma}_{i-1}\pi_{ij} \right)$$

with the abbreviation (9.5.13).

(c) Deduce (9.5.15).

(d) Check that (9.5.16) is correct.

9.5.3. Show that $\| H_j \| \leq 1$. Deduce that the eigenvalues of H_j (which are Ritz values of U) satisfy $|\lambda| \leq 1$. Thus they lie on or inside the unit circle.

9.6 The Unsymmetric Lanczos Process

As we have seen, the recurrence in the Arnoldi process becomes a short recurrence if A is Hermitian, in which case it is called the symmetric Lanczos process. The unsymmetric Lanczos process [143], which we will now introduce, produces short recurrences even for non-Hermitian A by giving up orthogonality. In the Hermitian case, the unsymmetric Lanczos process reduces to the symmetric Lanczos process. This and the next three sections draw substantially from [222].

A pair of sequences of vectors $u_1, u_2, \ldots, u_j$ and $w_1, w_2, \ldots, w_j$ is called *biorthonormal* if

$$\langle u_i, w_k \rangle = w_k^* u_i = \begin{cases} 0 & \text{if } i \neq k, \\ 1 & \text{if } i = k. \end{cases} \tag{9.6.1}$$

This is a much weaker property than orthonormality (Exercise 9.6.1). The unsymmetric Lanczos process builds a biorthonormal pair of sequences, each of which spans a Krylov subspace.

Let $A \in \mathbb{C}^{n \times n}$ be an arbitrary square matrix. The process begins with any two vectors $u_1, w_1 \in \mathbb{C}^n$ satisfying $\langle u_1, w_1 \rangle = 1$. After $j - 1$ steps, vectors $u_1, \ldots, u_j$ and $w_1, \ldots, w_j$ satisfying the biorthonormality property (9.6.1) will have been produced. They also satisfy

$$\text{span}\{u_1, \ldots, u_k\} = \mathcal{K}_k(A, v_1), \quad k = 1, \ldots, j,$$

and

$$\text{span}\{w_1, \ldots, w_k\} = \mathcal{K}_k(A^*, w_1), \quad k = 1, \ldots, j.$$

The vectors u_{j+1} and w_{j+1} are then produced as follows:

$$h_{ij} = \langle Au_j, w_i \rangle, \quad g_{ij} = \langle A^* w_j, u_i \rangle, \quad i = 1, \ldots, j, \tag{9.6.2}$$

$$\hat{u}_{j+1} = Au_j - \sum_{i=1}^{j} u_i h_{ij}, \tag{9.6.3}$$

$$\hat{w}_{j+1} = A^* w_j - \sum_{i=1}^{j} w_i g_{ij}, \tag{9.6.4}$$

$$\tau_{j+1} = \langle \hat{u}_{j+1}, \hat{w}_{j+1} \rangle, \tag{9.6.5}$$

and if $\tau_{j+1} \neq 0$, choose $h_{j+1,j}$ and $g_{j+1,j}$ so that $h_{j+1,j} \overline{g_{j+1,j}} = \tau_{j+1}$, and then set

$$u_{j+1} = \hat{u}_{j+1} / h_{j+1,j}, \quad w_{j+1} = \hat{w}_{j+1} / g_{j+1,j}. \tag{9.6.6}$$

Some of the basic properties of the unsymmetric Lanczos process are worked out in Exercise 9.6.2. To begin with, $\hat{u}_{j+1}$ is orthogonal to $w_1, \ldots, w_k$, and $\hat{w}_{j+1}$ is orthogonal to $u_1, \ldots, u_k$. Indeed, the definitions of h_{ij} and g_{ij} given by (9.6.2) are the unique ones for which this desired property is realized. Assuming that τ_{j+1} defined by (9.6.5) is not zero, there are many ways to choose $h_{j+1,1}$ and $g_{j+1,1}$ so that $h_{j+1,j} \overline{g_{j+1,j}} = \tau_{j+1}$. For any such choice, the vectors u_{j+1} and w_{j+1} satisfy $\langle u_{j+1}, w_{j+1} \rangle = 1$. If $\tau_{j+1} = 0$, the method breaks down. The consequences of breakdown will be discussed below.

It is easy to prove by induction that the unsymmetric Lanczos process produces vectors satisfying $\text{span}\{u_1, \ldots, u_j\} = \mathcal{K}_j(A, u_1)$ and $\text{span}\{w_1, \ldots, w_j\} = \mathcal{K}_j(A^*, w_1)$ for all j. Moreover, if at some step $Au_j \in \text{span}\{u_1, \ldots, u_j\}$, then $\hat{u}_{j+1} = 0$ (see Exercise 9.6.2). Thus the process is a Krylov process as defined in Section 9.2. It is, of course, also true that if $A^* w_j \in \text{span}\{w_1, \ldots, w_j\}$, then $\hat{w}_{j+1} = 0$. We can think of the unsymmetric Lanczos process as a Krylov process on A that produces the sequence (u_j) with the help of the auxiliary sequence (w_j), but it is equally well a Krylov process on A^* that produces the sequence (w_j) with the help of the auxiliary sequence (u_j). Better yet, we can think of it as a pair of mutually supporting Krylov processes. The process is perfectly symmetric with respect to the roles of A and A^*.

There are several ways in which a breakdown ($\tau_{j+1} = 0$) can occur. First, it can happen that $Au_j \in \text{span}\{u_1, \ldots, u_j\}$, causing $\hat{u}_{j+1} = 0$. This happens only when span$\{u_1, \ldots, u_j\}$ is invariant under A (Proposition 9.2.3). Since we can then extract j eigenvalues and eigenvectors of A from span$\{u_1, \ldots, u_j\}$, we call this a benign breakdown. Another good sort of breakdown occurs when $A^*w_j \in \text{span}\{w_1, \ldots, w_j\}$, causing $\hat{w}_{j+1} = 0$. In this case span$\{w_1, \ldots, w_j\}$ is invariant under A^*, so we can extract j eigenvalues and eigenvectors of A^* (left eigenvectors of A) from span$\{w_1, \ldots, w_j\}$. It can also happen that $\tau_{j+1} = \langle \hat{u}_{j+1}, \hat{w}_{j+1} \rangle = 0$ even though both $\hat{u}_{j+1}$ and $\hat{w}_{j+1}$ are nonzero. This is a serious breakdown. The surest remedy is to start over with different starting vectors. Perhaps more dangerous than a serious breakdown is a near breakdown. This happens when $\hat{u}_{j+1}$ and $\hat{w}_{j+1}$ are healthily far from zero but are nearly orthogonal, so that τ_{j+1} is nearly zero. This does not cause the process to stop, but it signals potential instability and inaccuracy of the results. Because of the risk of serious breakdowns or near breakdowns, it is best to keep the Lanczos runs fairly short and do implicit restarts frequently.[12]

We claimed at the beginning that the unsymmetric Lanczos process has short recurrences, which means that most of the coefficients h_{ij} and g_{ij} in (9.6.2) are zero. To see that this is so, we introduce notation similar to that used in previous sections. Suppose we have taken k steps of the unsymmetric Lanczos process. Let $U_k = \begin{bmatrix} u_1 & \cdots & u_k \end{bmatrix}$ and $W_k = \begin{bmatrix} w_1 & \cdots & w_k \end{bmatrix}$, and let $H_{k+1,k}$, H_k, $G_{k+1,k}$, and G_k denote the upper Hessenberg matrices built from the coefficients h_{ij} and g_{ij}. Then, reorganizing (9.6.3), (9.6.4), and (9.6.6), we find that

$$Au_j = \sum_{i=1}^{j+1} u_i h_{ij} \quad \text{and} \quad A^*w_j = \sum_{i=1}^{j+1} w_i g_{ij}, \quad j = 1, \ldots, k,$$

and consequently

$$AU_k = U_{k+1}H_{k+1,k} \quad \text{and} \quad A^*W_k = W_{k+1}G_{k+1,k}, \tag{9.6.7}$$

which can be rewritten as

$$AU_k = U_k H_k + u_{k+1}h_{k+1,k}e_k^T \tag{9.6.8}$$

and

$$A^*W_k = W_k G_k + w_{k+1}g_{k+1,k}e_k^T. \tag{9.6.9}$$

Biorthonormality implies that $W_k^*U_k = I_k$ and $W_k^*u_{k+1} = 0$. Therefore we can deduce from (9.6.8) that $H_k = W_k^*AU_k$. Similarly, from (9.6.9), we have $G_k = U_k^*A^*W_k$, so $G_k = H_k^*$. Since G_k and H_k are both upper Hessenberg, we deduce that they must in fact be tridiagonal. Thus most of the h_{ij} and g_{ij} are indeed zero, and those that are not zero are related. Let $\alpha_j = h_{jj}$, $\beta_j = h_{j+1,j}$, and $\gamma_j = h_{j,j+1}$. Then $\overline{\alpha_j} = g_{jj}$, $\overline{\beta_j} = g_{j,j+1}$, and $\overline{\gamma_j} = g_{j+1,j}$, and the recurrences (9.6.3) and (9.6.4) become

$$u_{j+1}\beta_j = Au_j - u_j\alpha_j - u_{j-1}\gamma_{j-1} \tag{9.6.10}$$

---

[12]An alternate approach is to circumvent the breakdown by a "look ahead" technique [90, 176].

and

$$w_{j+1}\overline{\gamma_j} = A^*w_j - w_j\overline{\alpha_j} - w_{j-1}\overline{\beta_{j-1}}, \tag{9.6.11}$$

respectively, where it is understood that $u_0\gamma_0 = 0 = w_0\overline{\beta_0}$.

These are elegant formulas. Unfortunately the process suffers from steady loss of biorthogonality, just as the Arnoldi process suffers from loss of orthogonality. If one wishes to keep biorthonormality intact, one must add a reorthogonalization at each step.

In the case $A^* = A$, the recurrences (9.6.10) and (9.6.11) are both driven by the same matrix A, (9.6.10) is identical in appearance to the symmetric Lanczos recurrence (9.4.9), and (9.6.11) has the same general look. If we start the unsymmetric Lanczos process with $w_1 = u_1$, then we get $w_{j+1} = u_{j+1}$ at every step, provided that we make the normalization (9.6.6) in such a way that $h_{j+1,j} = g_{j+1,j} > 0$, i.e., $\beta_j = \gamma_j > 0$. This is accomplished by taking $\beta_j = \gamma_j = \|\hat{u}_{j+1}\| = \|\hat{w}_{j+1}\|$. Then (9.6.10) and (9.6.11) become literally identical to each other and to the symmetric Lanczos recurrence (9.4.9). Thus the unsymmetric Lanczos process reduces to the symmetric Lanczos process in this case. The Arnoldi process and the unsymmetric Lanczos process are two very different extensions of the symmetric Lanczos process.

The Real Case

From this point on we will restrict our attention to real quantities. Thus we have $A \in \mathbb{R}^{n \times n}$ and $u_1, v_1 \in \mathbb{R}^n$. All quantities produced by the unsymmetric Lanczos process are then real, assuming that β_j and γ_j ($h_{j+1,j}$ and $g_{j+1,j}$) are always chosen to be real in (9.6.6). Let us take a closer look at the normalization step. If the real number $\tau_{j+1} = \langle \hat{u}_{j+1}, \hat{w}_{j+1}\rangle$ is not zero, then it is either positive or negative. In the former case we can satisfy $\beta_j\gamma_j = \tau_{j+1}$ by taking $\beta_j = \gamma_j = \sqrt{\tau_{j+1}}$; in the latter case we can take $\beta_j = \sqrt{-\tau_{j+1}} = -\gamma_j$. In either event we have $|\beta_j| = |\gamma_j|$, which implies that the matrix

$$H_k = \begin{bmatrix} \alpha_1 & \gamma_1 & & & & \\ \beta_1 & \alpha_2 & \gamma_2 & & & \\ & \beta_2 & \alpha_3 & \ddots & & \\ & & \ddots & \ddots & \gamma_{k-1} \\ & & & \beta_{k-1} & \alpha_k \end{bmatrix}$$

is pseudosymmetric (4.9.3). Thus it can be expressed as a product $H_k = T_k D_k$, where T_k is symmetric and tridiagonal and D_k is a signature matrix:

$$T_k = \begin{bmatrix} a_1 & b_1 & & & & \\ b_1 & a_2 & b_2 & & & \\ & b_2 & a_3 & \ddots & & \\ & & \ddots & \ddots & b_{k-1} \\ & & & b_{k-1} & a_k \end{bmatrix}, \quad D_k = \begin{bmatrix} d_1 & & & & \\ & d_2 & & & \\ & & d_3 & & \\ & & & \ddots & \\ & & & & d_k \end{bmatrix},$$

with $d_j = \pm 1$ for $j = 1, \ldots, k$. We also have $G_k = H_k^T = D_k T_k$.

Now let us see what the unsymmetric Lanczos process looks like in this new notation. The recurrences (9.6.10) and (9.6.11) become

$$u_{j+1} b_j d_j = A u_j - u_j a_j d_j - u_{j-1} b_{j-1} d_j \tag{9.6.12}$$

and

$$w_{j+1} d_{j+1} b_j = A^T w_j - w_j d_j a_j - w_{j-1} d_{j-1} b_{j-1}, \tag{9.6.13}$$

respectively. At the beginning of the jth step we will already have computed the quantities u_{j-1}, u_j, w_{j-1}, w_j, d_j, d_{j-1}, and b_{j-1}. Taking the inner product of (9.6.12) with w_j or of (9.6.13) with u_j and invoking biorthogonality, we find that

$$a_j = d_j^{-1} \langle A u_j, w_j \rangle = d_j^{-1} \langle A^T w_j, u_j \rangle = d_j^{-1} w_j^T A u_j,$$

in agreement with (9.6.2). Using this formula to compute a_j, we now have enough information to compute the right-hand sides of (9.6.12) and (9.6.13). As before, let $\hat{u}_{j+1}$ and $\hat{w}_{j+1}$ denote the vectors that result from these two computations, and let $\tau_{j+1} = \langle \hat{u}_{j+1}, \hat{w}_{j+1} \rangle$. The condition $\langle u_{j+1}, w_{j+1} \rangle = 1$ implies that $\tau_{j+1} = b_j^2 d_j d_{j+1}$, which can be used to determine b_j and d_{j+1}. If $\tau_{j+1} > 0$, we take $b_j = +\sqrt{\tau_{j+1}}$ and $d_{j+1} = d_j$; if $\tau_{j+1} < 0$, we take $b_j = +\sqrt{-\tau_{j+1}}$ and $d_{j+1} = -d_j$. We can then compute $u_{j+1} = \hat{u}_{j+1}/(b_j d_j)$ and $w_{j+1} = \hat{w}_{j+1}/(b_j d_{j+1})$. These considerations give the following algorithm.

Unsymmetric Lanczos process without reorthogonalization.
Start with initial vectors u_1 and w_1 satisfying $w_1^T u_1 = 1$.

$$
\begin{aligned}
&d_1 \leftarrow \pm 1 \\
&\text{for } j = 1, 2, 3, \ldots \\
&\quad\left[
\begin{aligned}
&y \leftarrow A u_j \\
&a_j \leftarrow d_j w_j^T y \\
&u_{j+1} \leftarrow y - u_j a_j d_j \\
&w_{j+1} \leftarrow A^T w_j - w_j d_j a_j \\
&\text{if } j > 1 \text{ then} \\
&\quad\left[
\begin{aligned}
&u_{j+1} \leftarrow u_{j+1} - u_{j-1} b_{j-1} d_j \\
&w_{j+1} \leftarrow w_{j+1} - w_{j-1} d_{j-1} b_{j-1}
\end{aligned}
\right. \\
&\tau_{j+1} \leftarrow w_{j+1}^T u_{j+1} \\
&\text{if } \tau_{j+1} = 0 \text{ then} \\
&\quad\left[\; \text{stop (breakdown)} \right. \\
&\text{if } \tau_{j+1} > 0 \text{ then} \\
&\quad\left[\; b_j \leftarrow \sqrt{\tau_{j+1}},\; d_{j+1} \leftarrow d_j \right. \\
&\text{if } \tau_{j+1} < 0 \text{ then} \\
&\quad\left[\; b_j \leftarrow \sqrt{-\tau_{j+1}},\; d_{j+1} \leftarrow -d_j \right. \\
&u_{j+1} \leftarrow u_{j+1}/(b_j d_j) \\
&w_{j+1} \leftarrow w_{j+1}/(d_{j+1} b_j)
\end{aligned}
\right.
\end{aligned}
\tag{9.6.14}
$$

It does not matter whether d_1 is taken to be $+1$ or -1. Reversing d_1 has no essential effect on the algorithm: All computed quantities are the same as before, except that minus signs are introduced in a systematic way.

If we want to keep the two sequences biorthogonal in practice, we must include reorthogonalization. Instead of rewriting the algorithm completely, we will just indicate the changes that would need to be made in (9.6.14). Immediately before the line that computes τ_{j+1}, insert the following commands:

$$s_{1:j} \leftarrow W_j^T u_{j+1}, \quad t_{1:j} \leftarrow U_j^T w_{j+1}$$
$$u_{j+1} \leftarrow u_{j+1} - U_j s_{1:j}$$
$$w_{j+1} \leftarrow w_{j+1} - W_j t_{1:j}$$
$$a_j \leftarrow a_j + s_j d_j$$

and insert updates $U_{j+1} \leftarrow \begin{bmatrix} U_j & u_{j+1} \end{bmatrix}$ and $W_{j+1} \leftarrow \begin{bmatrix} W_j & w_{j+1} \end{bmatrix}$ in the appropriate places.

Implicit Restarts

Either of the two implicit restart methods described in Section 9.2 can be applied to the unsymmetric Lanczos process. In this case we want to preserve the pseudosymmetry of $H_k = T_k D_k$, so we use transformations that preserve that structure. Thus the HR algorithm is the right version of GR to use here [106].

Let's look at the second restart method in some detail. In our current situation, the equations (9.6.8) and (9.6.9), which become

$$AU_k = U_k T_k D_k + u_{k+1} b_k d_k e_k^T \tag{9.6.15}$$

and

$$A^T W_k = W_k D_k T_k + w_{k+1} d_{k+1} b_k e_k^T \tag{9.6.16}$$

in our new notation, correspond to (9.3.12). We can use the HR algorithm, implemented as an HZ algorithm as described in Section 6.4, to diagonalize $T_k D_k$ and $D_k T_k$. Specifically, the HZ algorithm produces H, T, and D such that T is block diagonal, D is a signature matrix,

$$T = H^{-1} T_k H^{-T} \quad \text{and} \quad D = H^T D_k H.$$

Thus

$$TD = H^{-1}(T_k D_k)H \quad \text{and} \quad DT = H^T (D_k T_k) H^{-T}.$$

Both TD and DT are block diagonal with a 2×2 block for each complex-conjugate pair of eigenvalues and a 1×1 block for each real eigenvalue. The m eigenvalues that we wish to retain should be at the top of TD and DT. Notice that if they are not at the top, the necessary swaps are easily made in this case, because the matrices are block diagonal (not merely triangular). We have $D_k = HDH^T$ and $T_k = HTH^T$. If we wish to swap the real eigenvalues in positions (i, i) and (j, j) of TD and DT, we simply swap the corresponding entries of T and of D while at the same time swapping the ith and jth columns of H. Swapping complex-conjugate pairs is no harder, nor is swapping a complex pair with a real eigenvalue.

Multiply (9.6.15) on the right by H and (9.6.16) on the right by H^{-T} to obtain

$$A \tilde{U}_k = \tilde{U}_k TD + u_{k+1} \tilde{z}^T, \tag{9.6.17}$$

$$A^T \tilde{W}_k = \tilde{W}_k DT + w_{k+1} \tilde{y}^T, \qquad (9.6.18)$$

where $\tilde{U}_k = U_k H$, $\tilde{W}_k H^{-T}$, $\tilde{z}^T = b_k d_k e_k^T H$, and $\tilde{y}^T = d_{k+1} b_k e_k^T H^{-T}$. Notice that this transformation preserves biorthonormality: $\tilde{W}_k^T \tilde{U}_k = I$, $\tilde{W}_k^T u_{k+1} = 0$, and $\tilde{U}_k^T w_{k+1} = 0$. Moreover, $\tilde{y}^T$ is closely related to $\tilde{z}^T$. The equation $D_k = HDH^T$ implies that $H^{-T} = D_k HD$. Therefore $\tilde{y}^T = d_{k+1} \tilde{z}^T D$. Since $d_{k+1} = \pm 1$ and D is a signature matrix, this implies that the entries of $\tilde{y}$ and $\tilde{z}$ are the same up to signs. Thus if the first few entries of $\tilde{z}$ are small enough to be set to zero, indicating convergence and locking of an invariant subspace as in (9.3.17), then so are the first few entries of $\tilde{y}$. Thus, whenever we lock in a subspace that is invariant under A, we also lock in a subspace that is invariant under A^T. Locking was discussed in Section 9.2 and will not be revisited here.

We now restart by retaining only the first m columns of (9.6.17) and (9.6.18):

$$A\tilde{U}_m = \tilde{U}_m T_m D_m + u_{k+1} z^T, \qquad (9.6.19)$$

$$A^T \tilde{W}_m = \tilde{W}_m D_m T_m + w_{k+1} y^T, \qquad (9.6.20)$$

where $y^T = d_{k+1} z^T D_m$. In order to resume the nonsymmetric Lanczos process, we must transform these equations to a form like

$$A\hat{U}_m = \hat{U}_m \hat{T}_m \hat{D}_m + \hat{u}_{m+1} \hat{b}_m \hat{d}_m e_m^T,$$

$$A^T \hat{W}_m = \hat{W}_m \hat{D}_m \hat{T}_m + \hat{w}_{m+1} \hat{d}_{m+1} \hat{b}_m e_m^T,$$

in which z^T and y^T have been transformed to multiples of e_m^T. Let V_1 be an elimination matrix such that $z^T V_1 = \alpha e_m^T$. Assume further that V_1 has been constructed in such a way that $V_1^T D_m V_1 = \check{D}_m$ is another signature matrix. Such a V_1 can (almost always) be constructed as a product of hyperbolic and orthogonal transformations, essentially as described in the proof of Theorem 3.1.5. Since $V_1^{-T} = D_m V_1 \check{D}_m$, we also have $y^T V_1^{-T} = d_{k+1} z^T D_m^2 V_1 \check{D}_m = d_{k+1} \check{d}_m \alpha e_m^T$, so $y^T V_1^{-T} = \beta e_m^T$, where $\beta = d_{k+1} \check{d}_m \alpha$. Note that $|\beta| = |\alpha|$.

Let $\check{U}_m = \tilde{U}_m V_1$, $\check{W}_m = \tilde{W}_m V_1^{-T}$, and $\check{T}_m = V_1^{-1} T_m V_1^{-T}$. Then, multiplying (9.6.19) on the right by V_1 and (9.6.20) on the right by V_1^{-T}, we obtain

$$A\check{U}_m = \check{U}_m \check{T}_m \check{D}_m + u_{k+1} \alpha e_m^T \qquad (9.6.21)$$

and

$$A^T \check{W}_m = \check{W}_m \check{D}_m \check{T}_m + w_{k+1} \beta e_m^T. \qquad (9.6.22)$$

This is almost the form we want; the only problem is that $\check{T}_m$ is not tridiagonal, although it is symmetric. The final task of the restart is to return $\check{T}_m$ to tridiagonal form, row by row, from bottom to top, by a congruence transformation that preserves the signature-matrix form of $\check{D}_m$. Call the transforming matrix V_2, constructed so that $\hat{T}_m = V_2^{-1} \check{T}_m V_2^{-T}$ is tridiagonal and $\hat{D}_m = V_2^{-1} \check{D}_m V_2^{-T} = V_2^T \check{D}_m V_2$ is a signature matrix. V_2 can be constructed as a product of $m-2$ matrices essentially of the form described in Theorem 3.1.5. By construction, V_2 has the form

$$V_2 = \begin{bmatrix} \tilde{V}_2 & 0 \\ 0 & 1 \end{bmatrix},$$

and so $e_m^T V_2 = e_m^T$ and $e_m^T V_2^{-T} = e_m^T$. Multiplying (9.6.21) and (9.6.22) on the right by V_2 and V_2^{-T}, respectively, we obtain

$$A\hat{U}_m = \hat{U}_m \hat{T}_m \hat{D}_m + \hat{u}_{m+1} \hat{b}_m \hat{d}_m e_m^T, \tag{9.6.23}$$

$$A^T \hat{W}_m = \hat{W}_m \hat{D}_m \hat{T}_m + \hat{w}_{m+1} \hat{d}_{m+1} \hat{b}_m e_m^T, \tag{9.6.24}$$

where $\hat{U}_m = \breve{U}_m V_2$, $\hat{W}_m = \breve{W}_m V_2^{-T}$, $\hat{u}_{m+1} = u_{k+1}$, $\hat{w}_{m+1} = w_{k+1}$, $\hat{d}_m$ is the last diagonal entry in $\hat{D}_m$, $\hat{b}_m = \alpha/\hat{d}_m$, and $\hat{d}_{m+1} = \beta/\hat{b}_m = \pm 1$. These computations are consistent because $|\beta| = |\alpha|$. Notice that biorthonormality has been preserved all along: $\hat{W}_m^T \hat{U}_m = I$, $\hat{W}_m^T \hat{u}_{m+1} = 0$, and $\hat{U}_m^T \hat{w}_{m+1} = 0$.

Equations (9.6.23) and (9.6.24) are now in a form where we can pick up the unsymmetric Lanczos process at step $m + 1$. The implicit restart is complete.

An alternate way of generating the transformations V_1 and V_2 that saves a few flops and has a certain aesthetic appeal is sketched in Exercise 9.6.5.

We must not neglect to remind the reader that the unsymmetric Lanczos process can break down along the way. The implicit restart process can also break down, as it relies on hyperbolic transformations. It is therefore important to keep the Lanczos runs fairly short and restart frequently. This will guarantee that the probability of a breakdown on any given run is slight. Moreover, in the rare event of a breakdown, not much computational effort will have been invested in that particular short run. When a breakdown does occur, the process needs to be restarted somehow. One could start from scratch or think of some more clever way of continuing without losing everything. It is clear at least that any vectors that have converged and been locked in can be kept. We have not invested much thought in the question of emergency restarts because, in our experience with this and related algorithms, breakdowns are rare.

After a number of restarts, the Lanczos process will normally produce an invariant subspace of the desired dimension. Let's say that dimension is $\hat{m} \le m$. The final configuration is

$$AU_{\hat{m}} = U_{\hat{m}} T_{\hat{m}} D_{\hat{m}}, \qquad A^T W_{\hat{m}} = W_{\hat{m}} D_{\hat{m}} T_{\hat{m}}.$$

$\mathcal{R}(U_{\hat{m}})$ is invariant under A, and $\mathcal{R}(W_{\hat{m}})$ is invariant under A^T. Moreover, the right and left eigenvectors of A associated with $\mathcal{R}(U_{\hat{m}})$ are normally readily accessible in $U_{\hat{m}}$ and $W_{\hat{m}}$, respectively. This is so because T_m is normally block diagonal with a 1×1 block corresponding to each real eigenvalue of $T_{\hat{m}} D_{\hat{m}}$ and a 2×2 block corresponding to each complex-conjugate pair. If the (i, i) position of $T_{\hat{m}} D_{\hat{m}}$ houses a real eigenvalue of A, then the ith column of $U_{\hat{m}}$ is a right eigenvector of A associated with that eigenvalue, and the ith column of $W_{\hat{m}}$ is a left eigenvector. If $T_{\hat{m}} D_{\hat{m}}$ has a 2×2 block in positions i and $i + 1$, then columns i and $i + 1$ of $U_{\hat{m}}$ span the invariant subspace of A associated with that pair. Complex eigenvectors can be extracted easily. Similarly, complex left eigenvectors can be extracted from columns i and $i + 1$ of $W_{\hat{m}}$.

Since we have eigenvectors accessible, we can easily compute a residual for each eigenpair. This constitutes an a posteriori test of backward stability (Proposition 2.7.20). If the residual is not tiny, the results cannot be trusted. Since the Lanczos process is potentially unstable, this test is absolutely essential and should never be omitted.

Since the algorithm delivers both right and left eigenvectors, we can easily compute the condition number of each of the computed eigenvalues (2.7.4). If the residual is tiny and the condition number is not too big, the eigenvalue is guaranteed accurate.

Exercises

9.6.1.

(a) Let $u_1, u_2, \ldots, u_j$ and $w_1, w_2, \ldots, w_j$ be two sequences of vectors in $\mathbb{C}^n$, and let $U_j = \begin{bmatrix} u_1 & \cdots & u_j \end{bmatrix}$ and $W_j = \begin{bmatrix} w_1 & \cdots & w_j \end{bmatrix} \in \mathbb{C}^{n \times j}$. Show that the two sequences are biorthonormal if and only if $W_j^* U_j = I_j$.

(b) Given vectors $u_1, \ldots, u_n \in \mathbb{C}^n$, show that there exists a sequence $w_1, \ldots, w_n$ that is biorthonormal to $u_1, \ldots, u_n$ if and only if the matrix $U = \begin{bmatrix} u_1 & \cdots & u_n \end{bmatrix}$ is nonsingular, that is, the vectors $u_1, \ldots, u_n$ are linearly independent.

(c) State and prove a generalization of the result in part (b) that is valid for a set of j vectors $u_1, \ldots, u_j$, where $j \leq n$.

Since linear independence is a much weaker property than orthonormality, these results show that biorthonormality is much weaker than orthonormality.

9.6.2. This exercise works out some of the basic properties of the unsymmetric Lanczos process.

(a) If $\hat{u}_{j+1}$ is given by (9.6.3), show that $\hat{u}_{j+1}$ is orthogonal to w_k if and only if $h_{kj} = \langle Au_j, w_k \rangle$. Similarly, show that $\hat{w}_{j+1}$ is orthogonal to u_k if and only if $g_{kj} = \langle A^* w_j, u_k \rangle$.

(b) Show that if $\tau_{j+1} \neq 0$ in (9.6.5), and $h_{j+1,j}$ and $g_{j+1,j}$ are chosen so that $h_{j+1,j} \overline{g_{j+1,j}} = \tau_{j+1}$, then the vectors defined by (9.6.6) satisfy $\langle u_{j+1}, w_{j+1} \rangle = 1$, and the sequences $u_1, \ldots, u_{j+1}$ and $w_1, \ldots, w_{j+1}$ are biorthonormal.

(c) Prove by induction on j that (if $\tau_{j+1} \neq 0$ at every step) the unsymmetric Lanczos process (9.6.2)–(9.6.6) produces vectors satisfying $\text{span}\{u_1, \ldots, u_j\} = \mathcal{K}_j(A, u_1)$ and $\text{span}\{w_1, \ldots, w_j\} = \mathcal{K}_j(A^*, w_1)$ for $j = 1, 2, 3, \ldots$.

(d) Show that if $Au_j \in \text{span}\{u_1, \ldots, u_j\}$, then this fact and the relationships $\langle \hat{u}_{j+1}, w_k \rangle = 0, k = 1, \ldots, j$, imply that $\hat{u}_{j+1} = 0$.

(e) Show that if $Aw_j \in \text{span}\{w_1, \ldots, w_j\}$, then $\hat{w}_{j+1} = 0$.

9.6.3.

(a) Prove (9.6.7), (9.6.8), and (9.6.9).

(b) Show that $W_k^* U_k = I_k$, $W_k^* u_{k+1} = 0$, and $U_k^* w_{k+1} = 0$.

(c) Use (9.6.8) and (9.6.9) to show that $G_k = H_k^*$.

9.6.4. Verify that the unsymmetric Lanczos process reduces to the symmetric Lanczos process in the case $A^* = A$ if we start with $w_1 = u_1$ and take $\beta_j = \gamma_j = \|\hat{u}_{j+1}\|$ at each step.

9.6.5. This exercise sketches an economical way to build the transforming matrix $V_1 V_2$. The challenge is this: Given a diagonal matrix T_m, a signature matrix D_m, and $z \in \mathbb{C}^n$, construct V so that $\hat{T}_m = V^{-1} T_m V^{-T}$ is tridiagonal, $\hat{D}_m = V^{-1} D_m V^{-T} = V^T D_m V$ is a signature matrix, and $z^T V = \alpha e_m^T$. We will generate V as a product of essentially 2×2 transformations, each of which is either orthogonal or hyperbolic. It is not claimed that the construction will always work; it will break down whenever a hyperbolic transformation that is needed does not exist. The idea is to create zeros in z one at a time, at the same time keeping the T matrix from filling in.

(a) Show that there is a matrix $\hat{V}$, a product of $n-1$ essentially 2×2 matrices such that $z^T \hat{V} = \alpha e_m^T$ and $\hat{V}^T D_m \hat{V}$ is a signature matrix. The first transformation transforms z_1 to zero. It is hyperbolic if $d_1 = -d_2$ and orthogonal if $d_1 = d_2$.

(b) The matrix $\hat{V}^{-1} T_m \hat{V}^{-T}$ is far from tridiagonal. We seek to intersperse additional factors among the factors of $\hat{V}$ in such a way that the tridiagonal form is preserved. Show that after the first two factors of $\hat{V}$ have been applied to T_m, the resulting matrix is not tridiagonal, as it has a "bulge" at positions $(3, 1)$ and $(1, 3)$. Show that that bulge can be chased away by a single transformation acting in the $(1, 2)$ plane, which does not disturb the zeros that have been introduced into z.

(c) Now suppose we have chased that first "bulge" and now proceed to create one more zero in the z vector by applying the third factor of $\hat{V}$. Show that the corresponding transformation of the T matrix makes that matrix fail to be tridiagonal because of a bulge in positions $(4, 2)$ and $(2, 4)$. Show that that bulge can be removed from the matrix by an upward bulge-chase involving transformations in the $(2, 3)$ and $(1, 2)$ planes. These transformations do not disturb the zeros that have been created in z.

(d) Now suppose that the first j entries of z have been transformed to zero, the top $(j + 1) \times (j + 1)$ submatrix of the transformed T matrix is tridiagonal, and the bottom $(m - j - 1) \times (m - j - 1)$ submatrix of the transformed T is still diagonal. Show that the transformation that zeros out the $(j + 1)$th entry of the z matrix also introduces a "bulge" in positions $(j + 2, j)$ and $(j, j + 2)$. Show that this bulge can be chased from the matrix by an upward bulge-chase (an implicit HZ iteration in reverse) that does not disturb the zeros that have been introduced into z. The result is a transformed T matrix whose top $(j + 2) \times (j + 2)$ submatrix is tridiagonal and whose bottom $(m - j - 2) \times (m - j - 2)$ submatrix is still diagonal.

(e) What does the final step of the process look like?

Notice that this procedure is about the same as the procedure for returning a unitary matrix to Hessenberg form, sketched in Section 9.5. The methodology is not specific to the unitary and tridiagonal cases; it can be applied to implicit restarts by the second method in any context.

9.7   Skew-Hamiltonian Krylov Processes

In this and the next few sections we draw heavily from [222].

Hamiltonian and Related Structures

In [9] and [153] we studied large eigenvalue problems in which the smallest few eigenvalues of a Hamiltonian matrix are needed. Since these are also the largest eigenvalues of H^{-1}, a Krylov subspace method can be applied to H^{-1} to find them, provided that H^{-1} is accessible. Since H^{-1} inherits the Hamiltonian structure of H, we prefer to use a procedure that respects the Hamiltonian structure, in the interest of efficiency, stability, and accuracy. Eigenvalues of real Hamiltonian matrices occur in pairs $\{\lambda, -\lambda\}$ or quadruples $\{\lambda, -\lambda, \overline{\lambda}, -\overline{\lambda}\}$. A structure-preserving method will extract entire pairs and quadruples intact.

In situations where we have some prior information, we might prefer to shift before we invert. Specifically, if we know that the eigenvalues of interest lie near τ, we might prefer to work with $(H - \tau I)^{-1}$. Unfortunately, the shift destroys the structure, so we are lead to think of ways of effecting shifts without destroying the structure. One simple remedy is to work with the matrix

$$(H - \tau I)^{-1}(H + \tau I)^{-1},$$

which is not Hamiltonian but skew-Hamiltonian. It makes perfect sense that the shifts τ and $-\tau$ should be used together in light of the symmetry of the spectrum of H. If τ is neither real nor purely imaginary, we prefer to work with the skew-Hamiltonian matrix

$$(H - \tau I)^{-1}(H - \overline{\tau} I)^{-1}(H + \tau I)^{-1}(H + \overline{\tau} I)^{-1},$$

in order to stay within the real number system. If we want to apply either of these transformations, we would like to use a Krylov process that preserves skew-Hamiltonian structure.

If we wish to use a shift *and* keep the Hamiltonian structure, we can work with the Hamiltonian matrix

$$H^{-1}(H - \tau I)^{-1}(H + \tau I)^{-1}$$

or

$$H(H - \tau I)^{-1}(H + \tau I)^{-1},$$

for example.

Another possibility is to work with the Cayley transform

$$(H - \tau I)^{-1}(H + \tau I),$$

which is symplectic. To attack this one, we need a Krylov process that preserves symplectic structure.

In this and the next two sections we will develop Krylov processes that preserve these three related structures. We will begin with skew-Hamiltonian structure, which is easiest. Once we have developed a skew-Hamiltonian Lanczos process, we will find that we can derive Hamiltonian and symplectic Lanczos processes from it with amazingly little effort.

Recall from Exercise 2.2.17 that a similarity transformation preserves skew-Hamiltonian, Hamiltonian, or symplectic structure if the transforming matrix is symplectic.

As we observed in Section 9.2, Krylov processes effect partial similarity transformations. Therefore, a Krylov process can preserve any of the three structures if it produces vectors that are the columns of a symplectic matrix.

We ask therefore what special properties the columns of a symplectic matrix have. Recall that a matrix $S \in \mathbb{R}^{2n \times 2n}$ is symplectic if $S^T J S = J$, where J is given by

$$J = \begin{bmatrix} 0 & I \\ -I & 0 \end{bmatrix}.$$

Partitioning S as $S = \begin{bmatrix} S_1 & S_2 \end{bmatrix}$, where S_1 and S_2 are both $2n \times n$, we see immediately from the equation $S^T J S = J$ that $S_1^T J S_1 = 0$, $S_2^T J S_2 = 0$, and $S_1^T J S_2 = I$.

Recall that a subspace $\mathcal{S} \subseteq \mathbb{R}^{2n}$ is called *isotropic* if $x^T J y = 0$ for all x, $y \in \mathcal{S}$. A subspace $\mathcal{S} = \mathcal{R}(X)$ is isotropic if and only if $X^T J X = 0$. It follows that if $S \in \mathbb{R}^{2n \times 2n}$ is symplectic, then the space spanned by the first n columns of S is isotropic, and so is the space spanned by the last n columns. These subspaces are maximal, as the dimension of an isotropic subspace of $\mathbb{R}^{2n}$ cannot exceed n.

Proposition 9.7.1. *Let $\mathcal{S}$ be an isotropic subspace of $\mathbb{R}^{2n}$. Then $\dim(\mathcal{S}) \leq n$.*

Proof: Let $\mathcal{S}$ be an isotropic subspace of $\mathbb{R}^{2n}$ of dimension k, and let $q_1, \ldots, q_k$ be an orthonormal basis of $\mathcal{S}$. Then $q_1, \ldots, q_k, J q_1, \ldots, J q_k$ is an orthonormal set in $\mathbb{R}^{2n}$, so $2k \leq 2n$. Thus $k \leq n$. □

Now we come to the special property of skew-Hamiltonian matrices.

Proposition 9.7.2. *Let $B \in \mathbb{R}^{2n \times 2n}$ be skew-Hamiltonian, and let $u \in \mathbb{R}^{2n}$ be an arbitrary nonzero vector. Then the Krylov subspace $\mathcal{K}_j(B, u)$ is isotropic for all j.*

Proof: Since B is skew-Hamiltonian, one easily shows that all of its powers B^i, $i = 0, 1, 2, 3, \ldots$, are also skew-Hamiltonian. This means that $J B^i$ is skew symmetric. To establish isotropy of $\mathcal{K}_j(B, u)$, it suffices to prove that $(B^i u)^T J B^k u = 0$ for all $i \geq 0$ and $k \geq 0$. Since $B^T J = J B$, we have $(B^i u)^T J B^k u = u^T J B^{i+k} u$, which equals zero because $J B^{i+k}$ is skew symmetric. □

Proposition 9.7.2 shows that every Krylov process driven by a skew-Hamiltonian matrix automatically produces vectors that span isotropic subspaces. Thus we should have no trouble finding Krylov processes that preserve skew-Hamiltonian structure. Notice that this property guarantees that a skew-Hamiltonian Krylov process cannot be run to completion. Since an isotropic subspace cannot have dimension greater than n, we conclude that the Krylov space $\mathcal{K}_{n+1}(B, u_1)$ has dimension at most n. This implies that $\mathcal{K}_{n+1}(B, u_1) = \mathcal{K}_n(B, u_1)$ and $\mathcal{K}_n(B, u_1)$ is invariant under B. Thus a skew-Hamiltonian Krylov process must terminate with an invariant subspace in at most n iterations.

Skew-Hamiltonian Arnoldi Process

When we run the Arnoldi process on a skew-Hamiltonian matrix, we find that it does indeed produce isotropic subspaces. However, over the course of the iterations the isotropy gradually deteriorates in just the same way as orthogonality is lost if reorthogonalization is not used. In order to preserve isotropy, we must enforce it explicitly, as shown in the following algorithm.

Skew-Hamiltonian Arnoldi process.

$$
\begin{aligned}
&u_1 \leftarrow x/\|x\| \\
&U_1 \leftarrow [u_1] \\
&\text{for } j = 1, 2, 3, \ldots \\
&\qquad\left[\begin{aligned}
&u_{j+1} \leftarrow Bu_j \quad (B \text{ is skew-Hamiltonian}) \\
&h_{1:j,j} \leftarrow U_j^T u_{j+1} \\
&u_{j+1} \leftarrow u_{j+1} - U_j h_{1:j,j} \quad \text{(orthogonalize)} \\
&\delta_{1:j} \leftarrow U_j^T u_{j+1} \\
&u_{j+1} \leftarrow u_{j+1} - U_j \delta_{1:j} \quad \text{(reorthogonalize)} \\
&h_{1:j,j} \leftarrow h_{1:j,j} + \delta_{1:j} \\
&\gamma_{1:j} \leftarrow (JU_j)^T u_{j+1} \\
&u_{j+1} \leftarrow u_{j+1} - JU_j \gamma_{1:j} \quad \text{(enforce isotropy)} \\
&h_{j+1,j} \leftarrow \|u_{j+1}\| \\
&\text{if } h_{j+1,j} = 0 \\
&\qquad\left[\begin{aligned}
&\text{set flag } (\text{span}\{u_1,\ldots,u_j\} \text{ is invariant}) \\
&\text{exit}
\end{aligned}\right. \\
&u_{j+1} \leftarrow u_{j+1}/h_{j+1,j} \\
&U_{j+1} \leftarrow \begin{bmatrix} U_j & u_{j+1} \end{bmatrix}
\end{aligned}\right.
\end{aligned}
\tag{9.7.1}
$$

This is just like the ordinary Arnoldi process, except that the two lines

$$
\gamma_{1:j} \leftarrow (JU_j)^T u_{j+1}
$$
$$
u_{j+1} \leftarrow u_{j+1} - JU_j \gamma_{1:j}
$$

have been added. These lines ensure that u_{j+1} is orthogonal to $Ju_1, \ldots, Ju_j$, which is what isotropy means. In principle the numbers in $\gamma_{1:j}$ should all be zero, but in practice they are tiny numbers on the order of 10^{-15}. The tiny correction at each step ensures that the subspace stays isotropic to working precision. When this algorithm is run with restarts, it becomes the *skew-Hamiltonian implicitly restarted Arnoldi* (SHIRA) method [153]. The enforcement of isotropy is crucial to the efficiency of the algorithm. If it is not done, the eigenvalues will show up in duplicate. For example, if eight eigenvalues are wanted, four distinct eigenvalues will be produced, each in duplicate. (Recall that all eigenvalues of a skew-Hamiltonian matrix have multiplicity two or greater.) If, on the other hand, isotropy is enforced, eight distinct eigenvalues are produced.

Skew-Hamiltonian Lanczos Process

If the unsymmetric Lanczos process is applied to a skew-Hamiltonian matrix B, then the spaces $\mathcal{K}_j(B, u_1) = \text{span}\{u_1, \ldots, u_j\}$ are all isotropic, and so are the spaces $\mathcal{K}_j(B^T, w_1) = \text{span}\{w_1, \ldots, w_j\}$, because B^T is also skew-Hamiltonian.

Suppose we are able to run the algorithm for n steps, terminating in invariant subspaces $\mathcal{K}_n(B, u_1) = \text{span}\{u_1, \ldots, u_n\} = \mathcal{R}(U_n)$ and $\mathcal{K}_n(B^T, w_1) = \text{span}\{w_1, \ldots, w_n\} = \mathcal{R}(W_n)$.

Then $BU_n = U_n T_n D_n$ and $B^T W_n = W_n D_n T_n$, as the remainder terms are zero. Since we can go no further than this, let us make the following abbreviations: $U = U_n \in \mathbb{R}^{2n \times n}$, $W = W_n \in \mathbb{R}^{2n \times n}$, $T = T_n \in \mathbb{R}^{n \times n}$, and $D = D_n \in \mathbb{R}^{n \times n}$. Now we have

$$BU = U(TD), \tag{9.7.2}$$

$$B^T W = W(DT), \tag{9.7.3}$$

$$U^T W = I, \tag{9.7.4}$$

and by isotropy,

$$U^T J U = 0 \quad \text{and} \quad W^T J W = 0. \tag{9.7.5}$$

These equations represent only partial similarities. We can build full similarity transformations by exploiting the structure. Since $U^T W = I$, if we define a new matrix V by

$$V = -JW, \tag{9.7.6}$$

then we immediately have $U^T J V = I$. Since also $V^T J V = 0$ (isotropy condition inherited from W), the matrix $\begin{bmatrix} U & V \end{bmatrix}$ is symplectic.

Since B^T is skew-Hamiltonian, we have $JB^T = BJ$. Multiplying (9.7.3) by $-J$ on the left, we obtain

$$BV = V(DT). \tag{9.7.7}$$

Now we can combine (9.7.2) and (9.7.7) to obtain the full similarity transformation

$$B\begin{bmatrix} U & V \end{bmatrix} = \begin{bmatrix} U & V \end{bmatrix} \begin{bmatrix} TD & \\ & DT \end{bmatrix}. \tag{9.7.8}$$

Equation (9.7.8) holds only if we are able to reach the nth step with no breakdowns. Even if we do not get that far, we still obtain partial results. After k steps the Lanczos configuration is

$$BU_k = U_k T_k D_k + u_{k+1} b_k d_k e_k^T, \tag{9.7.9}$$

$$B^T W_k = W_k D_k T_k + w_{k+1} d_{k+1} b_k e_k^T. \tag{9.7.10}$$

If we multiply the second of these equations by $-J$ on the left, we obtain

$$BV_k = V_k D_k T_k + v_{k+1} d_{k+1} b_k e_k^T. \tag{9.7.11}$$

Equations (9.7.9) and (9.7.11) form the basis for our skew-Hamiltonian Lanczos process. The associated recurrences are

$$u_{j+1} b_j d_j = B u_j - u_j a_j d_j - u_{j-1} b_{j-1} d_j, \tag{9.7.12}$$

$$v_{j+1} d_{j+1} d_j = B v_j - v_j d_j a_j - v_{j-1} d_{j-1} b_{j-1}. \tag{9.7.13}$$

The coefficients are computed just as in the ordinary nonsymmetric Lanczos process (9.6.14), with w_j replaced by $J v_j$. Since the unsymmetric Lanczos process starts with vectors satisfying $u_1^T w_1 = 1$, the recurrences (9.7.12) and (9.7.13) must be initialized with vectors satisfying $u_1^T J v_1 = 1$. Then the vectors produced will automatically satisfy the properties of the columns of a symplectic matrix. One should always bear in mind, however, that roundoff errors will degrade the "symplecticness" of the vectors in practice, unless it is explicitly enforced at each step.

Exercises

9.7.1. Let H be a Hamiltonian matrix. In each of the following tasks, assume that the indicated inverse matrices exist.

(a) Prove that H^{-1} is Hamiltonian.

(b) Prove that $(H - \tau I)^{-1}(H + \tau I)^{-1}$ is skew-Hamiltonian.

(c) Prove that $H(H - \tau I)^{-1}(H + \tau I)^{-1}$ is Hamiltonian.

(d) Prove that $H^{-1}(H - \tau I)^{-1}(H + \tau I)^{-1}$ is Hamiltonian.

(e) Prove that the Cayley transform $(H - \tau I)^{-1}(H + \tau I)$ is symplectic.

9.7.2. Suppose that the skew-Hamiltonian Arnoldi process is run for n steps, culminating in an invariant subspace $\mathcal{K}_n(B, u_1) = \text{span}\{u_1, \ldots, u_n\}$. The final Arnoldi configuration is

$$BU = UH,$$

where $U = \begin{bmatrix} u_1 & \cdots & u_n \end{bmatrix}$ and $H \in \mathbb{R}^{n \times n}$ is upper Hessenberg.

(a) Let $S = \begin{bmatrix} U & -JU \end{bmatrix} \in \mathbb{R}^{2n \times 2n}$. Show that S is both orthogonal and symplectic.

(b) Let $C = S^{-1}BS$. Show that C has the form

$$C = \begin{bmatrix} H & N \\ 0 & H^T \end{bmatrix},$$

where $N^T = -N$.

Thus an orthogonal symplectic similarity transformation has been effected on B. If the Arnoldi process is carried only part way to completion, then a partial orthogonal, symplectic similarity transformation is carried out.

9.7.3. With notation as in (9.7.2) through (9.7.5), define $X = JU$.

(a) Show that $\begin{bmatrix} W & X \end{bmatrix}$ is symplectic.

(b) Starting from the equation $BU = U(TD)$, show that $B^T X = X(TD)$. Consequently

$$B^T \begin{bmatrix} W & X \end{bmatrix} = \begin{bmatrix} W & X \end{bmatrix} \begin{bmatrix} DT & \\ & TD \end{bmatrix}.$$

This is another symplectic similarity transformation, which is easily seen to be equivalent to (9.7.8)

(c) Show that $\begin{bmatrix} W & X \end{bmatrix} = \begin{bmatrix} U & V \end{bmatrix}^{-T}$.

9.7.4.

(a) Write a pseudocode algorithm like (9.6.14) for the skew-Hamiltonian Lanczos process (9.7.12), (9.7.13).

(b) Write down a version of the algorithm in which the "symplecticness" conditions $U_j^T J U_j = 0$, $V_j^T J V_j = 0$, and $U_j^T J V_j = I$ are explicitly enforced.

9.8 The Hamiltonian Lanczos Process

The Hamiltonian Lanczos process with restarts was first published in [28]. The version presented here is from [222].

Let $H \in \mathbb{R}^{2n \times 2n}$ be Hamiltonian. Then H^2 is skew-Hamiltonian, and we will take the skew-Hamiltonian Lanczos process for H^2 as our point of departure. From (9.7.12) and (9.7.13) we have

$$u_{j+1}b_j d_j = H^2 u_j - u_j a_j d_j - u_{j-1}b_{j-1}d_j, \tag{9.8.1}$$

$$v_{j+1}d_{j+1}b_j = H^2 v_j - v_j d_j a_j - v_{j-1}d_{j-1}b_{j-1}, \tag{9.8.2}$$

where the recurrences are started with a pair of vectors satisfying $u_1^T J v_1 = 1$. Define new vectors $\tilde{v}_j = Hu_j d_j$, $j = 1, 2, 3, \ldots$. Since $d_j = \pm 1$, we also have

$$\tilde{v}_j d_j = Hu_j. \tag{9.8.3}$$

Multiply (9.8.1) on the left by H and by d_j to obtain

$$\tilde{v}_{j+1}d_{j+1}b_j = H^2 \tilde{v}_j - \tilde{v}_j d_j a_j - \tilde{v}_{j-1}d_{j-1}b_{j-1}, \tag{9.8.4}$$

a recurrence that appears identical to (9.8.2). Now suppose we start (9.8.1) and (9.8.2) with vectors that satisfy both $u_1^T J v_1 = 1$ and $Hu_1 = v_1 d_1$. Then (9.8.2) and (9.8.4) have the same starting vector, and they are identical, so $\tilde{v}_j = v_j$ for every j. This means that we do not have to run both of the recurrences (9.8.1) and (9.8.2). We can just run (9.8.1) along with the supplementary condition (9.8.3). Noting also that the term $H^2 u_j$ can be replaced by $Hv_j d_j$, we obtain the *Hamiltonian Lanczos process*

$$u_{j+1}b_j = Hv_j - u_j a_j - u_{j-1}b_{j-1}, \tag{9.8.5}$$

$$v_{j+1}d_{j+1} = Hu_{j+1}. \tag{9.8.6}$$

Now we just need to give some thought to how the coefficients are computed. Multiplying (9.8.5) on the left by $v_j^T J$, we find that

$$a_j = -v_j^T J H v_j.$$

The other coefficients are produced in the normalization process. In practice, (9.8.5) and (9.8.6) take the form

$$\hat{u}_{j+1} = Hv_j - u_j a_j - u_{j-1}b_{j-1}, \tag{9.8.7}$$

$$\hat{v}_{j+1} = H\hat{u}_{j+1}. \tag{9.8.8}$$

Then $\hat{u}_{j+1} = u_{j+1}b_j$ and $\hat{v}_{j+1} = Hu_{j+1}b_j = v_{j+1}d_{j+1}b_j$. We can now determine b_j and d_{j+1} uniquely from the condition $u_{j+1}^T J v_{j+1} = 1$. Let $\pi = \hat{u}_{j+1}^T J \hat{v}_{j+1}$. Then we must have $\pi = d_{j+1}b_j^2$. If $\pi = 0$, we have a breakdown. If $\pi > 0$, we must have $d_{j+1} = 1$ and $b_j = \sqrt{\pi}$, and if $\pi < 0$, we must have $d_{j+1} = -1$ and $b_j = \sqrt{-\pi}$. We then have $u_{j+1} = \hat{u}_{j+1}/b_j$ and $v_{j+1} = \hat{v}_{j+1}/(d_{j+1}b_j)$. Now we have all we need for the next step.

Before we can write down the algorithm, we need to say a few words about picking starting vectors. u_1 and v_1 must satisfy both $v_1 d_1 = Hu_1$ and $u_1^T J v_1 = 1$. We can obtain vectors satisfying these conditions by picking a more or less arbitrary starting vector $\hat{u}_1$ and using the normalization process described in the previous paragraph. Thus we define $\hat{v}_1 = H\hat{u}_1$, $\pi = \hat{u}_1^T J \hat{v}_1$, and so on.

Now we are ready to write down the algorithm.

Hamiltonian Lanczos Process.

Start with initial vector u_1.

$$b_0 \leftarrow 0, \ u_0 \leftarrow 0$$
$$\text{for } j = 0, 1, 2, \ldots,$$

$$
\begin{array}{l}
\left\lceil \ \text{if } j > 0 \text{ then} \right. \\
\quad \left\lceil \ y \leftarrow Hv_j \right. \\
\quad \quad a_j \leftarrow -v_j^T Jy \\
\quad \left\lfloor \ u_{j+1} \leftarrow y - u_j a_j - u_{j-1} b_{j-1} \right. \\
\quad v_{j+1} \leftarrow Hu_{j+1} \\
\quad \pi \leftarrow u_{j+1}^T Jv_{j+1} \\
\quad \text{if } \pi = 0 \text{ then} \\
\quad \quad \left\lceil \ \text{stop (breakdown)} \right. \\
\quad \text{if } \pi > 0 \text{ then} \\
\quad \quad \left\lceil \ s \leftarrow \sqrt{\pi}, \ d_{j+1} \leftarrow 1 \right. \\
\quad \text{if } \pi < 0 \text{ then} \\
\quad \quad \left\lceil \ s \leftarrow \sqrt{-\pi}, \ d_{j+1} \leftarrow -1 \right. \\
\quad u_{j+1} \leftarrow u_{j+1}/s \\
\quad v_{j+1} \leftarrow v_{j+1}/(d_{j+1}s) \\
\quad \text{if } j > 0 \text{ then} \\
\quad \left\lfloor \ \left\lceil \ b_j \leftarrow s \right. \right.
\end{array}
\tag{9.8.9}
$$

This is the theoretical algorithm. In exact arithmetic it guarantees "symplecticness" of the vectors produced. In the real world of roundoff errors, the symplectic structure will gradually be lost unless it is explicitly enforced. This can be achieved by adding the lines

$$
\begin{aligned}
z &\leftarrow Ju_{j+1} \\
u_{j+1} &\leftarrow u_{j+1} - V_j(U_j^T z) + U_j(V_j^T z) \\
a_j &\leftarrow a_j + v_j^T z
\end{aligned}
\tag{9.8.10}
$$

immediately after the line $u_{j+1} \leftarrow y - u_j a_j - u_{j-1} b_{j-1}$. Here U_j and V_j have their usual meanings. This step ensures that new vector $\hat{u}_{j+1}$ satisfies $U_j^T J\hat{u}_{j+1} = 0$ and $V_j^T J\hat{u}_{j+1} = 0$ to working precision, so that we have $U_k^T JU_k = 0$ and $U_k^T JV_k = I_k$ at every step. Our experience has been that because of the very tight connection between the "u" and "v" vectors in this algorithm, the condition $V_k^T JV_k = 0$ takes care of itself.

Implicit Restarts

Rearranging (9.8.5) and (9.8.6), we obtain

$$
\begin{aligned}
Hv_j &= u_{j-1}b_{j-1} + u_j a_j + u_{j+1} b_j, \\
Hu_{j+1} &= v_{j+1}d_{j+1},
\end{aligned}
$$

which when packed into matrix equations gives

$$HV_k = U_k T_k + u_{k+1} b_k e_k^T, \qquad HU_k = V_k D_k \qquad (9.8.11)$$

after $k - 1/2$ steps. Here D_k is the signature matrix $D_k = \text{diag}\{d_1, \ldots, d_k\}$, and T_k is the symmetric tridiagonal matrix

$$T_k = \begin{bmatrix} a_1 & b_1 & & \\ b_1 & a_2 & \ddots & \\ & \ddots & \ddots & b_{k-1} \\ & & b_{k-1} & a_k \end{bmatrix},$$

as always. Equations (9.8.11) can also be written as the single equation

$$H \begin{bmatrix} U_k & V_k \end{bmatrix} = \begin{bmatrix} U_k & V_k \end{bmatrix} \begin{bmatrix} & T_k \\ D_k & \end{bmatrix} + u_{k+1} b_k e_{2k}^T.$$

The Hamiltonian matrix $\begin{bmatrix} & T_k \\ D_k & \end{bmatrix}$, which is familiar from Section 8.9, tells us how restarts should be done. Either of the restart methods that we have presented can be used. The filtering should be done by the product HR algorithm ($= HZ$ algorithm), as explained in Section 8.9.

Extracting Eigenvectors

After a number of restarts, the process will normally converge to an invariant subspace of some dimension m. The result has the form

$$H \begin{bmatrix} U_m & V_m \end{bmatrix} = \begin{bmatrix} U_m & V_m \end{bmatrix} \begin{bmatrix} & T_m \\ D_m & \end{bmatrix}, \qquad (9.8.12)$$

where D_m is a signature matrix and T_m is block diagonal. It is thus a simple matter to determine the eigenpairs of $\begin{bmatrix} & T_m \\ D_m & \end{bmatrix}$. Each eigenvector $\begin{bmatrix} x \\ y \end{bmatrix}$ of $\begin{bmatrix} & T_m \\ D_m & \end{bmatrix}$, corresponding to eigenvalue λ, yields an eigenvector $U_m x + V_m y$ of H corresponding to eigenvalue λ.

It is also a simple matter to get left eigenvectors. Let $W_m = J V_m$ and $X_m = -J U_m$ (consistent with notation used previously). Multiplying (9.8.12) by J on the left and using the fact that H is Hamiltonian, we immediately find that

$$H^T \begin{bmatrix} X_m & W_m \end{bmatrix} = \begin{bmatrix} X_m & W_m \end{bmatrix} \begin{bmatrix} 0 & T_m \\ D_m & 0 \end{bmatrix}.$$

Thus left eigenvectors can be obtained by the same method that we use to compute right eigenvectors.

Playing It Safe

Since we are able to get eigenvectors, we should always compute residuals as an a posteriori test of backward stability (Proposition 2.7.20). Because the Hamiltonian Lanczos process is potentially unstable, this test should never be skipped.

Since we have both left and right eigenvectors, we can also compute condition numbers of the computed eigenvalues (2.7.4). If the residual is tiny and the condition number is not too big, the eigenvalue is guaranteed to be accurate.

Exercises

9.8.1. The file `hamlan.m` encodes the Hamiltonian Lanczos process. Download this file and the driver program `hamlango.m` from the website and play with them. Try adjusting the matrix size and the number of Lanczos steps and see how this affects the results.

9.8.2. Download the files `rehamlan.m` and `rehamlango.m` and their auxiliary codes from the website. Figure out what they do and experiment with them (in Hamiltonian mode). For example, try adjusting `nmin`, `nmax`, and `nwanted`, and see how this affects the performance of the codes.

9.8.3. Suppose that $H \in \mathbb{R}^{2n \times 2n}$ is both Hamiltonian and skew symmetric.

(a) Show that H has the form
$$H = \begin{bmatrix} A & K \\ -K & A \end{bmatrix},$$
where $A^T = -A$ and $K^T = K$.

(b) Suppose that the skew symmetric Lanczos process (Exercise 9.4.11) is applied to H. Identify a pair of orthogonal isotropic Krylov subspaces that are generated by that algorithm.

(c) Compare and contrast the skew symmetric Lanczos process with the Hamiltonian Lanczos process derived in this section.

9.9   The Symplectic Lanczos Process

The symplectic Lanczos process was discussed in [17, 29, 30]. The version presented here is from [222].

If $S \in \mathbb{R}^{2n \times 2n}$ is symplectic, then $S + S^{-1}$ is skew-Hamiltonian (Proposition 1.4.4). Applying the skew-Hamiltonian Lanczos process (9.7.12) and (9.7.13) to $S + S^{-1}$, we have

$$u_{j+1}b_j d_j = (S + S^{-1})u_j - u_j a_j d_j - u_{j-1}b_{j-1}d_j, \tag{9.9.1}$$

$$v_{j+1}d_{j+1}b_j = (S + S^{-1})v_j - v_j d_j a_j - v_{j-1}d_{j-1}b_{j-1}, \tag{9.9.2}$$

where the recurrences are started with vectors satisfying $u_1^T J v_1 = 1$. Define new vectors $\tilde{v}_j = S^{-1}u_j d_j$, $j = 1, 2, 3, \ldots$. Since $d_j = \pm 1$, we also have

$$\tilde{v}_j d_j = S^{-1}u_j. \tag{9.9.3}$$

Multiply (9.9.1) by S^{-1} and by d_j to obtain

$$\tilde{v}_{j+1}d_{j+1}d_j = (S + S^{-1})\tilde{v}_j - \tilde{v}_j d_j a_j - \tilde{v}_{j-1}d_{j-1}b_{j-1}, \qquad (9.9.4)$$

which is identical to (9.9.2). Now suppose we start (9.9.1) and (9.9.2) with vectors that satisfy both $u_1^T J v_1 = 1$ and $S^{-1}u_1 = v_1 d_1$. Then (9.9.2) and (9.9.4) have the same starting vector, and they are identical, so that $\tilde{v}_j = v_j$ for every j. This means that we do not have to run both of the recurrences (9.9.1) and (9.9.2). We can just run (9.9.1) along with the supplementary condition (9.9.3). Noting also that the term $S^{-1}u_j$ can be replaced by $v_j d_j$, we obtain

$$u_{j+1}b_j = Su_j d_j + v_j - u_j a_j - u_{j-1}b_{j-1}, \qquad (9.9.5)$$
$$v_{j+1}d_{j+1} = S^{-1}u_{j+1}. \qquad (9.9.6)$$

We call this pair of recurrences the *symplectic Lanczos process*. The occurrence of S^{-1} is not a problem, as $S^{-1} = -JS^T J$.[13]

Now let us consider how to compute the coefficients. Multiplying (9.9.5) on the left by $v_j^T J$, we find that

$$a_j = -d_j v_j^T JSu_j.$$

The other coefficients are produced in the normalization process. In practice, (9.9.5) and (9.9.6) take the form

$$\hat{u}_{j+1} = Su_j d_j + v_j - u_j a_j - u_{j-1}b_{j-1}, \qquad (9.9.7)$$
$$\hat{v}_{j+1} = S^{-1}\hat{u}_{j+1}. \qquad (9.9.8)$$

Then $\hat{u}_{j+1} = u_{j+1}b_j$ and $\hat{v}_{j+1} = S^{-1}u_{j+1}b_j = v_{j+1}d_{j+1}b_j$. This is exactly the same situation as we had in the Hamiltonian case, and we can determine b_j and d_{j+1} by exactly the same method: Let $\pi = \hat{u}_{j+1}^T J \hat{v}_{j+1}$. Then we must have $\pi = d_{j+1}b_j^2$. If $\pi = 0$, we have a breakdown. If $\pi > 0$, we must have $d_{j+1} = 1$ and $b_j = \sqrt{\pi}$, and if $\pi < 0$, we must have $d_{j+1} = -1$ and $b_j = \sqrt{-\pi}$. We then have $u_{j+1} = \hat{u}_{j+1}/b_j$ and $v_{j+1} = \hat{v}_{j+1}/(d_{j+1}b_j)$.

The method for picking starting vectors is also the same as for the Hamiltonian case: We pick a more or less arbitrary starting vector $\hat{u}_1$ and use the normalization process described in the previous paragraph. Thus we define $\hat{v}_1 = S^{-1}\hat{u}_1$, $\pi = \hat{u}_1^T J \hat{v}_1$, and so on. We end up with starting vectors u_1 and v_1 that satisfy both $v_1 d_1 = S^{-1}u_1$ and $u_1^T J v_1 = 1$.

Putting these ideas together, we have the following algorithm.

---

[13]An equivalent set of recurrences is obtained by retaining (9.9.2), (9.9.3) and reorganizing them to get $v_{j+1}d_{j+1}b_j = Sv_j + u_j d_j - v_j d_j a_j - v_{j-1}d_{j-1}b_{j-1}$ and $u_{j+1}d_{j+1} = S^{-1}v_{j+1}$. These are the recurrences that were derived in [222].

Symplectic Lanczos process.
Start with initial vector u_1.

$$
\begin{aligned}
&b_0 \leftarrow 0, \ d_0 \leftarrow 0, \ v_0 \leftarrow 0 \\
&\text{for } j = 0, 1, 2, \ldots \\
&\left\lceil \ \text{if } j > 0 \text{ then} \right. \\
&\qquad \left\lceil \ y \leftarrow S u_j \right. \\
&\qquad \quad a_j \leftarrow -d_j(v_j^T J y) \\
&\qquad \left\lfloor \ u_{j+1} \leftarrow y d_j + v_j - u_j a_j - u_{j-1} b_{j-1} \right. \\
&\quad \ v_{j+1} \leftarrow S^{-1} u_{j+1} \\
&\quad \ \pi \leftarrow u_{j+1}^T J v_{j+1} \\
&\quad \ \text{if } \pi = 0 \text{ then} \\
&\qquad \left[\ \text{stop (breakdown)} \right. \\
&\quad \ \text{if } \pi > 0 \text{ then} \\
&\qquad \left[\ s \leftarrow \sqrt{\pi}, \ d_{j+1} \leftarrow 1 \right. \\
&\quad \ \text{if } \pi < 0 \text{ then} \\
&\qquad \left[\ s \leftarrow \sqrt{-\pi}, \ d_{j+1} \leftarrow -1 \right. \\
&\quad \ u_{j+1} \leftarrow u_{j+1}/s \\
&\quad \ v_{j+1} \leftarrow v_{j+1}/(d_{j+1} s) \\
&\quad \ \text{if } j > 0 \text{ then} \\
&\qquad \left[\ b_j \leftarrow s \right.
\end{aligned}
\tag{9.9.9}
$$

Just as for the Hamiltonian algorithm, the "symplecticness" of the vectors produced will decline over time unless we explicitly enforce it. To do this, insert exactly the same lines (9.8.10) as in the Hamiltonian algorithm.

Implicit Restarts

Rearranging (9.9.5) and (9.9.6), we obtain

$$
S u_j = u_{j-1} b_{j-1} d_j + u_j a_j d_j + u_{j+1} b_j d_j - v_j d_j,
$$

$$
S v_j = u_j d_j,
$$

which when packed into matrix equations gives

$$
S U_k = U_k(T_k D_k) + u_{k+1} b_{k+1} d_{k+1} e_k^T - V_k D_k, \qquad S V_k = U_k D_k \tag{9.9.10}
$$

after $k - 1/2$ steps. Here D_k is the signature matrix $D_k = \text{diag}\{d_1, \ldots, d_k\}$, and T_k is the same symmetric tridiagonal matrix as always. Equations (9.9.10) can also be written as the single equation

$$
S \begin{bmatrix} U_k & V_k \end{bmatrix} = \begin{bmatrix} U_k & V_k \end{bmatrix} \begin{bmatrix} T_k D_k & D_k \\ -D_k & 0 \end{bmatrix} + u_{k+1} b_{k+1} d_{k+1} e_k^T.
$$

The symplectic matrix $\begin{bmatrix} T_k D_k & D_k \\ -D_k & 0 \end{bmatrix}$, which is familiar from Section 8.9, tells us how restarts should be done. Either of the restart methods that we have presented can be used. The filtering should be done by the product HR algorithm ($= HZ$ algorithm), as explained in Section 8.9.

Extracting Eigenvectors

Once the process has converged to an invariant subspace of dimension m, we have

$$S \begin{bmatrix} U_m & V_m \end{bmatrix} = \begin{bmatrix} U_m & V_m \end{bmatrix} \begin{bmatrix} T_m D_m & D_m \\ -D_m & 0 \end{bmatrix}, \qquad (9.9.11)$$

where D_m is a signature matrix and T_m is block diagonal. The eigenvectors of $\begin{bmatrix} T_m D_m & D_m \\ -D_m & 0 \end{bmatrix}$ are easy to compute, and each such eigenvector $\begin{bmatrix} x \\ y \end{bmatrix}$ yields an eigenvector $U_m x + V_m y$ of S.

Left eigenvectors are also easily obtained. Recalling that

$$\begin{bmatrix} T_m D_m & D_m \\ -D_m & 0 \end{bmatrix}^{-1} = \begin{bmatrix} 0 & -D_m \\ D_m & D_m T_m \end{bmatrix},$$

multiply (9.9.11) on the right by this matrix and on the left by S^{-1} to obtain

$$S^{-1} \begin{bmatrix} U_m & V_m \end{bmatrix} = \begin{bmatrix} U_m & V_m \end{bmatrix} \begin{bmatrix} 0 & -D_m \\ D_m & D_m T_m \end{bmatrix}. \qquad (9.9.12)$$

Let $W_m = J V_m$ and $X_m = -J U_m$, as before. Then, multiplying (9.9.12) by J on the left and using the identity $J S^{-1} = S^T J$, we obtain

$$S^T \begin{bmatrix} -X_m & W_m \end{bmatrix} = \begin{bmatrix} -X_m & W_m \end{bmatrix} \begin{bmatrix} 0 & -D_m \\ D_m & D_m T_m \end{bmatrix}. \qquad (9.9.13)$$

From this we can get the right eigenvectors of S^T and hence the left eigenvectors of S.

The caveat about a posteriori testing that we gave for the Hamiltonian case is in equal force here. We should always compute residuals as a test of backward stability. Moreover, the presence of both right and left eigenvectors allows us to check the condition numbers of all the computed eigenvalues.

Exercises

9.9.1. The file `symplan.m` encodes the symplectic Lanczos process. Download this file and the driver program `symplango.m` from the website and play with them. Try adjusting the matrix size and the number of Lanczos steps and see how this affects the results. *Note:* The symplectic matrix that is used in this example was obtained by taking a Cayley transform of Hamiltonian matrix:
$$S = (H - \tau I)^{-1}(H + \tau I).$$
You can vary the value of τ and see how this affects the results. The eigenvalues and approximations that are shown in the plots are those of the Hamiltonian matrix H, not of the symplectic matrix S.

9.9.2. Download the files `rehamlan.m` and `rehamlango.m` and their auxiliary codes from the website. Figure out how to make them run in symplectic mode, then experiment with them. For example, try adjusting `nmin`, `nmax`, and `nwanted`, and see how this affects the performance of the codes. *Note:* Read the note to the previous exercise.

9.10 Product Krylov Processes

In this section we return to the theme of Chapter 8. Again we draw from Kressner [134].
Suppose we want to compute some eigenvalues of a large matrix presented as a product of
k matrices: $A = A_k A_{k-1} \cdots A_1$. The developments in this section hold for arbitrary k, but
we will consider the case $k = 3$ for illustration. This is done solely to keep the notation
under control. Thus let $A = A_3 A_2 A_1$. As in Chapter 8 we find it convenient to work with
the cyclic matrix

$$C = \begin{bmatrix} & & A_3 \\ A_1 & & \\ & A_2 & \end{bmatrix}. \tag{9.10.1}$$

Consider what happens when we apply the Arnoldi process to C with a special starting
vector of the form

$$\begin{bmatrix} u_1 \\ 0 \\ 0 \end{bmatrix}.$$

Multiplying this vector by C, we obtain

$$\begin{bmatrix} 0 \\ A_1 u_1 \\ 0 \end{bmatrix},$$

which is already orthogonal to the first vector. Thus only a normalization needs to be done:

$$\begin{bmatrix} 0 \\ v_1 \\ 0 \end{bmatrix} s_{11} = \begin{bmatrix} & & A_3 \\ A_1 & & \\ & A_2 & \end{bmatrix} \begin{bmatrix} u_1 \\ 0 \\ 0 \end{bmatrix},$$

where $s_{11} = \| A_1 u_1 \|_2$. On the second step we again get a vector that is already orthogonal
to the first two vectors, so only a normalization is necessary:

$$\begin{bmatrix} 0 \\ 0 \\ w_1 \end{bmatrix} t_{11} = \begin{bmatrix} & & A_3 \\ A_1 & & \\ & A_2 & \end{bmatrix} \begin{bmatrix} 0 \\ v_1 \\ 0 \end{bmatrix},$$

where $t_{11} = \| A_2 v_1 \|_2$. Finally, on the third step, we get a vector that has to be orthogonalized
against one of the previous vectors:

$$\begin{bmatrix} u_2 \\ 0 \\ 0 \end{bmatrix} h_{21} = \begin{bmatrix} & & A_3 \\ A_1 & & \\ & A_2 & \end{bmatrix} \begin{bmatrix} 0 \\ 0 \\ w_1 \end{bmatrix} - \begin{bmatrix} u_1 \\ 0 \\ 0 \end{bmatrix} h_{11}.$$

This step can be expressed more concisely as

$$u_2 h_{21} = A_3 w_1 - u_1 h_{11}.$$

The computations would be done just as in any Arnoldi step: $\hat{u}_2 \leftarrow A_3 w_1$, $h_{11} = u_1^* \hat{u}_2$,
$\hat{u}_2 \leftarrow \hat{u}_2 - u_1 h_{11}$, $h_{21} = \| \hat{u}_2 \|_2$, and $u_2 = \hat{u}_2 / h_{21}$. (In practice the orthogonalization would
be done twice.)

The pattern is clear. In each cycle of three Arnoldi steps a new "u," "v," and "w" vector will be produced, each of which has to be orthogonalized only against the previous "u," "v," or "w" vectors, respectively. For example, the w vector that is produced on the jth cycle is given by

$$
\begin{bmatrix} 0 \\ 0 \\ w_j \end{bmatrix} t_{jj} = \begin{bmatrix} & & A_3 \\ A_1 & & \\ & A_2 & \end{bmatrix} \begin{bmatrix} 0 \\ v_j \\ 0 \end{bmatrix} - \sum_{i=1}^{j-1} \begin{bmatrix} 0 \\ 0 \\ w_i \end{bmatrix} t_{ij}
$$

or, more concisely,

$$
w_j t_{jj} = A_2 v_j - \sum_{i=1}^{j-1} w_i t_{ij}.
$$

Similar formulas hold for v_j and u_{j+1}. Placed together, in order of execution, they are

$$
v_j s_{jj} = A_1 u_j - \sum_{i=1}^{j-1} v_i s_{ij},
$$

$$
w_j t_{jj} = A_2 v_j - \sum_{i=1}^{j-1} w_i t_{ij},
$$

$$
u_{j+1} h_{j+1,j} = A_3 w_j - \sum_{i=1}^{j} u_i h_{ij}.
$$

Rewrite these equations as

$$
A_1 u_j = \sum_{i=1}^{j} v_i s_{ij}, \qquad A_2 v_j = \sum_{i=1}^{j} w_i t_{ij}, \qquad A_3 w_j = \sum_{i=1}^{j+1} u_i h_{ij},
$$

and then assemble them into matrix equations. Suppose we have gone through k complete cycles. Then we have

$$
A_1 U_k = V_k S_k,
$$

$$
A_2 V_k = W_k T_k, \tag{9.10.2}
$$

$$
A_3 W_k = U_k H_k + u_{k+1} h_{k+1,k} e_k^T,
$$

where $U_k = \begin{bmatrix} u_1 & \cdots & u_k \end{bmatrix}$, for example, and the notation is mostly self-explanatory. Notice that S_k and T_k are upper triangular matrices, but H_k is upper Hessenberg. Equations (9.10.2) can be combined into the single equation

$$
\begin{bmatrix} & & A_3 \\ A_1 & & \\ & A_2 & \end{bmatrix} \begin{bmatrix} U_k & & \\ & V_k & \\ & & W_k \end{bmatrix}
$$

$$
= \begin{bmatrix} U_k & & \\ & V_k & \\ & & W_k \end{bmatrix} \begin{bmatrix} & & H_k \\ S_k & & \\ & T_k & \end{bmatrix} + \begin{bmatrix} u_{k+1} \\ \\ \end{bmatrix} h_{k+1,k} e_{3k}^T. \tag{9.10.3}
$$

If we wanted to rewrite the matrix of vectors

$$\begin{bmatrix} U_k & & \\ & V_k & \\ & & W_k \end{bmatrix}$$

so that the order of the vectors in the matrix was the same as the order in which they were computed, we would have to do a perfect shuffle of the columns. Then, to make (9.10.3) hold for the shuffled matrix, we would have to shuffle the rows and columns of the matrix of coefficients

$$\begin{bmatrix} & & H_k \\ S_k & & \\ & T_k & \end{bmatrix}.$$

This results in an upper Hessenberg matrix whose structure is illustrated by (8.3.4).

Combining the equations in (9.10.2), we get

$$(A_3 A_2 A_1) U_k = U_k (H_k T_k S_k) + u_{k+1} (h_{k+1,k} t_{kk} s_{kk}) e_k^T. \tag{9.10.4}$$

Since $H_k T_k S_k$ is upper Hessenberg, this shows that our Arnoldi process on C is automatically effecting an Arnoldi process on the product $A = A_3 A_2 A_1$. If we rewrite the equation $A_3 W_k = U_k H_k + u_{k+1} h_{k+1,k}$ in the style $A_3 W_k = U_{k+1} H_{k+1,k}$, we can write (9.10.2) as

$$(A_3 A_2 A_1) U_k = U_{k+1} (H_{k+1,k} T_k S_k).$$

Moreover, if we recombine (9.10.2) in different ways, we obtain

$$(A_1 A_3 A_2) V_k = V_{k+1} (S_{k+1} H_{k+1,k} T_k)$$

and

$$(A_2 A_1 A_3) W_k = W_{k+1} (T_{k+1} S_{k+1} H_{k+1,k}),$$

showing that we are also doing Arnoldi processes on the cycled products $A_1 A_3 A_2$ and $A_2 A_1 A_3$.

So far we have restricted our attention to the Arnoldi process, but other product Krylov processes can be built in the same fashion. For all such processes, an equation of the form (9.10.3) will hold, but the vectors produced need not be orthonormal. As we shall see below, the Hamiltonian Lanczos process (Section 9.8) can be viewed as a product Krylov process.

Implicit restarts can be done by either of the two restart methods that we have presented. Whichever method is used, the filtering can be done by an appropriate product GR algorithm (Section 8.3) to preserve the product structure.

Symmetric Lanczos Process for Singular Values

Consider the problem of finding the largest few singular values and associated singular vectors of B. This is equivalent to the eigenvalue problem for $B^* B$ and $B B^*$, so consider the cyclic matrix

$$C = \begin{bmatrix} 0 & B^* \\ B & 0 \end{bmatrix}.$$

If we run the symmetric Lanczos ($=$ Arnoldi) process on this matrix, we obtain, in analogy with (9.10.3),

$$\begin{bmatrix} & B^* \\ B & \end{bmatrix} \begin{bmatrix} V_k & \\ & U_k \end{bmatrix} = \begin{bmatrix} V_k & \\ & U_k \end{bmatrix} \begin{bmatrix} & H_k \\ T_k & \end{bmatrix} + \begin{bmatrix} v_{k+1} \\ \end{bmatrix} h_{k+1,k} e_{2k}^T. \quad (9.10.5)$$

By symmetry we have $H_k = T_k^*$, from which it follows immediately that both H_k and T_k are bidiagonal. (Moreover, they can be taken to be real. If we shuffle the rows and columns, we obtain a tridiagonal matrix with zeros on the main diagonal, like (8.4.3).) Letting

$$T_k = \begin{bmatrix} \alpha_1 & \beta_1 & & \\ & \alpha_2 & \ddots & \\ & & \ddots & \beta_{k-1} \\ & & & \alpha_k \end{bmatrix},$$

(9.10.5) gives the short recurrences

$$u_j \alpha_j = B v_j - u_{j-1} \beta_{j-1}$$

and

$$v_{j+1} \beta_j = B^* u_j - v_j \alpha_j,$$

which appeared first in [97]. These automatically effect symmetric Lanczos processes on $B^* B$ and $B B^*$.

They can be restarted by either of the methods of Section 9.2 [38, 132, 12]. Consider the second method, for example. After $2k$ steps we have

$$\begin{bmatrix} & B^* \\ B & \end{bmatrix} \begin{bmatrix} V_k & \\ & U_k \end{bmatrix} = \begin{bmatrix} V_k & \\ & U_k \end{bmatrix} \begin{bmatrix} & T_k^T \\ T_k & \end{bmatrix} + \begin{bmatrix} v_{k+1} \\ 0 \end{bmatrix} \beta_k e_{2k}^T. \quad (9.10.6)$$

If we want to restart with $2m$ vectors, we compute the SVD of T_k: $T_k = P_k \Sigma_k Q_k^T$, where the singular values in Σ_k are ordered so that the m that we want to keep (the m biggest ones in this case) are at the top. Then

$$\begin{bmatrix} & T_k^T \\ T_k & \end{bmatrix} = \begin{bmatrix} Q_k & \\ & P_k \end{bmatrix} \begin{bmatrix} & \Sigma_k^T \\ \Sigma_k & \end{bmatrix} \begin{bmatrix} Q_k^T & \\ & P_k^T \end{bmatrix}.$$

Substituting this into (9.10.6) and multiplying on the right by $\operatorname{diag}\{Q_k, P_k\}$, we obtain

$$\begin{bmatrix} & B^* \\ B & \end{bmatrix} \begin{bmatrix} \tilde{V}_k & \\ & \tilde{U}_k \end{bmatrix} = \begin{bmatrix} \tilde{V}_k & \\ & \tilde{U}_k \end{bmatrix} \begin{bmatrix} & \Sigma_k^T \\ \Sigma_k & \end{bmatrix} + \begin{bmatrix} v_{k+1} \\ 0 \end{bmatrix} \beta_k \begin{bmatrix} 0 & \tilde{z}^T \end{bmatrix},$$
$$(9.10.7)$$

where $\tilde{V}_k = V_k Q_k$, $\tilde{U}_k = U_k P_k$, and $\tilde{z}^T = e_k^T P_k$. If we do a perfect shuffle of the vectors in (9.10.7), this becomes an instance of (9.3.13). Now let $\tilde{V}_m$ and $\tilde{U}_m$ denote the first m columns of $\tilde{V}_k$ and $\tilde{U}_k$, respectively. Let z denote the first m entries of $\tilde{z}$, and let Σ_m denote the upper left-hand $m \times m$ submatrix of Σ_k. Σ_m contains the singular values of T_k that we wish to retain. Keeping $2m$ columns of (9.10.7), we have

$$\begin{bmatrix} & B^* \\ B & \end{bmatrix} \begin{bmatrix} \tilde{V}_m & \\ & \tilde{U}_m \end{bmatrix} = \begin{bmatrix} \tilde{V}_m & \\ & \tilde{U}_m \end{bmatrix} \begin{bmatrix} & \Sigma_m^T \\ \Sigma_m & \end{bmatrix} + \begin{bmatrix} v_{k+1} \\ 0 \end{bmatrix} \beta_k \begin{bmatrix} 0 & z^T \end{bmatrix},$$
$$(9.10.8)$$

which is (upon shuffling) an instance of (9.3.14). Now let X_m and Y_m be orthogonal matrices such that $z^T X_m^T = \gamma e_m^T$ (for some γ) and the matrix $\hat{B}_m = X_m \Sigma_m Y_m^T$ is bidiagonal. This is achieved by starting with an orthogonal matrix X_{m1} such that $z^T X_{m1}^T = \alpha e_m^T$, then transforming $X_{m1}\Sigma_m$ to bidiagonal form from bottom to top. (Here is one viewpoint. Do a perfect shuffle of the rows and columns of

$$\begin{bmatrix} & (X_{m1}\Sigma_m)^T \\ X_{m1}\Sigma_m & \end{bmatrix}.$$

Then reduce the resulting matrix to upper Hessenberg form from bottom to top, as described in Section 3.4, using orthogonal transformations. The result will be tridiagonal by symmetry, with zeros on the main diagonal. When we deshuffle this tridiagonal matrix, we get

$$\begin{bmatrix} & \hat{B}_m^T \\ \hat{B}_m & \end{bmatrix},$$

where $\hat{B}_m$ is bidiagonal.)

Since $\hat{B}_m = X_m \Sigma_m Y_m^T$, we have

$$\begin{bmatrix} & \Sigma_m^T \\ \Sigma_m & \end{bmatrix} = \begin{bmatrix} Y_m^T & \\ & X_m^T \end{bmatrix}\begin{bmatrix} & \hat{B}_m^T \\ \hat{B}_m & \end{bmatrix}\begin{bmatrix} Y_m & \\ & X_m \end{bmatrix}.$$

Substituting this into (9.10.8) and multiplying on the right by $\begin{bmatrix} Y_m^T & \\ & X_m^T \end{bmatrix}$, we obtain

$$\begin{bmatrix} & B^* \\ B & \end{bmatrix}\begin{bmatrix} \hat{V}_m & \\ & \hat{U}_m \end{bmatrix} = \begin{bmatrix} \hat{V}_m & \\ & \hat{U}_m \end{bmatrix}\begin{bmatrix} & \hat{B}_m^T \\ \hat{B}_m & \end{bmatrix} + \begin{bmatrix} \hat{v}_{m+1} \\ 0 \end{bmatrix}\hat{\beta}_m \begin{bmatrix} 0 & e_m^T \end{bmatrix},$$

$$(9.10.9)$$

where $\hat{V}_m = \tilde{V}_m Y_m^T$, $\hat{U}_m = \tilde{U}_m X_m^T$, $\hat{v}_{m+1} = v_{k+1}$, and $\hat{\beta}_m = \beta_k \gamma$. Equation (9.10.9) is (upon shuffling) an instance of (9.3.15), and it is just what we need to restart the Lanczos process.

In the description here, everything is done twice. In an actual implementation, just the following needs to be done: First the SVD $T_k = P_k \Sigma_k Q_k^T$ is computed. The vector z^T is obtained as the first m entries of the last row of P_k, and Σ_m is obtained as a submatrix of Σ_k. Then the transformation $\hat{B}_m = X_m \Sigma_m Y_m^T$ with $z^T X_m^T = \gamma e_m^T$ is done. $\tilde{U}_m$ and $\tilde{V}_m$ are obtained by computing the first m columns of the transformations $\tilde{U}_k = U_k P_k$ and $\tilde{V}_k = V_k Q_k$, respectively, and then $\hat{U}_m = \tilde{U}_m X_m^T$ and $\hat{V}_m = \tilde{V}_m Y_m^T$ are computed.

The method we have just described is just the thick restart method of Wu and Simon [233], taking the cyclic structure of the matrix into account.

The Hamiltonian Lanczos Process Revisited

Since the Hamiltonian Lanczos process for H was derived from the skew-Hamiltonian Lanczos process for the product HH, we should not be surprised that this algorithm can also be derived as a product Krylov process. Such a process would satisfy an equation analogous to (9.10.3), which in this case would have the form

$$\begin{bmatrix} H & \\ & H \end{bmatrix}\begin{bmatrix} U_k & \\ & V_k \end{bmatrix} = \begin{bmatrix} U_k & \\ & V_k \end{bmatrix}\begin{bmatrix} & T_k \\ D_k & \end{bmatrix} + \begin{bmatrix} u_{k+1} \\ \end{bmatrix}\beta_k e_{2k}^T. \quad (9.10.10)$$

We use the symbols D_k and T_k here, because they will turn out to have the same form as the D_k and T_k from Sections 9.7, 9.8, and 9.9. However, all we know at this point is that D_k is upper triangular and T_k is upper Hessenberg. Equation (9.10.10) can be written more compactly as

$$HU_k = V_k D_k, \qquad HV_k = U_k T_k + u_{k+1} \beta_k e_k^T \qquad (9.10.11)$$

or

$$H \begin{bmatrix} U_k & V_k \end{bmatrix} = \begin{bmatrix} U_k & V_k \end{bmatrix} \begin{bmatrix} & T_k \\ D_k & \end{bmatrix} + u_{k+1} t_{k+1,k} e_{2k}^T. \qquad (9.10.12)$$

We can preserve Hamiltonian structure by building vectors $u_1, \ldots, u_k$ and $v_1, \ldots, v_k$ that are columns of a symplectic matrix. If we rewrite the second equation of (9.10.11) in the style $HV_k = U_{k+1} T_{k+1,k}$, we have

$$H^2 U_k = U_{k+1} T_{k+1,k} D_k \quad \text{and} \quad H^2 V_k = V_{k+1} D_{k+1} T_{k+1,k},$$

and so $\mathcal{R}(U_k) = \mathcal{K}_k(H^2, u_1)$ and $\mathcal{R}(V_k) = \mathcal{K}_k(H^2, v_1)$. Thus we have isotropy ($U_k^T J U_k = 0$ and $V_k^T J V_k = 0$) automatically by Proposition 9.7.2. The one other property we need is

$$U_k^T J V_k = I_k. \qquad (9.10.13)$$

Equating the jth columns in (9.10.11), we have

$$Hu_j = \sum_{i=1}^{j} v_i d_{ij}, \qquad Hv_j = \sum_{i=1}^{j+1} u_i t_{ij}, \qquad (9.10.14)$$

which gives the recurrences

$$u_{j+1} t_{j+1,j} = Hv_j - \sum_{i=1}^{j} u_i t_{ij}, \qquad (9.10.15)$$

$$v_{j+1} d_{j+1,j+1} = Hu_{j+1} - \sum_{i=1}^{j} v_i d_{i,j+1}. \qquad (9.10.16)$$

It is not hard to show that the coefficients t_{ij} and $d_{i,j+1}$ can be chosen in such a way that (9.10.13) holds. Moreover, it can be done in such a way that $d_{jj} = \pm 1$ for all j (assuming no breakdowns).

Once we have achieved this, the symplectic structure of $\begin{bmatrix} U_k & V_k \end{bmatrix}$ in (9.10.12) can be used to prove that the small matrix

$$\begin{bmatrix} & T_k \\ D_k & \end{bmatrix}$$

must be Hamiltonian. Therefore D_k and T_k are symmetric, D_k is diagonal (and even a signature matrix), and T_k is tridiagonal. Almost all of the coefficients in (9.10.15) and (9.10.16) are zero, and the process that we have derived is exactly the Hamiltonian Lanczos process of Section 9.8. The details are left as an exercise.

9.11   Block Krylov Processes

Krylov processes can have difficulties with matrices that have multiple eigenvalues. For example, consider a Hermitian matrix $A \in \mathbb{C}^{n \times n}$ whose largest eigenvalue λ_1 has multiplicity greater than one. If we run the symmetric Lanczos process on A, our approximate eigenvalues are taken from a symmetric properly tridiagonal matrix T_k, which cannot have multiple eigenvalues (Exercise 9.11.2). Therefore we must find and lock out one eigenvector associated with λ_1 before we can even begin to search for the rest of the eigenspace associated with λ_1. In fact, in exact arithmetic we will never find the rest of the eigenspace (Exercise 9.11.3), so we are dependent on roundoff errors to help us find it. Clearly there is a danger of overlooking some wanted eigenvectors, particularly if the multiplicity is high.

A remedy for this problem is to use a block Krylov process. We will illustrate by discussing the block Arnoldi process. Throughout this book the lowercase symbol u has always stood for a vector. If we wanted to refer to a matrix with several columns, we always used an uppercase letter. In this section we will violate that convention in the interest of avoiding a notational conflict. Our block Arnoldi process begins with a u_1 that is not a vector but a matrix with b columns, where b is the *block size*. We can take the columns to be orthonormal; for example, we can get u_1 by picking an $n \times b$ matrix at random and orthonormalizing its columns by a QR decomposition.

After $j - 1$ steps we will have $u_1, \ldots, u_j$, each consisting of b orthonormal columns such that the jb columns of $u_1, \ldots, u_j$ taken together form an orthonormal set. We now describe the jth step. In analogy with (9.4.1) we define coefficients h_{ij} by

$$h_{ij} = u_i^* A u_j, \qquad i = 1, \ldots, j. \tag{9.11.1}$$

Each h_{ij} is now a $b \times b$ matrix. In analogy with (9.4.2) we define

$$\hat{u}_{j+1} = A u_j - \sum_{i=1}^{j} u_i h_{ij}, \tag{9.11.2}$$

where $\hat{u}_{j+1}$ is a matrix with b columns. We cannot expect that the columns of $\hat{u}_{j+1}$ will be orthogonal, but it is easy to show that they are orthogonal to all of the columns of $u_1, \ldots, u_j$. In practice one would do this orthonormalization step twice, as illustrated in algorithm (9.4.7). The process of normalizing $\hat{u}_{k+1}$ is more complicated than what is shown in (9.4.3) and (9.4.4) but hardly difficult. Instead of taking the norm and dividing, we do a condensed QR decomposition:

$$\hat{u}_{j+1} = u_{j+1} h_{j+1,j}. \tag{9.11.3}$$

Here u_{j+1} is $n \times b$ with orthonormal columns, and $h_{j+1,j}$ is $b \times b$ and upper triangular. This completes the description of a block Arnoldi step.

Algorithm (9.4.7) serves as a description of the block Arnoldi process, except that the normalization step has to be changed to (9.11.3). Equations like

$$A U_k = U_k H_k + u_{k+1} h_{k+1,k} e_k^T$$

continue to hold if we reinterpret e_k to be the last b columns of the $kb \times kb$ identity matrix. H_k is now a block upper Hessenberg matrix. Since the subdiagonal blocks $h_{j+1,j}$ are upper triangular, H_k is in fact b-Hessenberg. (Everything below the bth subdiagonal is zero.) Since H_k is not Hessenberg in the scalar sense, it can have multiple eigenvalues, and eigenvalues of A with geometric multiplicity as large as b can be found and extracted together. Implicit restarts can be done in much the same way as they are for the scalar case.

We have described a block Arnoldi process. Other block Krylov processes can be developed in the same way.

Exercises

9.11.1. The matrix generated by the MATLAB command

```
A = delsq(numgrid('S',40),
```

a discrete negative Laplacian on a square, has numerous repeated eigenvalues. Download the block symmetric Lanczos code `blocklan.m` from the website. Use it to compute a few of the smallest eigenvalues of A by the block Lanczos process applied to A^{-1}. Try block sizes 1, 2, and 3. Notice that with block sizes 2 and greater, all of the small repeated eigenvalues are found, while with block size 1, some of the repeated eigenvalues are overlooked.

9.11.2. Let $H \in \mathbb{C}^{k \times k}$ be a properly upper Hessenberg matrix.

(a) Show that for any $\mu \in \mathbb{C}$, $H - \mu I$ has rank at least $k - 1$.

(b) Show that every eigenvalue of H has geometric multiplicity 1. Deduce that H has one Jordan block associated with each eigenvalue.

(c) Now suppose that H is also Hermitian, and hence tridiagonal. Show that each eigenvalue has algebraic multiplicity 1. Therefore H has k distinct eigenvalues.

9.11.3. Let $A \in \mathbb{C}^{n \times n}$ be a semisimple matrix with j distinct eigenvalues $\lambda_1, \ldots, \lambda_j$, where $j < n$. Let $\mathcal{S}_i$ denote the eigenspace associated with λ_i, $i = 1, \ldots, j$.

(a) Show that each $x \in \mathbb{C}^n$ can be expressed uniquely as

$$x = q_1 + q_2 + \cdots + q_j,$$

where $q_i \in \mathcal{S}_i$, $i = 1, \ldots, j$.

(b) Show that every Krylov space $\mathcal{K}_i(A, x)$ satisfies

$$\mathcal{K}_i(A, x) \subseteq \text{span}\{q_1, \ldots, q_j\}.$$

Therefore any Krylov process starting with x sees only multiples of q_i from the eigenspace $\mathcal{S}_i$. If λ_i is a multiple eigenvalue, the complete eigenspace will never be found (in exact arithmetic).

(c) Show that this problem is not alleviated by implicit restarts.

9.11.4. Think about how the results of Exercise 9.11.3 might be extended to defective matrices.

9.11.5. Consider the block Arnoldi process applied to a matrix $A \in \mathbb{R}^{2n \times 2n}$ that is both skew symmetric ($A^T = -A$) and Hamiltonian ($(JA)^T = JA$).

(a) What simplifications occur in (9.11.2) due to the fact that A is skew symmetric?

(b) Show that $JA = AJ$, where $J = \begin{bmatrix} 0 & I \\ -I & 0 \end{bmatrix}$, as always.

(c) Now take block size $b = 2$ and consider the block Arnoldi process with starting vectors $u_1 = \begin{bmatrix} q_1 & -Jq_1 \end{bmatrix}$, where $q_1 \in \mathbb{R}^{2n}$ is a starting vector chosen so that $\| q_1 \| = 1$. Show that the columns of u_1 are orthonormal.

(d) Consider the first step $\hat{u}_2 = Au_1 - u_1 h_{11}$. Show that $h_{11} = \begin{bmatrix} 0 & -\alpha_1 \\ \alpha_1 & 0 \end{bmatrix}$, where $\alpha_1 = q_1^T J A q_1$. Letting $\hat{u}_2 = \begin{bmatrix} \hat{q}_2 & v \end{bmatrix}$, determine formulas for $\hat{q}_2$ and v, and show that $v = -J\hat{q}_2$.

(e) In the QR decomposition $\hat{u}_2 = u_2 h_{21}$, show that

$$u_2 = \begin{bmatrix} q_2 & -Jq_2 \end{bmatrix} \quad \text{and} \quad h_{21} = \begin{bmatrix} \beta_1 & 0 \\ 0 & \beta_1 \end{bmatrix},$$

where $\beta_1 = \| \hat{q}_2 \|$ and $q_2 = \hat{q}_2 / \beta_1$.

(f) Using the equations

$$u_{j+1} h_{j+1,j} = \hat{u}_{j+1} = A u_j - u_j h_{jj} - u_{j-1} h_{j-1,j},$$

prove by induction on j that $\hat{u}_j = \begin{bmatrix} \hat{q}_j & -J\hat{q}_j \end{bmatrix}$, $u_j = \begin{bmatrix} q_j & -Jq_j \end{bmatrix}$,

$$h_{jj} = \begin{bmatrix} 0 & -\alpha_j \\ \alpha_j & 0 \end{bmatrix}, \quad \text{and} \quad h_{j+1,j} = -h_{j,j+1} = \begin{bmatrix} \beta_j & 0 \\ 0 & \beta_j \end{bmatrix}$$

for all j, where $\alpha_j = q_j^T J A q_j$, $\beta_j = \| \hat{q}_{j+1} \|$, and the vectors q_j are generated by the recurrence

$$q_{j+1} \beta_j = \hat{q}_{j+1} = A q_j + J q_j \alpha_j + q_{j-1} \beta_{j-1}.$$

This is a *Hamiltonian skew symmetric Lanczos process*.

(g) Let $Q_j = \begin{bmatrix} q_1 & \cdots & q_j \end{bmatrix}$ and $\mathcal{Q}_j = \mathcal{R}(Q_j)$. Show that $\mathcal{Q}_j$ is isotropic. Show that the $2n \times 2j$ matrix $U_j = \begin{bmatrix} Q_j & -JQ_j \end{bmatrix}$ is symplectic, in the sense that $U_j^T J U_j = J_{2j}$.

(h) Show that

$$AQ_j = Q_j B_j - JQ_j A_j + q_{j+1} \beta_j e_j^T,$$

where

$$
B_j = \begin{bmatrix} 0 & -\beta_1 & & \\ \beta_1 & 0 & \ddots & \\ & \ddots & \ddots & -\beta_{j-1} \\ & & \beta_{j-1} & 0 \end{bmatrix} \quad \text{and} \quad A_j = \begin{bmatrix} \alpha_1 & 0 & & \\ 0 & \alpha_2 & \ddots & \\ & \ddots & \ddots & 0 \\ & & 0 & \alpha_j \end{bmatrix}.
$$

(i) Show that

$$
A \begin{bmatrix} Q_j & -JQ_j \end{bmatrix} = \begin{bmatrix} Q_j & -JQ_j \end{bmatrix} \begin{bmatrix} B_j & -A_j \\ A_j & B_j \end{bmatrix} + R_j,
$$

where

$$
R_j = q_{j+1}\beta_j \begin{bmatrix} e_j^T & 0 \end{bmatrix} - J q_{j+1}\beta_j \begin{bmatrix} 0 & e_j^T \end{bmatrix}.
$$

Bibliography

[1] H. D. I. Abarbanel, R. Brown, and M. B. Kennel. Local Lyapunov exponents computed from observed data. *Chaos*, 2:343–365, 1992. (Cited on pp. 316, 317.)

[2] N. Abdelmalek. Round off error analysis for Gram-Schmidt method and solution of linear least squares problems. *BIT*, 11:345–368, 1971. (Cited on p. 371.)

[3] L. Ahlfors. *Complex Analysis*. McGraw–Hill, New York, Third edition, 1979. (Cited on p. 91.)

[4] M. Ahues and F. Tisseur. A New Deflation Criterion for the QR Algorithm. Technical Report UT-CS-97-353, University of Tennessee, Knoxville, TN, 1997. LAPACK working note 122. (Cited on p. 204.)

[5] G. Ammar, W. Gragg, and L. Reichel. On the eigenproblem for orthogonal matrices. In *Proceedings of the 25th IEEE Conference on Decision and Control*, Athens, Greece, 1986, pp. 1963–1966. (Cited on p. 341.)

[6] G. S. Ammar, L. Reichel, and D. C. Sorensen. An implementation of a divide and conquer algorithm for the unitary eigenproblem. *ACM Trans. Math. Software*, 18:292–307, 1992. (Cited on p. 341.)

[7] G. S. Ammar, L. Reichel, and D. C. Sorensen. Corrigendum: Algorithm 730: An implementation of a divide and conquer algorithm for the unitary eigenproblem. *ACM Trans. Math. Software*, 20:161, 1994. (Cited on p. 341.)

[8] E. Anderson, Z. Bai, C. Bischof, S. Blackford, J. Demmel, J. Dongarra, J. Du Croz, A. Greenbaum, S. Hammerling, A. McKenney, and D. Sorensen. *LAPACK Users' Guide*. Software Environ. Tools 9, SIAM, Philadelphia, Third edition, 1999. www.netlib.org/lapack/. (Cited on pp. 103, 131, 162, 194, 198.)

[9] T. Apel, V. Mehrmann, and D. Watkins. Structured eigenvalue methods for the computation of corner singularities in 3D anisotropic elastic structures. *Comput. Methods Appl. Mech. Engrg.*, 191:4459–4473, 2002. (Cited on p. 398.)

[10] V. I. Arnold. *Mathematical Methods of Classical Mechanics*. Springer-Verlag, New York, Second edition, 1989. (Cited on p. 17.)

[11] W. E. Arnoldi. The principle of minimized iterations in the solution of the matrix eigenvalue problem. *Quart. Appl. Math.*, 9:17–29, 1951. (Cited on pp. 360, 370.)

[12] J. Baglama and L. Reichel. Augmented implicitly restarted Lanczos bidiagonalization methods. *SIAM J. Sci. Comput.*, 27:19–42, 2005. (Cited on p. 413.)

[13] Z. Bai and J. Demmel. On a block implementation of the Hessenberg multishift QR iteration. *Internat. J. High Speed Comput.*, 1:97–112, 1989. (Cited on pp. 177, 198, 199, 265.)

[14] Z. Bai and J. Demmel. On swapping diagonal blocks in real Schur form. *Linear Algebra Appl.*, 186:73–95, 1993. (Cited on p. 194.)

[15] Z. Bai, J. Demmel, J. Dongarra, A. Ruhe, and H. van der Vorst, editors. *Templates for the Solution of Algebraic Eigenvalue Problems: A Practical Guide.* Software Environ. Tools 11, SIAM, Philadelphia, 2000. (Cited on pp. 119, 198.)

[16] Z. Bai, J. Demmel, and A. Mc Kenney. On computing condition numbers for the nonsymmetric eigenvalue problem. *ACM Trans. Math. Software*, 19:202–223, 1993. (Cited on pp. 103, 194.)

[17] G. Banse. *Symplektische Eigenwertverfahren zur Lösung zeitdiskreter optimaler Steuerungsprobleme.* Ph.D. thesis, University of Bremen, Bremen, Germany, 1995. (Cited on pp. 209, 406.)

[18] G. Banse and A. Bunse-Gerstner. A condensed form for the solution of the symplectic eigenvalue problem. In U. Helmke, R. Menniken, and J. Sauer, editors, *Systems and Networks: Mathematical Theory and Applications*, pp. 613–616. Akademie Verlag, Berlin, 1994. (Cited on p. 209.)

[19] I. Bar-On and M. Paprzycki. High performance solution of the complex symmetric eigenproblem. *Numer. Algorithms*, 18:195–208, 1998. (Cited on p. 119.)

[20] I. Bar-On and V. Ryaboy. Fast diagonalization of large and dense complex symmetric matrices, with applications to quantum reaction dynamics. *SIAM J. Sci. Comput.*, 18:1412–1435, 1997. (Cited on p. 119.)

[21] R. H. Bartels and G. W. Stewart. Solution of the equation $AX + XB = C$. *Comm. ACM*, 15:820–826, 1972. (Cited on pp. 194, 196, 197.)

[22] F. L. Bauer. Das Verfahren der Treppeniteration und verwandte Verfahren zur Lösung algebraischer Eigenwertprobleme. *Z. Angew. Math. Phys.*, 8:214–235, 1957. (Cited on p. 159.)

[23] F. L. Bauer. On modern matrix iteration processes of Bernoulli and Graefe types. *J. ACM*, 5:246–257, 1958. (Cited on p. 159.)

[24] F. L. Bauer and C. T. Fike. Norms and exclusion theorems. *Numer. Math*, 2:137–141, 1960. (Cited on pp. 94, 108.)

[25] C. Beattie, M. Embree, and J. Rossi. Convergence of restarted Krylov subspaces to invariant subspaces. *SIAM J. Matrix Anal. Appl.*, 25:1074–1109, 2004. (Cited on p. 365.)

[26] C. A. Beattie, M. Embree, and D. C. Sorensen. Convergence of polynomial restart Krylov methods for eigenvalue computations. *SIAM Rev.*, 47:492–515, 2005. (Cited on p. 365.)

[27] P. Benner, R. Byers, V. Mehrmann, and H. Xu. Numerical computation of deflating subspaces of skew-Hamiltonian/Hamiltonian pencils. *SIAM J. Matrix Anal. Appl.*, 24:165–190, 2002. (Cited on p. 319.)

[28] P. Benner and H. Fassbender. An implicitly restarted symplectic Lanczos method for the Hamiltonian eigenvalue problem. *Linear Algebra Appl.*, 263:75–111, 1997. (Cited on p. 403.)

[29] P. Benner and H. Faßbender. The symplectic eigenvalue problem, the butterfly form, the SR algorithm, and the Lanczos method. *Linear Algebra Appl.*, 275–276:19–47, 1998. (Cited on pp. 209, 406.)

[30] P. Benner and H. Faßbender. An implicitly restarted symplectic Lanczos method for the symplectic eigenvalue problem. *SIAM J. Matrix Anal. Appl.*, 22:682–713, 2000. (Cited on p. 406.)

[31] P. Benner, H. Fassbender, and D. S. Watkins. Two connections between the SR and HR eigenvalue algorithms. *Linear Algebra Appl.*, 272:17–32, 1998. (Cited on pp. 339, 341.)

[32] P. Benner, H. Fassbender, and D. S. Watkins. SR and SZ algorithms for the symplectic (butterfly) eigenproblem. *Linear Algebra Appl.*, 287:41–76, 1999. (Cited on p. 209.)

[33] P. Benner and D. Kressner. Fortran 77 subroutines for computing the eigenvalues of Hamiltonian matrices. *ACM Trans. Math. Software*, 32:352–373, 2006. www. tu-chemnitz.de/mathematik/hapack/. (Cited on pp. 208, 324.)

[34] P. Benner, D. Kressner, and V. Mehrmann. Skew-Hamiltonian and Hamiltonian eigenvalue problems: Theory, algorithms, and applications. In Z. Drmac, M. Marusic, and Z. Tutek, editors, *Proceedings of the Conference on Applied Mathematics and Scientific Computing*, pp. 3–39. Springer-Verlag, New York, 2005. (Cited on p. 208.)

[35] P. Benner, V. Mehrmann, and H. Xu. A numerically stable, structure preserving method for computing the eigenvalues of real Hamiltonian or symplectic pencils. *Numer. Math.*, 78:329–358, 1998. (Cited on p. 322.)

[36] C. Bischof and C. Van Loan. The WY representation for products of Householder matrices. *SIAM J. Sci. Stat. Comput.*, 8:s2–s13, 1987. (Cited on p. 131.)

[37] Å. Björck and G. H. Golub. Numerical methods for computing angles between linear subspaces. *Math. Comp.*, 27:579–594, 1973. (Cited on p. 84.)

[38] Å. Björck, E. Grimme, and P. Van Dooren. An implicit shift bidiagonalization algorithm for ill-posed systems. *BIT*, 34:510–534, 1994. (Cited on p. 413.)

[39] L. S. Blackford, J. Choi, A. Cleary, E. D'Azevedo, J. Demmel, I. Dhillon, J. Dongarra, S. Hammarling, G. Henry, A. Petitet, K. Stanley, D. Walker, and R. C. Whaley. *ScaLAPACK Users' Guide.* Software Environ. Tools 4, SIAM, Philadelphia, 1997. www.netlib.org/scalapack/. (Cited on p. 199.)

[40] B. Bohnhorst. *Ein Lanczos-ähnliches Verfahren zur Lösung des unitären Eigenwert-problems.* Ph.D. thesis, University of Bielefeld, Bielefeld, Germany, 1993. (Cited on p. 380.)

[41] A. Bojanczyk, G. H. Golub, and P. Van Dooren. The periodic Schur decomposition; Algorithm and applications. In *Proc. SPIE* 1770, 1992, pp. 31–42. (Cited on p. 319.)

[42] A. Bojanczyk, N. J. Higham, and H. Patel. Solving the indefinite least squares problem by hyperbolic QR factorization. *SIAM J. Matrix Anal. Appl.*, 24:914–931, 2003. (Cited on p. 122.)

[43] A. Bojanczyk and P. Van Dooren. Reordering diagonal blocks in real Schur form. In G. H. Golub, M. S. Moonen, and B. L. R. De Moor, editors, *Linear Algebra for Large Scale and Real-Time Applications.* Kluwer Academic Publishers, Amsterdam, 1993. citeseer.ist.psu.edu/bojanczyk93reordering.html. (Cited on p. 194.)

[44] K. Braman, R. Byers, and R. Mathias. The multishift QR algorithm. Part I: Maintaining well-focused shifts and level 3 performance. *SIAM J. Matrix Anal. Appl.*, 23:929–947, 2002. (Cited on pp. 200, 201, 202.)

[45] K. Braman, R. Byers, and R. Mathias. The multishift QR algorithm. Part II: Aggressive early deflation. *SIAM J. Matrix Anal. Appl.*, 23:948–973, 2002. (Cited on pp. 203, 204, 271.)

[46] R. Brawer and M. Pirovino. The linear algebra of the Pascal matrix. *Linear Algebra Appl.*, 174:13–23, 1992. (Cited on p. 347.)

[47] M. A. Brebner and J. Grad. Eigenvalues of $Ax = \lambda Bx$ for real symmetric matrices A and B computed by reduction to pseudosymmetric form and the HR process. *Linear Algebra Appl.*, 43:99–118, 1982. (Cited on p. 206.)

[48] F. Brenti. Combinatorics and total positivity. *J. Combin. Theory Ser. A*, 71:175–218, 1995. (Cited on p. 346.)

[49] W. Bunse and A. Bunse-Gerstner. *Numerische lineare Algebra.* B. G. Teubner, Stuttgart, 1985. (Cited on pp. 12, 13.)

[50] A. Bunse-Gerstner. An analysis of the HR algorithm for computing the eigenvalues of a matrix. *Linear Algebra Appl.*, 35:155–173, 1981. (Cited on p. 206.)

[51] A. Bunse-Gerstner and L. Elsner. Schur parameter pencils for the solution of the unitary eigenproblem. *Linear Algebra Appl.*, 154–156:741–778, 1991. (Cited on pp. 341, 380, 382.)

[52] A. Bunse-Gerstner and C. He. On a Sturm sequence of polynomials for unitary Hessenberg matrices. *SIAM J. Matrix Anal. Appl.*, 16:1043–1055, 1995. (Cited on p. 341.)

[53] A. Bunse-Gerstner and V. Mehrmann. A symplectic QR-like algorithm for the solution of the real algebraic Riccati equation. *IEEE Trans. Automat. Control*, AC-31:1104–1113, 1986. (Cited on pp. 120, 207, 339.)

[54] H. J. Buurema. *A Geometric Proof of Convergence for the QR Method*. Ph.D. thesis, Rijksuniversiteit te Groningen, Groningen, The Netherlands, 1970. (Cited on p. 159.)

[55] R. Byers. A Hamiltonian QR algorithm. *SIAM J. Sci. Stat. Comput.*, 7:212–229, 1986. (Cited on p. 208.)

[56] T. F. Chan. An improved algorithm for computing the singular value decomposition. *ACM Trans. Math. Software*, 8:72–83, 1982. (Cited on pp. 306, 307.)

[57] D. Chu, X. Liu, and V. Mehrmann. A numerically strongly stable method for computing the Hamiltonian Schur form. *Numer. Math.*, 105:375–412, 2007. (Cited on pp. 208, 324, 326, 328.)

[58] K.-W. E. Chu. The solution of the matrix equations $AXB - CXD = E$ and $(YA - DZ, YX - BZ) = (E, F)$. *Linear Algebra Appl.*, 93:93–105, 1987. (Cited on p. 260.)

[59] J. K. Cullum and R. A. Willoughby. *Lanczos Algorithms for Large Symmetric Eigenvalue Computations*. Birkhäuser Boston, Cambridge, MA, 1985. (Cited on pp. 119, 373.)

[60] J. K. Cullum and R. A. Willoughby. A QL procedure for computing the eigenvalues of complex symmetric tridiagonal matrices. *SIAM J. Matrix Anal. Appl.*, 17:83–109, 1996. (Cited on p. 119.)

[61] J. J. M. Cuppen. A divide and conquer method for the symmetric tridiagonal eigenproblem. *Numer. Math.*, 36:177–195, 1981. (Cited on p. 205.)

[62] J. W. Daniel, W. B. Gragg, L. Kaufman, and G. W. Stewart. Reorthogonalization and stable algorithms for updating the Gram–Schmidt QR factorization. *Math. Comp.*, 30:772–795, 1976. (Cited on p. 371.)

[63] R. J. A. David. *Algorithms for the Unitary Eigenvalue Problem*. Ph.D. thesis, Washington State University, Pullman, WA, 2007. (Cited on pp. 342, 375, 382, 385, 387.)

[64] R. J. A. David and D. S. Watkins. Efficient implementation of the multishift QR algorithm for the unitary eigenvalue problem. *SIAM J. Matrix Anal. Appl.*, 28:623–633, 2006. (Cited on pp. 342, 344.)

[65] R. J. A. David and D. S. Watkins. An inexact Krylov–Schur algorithm for the unitary eigenvalue problem. *Linear Algebra Appl.*, 2007, to appear. (Cited on pp. 382, 385, 387.)

[66] C. Davis and W. M. Kahan. Some new bounds on perturbations of subspaces. *Bull. Amer. Math. Soc.*, 75:863–868, 1969. (Cited on p. 84.)

[67] C. Davis and W. M. Kahan. The rotation of eigenvectors by a perturbation. III. *SIAM J. Numer. Anal.*, 7:1–46, 1970. (Cited on p. 84.)

[68] D. Day. How the QR Algorithm Fails to Converge and How to Fix It. Technical Report 96-0913J, Sandia National Laboratory, Albuquerque, NM, 1996. (Cited on p. 199.)

[69] B. De Moor and P. Van Dooren. Generalizations of the singular value and QR decompositions. *SIAM J. Matrix Anal. Appl.*, 13:993–1014, 1992. (Cited on p. 316.)

[70] B. De Moor and H. Zha. A tree of generalizations of the ordinary singular value decomposition. *Linear Algebra Appl.*, 147:469–500, 1991. (Cited on p. 316.)

[71] J. Della-Dora. Numerical linear algorithms and group theory. *Linear Algebra Appl.*, 10:267–283, 1975. (Cited on p. 159.)

[72] J. Demmel. On condition numbers and the distance to the nearest ill-posed problem. *Numer. Math.*, 51:251–289, 1987. (Cited on p. 107.)

[73] J. Demmel and B. Kågström. The generalized Schur decomposition of an arbitrary pencil $A - \lambda B$: Robust software with error bounds and applications. *ACM Trans. Math. Software*, 19:160–201, 1993. (Cited on pp. 234, 252.)

[74] J. Demmel and W. Kahan. Accurate singular values of bidiagonal matrices. *SIAM J. Sci. Stat. Comput.*, 11:873–912, 1990. (Cited on pp. 308, 347.)

[75] J. Demmel and K. Veselić. Jacobi's method is more accurate than QR. *SIAM J. Matrix Anal. Appl.*, 13:1204–1245, 1992. (Cited on p. 205.)

[76] J. J. Dongarra and D. C. Sorensen. A fully parallel algorithm for the symmetric eigenvalue problem. *SIAM J. Sci. Stat. Comput.*, 8:s139–s154, 1987. (Cited on p. 205.)

[77] A. A. Dubrulle. The Multishift QR Algorithm—Is It Worth the Trouble? Technical Report G320-3558x, IBM Corp., Palo Alto, CA, 1991. (Cited on p. 199.)

[78] J. Durbin. The fitting of time-series models. *Rev. Inst. Internat. Statist.*, 28:233–243, 1960. (Cited on p. 380.)

[79] A. Edelman and G. Strang. Pascal matrices. *Amer. Math. Monthly*, 111:189–197, 2004. (Cited on p. 347.)

[80] E. Elmroth and F. Gustavson. High-performance library software for QR factorization. In *Applied Parallel Computing. New Paradigms for HPC in Industry and Academia*, Lecture Notes in Comput. Sci. 1947, pp. 53–63. Springer-Verlag, New York, 2001. (Cited on p. 134.)

[81] L. Elsner. On some algebraic problems in connection with general eigenvalue problems. *Linear Algebra Appl.*, 26:123–138, 1979. (Cited on pp. 12, 13.)

[82] M. A. Epton. Methods for the solution of $AXD - BXC = E$ and its applications in the numerical solution of implicit ordinary differential equations. *BIT*, 20:341–345, 1980. (Cited on p. 260.)

[83] D. K. Fadeev and V. N. Fadeeva. *Computational Methods of Linear Algebra*. Freeman, San Francisco, 1963. (Cited on pp. 159, 360.)

[84] H. Fassbender. *Numerische Verfahren zur diskreten trigonometrischen Polynomapproximation*. Ph.D. thesis, University of Bremen, Bremen, Germany, 1993. (Cited on p. 380.)

[85] H. Fassbender. *Symplectic Methods for the Symplectic Eigenproblem*. Kluwer Academic/Plenum Publishers, Amsterdam/New York, 2000. (Cited on p. 209.)

[86] H. Fassbender, D. S. Mackey, N. Mackey, and H. Xu. Hamiltonian square roots of skew-Hamiltonian matrices. *Linear Algebra Appl.*, 287:125–159, 1999. (Cited on p. 142.)

[87] K. V. Fernando and B. N. Parlett. Accurate singular values and differential qd algorithms. *Numer. Math.*, 67:191–229, 1994. (Cited on pp. 169, 205, 347.)

[88] U. Flaschka, V. Mehrmann, and D. Zywietz. An analysis of structure preserving methods for symplectic eigenvalue problems. *RAIRO Automatique Productique Informatique Industrielle*, 25:165–190, 1991. (Cited on p. 209.)

[89] J. G. F. Francis. The QR transformation, parts I and II. *Computer J.*, 4:265–272, 332–345, 1961. (Cited on pp. 159, 177, 199.)

[90] R. W. Freund, M. H. Gutknecht, and N. M. Nachtigal. An implementation of the look-ahead Lanczos algorithm for non-Hermitian matrices. *SIAM J. Sci. Comput.*, 14:137–158, 1993. (Cited on p. 390.)

[91] F. R. Gantmacher. *The Theory of Matrices, Vol.* I. Chelsea Publishing, New York, 1959. (Cited on p. 63.)

[92] F. R. Gantmacher. *The Theory of Matrices, Vol.* II. Chelsea Publishing, New York, 1959. (Cited on p. 234.)

[93] M. Gasca and C. A. Micchelli, editors. *Total Positivity and Its Applications*. Kluwer Academic Publishers, Dordrecht, The Netherlands, 1996. (Cited on p. 346.)

[94] W. Gautschi. On generating Gaussian quadrature rules. In G. Hämmerlin, editor, *Numerische Integration*, pp. 147–154. Birkhäuser-Verlag, Basel, 1979. (Cited on p. 316.)

[95] L. Giraud, J. Langou, M. Rozložník, and J. van den Eshof. Rounding error analysis of the classical Gram–Schmidt orthogonalization process. *Numer. Math.*, 101:87–100, 2005. (Cited on p. 371.)

[96] G. Golub. Least squares, singular values, and matrix approximations. *Aplikace Mathematiky*, 13:44–51, 1968. (Cited on p. 307.)

[97] G. Golub and W. Kahan. Calculating the singular values and pseudo-inverse of a matrix. *SIAM J. Numer. Anal.*, 2:205–224, 1965. (Cited on pp. 304, 306, 413.)

[98] G. Golub and C. Reinsch. Singular value decomposition and least squares solutions. *Numer. Math.*, 14:403–420, 1970. (Cited on p. 304.)

[99] G. Golub, K. Sølna, and P. Van Dooren. Computing the SVD of a general matrix product/quotient. *SIAM J. Matrix Anal. Appl.*, 22:1–19, 2000. (Cited on p. 317.)

[100] G. Golub and C. F. Van Loan. *Matrix Computations*. Johns Hopkins University Press, Baltimore, Third edition, 1996. (Cited on p. 351.)

[101] G. Golub and J. H. Welsh. Calculation of Gauss quadrature rules. *Math. Comp.*, 23:221–230, 1969. (Cited on p. 316.)

[102] W. B. Gragg. Positive definite Toeplitz matrices, the Arnoldi process for isometric operators, and Gaussian quadrature on the unit circle. In E. S. Nikolaev, editor, *Numerical Methods in Linear Algebra*, pp. 16–23. Moscow University Press, Moscow, 1982 (in Russian). (Cited on pp. 380, 382, 428.)

[103] W. B. Gragg. The QR algorithm for unitary Hessenberg matrices. *J. Comput. Appl. Math.*, 16:1–8, 1986. (Cited on p. 341.)

[104] W. B. Gragg. Positive definite Toeplitz matrices, the Arnoldi process for isometric operators, and Gaussian quadrature on the unit circle. *J. Comput. Appl. Math.*, 46:183–198, 1993; English translation of [102]. (Cited on pp. 380, 382.)

[105] W. B. Gragg and L. Reichel. A divide and conquer algorithm for the unitary and orthogonal eigenproblems. *Numer. Math.*, 57:695–718, 1990. (Cited on p. 341.)

[106] E. J. Grimme, D. C. Sorensen, and P. Van Dooren. Model reduction of state space systems via an implicitly restarted Lanczos method. *Numer. Algorithms*, 12:1–31, 1996. (Cited on p. 393.)

[107] M. Gu, R. Guzzo, X.-B. Chi, and X.-Q. Cao. A stable divide and conquer algorithm for the unitary eigenproblem. *SIAM J. Matrix Anal. Appl.*, 25:385–404, 2003. (Cited on p. 341.)

[108] J. B. Haag and D. S. Watkins. QR-like algorithms for the nonsymmetric eigenvalue problem. *ACM Trans. Math. Software*, 19:407–418, 1993. (Cited on pp. 176, 177.)

[109] C. R. Hadlock. *Field Theory and Its Classical Problems*. The Carus Mathematical Monographs, Mathematical Association of America, Washington, DC, 1978. (Cited on p. 151.)

[110] J. Handy. *The Cache Memory Book*. Academic Press, Second edition, 1998. (Cited on p. 130.)

[111] J. J. Hench and A. J. Laub. Numerical solution of the discrete-time periodic Riccati equation. *IEEE Trans. Automat. Control*, 39:1197–1210, 1994. (Cited on pp. 296, 300, 303, 319, 323.)

[112] G. Henry, D. Watkins, and J. Dongarra. A parallel implementation of the nonsymmetric QR algorithm for distributed memory architectures. *SIAM J. Sci. Comput.*, 24:284–311, 2002. (Cited on p. 199.)

[113] M. R. Hestenes and E. Stiefel. Methods of conjugate gradients for solving linear systems. *J. Res. Nat. Bur. Standards*, 49:409–436, 1952. (Cited on p. 360.)

[114] N. J. Higham. *Accuracy and Stability of Numerical Algorithms*. SIAM, Philadelphia, Second edition, 2002. (Cited on pp. 94, 194.)

[115] D. Hinrichsen and N. K. Son. Stability radii of linear discrete-time systems and symplectic pencils. *Int. J. Robust Nonlinear Control*, 1:79–97, 1991. (Cited on p. 208.)

[116] W. Hoffmann. Iterative algorithms for Gram–Schmidt orthogonalization. *Computing*, 41:335–348, 1989. (Cited on p. 371.)

[117] R. A. Horn and C. A. Johnson. *Matrix Analysis*. Cambridge University Press, Cambridge, UK, 1985. (Cited on p. 63.)

[118] H. Hotelling. Relation between two sets of variates. *Biometrika*, 28:322–377, 1936. (Cited on p. 83.)

[119] A. S. Householder. *The Theory of Matrices in Numerical Analysis*. Blaisdell, New York, 1964; reprinted by Dover, New York, 1975. (Cited on p. 159.)

[120] C. G. J. Jacobi. Über ein leichtes Verfahren die in der Theorie der Säculärstörungen vorkommenden Gleichungen numerisch aufzulösen. *J. Reine Angew. Math.*, 30:51–94, 1846. (Cited on p. 205.)

[121] C. Jordan. Essai sur la géométrie à n dimensions. *Bull. Soc. Math. France*, 3:103–174, 1875. (Cited on p. 83.)

[122] B. Kågström. A direct method for reordering eigenvalues in the generalized real Schur form of a regular matrix pair (A, B). In M. S. Moonen, G. H. Golub, and B. L. R. De Moor, editors, *Linear Algebra for Large Scale and Real-Time Applications*, pp. 195–218. Kluwer Academic Publishers, Dordrecht, The Netherlands, 1993. (Cited on pp. 260, 261, 285.)

[123] B. Kågström and D. Kressner. Multishift variants of the QZ algorithm with aggressive early deflation. *SIAM J. Matrix Anal. Appl.*, 29:199–227, 2006. (Cited on pp. 246, 252.)

[124] B. Kågström and P. Poromaa. LAPACK-style algorithms and software for solving the generalized Sylvester equation and estimating the separation between regular matrix pairs, *ACM Trans. Math. Software*, 22:78–103, 1996. (Cited on p. 260.)

[125] B. Kågström and L. Westin. Generalized Schur methods with condition estimators for solving the generalized Sylvester equation. *IEEE Trans. Automat. Control*, 34:745–751, 1989. (Cited on p. 260.)

[126] W. Kahan. Accurate Eigenvalues of a Symmetric Tridiagonal Matrix. Technical Report CS 41, Computer Science Department, Stanford University, Palo Alto, CA, 1966. (Cited on p. 308.)

[127] S. Kaniel. Estimates for some computational techniques in linear algebra. *Math. Comp.*, 20:369–378, 1966. (Cited on p. 351.)

[128] S. Karlin. *Total Positivity*. Stanford University Press, Palo Alto, CA, 1968. (Cited on p. 346.)

[129] L. Kaufman. The LZ-algorithm to solve the generalized eigenvalue problem. *SIAM J. Numer. Anal.*, 11:997–1024, 1974. (Cited on pp. 242, 246.)

[130] L. Kaufman. Some thoughts on the QZ algorithm for solving the generalized eigenvalue problem. *ACM Trans. Math. Software*, 3:65–75, 1977. (Cited on p. 246.)

[131] P. Koev. Accurate eigenvalues and SVDs of totally nonnegative matrices. *SIAM J. Matrix Anal. Appl.*, 27:1–23, 2005. (Cited on p. 347.)

[132] E. Kokiopoulou, C. Bekas, and A. Gallopoulos. Computing smallest singular triplets with implicitly restarted Lanczos bidiagonalization. *Appl. Numer. Math.*, 49:39–61, 2004. (Cited on p. 413.)

[133] D. Kressner. An efficient and reliable implementation of the periodic QZ algorithm. In S. Bittanti and P. Colaneri, editors, *Periodic Control Systems* 2001 (Proceedings of the IFAC Workshop on Periodic Control Systems, 2001). Elsevier, New York, 2002. www.math.tu-berlin.de/~kressner/. (Cited on p. 319.)

[134] D. Kressner. *Numerical Methods for General and Structured Eigenproblems*. Springer, New York, 2005. (Cited on pp. 202, 208, 239, 320, 410.)

[135] D. Kressner. On the use of larger bulges in the QR algorithm. *Electron. Trans. Numer. Anal.*, 20:50–63, 2005. (Cited on p. 271.)

[136] D. Kressner. The periodic QR algorithm is a disguised QR algorithm. *Linear Algebra Appl.*, 417:423–433, 2006. (Cited on pp. 297, 300, 305.)

[137] D. Kressner. The effect of aggressive early deflation on the convergence of the QR algorithm. *SIAM J. Matrix Anal. Appl.*, 2007, to appear. (Cited on pp. 145, 204, 216.)

[138] A. N. Krylov. On the numerical solution of the equation by which in technical questions frequencies of small oscillations of material systems are determined, *Izv. Akad. Nauk SSSR. Otd. Mat. Estest.*, 7:491–539, 1931 (in Russian). (Cited on p. 360.)

[139] V. N. Kublanovskaya. On some algorithms for the solution of the complete eigenvalue problem. *USSR Comput. Math. Math. Phys.*, 3:637–657, 1961. (Cited on p. 159.)

[140] V. N. Kublanovskaya. On a method of solving the complete eigenvalue problem for a degenerate matrix. *USSR Comput. Math. Math. Phys.*, 6:1–14, 1968. (Cited on pp. 56, 63.)

[141] P. Lancaster and L. Rodman. *The Algebraic Riccati Equation*. Oxford University Press, London, 1995. (Cited on pp. 207, 208.)

[142] P. Lancaster and M. Tismenetsky. *The Theory of Matrices*. Academic Press, New York, Second edition, 1985. (Cited on pp. 63, 213.)

[143] C. Lanczos. An iteration method for the solution of the eigenvalue problem of linear differential and integral operators. *J. Res. Nat. Bur. Stand.*, 45:255–282, 1950. (Cited on pp. 206, 360, 372, 388.)

[144] B. Lang. Effiziente Orthogonaltransformationen bei der Eigen- und Singulärwertzerlegung. Habilitationsschrift, Universität Wuppertal, Wuppertal, Germany, 1997. (Cited on p. 200.)

[145] B. Lang. Using level 3 BLAS in rotation-based algorithms. *SIAM J. Sci. Comput.*, 19:626–634, 1998. (Cited on p. 205.)

[146] A. J. Laub. A Schur method for solving algebraic Riccati equations. *IEEE Trans. Automat. Control*, AC-24:913–921, 1979. (Cited on p. 207.)

[147] R. B. Lehoucq, D. C. Sorensen, and C. Yang. *ARPACK Users' Guide: Solution of Large-Scale Eigenvalue Problems with Implicitly Restarted Arnoldi Methods*. SIAM, Philadelphia, 1998. Software at www.caam.rice.edu/software/ARPACK/. (Cited on pp. 362, 368, 372, 378.)

[148] N. Levinson. The Wiener RMS error criterion in filter design and prediction. *J. Math. Phys.*, 25:261–278, 1947; reprinted as an appendix to [229]. (Cited on p. 380.)

[149] D. S. Mackey and N. Mackey. On the Determinant of Symplectic Matrices. Technical Report, Numerical Analysis Report No. 422, Manchester Centre for Computational Mathematics, Manchester, UK, 2003. (Cited on p. 11.)

[150] D. S. Mackey, N. Mackey, and F. Tisseur. Structured tools for structured matrices. *Electron. J. Linear Algebra*, 10:106–145, 2003. (Cited on p. 120.)

[151] D. S. Mackey, N. Mackey, and F. Tisseur. G-reflectors: Analogues of Householder transformations in scalar product spaces. *Linear Algebra Appl.*, 385:187–213, 2004. (Cited on p. 118.)

[152] V. L. Mehrmann. *The Autonomous Linear Quadratic Control Problem*. Lecture Notes in Control and Inform. Sci. 163. Springer-Verlag, New York, 1991. (Cited on p. 208.)

[153] V. Mehrmann and D. Watkins. Structure-preserving methods for computing eigenpairs of large sparse skew-Hamiltoninan/Hamiltonian pencils. *SIAM J. Sci. Comput.*, 22:1905–1925, 2001. (Cited on pp. 398, 400.)

[154] G. Meurant. *The Lanczos and Conjugate Gradient Algorithms: From Theory to Finite Precision Computations.* Software Environ. Tools 19, SIAM, Philadelphia, 2006. (Cited on p. 373.)

[155] G. S. Minimis and C. C. Paige. Implicit shifting in the QR and related algorithms. *SIAM J. Matrix Anal. Appl.*, 12:385–400, 1991. (Cited on p. 185.)

[156] C. B. Moler and G. W. Stewart. An algorithm for generalized matrix eigenvalue problems. *SIAM J. Numer. Anal.*, 10:241–256, 1973. (Cited on pp. 242, 246.)

[157] R. B. Morgan. On restarting the Arnoldi method for large nonsymmetric eigenvalue problems. *Math. Comp.*, 65:1213–1230, 1996. (Cited on p. 368.)

[158] S. Oliveira and D. Stewart. Exponential splittings of products of matrices and and accurately computing singular values of long products. *Linear Algebra Appl.*, 309:175–190, 2000. (Cited on p. 317.)

[159] E. E. Osborne. On pre-conditioning of matrices. *J. ACM*, 7:338–345, 1960. (Cited on p. 198.)

[160] M. Overton. *Numerical Computing with IEEE Floating Point Arithmetic.* SIAM, Philadelphia, 2001. (Cited on pp. 162, 292.)

[161] C. C. Paige. *The Computation of Eigenvalues and Eigenvectors of Very Large Sparse Matrices.* Ph.D. thesis, London University, London, 1971. (Cited on pp. 351, 373.)

[162] C. C. Paige. Error analysis of the Lanczos algorithm for tridiagonalizing a symmetric matrix. *J. Inst. Math. Appl.*, 18:341–349, 1976. (Cited on p. 373.)

[163] C. C. Paige. Accuracy and effectiveness of the Lanczos algorithm for the symmetric eigenproblem. *Linear Algebra Appl.*, 34:235–258, 1980. (Cited on p. 373.)

[164] C. C. Paige and M. A. Saunders. Toward a generalized singular value decomposition. *SIAM J. Numer. Anal.*, 18:398–405, 1981. (Cited on p. 84.)

[165] C. C. Paige and C. Van Loan. A Schur decomposition for Hamiltonian matrices. *Linear Algebra Appl.*, 41:11–32, 1981. (Cited on p. 324.)

[166] C. C. Paige and M. Wei. History and generality of the CS decomposition. *Linear Algebra Appl.*, 208/209:303–326, 1994. (Cited on p. 84.)

[167] T. Pappas, A. Laub, and N. Sandell. On the numerical solution of the discrete-time algebraic Riccati equation. *IEEE Trans. Automat. Control*, AC-25:631–641, 1980. (Cited on p. 208.)

[168] B. N. Parlett. Global convergence of the basic QR algorithm on Hessenberg matrices. *Math. Comp.*, 22:803–817, 1968. (Cited on p. 224.)

[169] B. N. Parlett. *The Symmetric Eigenvalue Problem.* Classics in Appl. Math. 20, SIAM, Philadelphia, 1997 (originally Prentice–Hall, 1980). (Cited on pp. 205, 371, 373.)

[170] B. N. Parlett. The new qd algorithms. In A. Iserles, editor, *Acta Numerica 4*, pp. 459–491. Cambridge University Press, Cambridge, UK, 1995. (Cited on pp. 169, 205, 347.)

[171] B. N. Parlett and I. S. Dhillon. Relatively robust representations of symmetric tridiagonals. *Linear Algebra Appl.*, 309:121–151, 2000. (Cited on p. 206.)

[172] B. N. Parlett and W. G. Poole, Jr. A geometric theory for the QR, LU and power iterations. *SIAM J. Numer. Anal.*, 10:389–412, 1973. (Cited on p. 159.)

[173] B. N. Parlett and J. Le. Forward instability of tridiagonal QR. *SIAM J. Matrix Anal. Appl.*, 14:279–316, 1993. (Cited on p. 191.)

[174] B. N. Parlett and C. Reinsch. Balancing a matrix for calculation of eigenvalues and eigenvectors. *Numer. Math.*, 13:293–304, 1969. (Cited on p. 198.)

[175] B. N. Parlett and D. S. Scott. The Lanczos algorithm with selective orthogonalization. *Math. Comp.*, 33:217–138, 1979. (Cited on p. 373.)

[176] B. N. Parlett, D. R. Taylor, and Z. S. Liu. A look-ahead Lanczos algorithm for nonsymmetric matrices. *Math. Comp.*, 44:105–124, 1985. (Cited on p. 390.)

[177] T. J. Rivlin. *Chebyshev Polynomials, From Approximation Theory to Algebra and Number Theory*. Wiley-Interscience, New York, Second edition, 1990. (Cited on p. 352.)

[178] H. Rutishauser. Der Quotienten-Differenzen-Algorithmus. *Z. Angew. Math. Phys.*, 5:233–251, 1954. (Cited on pp. 159, 168.)

[179] H. Rutishauser. *Der Quotienten-Differenzen-Algorithmus*. Mitt. Inst. angew. Math. ETH 7. Birkhäuser, Basel, 1957. (Cited on pp. 159, 168.)

[180] H. Rutishauser. Solution of eigenvalue problems with the LR-transformation. *Nat. Bur. Standards Appl. Math. Ser.*, 49:47–81, 1958. (Cited on pp. 159, 168.)

[181] H. Rutishauser. Bestimmung der Eigenwerte orthogonaler Matrizen. *Numer. Math.*, 9:104–108, 1966. (Cited on p. 341.)

[182] H. Rutishauser. *Lectures on Numerical Mathematics*. Birkhäuser, 1990. (Cited on pp. 159, 169, 205, 347.)

[183] Y. Saad. On the rates of convergence of the Lanczos and the block-Lanczos methods. *SIAM J. Numer. Anal.*, 17:687–706, 1980. (Cited on p. 351.)

[184] Y. Saad. Chebyshev acceleration techniques for solving nonsymmetric eigenvalue problems. *Math. Comp.*, 42:567–588, 1984. (Cited on p. 368.)

[185] Y. Saad. *Numerical Methods for Large Eigenvalue Problems*. Manchester University Press, Manchester, UK, 1992; available for free online at www-users.cs.umn.edu/~saad/books.html. (Cited on p. 351.)

[186] R. Schreiber and C. Van Loan. A storage-efficient WY representation for products of Householder transformations. *SIAM J. Sci. Stat. Comput.*, 10:53–57, 1989. (Cited on p. 131.)

[187] H. Shapiro. The Weyr characteristic. *Amer. Math. Monthly*, 108:919–929, 1999. (Cited on p. 60.)

[188] V. Sima. Accurate computation of eigenvalues and real Schur form of 2×2 real matrices. In *Proceedings of the Second NICONET Workshop on "Numerical Control Software: SLICOT, a Useful Tool in Industry,"* INRIA Rocquencourt, France, 1999, pp. 81–86. (Cited on p. 162.)

[189] H. D. Simon. Analysis of the symmetric Lanczos algorithm with reorthogonalization methods. *Linear Algebra Appl.*, 61:101–131, 1984. (Cited on p. 373.)

[190] R. A. Smith. The condition numbers of the matrix eigenvalue problem. *Numer. Math.*, 10:232–240, 1967. (Cited on p. 232.)

[191] D. C. Sorensen. Implicit application of polynomial filters in a k-step Arnoldi method. *SIAM J. Matrix Anal. Appl.*, 13:357–385, 1992. (Cited on pp. 362, 368, 372.)

[192] A. Stathopoulos, Y. Saad, and K. Wu. Dynamic thick restarting of the Davidson, and the implicitly restarted Arnoldi methods. *SIAM J. Sci. Comput.*, 19:227–245, 1998. (Cited on p. 368.)

[193] D. Stewart. A new algorithm for the SVD of a long product of matrices and the stability of products. *Electron. Trans. Numer. Anal.*, 5:29–47, 1997. (Cited on p. 317.)

[194] G. W. Stewart. Error bounds for approximate invariant subspaces of closed linear operators. *SIAM J. Numer Anal.*, 8:796–808, 1971. (Cited on pp. 96, 101.)

[195] G. W. Stewart. On the sensitivity of the eigenvalue problem $Ax = \lambda Bx$. *SIAM J. Numer. Anal.*, 9:669–686, 1972. (Cited on p. 259.)

[196] G. W. Stewart. Error and perturbation bounds for subspaces associated with certain eigenvalue problems. *SIAM Rev.*, 15:727–764, 1973. (Cited on pp. 96, 101, 259.)

[197] G. W. Stewart. On the perturbation of pseudo-inverses, projections, and linear least squares problems. *SIAM Rev.*, 19:634–662, 1977. (Cited on p. 84.)

[198] G. W. Stewart. A Krylov–Schur algorithm for large eigenproblems. *SIAM J. Matrix Anal. Appl.*, 23:601–614, 2001. (Cited on pp. 365, 366, 368, 372.)

[199] G. W. Stewart and Ji-guang Sun. *Matrix Pertubation Theory*. Academic Press, New York, 1990. (Cited on p. 96.)

[200] M. Stewart. An Error Analysis of a Unitary Hessenberg QR Algorithm. Technical Report TR-CS-98-11, Department of Computer Science, Australian National University, Canberra, Australia, 1998. http://eprints.anu.edu.au/archive/00001557/. (Cited on p. 341.)

[201] G. Szegő. *Orthogonal Polynomials*, AMS Colloquium Publications 23. American Mathematical Society, Providence, RI, Fourth edition, 1975. (Cited on p. 380.)

[202] A. E. Taylor and D. C. Lay. *Introduction to Functional Analysis*. Krieger, Malabar, FL, Second edition, 1986. (Cited on pp. 36, 71, 218.)

[203] F. Tisseur. Tridiagonal-diagonal reduction of symmetric indefinite pairs. *SIAM J. Matrix Anal. Appl.*, 26:215–232, 2004. (Cited on p. 122.)

[204] L. N. Trefethen and M. Embree. *Spectra and Pseudospectra, The Behavior of Non-normal Matrices and Operators*. Princeton University Press, Princeton, NJ, 2005. (Cited on p. 91.)

[205] R. A. van de Geijn. Deferred shifting schemes for parallel QR methods. *SIAM J. Matrix Anal. Appl.*, 14:180–194, 1993. (Cited on p. 200.)

[206] P. Van Dooren. The computation of Kronecker's canonical form of a singular pencil. *Linear Algebra Appl.*, 27:103–141, 1979. (Cited on pp. 234, 252.)

[207] C. F. Van Loan. A general matrix eigenvalue algorithm. *SIAM J. Numer. Anal.*, 12:819–834, 1975. (Cited on p. 318.)

[208] C. F. Van Loan. A symplectic method for approximating all the eigenvalues of a Hamiltonian matrix. *Linear Algebra Appl.*, 61:233–252, 1984. (Cited on pp. 140, 320.)

[209] R. S. Varga. *Geršgorin and His Circles*. Springer, New York, 2004. (Cited on p. 92.)

[210] V. V. Voyevodin. A method for the solution of the complete eigenvalue problem. *Zh. Vych. Math.*, 2:15–24, 1962. (Cited on p. 159.)

[211] R. C. Ward. The combination shift QZ algorithm. *SIAM J. Numer. Anal.*, 12:835–853, 1975. (Cited on p. 253.)

[212] R. C. Ward. Balancing the generalized eigenvalue problem. *SIAM J. Sci. Stat. Comput.*, 2:141–152, 1981. (Cited on p. 246.)

[213] R. C. Ward and L. J. Gray. Eigensystem computation for skew-symmetric and a class of symmetric matrices. *ACM Trans. Math. Software*, 4:278–285, 1978. (Cited on p. 316.)

[214] D. S. Watkins. Understanding the QR algorithm. *SIAM Rev.*, 24:427–440, 1982. (Cited on p. 159.)

[215] D. S. Watkins. Some perspectives on the eigenvalue problem. *SIAM Rev.*, 35:430–471, 1993. (Cited on pp. 373, 380, 382.)

[216] D. S. Watkins. Forward stability and transmission of shifts in the QR algorithm. *SIAM J. Matrix Anal. Appl.*, 16:469–487, 1995. (Cited on pp. 265, 273.)

[217] D. S. Watkins. The transmission of shifts and shift blurring in the QR algorithm. *Linear Algebra Appl.*, 241–243:877–896, 1996. (Cited on pp. 177, 265, 271.)

[218] D. S. Watkins. Unitary orthogonalization processes. *J. Comput. Appl. Math.*, 86:335–345, 1997. (Cited on p. 380.)

[219] D. S. Watkins. Bulge exchanges in algorithms of QR type. *SIAM J. Matrix Anal. Appl.*, 19:1074–1096, 1998. (Cited on pp. 265, 289.)

[220] D. S. Watkins. Performance of the QZ algorithm in the presence of infinite eigenvalues. *SIAM J. Matrix Anal. Appl.*, 22:364–375, 2000. (Cited on pp. 250, 253, 255, 265.)

[221] D. S. Watkins. *Fundamentals of Matrix Computations*. Wiley, New York, Second edition, 2002. (Cited on pp. ix, 9, 10, 15, 117, 132, 206, 371.)

[222] D. S. Watkins. On Hamiltonian and symplectic Lanczos processes. *Linear Algebra Appl.*, 385:23–45, 2004. (Cited on pp. 388, 398, 403, 406, 407.)

[223] D. S. Watkins. Bulge Exchanges in GZ Algorithms for the Standard and Generalized Eigenvalue Problems. Unpublished manuscript, 2005. (Cited on p. 265.)

[224] D. S. Watkins. Product eigenvalue problems. *SIAM Rev.*, 47:3–40, 2005. (Cited on p. 291.)

[225] D. S. Watkins. A case where balancing is harmful. *Electron. Trans. Numer. Anal.*, 23:1–4, 2006. (Cited on p. 198.)

[226] D. S. Watkins. On the reduction of a Hamiltonian matrix to Hamiltonian Schur form. *Electron. Trans. Numer. Anal.*, 23:141–157, 2006. (Cited on pp. 208, 324, 326, 328.)

[227] D. S. Watkins and L. Elsner. Convergence of algorithms of decomposition type for the eigenvalue problem. *Linear Algebra Appl.*, 143:19–47, 1991. (Cited on pp. 199, 213, 231.)

[228] D. S. Watkins and L. Elsner. Theory of decomposition and bulge-chasing algorithms for the generalized eigenvalue problem. *SIAM J. Matrix Anal. Appl.*, 15:943–967, 1994. (Cited on pp. 241, 246.)

[229] N. Wiener. *Extrapolation, Interpolation, and Smoothing of Stationary Time Series.* MIT Press, Cambridge, MA, 1949. (Cited on p. 431.)

[230] J. H. Wilkinson. *The Algebraic Eigenvalue Problem,*. Clarendon Press, Oxford, UK, 1965. (Cited on pp. 115, 159, 177, 206.)

[231] J. H. Wilkinson. Kronecker's canonical form and the QZ algorithm. *Linear Algebra Appl.*, 28:285–303, 1979. (Cited on p. 234.)

[232] J. H. Wilkinson and C. Reinsch, editors. *Handbook for Automatic Computation, Volume* II, *Linear Algebra*. Springer-Verlag, New York, 1971. (Cited on p. 198.)

[233] K. Wu and H. Simon. Thick-restart Lanczos method for large symmetric eigenvalue problems. *SIAM J. Matrix Anal. Appl.*, 22:602–616, 2000. (Cited on pp. 365, 366, 368, 414.)

Index